Lecture Notes in Computer Science 12156

More information about this series at http://www.springer.com/series/7407

Andrea Werneck Richa ·
Christian Scheideler (Eds.)

Structural Information and Communication Complexity

27th International Colloquium, SIROCCO 2020
Paderborn, Germany, June 29 – July 1, 2020
Proceedings

 Springer

Editors
Andrea Werneck Richa
Computer Science, CIDSE
Arizona State University
Tempe, AZ, USA

Christian Scheideler
Department of Computer Science
Paderborn University
Paderborn, Germany

ISSN 0302-9743   ISSN 1611-3349   (electronic)
Lecture Notes in Computer Science
ISBN 978-3-030-54920-6   ISBN 978-3-030-54921-3   (eBook)
https://doi.org/10.1007/978-3-030-54921-3

LNCS Sublibrary: SL1 – Theoretical Computer Science and General Issues

This Springer imprint is published by the registered company Springer Nature Switzerland AG
The registered company address is: Gewerbestrasse 11, 6330 Cham, Switzerland

Preface

The papers in this volume were presented at the 27th International Colloquium on Structural Information and Communication Complexity (SIROCCO 2020) which was supposed to be held from June 29 to July 1, 2020, in Paderborn, Germany. However, due to the Corona virus pandemic, it took place virtually at the same time.

SIROCCO is devoted to the study of the interplay between structural knowledge, communication, and computing in decentralized systems of multiple communicating entities. Special emphasis is given to innovative approaches leading to better understanding of the relationship between computing and communication.

This time, we had 41 submissions, where 39 of them were regular papers and 2 were brief announcements. Each submission was reviewed by at least three Program Committee (PC) members with the help of external reviewers, and the committee decided to accept 19 submissions as regular papers. In addition to these, this volume includes invited papers from two keynote speakers and their co-authors. We accepted four of the contributed submissions as brief announcements; while these do not appear in these proceedings, they had short presentations at the conference. Of the regular papers, "Ants on a Plane" by Abhinav Aggarwal and Jared Saia received the Best Paper Award, and "Local Gathering of Mobile Robots in Three Dimensions" by Michael Braun, Jannik Castenow, and Friedhelm Meyer auf der Heide won the Best Student Paper Award.

We would like to thank the authors who submitted their work to SIROCCO this year and the PC members and subreviewers for their valuable and insightful reviews and comments. We would also like to thank the keynote speakers Petra Berenbrink, Mohsen Ghaffari, and Jared Saia for their insightful talks, and Amos Korman for his featured talk as the recipient of the 2020 SIROCCO Innovation in Distributed Computing Prize. The SIROCCO Steering Committee, chaired by Magnús M. Halldórsson, provided help and guidance throughout the process. The EasyChair system was used to handle the submission of papers and to manage the review process. Without all of these people it would not have been possible to produce these proceedings and the great conference program.

June 2020

Andrea Werneck Richa
Christian Scheideler

Laudatio: 2020 SIROCCO Prize for Innovation in Distributed Computing

It is a pleasure to award the 2020 SIROCCO Prize for Innovation in Distributed Computing to Amos Korman. Amos has been one of the main contributors to distributed network computing since the early 2000s, with outstanding contributions to informative labeling schemes and dynamic networks, as witnessed by his SIROCCO papers [1, 5]. With Shay Kutten and David Peleg, he is the initiator of the whole field of proof-labeling schemes, an elegant and powerful mechanism for the design of fault-tolerant distributed algorithms, including self-stabilizing algorithms, which experienced significant developments since the conference version of their seminal paper [2] appeared in PODC 2005.

However, the Award Committee wishes to award the prize for another radically innovative line of research that Amos launched in 2010. This research aims at using tools from distributed algorithms design and analysis to the study of biological ensembles, such as ant colonies, schools of fish, cellular systems, flocks of birds, etc. Its guiding principle is that, by modeling the behaviors of biological ensembles using tools from distributed computing, one can derive bounds on the capabilities of each individual in these ensembles, regarding communication and computation abilities. In 2012, Amos Korman published his first two contributions in this field [3, 4], on collaborative search with implications for biology. These papers had a great impact on the distributed computing community, opening wide avenues for research.

In 2013, Amos with his colleagues set up the Workshop on Biological Distributed Algorithms (BDA). Since then, this workshop has been organized yearly, and successfully gathered computer scientists and biologists interested in computational aspects of biological systems. Hence, not only did Amos author influential papers, but also he was at the origin of the set up of an entire community of scientists.

Showing the impact of distributed computing not only on the theory and practice of artificial systems (networks, distributed systems and architectures, etc.), but also on other disciplines, is an important contribution to augmenting the visibility of this field of research. Amos has been one of the pioneers in noticing the importance of distributed computing for life sciences.

We award the 2020 SIROCCO Prize for Innovation in Distributed Computing to Amos Korman for his pioneering contributions to distributed computing methods for system biology.

The 2020 Award Committee[1]

Shantanu Das	Aix-Marseille University, France
Andrzej Pelc (Chair)	Université du Qubec en Outaouais, Canada
Boaz Patt-Shamir	Tel Aviv University, Israel

---

[1] We wish to thank the nominators for the nomination and for contributing heavily to this text.

Zvi Lotker   Ben-Gurion University of the Negev, Israel
Michel Raynal   Irisa, France
Jukka Suomela   Aalto University, Finland

Selected Publications Related to Amos Korman's Contribution:

1. Amos Korman and Shay Kutten:
   "Labeling Schemes with Queries"
   SIROCCO 2007: 109–123.
2. Amos Korman, Shay Kutten, and David Peleg:
   "Proof Labeling Schemes"
   Distributed Computing 22(4): 215–233 (2010).
3. Ofer Feinerman and Amos Korman:
   "Memory Lower Bounds for Randomized Collaborative Search and Applications to Biology"
   DISC 2012: 61–75.
4. Ofer Feinerman, Amos Korman, Zvi Lotker, and Jean-Sébastien Sereni:
   "Collaborative Search on the Plane without Communication"
   PODC 2012: 77–86.
5. Ofer Feinerman and Amos Korman:
   "Clock Synchronization and Estimation in Highly Dynamic Networks: An Information Theoretic Approach"
   SIROCCO 2015: 16-30.
6. Amos Korman and Yoav Rodeh:
   "Parallel Search with No Coordination"
   SIROCCO 2017: 195–211.
7. Ofer Feinerman and Amos Korman:
   "The ANTS problem"
   Distributed Computing 30(3): 149–168 (2017).
8. Lucas Boczkowski, Amos Korman, and Emanuele Natale:
   "Minimizing Message Size in Stochastic Communication Patterns: Fast Self-Stabilizing Protocols with 3 bits"
   Distributed Computing 32(3): 173–191 (2019).

Organization

Program Committee Chairs

Andrea Werneck Richa   Arizona State University, USA
Christian Scheideler   Paderborn University, Germany

Program Committee

Petra Berenbrink	University of Hamburg, Germany
Borzoo Bonakdarpour	Iowa State University, USA
Yuval Emek	Technion, Israel
Shantanu Das	Aix-Marseille University, France
Thomas Erlebach	University of Leicester, UK
Panagiota Fatourou	University of Crete, Greece
Olga Goussevskaia	Federal University of Minas Gerais, Brazil
Taisuke Izumi	Nagoya Institute of Technology, Japan
Valerie King	University of Victoria, Canada
Frederik Mallmann-Trenn	King's College London, UK
Lata Narayanan	Concordia University, Canada
Calvin Newport	Georgetown University, USA
Rotem Oshman	Tel Aviv University, Israel
Matthew Patitz	University of Arkansas, USA
Maria Potop-Butucaru	Sorbonne University, LIP 6, France
Andrea Werneck Richa (Co-chair)	Arizona State University, USA
Peter Robinson	City University of Hong Kong, Hong Kong, China
Christian Scheideler (Co-chair)	Paderborn University, Germany
Stefan Schmid	University of Vienna, Austria
Nodari Sitchinava	University of Hawaii at Manoa, USA
Jara Uitto	Aalto University, Finland
André van Renssen	The University of Sydney, Australia
Roger Wattenhofer	ETH Zurich, Switzerland
Maxwell Young	Mississippi State University, USA

Subreviewers SIROCCO 2020

Klaus-Tycho Förster   Jan Studený
Joshua Daymude   Anissa Lamani
Daniel Warner   Soumyottam Chatterjee
Orr Fischer   Hirotaka Ono

Satoshi Takabe
Orr Fischer
Joshua Daymude
Cristina Gava
Martin P. Seybold
William K. Moses Jr.
John Pfeifer
Daniel Hader
Omer Wasim
Milutin Brankovic
Othon Michail
Michael Feldmann
Christina Kolb
Eric Severson
Alexander Setzer

Damien Imbs
Josef Widder
Sampson Wong
Trent Rogers
Debasish Pattanayak
Giuseppe Antonio Di Luna
Cristina Gava
Jérémie Chalopin
Ulrich Schmid
Daniel Hader
Hung Viet Le
Kristian Hinnenthal
Yukiko Yamauchi
Patrick Eades

Contents

Multi-agent Systems

Communication Complexity

Game Theory

Invited Papers

Network Decomposition and Distributed Derandomization (Invited Paper)

Mohsen Ghaffari$^{(\boxtimes)}$

ETH Zurich, Zurich, Switzerland
ghaffari@inf.ethz.ch

Abstract. We overview a recent line of work [Rozhoň and Ghaffari at STOC 2020; Ghaffari, Harris, and Kuhn at FOCS 2018; and Ghaffari, Kuhn, and Maus at STOC 2017], which proved that any (locally-checkable) graph problem that admits an efficient randomized distributed algorithm also admits an efficient deterministic distributed algorithm, thereby resolving several central and decades-old open problems in distributed graph algorithms. We present a short and self-contained version of the proofs, and conclude by discussing several related questions that remain open.

This article accompanies a keynote talk of the author at the International Colloquium on Structural Information and Communication Complexity (SIROCCO) 2020. The writing is based on [24,28,45] and primarily targets non-experts.

Keywords: Distributed graph algorithms · Derandomization · Network decomposition · LOCAL model

1 Introduction

Understanding the gap between the power of randomized and deterministic algorithms is one of the fundamental questions in the theory of computation. Instances of this question have been studied extensively in various settings, e.g., the P vs BPP question in centralized computation, the NC vs RNC question in parallel computation, the L vs RL in log-space bounded computation, etc.

For over three decades, a similar question remained open for *distributed graph algorithms*, and various special cases of it were at the center of the community's research. Here, we briefly review the recent work that resolved this question [24, 28, 45], and we outline some of the related problems that remain open.

The Model. We consider a standard synchronous message-passing model for distributed computing on a network, often referred to as the LOCAL model [36,37]. The network is abstracted as an n-node undirected graph $G = (V, E)$.

We acknowledge support from the European Research Council (ERC), under the European Union's Horizon 2020 research and innovation programme (grant agreement No. 853109).

A. W. Richa and C. Scheideler (Eds.): SIROCCO 2020, LNCS 12156, pp. 3–18, 2020.
https://doi.org/10.1007/978-3-030-54921-3_1

There is one processor on each network node. The processors can communicate with each other in synchronous rounds, where per round each processor/node can send one message to each of its neighbors in G. In the basic version of the model, we assume no upper bound on the message sizes. A variant, called the CONGEST model [44], allows only $O(\log n)$-bit messages. At the very beginning, nodes do not know the topology of the graph G, except for knowing their own neighbors, and potentially estimates of some global parameters, e.g., a (polynomially-tight) upper bound on the network size $n = |V|$. In the deterministic version of the model, we also assume that the nodes have unique identifiers, typically of $O(\log n)$-bit length. At the end of the algorithm, each node should know its own part of the output, e.g., if we are coloring the vertices of G, each processor should know its own color. Finally, the focus is on the communication and thus we measure mainly the *round complexity*, i.e., the number of rounds until all nodes terminate. That is, the model allows the processors to perform arbitrary computation in between the rounds, given the information that they have.

Thanks to the relaxations on the message size and computation, the model admits a clean mathematical interpretation: all that a node can compute in r rounds, for any value r, is exactly functions of the information initially in nodes within its distance r. That is, for instance, any deterministic r-round algorithm is exactly a mapping from r-hop neighborhood topologies to the output. Hence, in this sense, determining the round complexity of a graph problem in the LOCAL model characterizes the *locality* of that problem.

Deterministic vs Randomized—State of the Art. In a spirit similar to viewing centralized algorithms with poly(n) time complexity as efficient, or parallel algorithms with poly$(\log n)$ time complexity as efficient, in distributed graph algorithms, it has become standard to consider algorithms with poly$(\log n)$ round complexity as *efficient*.

Starting in the 1980s, it was discovered that several key problems admit efficient randomized algorithms in the LOCAL model. For instance (Las Vegas) randomized algorithms[1] with $O(\log n)$ round complexity were presented for problems such as maximal independent set, maximal matching, $(\Delta + 1)$-vertex coloring, and $(2\Delta - 1)$ edge coloring [1,35,39]. Shortly after, and naturally, researchers asked whether these problems admit also efficient deterministic distributed algorithms. A prominent example is Linial's famous open question[2] about the Maximal Independent Set (MIS) problem [36,37], where he asked *"can it [MIS] always be found [deterministically] in polylogarithmic time?"* Important progress was made in the work of Awerbuch, Goldberg, Luby, and Plotkin [3] and the follow up by Panconesi and Srinivasan [43], which led to deterministic algorithms with complexity $2^{O(\sqrt{\log n})}$. This is a complexity much better than any polynomial in

---

[1] These are algorithms that always output a correct solution, and they terminate within the promised time bound *with high probability*, i.e., with probability at least $1 - 1/n^c$ for a desirably large fixed constant $c > 1$.

[2] For instance, it was called *"probably the most outstanding open problem in the area"*, by Barenboim and Elkin [9].

n but still much higher than the desired poly$(\log n)$ bound. Over the years, there was progress on efficient deterministic algorithms for a number of these problems, e.g., maximal matching [19,33], relaxations of vertex coloring with larger number of colors [8], and edge coloring [20,29,31,34]. However, several problems—notably including MIS and $(\Delta+1)$ vertex coloring—remained open, with the best known complexity remaining at the aforementioned $2^{O(\sqrt{\log n})}$ bound. See also the book of Barenboim and Elkin [9, Open Problems Chapter], which highlights a number of such open problems (as of 2013).

Deterministic vs Randomized—Generalized Question. Generalizing these, one can ask a much broader question [28]: Is it the case that, in the LOCAL model, the class of all *locally-checkable* problems that admit an efficient randomized algorithm (known as P-RLOCAL) is equal to the class of all locally checkable problems that admit an efficient deterministic algorithm (known as P-LOCAL)? In short, this is to ask whether

$$\text{P-LOCAL} \overset{?}{=} \text{P-RLOCAL}$$

As mentioned above, we interpret efficient algorithms as those with poly$(\log n)$ round complexity. Also, locally-checkable means graph problems for which a proposed solution can be checked deterministically in constant rounds, as defined by Naor and Stockmeyer [41], or even poly$(\log n)$ rounds. If the solution is invalid, at least one node should know. Notice that all the problems mentioned above are locally checkable, e.g., in vertex coloring, it suffices if each node compares its colors with just its neighbors, which can be done in one round. Also, we note that this restriction to locally-checkable problems is necessary, as otherwise the answer is clearly negative and there are problems that have even zero round randomized algorithms and no efficient deterministic algorithm[3].

Motivations. The above deterministic vs randomized question is a central problem in the area. In fact, the 2013 book of Barenboim and Elkin [9] even goes on to assert that *"Perhaps the most fundamental open problem in this field is to understand the power and limitations of randomization"*. Similar to its counterparts in other computational settings, this question is motivated by complexity-theoretic considerations as well as practical desires for deterministic guarantees. But besides these standard reasons, there is also a modern motivation, which is unique to the distributed setting, as we review next.

A 2016 result of Chang et al. [13] showed that to improve the randomized algorithms for several key problems– MIS, vertex coloring, etc–we *provably* need faster deterministic algorithms. Concretely, they showed that the randomized complexity of any problem on n-node graphs is at least its deterministic complexity in $\sqrt{\log n}$-node graphs. Considering the state of the art randomized algorithms this showed that any faster randomized algorithm would imply a faster deterministic algorithm. So, even if we are interested in only randomized

---

[3] For instance, the problem of marking $(1 \pm o(1))\sqrt{n}$ nodes can be solved in zero rounds randomized and needs $\tilde{\Omega}(\sqrt{n})$ rounds deterministically. Cf. [24, Sect. 7.2].

algorithms, we need to understand and improve deterministic algorithms. See [13,15,45] for quantitative statements for various problems.

Resolution. The question mentioned above was recently resolved in affirmative [24,28,45], proving that P-LOCAL=P-RLOCAL. That is, any locally checkable problem that admits an efficient randomized algorithm also admits an efficient deterministic algorithm. Concretely, this came in two parts: (A) The work in [24,28] showed that obtaining a poly($\log n$)-round deterministic algorithm for network decomposition (a classic problem that we will review soon) would imply P-LOCAL=P-RLOCAL, and (B) the work in [45] provided a poly($\log n$)-round deterministic algorithm for network decomposition.

Implications and Applications. The above result lead to poly($\log n$)-round deterministic algorithms for several well-studied graph problems, and thus resolved a number of central open problems in distributed graph algorithms. Just the network decomposition algorithm of [45], on its own, solved several special cases: for instance, it implies a poly($\log n$) round algorithms for MIS— thus resolving Linial's question—and $\Delta + 1$ coloring. These algorithms can be extended also to the CONGEST model, with $O(\log n)$-bit messages [7,12,45]. With the general derandomization P-LOCAL=P-RLOCAL, the list goes much further, and includes problems such as Lovász Local Lemma and its complexity-theoretic implications, packing and covering integer linear programs, defective and frugal coloring, forest decomposition, low out-degree edge orientations, spanners, etc. We refer the interested reader to [45, Sect. 3] for a coverage of these corollaries, as well as implications for *massively parallel computation.*

Roadmap. In the following sections, we first start with a warm up that allows us to organically introduce the concept of *network decomposition* and how they help for distributed algorithms. We then present (a weaker variant of) the efficient deterministic network decomposition algorithm of [45] in Sect. 3, and discuss the derandomization method of [24,28] in Sect. 4. These two sections provide a short and self-contained proof of P-LOCAL=P-RLOCAL. In the last section, we discuss several related problems that remain open.

2   Warm Up: MIS and Network Decomposition

Here, as a warm up, we discuss the MIS problem and its randomized algorithm and then, we explain how we naturally arrive at the concept of network decomposition, as introduced by Awerbuch et al. [3], starting with the goal of an efficient deterministic distributed MIS algorithm.

The MIS Problem. The Maximal Independent Set (MIS) problem asks for selecting a subset $S \subseteq V$ of vertices such that no two nodes in S are adjacent (*independence*), and moreover we cannot add any more node to S, i.e., any node $v \in V \setminus S$ has a neighbor in S (*maximality*).

Randomized Algorithm. There is a simple and celebrated[4] $O(\log n)$-round randomized algorithm for MIS from the 1980s, due to Luby [39] and Alon,

---

[4] See the Dijkstra prize of 2016, and its citation.

Babai, and Itai [1]. We start with an empty S and iteratively add independent vertices to it. Per iteration, each (remaining) node v picks a real random number[5] $r_v \in [0, 1]$ and then we add to S all nodes v that have a number strictly larger than all of their neighbors. We then remove all the (newly added) nodes of S and their neighbors from V, and proceed to the next iteration. The analysis shows that the algorithm terminates with high probability in $O(\log n)$ iterations, as per iteration half of the remaining edges are removed in expectation. See [46].

Deterministic Algorithm and Introducing Network Decomposition. If we were in the easy case of a small diameter network—i.e., if the network graph G has diameter D that is at most poly$(\log n)$—then we could solve the MIS problem via a simple topology-gathering approach: we would collect the topology of the entire network, including the node identifiers and all the edges between them, into one leader node in $O(D)$ rounds, then compute an MIS there via the simple sequential greedy MIS algorithm, and finally broadcast the computed MIS set to all nodes in $O(D)$ rounds. The interesting case is when G has a large diameter.

Wishfully thinking, if we could partition say at least half of the nodes into non-adjacent and vertex-disjoint clusters, each of diameter at most $D = $ poly$(\log n)$, then we could also compute an MIS of all the nodes in these clusters by repeating the above strategy in each cluster. Since clusters are non-adjacent, they can compute their independent sets simultaneously, all in parallel. Then, we would remove all these clustered vertices from the network, as well as all non-clustered vertices that are adjacent to a node in the computed independent set. This way, we would have removed at least $1/2$ of the vertices of the graph. So, by repeating the same procedure $\log n$ times, each time clustering at least half of the remaining nodes into non-adjacent and vertex-disjoint clusters of diameter at most D and performing topology-gatherin in each cluster, we would at the end obtain an MIS of the entire graph. This would have complexity $O(D \log n) = $ poly$(\log n)$, i.e., it would be an efficient deterministic algorithm for MIS.

The only missing piece in the above wishful thinking is this: where do we get such a nice clustering? This is basically what we call *network decomposition*, as introduced first by Awerbuch, Goldberg, Luby, and Plotkin [3], and as we define formally next. Before that, it is worth commenting that, even though the above approach is in the LOCAL model and uses large messages for topology gathering, one can use network decomposition to solve MIS and $\Delta + 1$ vertex coloring also in the CONGEST model [7, 12, 45].

Defining Network Decomposition. A network decomposition with C colors and D diameter is a partitioning of the vertices into disjoint clusters, each colored with a color from $\{1, 2, \ldots, C\}$, such that each cluster has diameter at most D and any two adjacent clusters—i.e., two clusters that contain adjacent vertices—have different colors.

In the above wishful thinking, in each iteration, we would get the clusters of one color of the network decomposition. Each cluster would have diameter

---

[5] A number with $O(\log n)$ bits of precision would also suffice.

at most D. And since we repeat $\log n$ times, our decomposition would have $C = \log n$ colors.

A subtlety worth point out is in the definition of the diameter for each cluster. If we define the diameter of a cluster as the maximum distance of any of its two vertices in the subgraph induced by the nodes of the cluster, this is known as a *strong-diameter* network decomposition. In contrast, if we define the diameter of the cluster as the the maximum distance of any of its two vertices in the original graph G (i.e., when the path is allowed to exit the cluster), this is known as a *weak-diameter* network decomposition. For the applications in the LOCAL model, e.g., as exemplified above for the MIS problem, even the version with weak-diameter is sufficient.

Known Results about Network Decomposition. It is well-known that every n-node graph admits a network decomposition with $C = \log n$ colors and $D = \log n$ strong-diameter, and there is a simple sequential algorithm to build such a decomposition (often called *ball-carving*) [4]. A randomized algorithm of Linial and Sak computes a decomposition with these parameters but weak-diameter in $O(\log n)$ rounds of the LOCAL model, and it can be changed to a poly($\log n$)-round algorithm that computes a decomposition with these parameters and strong-diameter, using a transformation of Awerbuch et al. [2]. Recently, Elkin and Neiman [17] presented an adaptation of the randomized approach of [40], which computes a strong-diameter network decomposition with the above ideal bounds in $O(\log^2 n)$ rounds of the CONGEST model. On the deterministic side, a classic LOCAL-model deterministic algorithm of Panconesi and Srinivasan computes a decomposition with $C = \log n$ colors and $D = \log n$ strong-diameter in $2^{O(\sqrt{\log n})}$ rounds [43], which itself was an improvement on the $2^{O(\sqrt{\log n \log \log n})}$ round algorithm of Awerbuch et al. [3]. This $2^{O(\sqrt{\log n})}$ round complexity remained the state of the art for over 25 years, until the efficient deterministic distributed construction of [45], which we discuss next.

3   Efficient Deterministic Network Decomposition

Now, we discuss the efficient deterministic algorithm of [45] for constructing a network decomposition. We discuss only a weaker variant with $C = O(\log n)$ colors and weak-diameter $D = \text{poly}(\log n)$. Two remarks are in order: (A) It is well-known how to improve these to the ideal bound of $O(\log n)$ and strong-diameter, with only a poly($\log n$) factor increase in the round complexity. See [2,45]. (B) This weaker statement is still sufficient for poly($\log n$)-round deterministic algorithms in the LOCAL model, e.g., for MIS and the efficient derandomization that we discuss in the next section.

Algorithm Outline. We describe an algorithm that finds a partial clustering that clusters at least half of the vertices into non-adjacent clusters each with weak-diameter at most $D = O(\log^3 n)$. Applying this once gives the clusters of the first color of the decomposition. This is what we call one *iteration*. By repeating the algorithm for $\log n$ iterations, each time on the remaining nodes, we get the $\log n$ colors of the decomposition.

Outline of One Iteration. The iteration starts with a trivial clustering where each node is one cluster. Unfortunately, in this clustering, all edges are *bad*, i.e., they connect two different clusters. We adjust the clusters gradually, growing or shrinking them appropriately, such that at the end no bad edge remains.

The iteration has b phases, where $b = O(\log n)$ denotes the number of bits in the identifiers. Each phase corresponds to one bit. The objective is that, at the end of iteration i, there should be no (bad) edge connecting two clusters whose identifiers (the identifier of the original node in the cluster) differ in the i least significant bits. Hence, at the very end, after phase $i = b$, there is no edge connecting two different clusters.

Each Phase. In the first phase, we call a cluster (and its nodes) red if the least significant identifier bit is 1 and blue otherwise. Generally, in phase i, we call a cluster (and its nodes) *red* if the ith least significant identifier bit of the cluster is 1 and *blue* otherwise (completely forgetting the colors of the previous phases). Each phase is made of $O(\log^2 n)$ *steps*. Per step, each red node proposes to join an arbitrarily chosen neighboring blue cluster, if there is one. A blue cluster either accepts all the request, which happens when the number of request is at least a $1/(2b)$ factor of the number of nodes in the cluster, or it denies all of them. A red node whose request is denied is *killed* (and loses its color). When a vertex v is *killed*, it will not be clustered in this iteration. A red node whose request is accepted becomes blue and joins the accepting cluster, and adopts the identifier of that cluster as its own cluster identifier. A red node who did not make a request, as it did not have a blue neighbor, remains in its cluster as a red node, for this step. Notice the asymmetry as the red nodes act individually whereas blue clusters act as a cluster. As mentioned, we perform this procedure for $O(\log^2 n)$ steps, and then proceed to the next phase, which has a new coloring, based on the next (cluster) identifier bit.

Analysis–the First Phase. Let us start with the first phase. We observe two properties: (A) per step that a blue cluster accepts requests, its size grows by at least a $(1 + 1/(2b))$ factor. Hence, the cluster can grow at most $O(b \log n) = O(\log^2 n)$ steps in this phase, as beyond that it would have more than n vertices. Hence, all blue clusters stop growing within $O(\log^2 n)$ steps. After $O(\log^2 n)$ steps, i.e., at the end of the first phase, there is no edge between a red cluster and a blue cluster. (B) The first time that a blue cluster denies the requests, it will never grow again in this phase, because all adjacent nodes are either blue or killed (and thus uncolored for the rest of this iteration). When the cluster stops growing, it kills a number of red nodes up to at most a $1/(2b)$ fraction of the number of (blue) nodes in this cluster. Hence, overall in this phase, at most a $1/(2b)$ fraction of nodes are killed.

Analysis–All Phases. Generalizing these observations, we now argue the three properties of this construction, for all phases:

– **No bad edges left:** Generalizing observation (A) above, we can argue, with an induction, that at the end of the ith phase, we have no (bad) edge between clusters whose IDs differ in the i least significant bits. Therefore, after all the b phases, no bad edge remains and the clusters are non-adjacent.

- **At most half killed:** Per phase, we kill at most a $1/(2b)$ fraction of the nodes. Hence, the fraction of nodes that are not killed during b phases—i.e., remain clustered—is at least $(1 - 1/(2b))^b \geq 1/2$.

- **Small cluster diameter:** Finally, the weak-diameter of each cluster grows by additive $O(1)$ per step of each phase, and thus by additive $O(\log^2 n)$ per phase. Hence, even after all the $b = O(\log n)$ phases, the weak-diameter is at most $O(\log^3 n)$. This also allows us to perform each one step in at most $O(\log^3 n)$ rounds, as each blue cluster can aggregate the number of requests and decide to accept or deny them all.

4   Distributed Derandomization

In the warm up section, we used network decomposition to solve MIS deterministically. We basically simulated the sequential greedy MIS algorithm, by working through the colors of the decomposition. We now discuss an approach of [24,28] that generalizes this and obtains an efficient distributed derandomization for all locally checkable problems, thus proving P-LOCAL=P-RLOCAL.

Theorem 1. P-LOCAL=P-RLOCAL

Proof (Proof Sketch). Consider a locally checkable problem $\mathcal{P}$ for which any solution can be checked deterministically in $p(n) \in \text{poly}(\log n)$ rounds. Let $\mathcal{A}$ be a randomized algorithm that solves $\mathcal{P}$ with high probability—i.e. with probability at least $1 - 1/n^c$ for a constant $c > 1$—in $r(n) \in \text{poly}(\log n)$ rounds. We exhibit a deterministic algorithm that solves $\mathcal{P}$ in $r'(n) \in \text{poly}(\log n)$ rounds.

Considering algorithm $\mathcal{A}$ and the local checkability of $\mathcal{P}$, we thus have an algorithm $\mathcal{A}'$ with round complexity $R(n) = p(n) + r(n)$ where each node v, besides its output to problem $\mathcal{P}$, checks its $p(n)$-hop neighborhood and sets an indicator variable $f_v \in \{0,1\}$ equal to 1 if and only if one of the conditions of $\mathcal{P}$ is violated. Hence, the event $(\forall v \in V, f_v = 0)$ deterministically indicates that the outputs provide a correct solution for problem $\mathcal{P}$. Moreover, $\mathbb{E}[\sum_{v \in V} f_v] \leq \sum_{v \in V} 1/n^c = n/n^c < 1$.

A Sequential Deterministic Local Algorithm for $\mathcal{P}$. First, let us devise a certain sequential and local algorithm (formally, an SLOCAL-model algorithm in the terminology of [28]) for $\mathcal{P}$ via a simple method of conditional expectation. Consider an arbitrary ordering of the vertices $u_1, u_2, \ldots, u_n$. We examine the vertices one by one, according to this order. When examining vertex u_1, we fix its randomness in algorithm $\mathcal{A}'$ in a way that the conditional expectation of the $\mathbb{E}[\sum_{v \in V} f_v]$ does not increase. Notice that u_1 influences f_v only for nodes v in its $R(n)$-hop neighborhood. Thus, we can fix the randomness of u_1 by reading only its $2R(n)$-hop neighborhood. We similarly fix the randomness of all other nodes. Each time, when examining vertex u_i, we read the $2R(n)$-hop neighborhood of u_i, including the randomness already fixed there, and then we fix the randomness of u_i in a manner that the conditional expectation of the $\mathbb{E}[\sum_{v \in V} f_v]$ does not increase. As a result, we have a deterministic sequential algorithm for which it still holds that $\mathbb{E}[\sum_{v \in V} f_v] < 1$. Since we have a deterministic algorithm and $\sum_{v \in V} f_v$ must be a non-negative integer, this means $\sum_{v \in V} f_v = 0$. That is, this sequential algorithm deterministically solves $\mathcal{P}$. Moreover, this sequential algorithm is local in the sense each value is determined by examining just a small neighborhood around it.

A Distributed Deterministic LOCAL algorithm for $\mathcal{P}$. Now, we describe a deterministic distributed LOCAL algorithm that simulates the above deterministic sequential local algorithm, using network decomposition. This is fairly similar to how we simulate the sequential greedy process for computing the MIS in the LOCAL model, using network decomposition.

Concretely, first compute a network decomposition of the graph $G^{2R(n)}$— where we connect any two vertices with distance at most $2R(n)$. This can be done in $R \operatorname{poly}(\log n) = \operatorname{poly}(\log n)$ rounds, using the network decomposition of the previous section. Then, we examine the colors of this network decomposition one by one. When examining one color, each cluster works separately, and aggregates the $R(n)$-hop topology around the cluster, including the already fixed randomness in that neighborhood. Then, the cluster simulates the above sequential process to fix the randomness of the nodes of the cluster. All clusters of the same color can work simultaneously, as the indicator variables f_v that they influence are disjoint—since the clusters of the same color are more than $2R(n)$ hops apart and each influences at most $R(n)$-hops far.   □

5   Open Problems

We have seen a $\operatorname{poly}(\log n)$-round algorithm for network decomposition in the LOCAL model. We also discussed that in the LOCAL model, efficient algorithms do not need randomness, for locally checkable problems, at least in the standard (coarse-grained) interpretation of efficiency as $\operatorname{poly}(\log n)$ round complexity. More concretely, there is an efficient distributed derandomization method, which shows that any $\operatorname{poly}(\log n)$-round randomized LOCAL algorithm for any problem whose solution can be checked deterministically in $\operatorname{poly}(\log n)$ rounds of the LOCAL model can be transformed to a $\operatorname{poly}(\log n)$-round deterministic LOCAL algorithm for the same problem. These two results have numerous implications and corollaries, and resolve several central open problems in distributed graph algorithms. See [45, Sect. 3]. In this last section, we discuss some of the related problems that remain open. Of course, this is by no-means exhaustive; we focus on questions that closely relate to the material discussed in this article.

5.1   Open Problems: LOCAL Model

(?) **Fine-Grained Polylogarithmic Complexity.** Now that the coarse-grained version of the deterministic vs. randomized question is resolved, we can look into the problem in a more fine-grained way. Concretely, an obvious question is to improve the complexity of the network decomposition and obtain a smaller exponent in the poly-logarithm. The algorithm of Rozhoň and Ghaffari [45] has an $O(\log^7 n)$ round complexity in the LOCAL model. As an ideal target, we can ask

Open Problem 1. *Is there a deterministic LOCAL model algorithm with round complexity $O(\log^2 n)$ for computing a network decomposition with $O(\log n)$ colors and $O(\log n)$ cluster diameter?*

A similar fine-grained question can be asked about a number of other central problems whose best known solution relies on network decomposition, e.g., MIS, $(\Delta + 1)$- coloring, Lovász Local Lemma, etc. Indeed, some of these might be more plausible ground for faster algorithms, because their algorithm does not necessarily need to build on network decomposition.

Open Problem 2. *Is there a deterministic LOCAL model algorithm with round complexity $O(\log n)$ for the maximal independent set problem? If not, how about for $(\Delta + 1)$ vertex coloring?*

The algorithm of Rozhon and Ghaffari [45] for MIS and $\Delta + 1$ coloring has an $O(\log^7 n)$ round complexity in the LOCAL model.

We note that even for the simplification of the MIS problem to the maximal matching problem (e.g., the special case of MIS on line graphs), currently the best known algorithm, due to Fischer [18, 19], has complexity $O(\log^2 \Delta \log n) = O(\log^3 n)$. Notably, this is one of the classic problems for which the best known solution does not rely on network decomposition. Maximal matching might be a better starting point for addressing Open Problem 2. See also [25, 31, 34] for some other problems with poly$(\log n)$ deterministic round complexity, for which the best known solution does not depend on network decomposition and there is room for improvement in poly-logarithmic bound.

Most notably, the gap for the coloring problem is much wider. While for MIS and maximal matching, deterministic algorithms cannot go much further below an $O(\log n)$ round complexity, thanks to a recent $\Omega(\log n/\log \log n)$ lower bound of Balliu et al. [5], for the $(\Delta + 1)$ coloring, there is no such obstacle known and (significantly) sublogarithmic complexities are plausible. The best known lower bound for the round complexity of $(\Delta + 1)$ coloring remains at $\Omega(\log^* n)$ [36, 37], even when restricted to deterministic algorithms.

Finally, one can view this fine-grained deterministic vs. randomized complexity through a complexity landscape lens and ask the following:

Open Problem 3. *What is the largest gap possible between the deterministic and randomized complexity of a locally checkable problem, in n-node graphs?*

Results of [45], as discussed in previous section, show that the gap is always at most $O(\log^7 n)$. On the other hand, there are problems with an $\Omega(\log n/\log \log n)$ factor gap. For instance, the sinkless orientation problem has randomized complexity $\Theta(\log \log n)$ [11, 31] and deterministic complexity $\Theta(\log \log n)$ [13, 31]. This gap was also lifted recently to higher complexities, by Balliu et al. [6], who showed locally checkable problems with $\Theta(\log \log n \cdot \log n)$ randomized complexity and $\Theta(\log^2 n)$ deterministic complexity. To the best of our knowledge, this $\Omega(\log n/\log \log n)$ gap remains the largest known separation.

(?) **Low-Diameter Ordering.** An issue with using network decomposition (for MIS, $\Delta + 1$ coloring, etc) is that, even if we can construct it very fast—e.g., in $O(\log n)$ rounds, as provided by a variant of the randomized algorithm of Linial and Saks [38]—still using the decomposition in the standard way requires $\tilde{\Omega}(\log^2 n)$ rounds. Here, $\tilde{\Omega}$ suppresses $\log \log n$ factors. The reason for the lower bound is as follows. In the standard way, we process the color classes one by one and then spend time proportional to the maximum cluster diameter per color class. As pointed out by Linial and Saks [38], in a decomposition with k colors, the cluster diameter has to be at least $\Omega(\log_k n)$—simply because there are graphs of chromatic number below $2k$ and girth $\Omega(\log_k n)$ and a better cluster diameter would contradict the chromatic number, since each lower diameter cluster is a forest and admits a 2-coloring. They also prove that in such a decomposition, the cluster-diameter is also at least $\Omega(n^{1/k})$ [38, theorem 3.1]. Considering the possibilities of k yields the aforementioned $\Omega(\log^2 n)$ lower bound[6].

One way of going around this issue, and approaching an $O(\log n)$ complexity, is to use a different helper tool for transforming sequential local algorithms to distributed LOCAL model algorithms. A concrete suggestion, closely related to network decomposition, is the *low-diameter ordering* problem introduced by Ghaffari, Kuhn, and Maus [28]. A d-*diameter* ordering is a numbering of the vertices with distinct values such that for any path with monotonically increasing numbers, the weak diameter is at most d. One can see that given a d-diameter ordering, we can solve for instance MIS in $O(d)$ rounds of the LOCAL model. Network decompositions with $O(\log n)$ colors and $O(\log n)$ cluster diameter imply an $O(\log^2 n)$-diameter ordering: set the numbering for each node v to be the tuple $(color_v, ID_v)$, and use lexicogrpahic comparisons. But this remains the best known. Already understanding the best existential bound on the low-diameter ordering—e.g., whether there is always an ordering with $O(\log n)$ diameter—is interesting. More generally, we can ask the following:

Open Problem 4. *What is the best existential bound for low-diameter ordering, i.e., what is the smallest function $d(n)$ such that any n-node graph admits a $d(n)$-diameter ordering? Also, is there a distributed algorithm with complexity $O(d(n))$ to build such an ordering?*

(?) **Shared Randomness.** Stepping back from the above fine-grained questions, a closely related coarse-grained question is worth emphasizing: While we now have a reasonable understanding of the power of private randomness, a similar questions can be asked (and remains open, to the best of our knowledge) regarding shared randomness in distributed graph algorithms.

Open Problem 5. *Is there an efficient distributed mechanism for removing shared randomness from algorithms? Concretely, consider an arbitrary locally*

---

[6] One particular question, related to this topic, that remains open to the best of our knowledge is that of computing a *minimal coloring*, that is, a coloring where each node with color i has a neighbor with color j for every $j < i$. With randomized network decomposition [38], we can compute such a coloring in $O(\log^2 n)$ rounds, with high probability. This remains the best known algorithm for minimal coloring. The problem has an $\Omega(\log n / \log \log n)$ round lower bound [21].

checkable problem P, whose solution can be checked deterministically in $O(1)$ rounds, and such that P admits a randomized LOCAL-*model algorithm with shared randomness with a* poly$(\log n)$ *round complexity in any n-node graph. Can we solve P deterministically in* poly$(\log n)$ *rounds of the* LOCAL *model?*

A somewhat related note: one can see, via a simple application of the probabilistic method that generalizes a classic argument of Newman [42] for two-party protocols, that we need at most $O(\log n)$ bits, if any, of shared randomness in distributed algorithms[7]. See also [27] for other related observations.

5.2 Open Problems: CONGEST Model

(?) **Strong Diameter Decomposition in the** CONGEST **Model.** The network decomposition algorithm of [45] also works in the CONGEST model with poly$(\log n)$ round complexity, where per round each node can send only $O(\log n)$ bits to each neighbor. The provided structure has clusters with weak-diameter poly$(\log n)$, meaning that any two vertices of the same cluster are within distance poly$(\log n)$, according to distances in the base graph. In the LOCAL model, this can be refined to a strong-diameter network decomposition: $O(\log n)$-colored clusters, each inducing a subgraph with diameter $O(\log n)$ [45]. In the CONGEST model, that question remains open.

Open Problem 6. *Devise a* poly$(\log n)$-*round deterministic algorithm in the* CONGEST *model that computes a strong-diameter network decomposition with* poly$(\log n)$ *parameters, that is, clusters colored with* poly$(\log n)$ *colors, where adjacent clusters have different colors, and such that the subgraph induced by each cluster has diameter* poly$(\log n)$.

We note that while this strong-diameter decomposition question remains open, the algorithm of [45] provides something functionally close to it, which appears to be sufficient for all known applications of network decomposition in the CONGEST model. In particular, for each cluster, the algorithm provides also a Steiner tree where the vertices of the cluster are terminal nodes. These Steiner trees have poly$(\log n)$-diameter and each edge of the graph appears in at most poly$(\log n)$ Steiner trees. This way, we can perform communications such as broadcast/convergecast in different clusters simultaneously, via their Steiner tree, with only a poly$(\log n)$ round complexity overhead. However, having a strong-diameter network decomposition, as described in Open Problem 6, would be more desirable, at least aesthetically (even if there is no extra application).

(?) **Usage in Shattering in the** CONGEST **Model.** Another extra property to desire for network decomposition in the CONGEST model has to do with their usage in the *shattering* framework [10,22,23] and the length of the identifiers. The algorithm of [45] assumes that nodes have $O(\log n)$-bit identifiers.

---

[7] This was observed in conversations with Fabian Kuhn.

In the LOCAL model, one can relax this much further and allow even S-bit identifiers where $\log^* s = \text{poly}(\log n)$, see [45, Remark 2.10]. However, that extension does not work in the CONGEST model (one can relax to $S = \text{poly}(\log n)$ bits, paying proportionally in the round complexity). In the shattering framework, [10, 22, 23], after running some randomized algorithm, we are left with connected components, each of which has only $N = O(\log n)$ nodes (or $\text{poly}(\Delta) \log n$ nodes, with some extra ruling set property, which allows us to move to a case similar to the case with $O(\log n)$ nodes). This opens the road to running on each component a deterministic algorithm with complexity $\text{poly}(\log N) = O(\log \log n)$, as would be suggested by [45]. However, there is a catch: The network decomposition algorithm assumes $O(\log N)$-bit identifiers on N-node graphs and we have only N-bit identifiers (inherited from the basic problem, before shattering). Hence, we cannot directly apply the algorithm of [45], in the CONGEST model.[8]

Open Problem 7. *Consider a CONGEST model with S-bit identifiers, for $S = \Omega(\log n)$, and $O(S)$-bit messages. Can we devise a $\text{poly}(\log n)$-round deterministic CONGEST model algorithm for network decomposition with $\text{poly}(\log n)$ parameters, so long as $\log^* S \leq \text{poly}(\log n)$ or even just $S \leq n$.*

Without resolving this, or finding a way around it, it is unclear how to obtain CONGEST-model randomized algorithms that enjoy from the new $\text{poly}(\log n)$-round deterministic network decomposition, and have a $\text{poly}(\log \log n)$ round complexity in the part after shattering. As of now, to the best of our knowledge, the state of the art randomized MIS algorithm in the CONGEST model remains at complexity $O(\log \Delta \sqrt{\log \log n} + 2^{O(\sqrt{\log \log n})})$ [30] and the state of the art randomized $(\Delta + 1)$-coloring in the CONGEST model remains at complexity $O(\log \Delta + 2^{O(\sqrt{\log \log n})})$ [23]. Both of these have terms that are clearly remnants of the $2^{O(\sqrt{\log n})}$-complexity network decomposition [23, 43]. And they have clear gaps to the corresponding LOCAL-model bounds, which are $O(\log \Delta + \text{poly}(\log \log n))$ [22] and $\text{poly}(\log \log n)$ [14, 15], respectively.

(?) **Simpler MIS Algorithms, without Derandomization.** As noted before, the network decomposition algorithm of [45] works also in the CONGEST model and when put together with the CONGEST-model deterministic MIS algorithm of Censor-Hillel et al. [12], leads to a $\text{poly}(\log n)$-round deterministic MIS algorithm in the CONGEST model. However, there is still one extra property to desire. The algorithm of [12] is far from simple and works by fixing the randomness of the randomized algorithm of [22], bit by bit, via conditional expectation (after observing that pairwise independence is sufficient per round, which implies that we need to fix only $O(\log n)$ bits): concretely, one uses a pessimistic estimator on the expected number of bad events, if we set the bit to be either 0 or 1, and then we choose the bit that does not increase the estimator, via gathering the two values over the entire network. This also requires the nodes to perform some

---

[8] If the CONGEST model was relaxed to allow $O(\log^2 n)$-bit messages, this issue would go away as then we could perform a renaming in each component similar to [45, Remark 2.10], since all relevant identifiers would fit in one message.

complex expectation calculations, based on the identifiers of their neighbors and the remaining space of randomness.

Given that MIS is a simple and natural problem, one can hope to see a more direct deterministic algorithm for MIS in the CONGEST model, that does not rely in this way on derandomizing some randomized algorithm. Hopefully, this could also provide a simpler and more practical deterministic algorithm. Note that it suffices to solve the problem in low-diameter graphs, with diameter $\mathrm{poly}(\log n)$, as this can then be extended to all graphs via network decomposition.

In this direction, one concrete result worth mentioning, and perhaps a starting point for algorithmic ideas, is something that can be obtained via a classic PRAM algorithm of Goldberg and Spencer [32]. This is simple and combinatorial algorithm. It appears[9] that one can obtain a distributed $\mathrm{poly}(\log n)$-round deterministic MIS by following this approach, using $O(\Delta^2 + \log n)$-bit messages. This is still much less than the message size needed in the topology-gathering approach (as described in Sect. 2), in graphs with reasonably small degrees.

We also note that there are also similar efficient deterministic CONGEST-model algorithms for other problems, including $\Delta + 1$ coloring (and even $degree + 1$ list coloring), and dominating set and set cover approximations. See [45, Corollary 3.17 & 3.18] and [16,26]. All of these rely on explicit derandomization of some randomized algorithm. Having a more direct algorithm could be valuable.

Acknowledgement. I am grateful to David Harris, Fabian Kuhn, Yannic Maus, and Václav Rozhoň for the joint work in [24,28,45] and discussions.

References

1. Alon, N., Babai, L., Itai, A.: A fast and simple randomized parallel algorithm for the maximal independent set problem. J. Algorithms **7**(4), 567–583 (1986)
2. Awerbuch, B., Berger, B., Cowen, L., Peleg, D.: Fast network decompositions and covers. J. Parallel Distrib. Comput. **39**(2), 105–114 (1996)
3. Awerbuch, B., Goldberg, A.V., Luby, M., Plotkin, S.A.: Network decomposition and locality in distributed computation. In: Proceedings of 30th IEEE Symposium on Foundations of Computer Science (FOCS), pp. 364–369 (1989)
4. Awerbuch, B., Peleg, D.: Sparse partitions. In: Proceedings of 31st IEEE Symposium on Foundations of Computer Science (FOCS), pp. 503–513 (1990)
5. Balliu, A., Brandt, S., Hirvonen, J., Olivetti, D., Rabie, M., Suomela, J.: Lower bounds for maximal matchings and maximal independent sets. In: Proceedings of the Symposium on Foundations of Computer Science (FOCS) (2019)
6. Balliu, A., Brandt, S., Olivetti, D., Suomela, J.: How much does randomness help with locally checkable problems? In: Proceedings of the Symposium on Principles of Distributed Computing (PODC), pp. to appear, arXiv:1902.06803 (2020)
7. Bamberger, P., Kuhn, F., Maus, Y.: Efficient deterministic distributed coloring with small bandwidth. In: Proceedings of the Symposium on Principles of Distributed Computing (PODC), pp. to appear, arXiv:1912.02814 (2020)

---

[9] This should be taken with a grain of salt; we have not checked the details thoroughly.

8. Barenboim, L., Elkin, M.: Deterministic distributed vertex coloring in polylogarithmic time. In: Proceedings of 29th Symposium on Principles of Distributed Computing (PODC), pp. 410–419 (2010)

9. Barenboim, L., Elkin, M.: Distributed graph coloring: fundamentals and recent developments. Synthesis Lect. Distrib. Comput. Theory **4**(1), 1–177 (2013)

10. Barenboim, L., Elkin, M., Pettie, S., Schneider, J.: The locality of distributed symmetry breaking. J. ACM **63**(3), 1–45 (2016)

11. Brandt, S., et al.: A lower bound for the distributed Lovász local lemma. In: Proceedings of the Symposium on Theory of Computation (STOC), pp. 479–488 (2016)

12. Censor-Hillel, K., Parter, M., Schwartzman, G.: Derandomizing local distributed algorithms under bandwidth restrictions. In: Proceedings of the Symposium on DIStributed Computing (DISC) (2017)

13. Chang, Y.J., Kopelowitz, T., Pettie, S.: An exponential separation between randomized and deterministic complexity in the LOCAL model. In: Proceedings of 57th IEEE Symposium on Foundations of Computer Science (FOCS) (2016)

14. Chang, Y.J., Li, W., Pettie, S.: An optimal distributed $(\Delta+1)$-coloring algorithm? In: Proceedings of 50th ACM Symposium on Theory of Computing (STOC) (2018)

15. Chang, Y.J., Li, W., Pettie, S.: Distributed $(\Delta+1)$-coloring viaultrafast graph shattering. SIAM J. Comput. **49**(3), 497–539 (2020)

16. Deurer, J., Kuhn, F., Maus, Y.: Deterministic distributed dominating set approximation in the congest model. In: Proceedings of the Symposium on Principles of Distributed Computing (PODC), pp. 94–103 (2019)

17. Elkin, M., Neiman, O.: Distributed strong diameter network decomposition. In: Proceedings of the Symposium on Principles of Distributed Computing (PODC) (2016)

18. Fischer, M.: Improved Deterministic Distributed Matching via Rounding. In: Proceedings of the Symposium on DIStributed Computing (DISC), pp. 1–15 (2017)

19. Fischer, M.: Improved deterministic distributed matching via rounding. Distrib. Comput. 1–13 (2018)

20. Fischer, M., Ghaffari, M., Kuhn, F.: Deterministic distributed edge-coloring via hypergraph maximal matching. In: Proceedings of the Symposium on Foundations of Computer Science (FOCS) (2017)

21. Gavoille, C., Klasing, R., Kosowski, A., Kuszner, L., Navarra, A.: On the complexity of distributed graph coloring with local minimality constraints. Netw. Int. J. **54**(1), 12–19 (2009)

22. Ghaffari, M.: An improved distributed algorithm for maximal independent set. In: Proceedings of the Symposium on Discrete Algorithms (SODA), pp. 270–277 (2016)

23. Ghaffari, M.: Distributed maximal independent set using small messages. In: Proceedings of the Symposium on Discrete Algorithms (SODA), pp. 805–820 (2019)

24. Ghaffari, M., Harris, D., Kuhn, F.: On derandomizing local distributed algorithms. In: Proceedings of the Symposium on Foundations of Computer Science (FOCS), pp. 662–673 (2018)

25. Ghaffari, M., Hirvonen, J., Kuhn, F., Maus, Y., Suomela, J., Uitto, J.: Improved distributed degree splitting and edge coloring. In: Proceedings of the Symposium on DIStributed Computing (DISC) (2017)

26. Ghaffari, M., Kuhn, F.: Derandomizing distributed algorithms with small messages: Spanners and dominating set. In: 32nd International Symposium on Distributed Computing (DISC 2018) (2018)

27. Ghaffari, M., Kuhn, F.: On the use of randomness in local distributed graph algorithms. In: Proceedings of the Symposium on Principles of Distributed Computing (PODC), pp. 290–299 (2019)
28. Ghaffari, M., Kuhn, F., Maus, Y.: On the complexity of local distributed graph problems. In: Proceedings of the Symposium on Theory of Computation (STOC), pp. 784–797 (2017)
29. Ghaffari, M., Kuhn, F., Maus, Y., Uitto, J.: Deterministic distributed edge-coloring with fewer colors. In: Proceedings of the Symposium on Theory of Computation (STOC), pp. 418–430 (2018)
30. Ghaffari, M., Portmann, J.: Improved network decompositions using small messages with applications on mis, neighborhood covers, and beyond. In: 33rd International Symposium on Distributed Computing (DISC 2019) (2019)
31. Ghaffari, M., Su, H.H.: Distributed degree splitting, edge coloring, and orientations. In: Proceedings of the Twenty-Eighth Annual ACM-SIAM Symposium on Discrete Algorithms, pp. 2505–2523. Society for Industrial and Applied Mathematics (2017)
32. Goldberg, M., Spencer, T.: Constructing a maximal independent set in parallel. SIAM J. Discrete Math. **2**(3), 322–328 (1989)
33. Hańćkowiak, M., Karoński, M., Panconesi, A.: On the distributed complexity of computing maximal matchings. SIAM J. Discrete Math. **15**(1), 41–57 (2001)
34. Harris, D.G.: Distributed local approximation algorithms for maximum matching in graphs and hypergraphs. In: Proceedings of the Symposium on Foundations of Computer Science (FOCS), pp. 700–724 (2019)
35. Israeli, A., Itai, A.: A fast and simple randomized parallel algorithm for maximal matching. Inf. Process. Lett. **22**(2), 77–80 (1986)
36. Linial, N.: Distributive graph algorithms - global solutions from local data. In: Proceedings of the Symposium on Foundations of Computer Science (FOCS), pp. 331–335 (1987)
37. Linial, N.: Locality in distributed graph algorithms. SIAM J. Comput. **21**(1), 193–201 (1992)
38. Linial, N., Saks, M.: Low diameter graph decompositions. Combinatorica **13**(4), 441–454 (1993)
39. Luby, M.: A simple parallel algorithm for the maximal independent set problem. SIAM J. Comput. **15**, 1036–1053 (1986)
40. Miller, G.L., Peng, R., Xu, S.C.: Parallel graph decompositions using random shifts. In: Proceedings of the Symposium on Parallel Algorithms and Architecture (SPAA), pp. 196–203 (2013)
41. Naor, M., Stockmeyer, L.: What can be computed locally? SIAM J. Comput. **24**(6), 1259–1277 (1995)
42. Newman, I.: Private vs common random bits in communication complexity. Inf. Process. Lett. **39**(2), 67–71 (1991)
43. Panconesi, A., Srinivasan, A.: Improved distributed algorithms for coloring and network decomposition problems. In: Proceedings of the Symposium on Theory of Computation (STOC), pp. 581–592 (1992)
44. Peleg, D.: Distributed Computing: A Locality-Sensitive Approach. SIAM, Philadelphia (2000)
45. Rozhoň, V., Ghaffari, M.: Polylogarithmic-time deterministic network decomposition and distributed derandomization. In: Proceedings of the Symposium on Theory of Computation (STOC), pp. to appear, arXiv:1907.10937 (2020)
46. Windsor, A.: A simple proof that finding a maximal independent set in a graph is in NC. Inf. Process. Lett. **92**(4), 185–187 (2004)

Resource Burning for Permissionless Systems (Invited Paper)

Diksha Gupta[1(✉)], Jared Saia[1], and Maxwell Young[2]

[1] Department of Computer Science, University of New Mexico,
Albuquerque, NM, USA
{dgupta,saia}@cs.unm.edu

[2] Department of Computer Science and Engineering, Mississippi State University,
Mississippi State, MS, USA
myoung@cse.msstate.edu

Abstract. Proof-of-work puzzles and CAPTCHAS consume enormous amounts of energy and time. These techniques are examples of resource burning: verifiable consumption of resources solely to convey information.

Can these costs be eliminated? It seems unlikely since resource burning shares similarities with "money burning" and "costly signaling", which are foundational to game theory, biology, and economics. Can these costs be reduced? Yes, research shows we can significantly lower the asymptotic costs of resource burning in many different settings.

In this paper, we survey the literature on resource burning; take positions based on predictions of how the tool is likely to evolve; and propose several open problems targeted at the theoretical distributed-computing research community.

> *"It's not about money, it's about sending a message."*
> The Joker [107]

1 Introduction

In 1993, Dwork and Naor proposed using computational puzzles to combat spam email [43]. In the ensuing three decades, *resource burning*—verifiable consumption of resources—has become a well-established tool in distributed security. The resource consumed has broadened to include not just computational power, but also communication capacity, computer memory, and human effort.

The rise of *permissionless systems* has coincided with the recent increase in popularity of resource burning. In permissionless systems, any participant—represented by a virtual identifier (***ID***) in the system—is free to join and depart without scrutiny, while enjoying a high degree of anonymity. For example, an ID might be an IP address, a digital wallet, or a username.

In this setting, security challenges arise from the inability to link an ID to the corresponding user. A single malicious user may create a large number of

---

This work is supported by the National Science Foundation grants NSF-CNS-1816250 and NSF-CNS-1816076.

A. W. Richa and C. Scheideler (Eds.): SIROCCO 2020, LNCS 12156, pp. 19–44, 2020.
https://doi.org/10.1007/978-3-030-54921-3_2

accounts on a social media platform to wield greater influence; or present itself as multiple clients to disproportionately consume resources provided by the system; or inject many IDs in a peer-to-peer network to gain control over routing and content. This malicious behavior is referred to as the ***Sybil attack***, originally described by Douceur [41].

Such attacks are possible because users are not "ID-bounded" in a permissionless system; that is, there is no cost, and therefore no limit, to the number of IDs that the attacker (***adversary***) can generate. However, the adversary is often "resource-bounded", even if this bound is unknown. In particular, it may be constrained, for example, in the number of machines it controls, or total channel capacity to which it has access. Resource burning leverages this constraint, forcing IDs to prove their distinct provenance by producing work that no single attacker can perform.

Paper Overview. Resource burning is a critical tool for defending permissionless systems. In support of this claim, we survey an assortment of topics: distributed ledgers, application-layer distributed denial-of-service (DDoS) attacks, review spam, and secure distributed hash tables (DHTs). Using these examples, we highlight how results in these different areas have converged upon resource burning as a critical ingredient for achieving security; this is summarized in Table 1.

Table 1. Summary of the domains surveyed, along with the corresponding resources, and core functionality that is secured by resource burning. We also make conjectures on the algorithmic spend rate. Here, T is the adversary's spend rate; J_G is the join rate for good IDs; and P_G is the posting rate of good IDs. We elaborate on these notions in Sect. 2.5. The $\tilde{O}$ notation omits polylogarithmic factors.

Domain	Primary resource consumed	Mechanism	Enabled functionality	Conjectured cost
Blockchains	CPU	CPU puzzles	Distributed ledger	$O(\sqrt{TJ_G} + J_G)$
DHTs	CPU	CPU puzzles	Decentralized storage and search	$\tilde{O}(\sqrt{TJ_G} + J_G)$
DDoS attacks	Bandwidth/CPU	Messages/CPU puzzles	Fair allocation of server resources	No conjecture
Review spam	Human time	CAPTCHAS	Trusted consumer recommendations	$\tilde{O}(T^{2/3} + P_G)$

As prelude to this survey, we predict how resource-burning may evolve, and how systems may adapt to this technique. These predictions are distilled in four position statements below.

> **Position 1.** *Resource burning is a fundamental tool for defending permissionless systems.*

PoW and CAPTCHAs have been around now for decades, persisting despite concerns over scalability, resource consumption, security guarantees, and predicted obsolescence (see discussion under Position 2 and Sect. 3). The continued practical success of resource burning aligns with theoretical justification from game-theoretic results on "money-burning" and "costly signaling" (Sect. 2.1). Given the increasing popularity of permissionless systems, and the need to defend them, resource burning will likely only increase in prevalence.

> **Position 2.** *Resource burning must be optimized.*

In May 2020, the annual energy consumption of Bitcoin was 57.92 terawatt-hours of electricity per year, which is comparable to the annual electricity consumption of Bangladesh; Ethereum was 7.9 terawatt-hours, comparable to that of Angola [39]. In 2012, humans spent an estimated $150,000$ hours per day solving CAPTCHAS [114,137]. The rise of permissionless systems will likely only increase these rates of resource burning.

On the positive side, recent theoretical results suggest that resource burning can be analyzed and optimized just like any other computational resource [59,61]. But there is significant work needed to: (1) develop a theory of resource burning focused on distributed security; and to (2) translate this theory into practical resource savings.

In this paper, we discuss current theoretical work on reducing resource-burning rates across multiple application: blockchains (Sect. 3); DHTs (Sect. 4); application-layer DDoS attacks (Sect. 5); and review spam (Sect. 6).

> **Position 3.** *Reducing from permissionless to permissioned systems is important.*

Four decades of research have resulted in efficient and reliable algorithms for permissioned networks. We should leverage these results when addressing problems in new permissionless systems. One way to do this is to develop tools, based on resource burning, that bound the fraction of IDs controlled by the adversary (***bad IDs***) in permissionless systems. In Sect. 3, we discuss results on the problem GENID, which provides this bound for static, permissionless networks; and DEFID which does so for permissionless networks with churn.

In Sects. 5 and 6, we discuss the threats posed by application-layer denial-of-service (DDoS) attacks and review spam. Neither problem aligns perfectly with a permissionless model. For example, servers are under administrative control, and online review systems often require credentials for account creation. However, these systems still remain vulnerable to malicious participants that are difficult to identify, and who monopolize system resources. We define a *hybrid* system model as one that contains both permissioned and permissionless properties. We note that any tools designed for permissionless systems will also work for hybrid systems. However, we would expect to be able to develop more efficient techniques to adapt tools from permissioned systems to these hybrid systems.

> **Position 4.** *Theoretical guarantees should hold indepen-*
> *dently of the resource burned. Research should focus on*
> *both domain-specific and domain-generic problems.*

As theoreticians, we should generalize as much as possible. Algorithms that use resource burning should require a certain "cost" that specifies the amount of the resource to be consumed, but should allow for that resource to be anything: computation, computer memory, bandwidth, human effort, or some other resource yet to be defined. As much as possible, theoretical results should be stated in terms of this cost, irrespective of the resource consumed. This ensures our theoretical results will continue to be relevant, even as underlying technologies providing verifiable resource burning may change.

Additionally, a key research focus should be on problems that generalize across multiple domains. In this paper, we describe two examples: GENID and DEFID (Sect. 3.1). Our remaining three examples are domain-specific. We believe it is important to work on both types of problems.

2 Background and Preliminaries

Resource burning has found application in various areas of computer security; indeed, its use was proposed by Douceur [41] as a defense against the Sybil attack [40,75,100,106]. However, resource burning has a broader history, with similar ideas appearing in several other scientific domains.

In Sect. 2.1, we present this background. In Sects. 2.2, 2.3, and 2.4, we elaborate on the notion of resource burning. Finally, in Sect. 2.5, we describe a general problem model that provides a unifying set of assumptions and terminology used throughout this document.

2.1 Game Theory, Biology and Economics

Resource burning is analogous to what is referred to as *money burning* in the game theory literature. To the best of our knowledge, the first significant algorithmic game theory study of money burning, due to Hartline and Roughgarden, analyzed the use of money burning in mechanism design [61]. Their main result is a near-optimal mechanism for multi-unit auctions, where the quantity optimized is *social welfare* or the sum of utilities of all players. They also give results showing that, under certain conditions, an auction utilizing money burning can obtain a $\Omega\left(\frac{1}{1+\log(n/k)}\right)$ fraction of the optimal social welfare, where the auction consists of n bidders who are bidding for k units. They conclude that *"the cost of implementing money-burning ... is relatively modest, provided an optimal money-burning mechanism is used"*.

Money burning is also known as *costly signaling* in the game theory literature, and it has two main uses in this context. First, it can signal commitment to a certain action, as is illustrated in the "lunch" game[1] [29,68]. Second, it can signal

---

[1] This is equivalent to what is referred to as the "battle of the sexes" game in [29].

the "type" of a player, as is in the"college" game [29]. We present these two games below.

Lunch Game. Two friends want to eat lunch together, but the first friend prefers option A and the second prefers option B. They each obtain payoff of −1 if they choose different locations. If they both pick option A, they obtain payoffs of 10 and 1 respectively. Conversely, if they both pick option B, they obtain payoffs of 1 and 10.

Now, if the first friend verifiably burns money equal to 1 unit of utility prior to playing the game, this signals a commitment to their preferred option, since if they were to choose the unpreferable option, their utility would now be at most 0. Thus, they would not have played the game. In this way, a friend who burns money can expect higher utility.

College Game. Each student is one of two types: *smart* or *daft*. Each student is considering college and can choose either the action *attend* or *not attend*. A smart student pays a cost of 1 (in terms of time and effort) to attend college, and a daft student pays a cost of 3 to attend college. We assume that the decision of the student to attend college is publicly known, but that otherwise, college has no impact: daft students stay daft even after attending.[2]

An employer wants to hire smart students. If the employer hires a smart student, their benefit is 2, and if they hire a daft student, their cost is 2. If a student is hired by the employer, they have a benefit of 2, and if they are not hired, they have a benefit of 0.

It is easy to verify that the following is a Nash equilibrium for this game:

○ Smart students attend college.
○ Daft students do not attend college.
○ The employer hires only students that attend college.

Here, smart students all choose to attend, *even though college has no intrinsic benefit*. Thus, the choice to attend college is a costly signal made by the smart students, and college itself is an example of resource burning.

If the option to attend college were removed from the game, and the fraction of smart students were less than 1/2, then a Nash equilibrium would be for the employer to never hire. In this case, the overall social welfare—the sum of expected benefits to all players—would decrease.

Biology. Costly signaling is a well-known phenomena in biology. A relevant example from animal behavior is *stotting*, in which quadrupeds, such as deer and gazelles, repeatedly jump high into the air. This is often done in view of a predator, suggesting that stotting is a costly signal to the predator that the prey is too healthy to catch [49]. Other examples occur in sexual-selection, where the use of plumage, large antlers, and loud cries are a costly signal of fitness [156].

Economics. In 1912, the economist Thostein Veblen coined the term "conspicuous consumption" to describe costly signaling used by people to advertise both

---

[2] On the positive side, smart students stay smart!.

wealth and leisure. For example, Veblen writes, *"The walking stick serves the purpose of an advertisement that the bearer's hands are employed otherwise than in useful effort, and it therefore has utility as an evidence of leisure"* [134]. Decades of economic studies suggest that conspicuous consumption is a critical part of historical and modern economies [98,113,118,119,131]. For example, Sundie et al write: *"Although showy spending is often perceived as wasteful, frivolous, and even narcissistic, an evolutionary perspective suggests that blatant displays of resources may serve an important function, namely, as a communication strategy designed to gain reproductive reward"*[131].

2.2   What is Resource Burning?

We define **resource burning** as the verifiable consumption of a resource. In particular, it is computationally easy to verify both the consumption of the resource, and also the ID that consumed the resource [6]. Below we describe several resource-burning techniques.

Proof-of-Work (PoW). PoW is arguably the current, best-known example of resource burning. Here, the resource is computational power. Proof-of-work has been proposed for spam-prevention [43,85,90]; blockchains [103]; and defense against Sybil attacks [10,88].

CAPTCHAs. A *completely automated public Turing test to tell computers and humans apart*, or a CAPTCHA, is a resource-burning tool where the resource is human effort [147]. CAPTCHAs may be based around text, images, or audio; however, several design and usability issues exist [148].

Proof-of-Space. Proof-of-space requires a prover to demonstrate utilization of a certain amount of storage space [1,13,42,44]. This approach is foundational for Spacemint cryptocurrency [111]. Like PoW, proof of space demonstrates the consumption of a certain amount of a physical resource, but can require less electrical power. A related proposal is "Proof of Space-Time" [102], which demands proof of consumption of a certain amount of storage space for a certain amount of time.

Resource Testing. Resource testing requires a prover to demonstrate utilization of a radio channel [55,56,101].[3] Consider a wireless setting where each device has a single radio that provides access to one of several channels. Thus, an adversary representing two bad IDs, but with a single device, can only listen to one channel at a time. A base station can assign each ID to separate channels; send a random message on one of these channels chosen randomly; and demand that the message be echoed back by the corresponding ID. Since the adversary can only listen to a one channel at a time, it will fail this test with probability at least $1/2$.

---

[3] Resource burning refers to the game-theoretic money burning technique; resource testing refers to that technique specifically applied in the wireless domain.

2.3   What is *not* Resource Burning

Proof-of-Stake (PoS) is a defense for permissionless systems, wherein security relies on the adversary holding a minority stake in an abstract finite resource [2]. It has been proposed primarily for cryptocurrency systems (Sect. 3). When making a group decision, PoS ensures that each ID has voting weight proportional to the amount of cryptocurrency that ID holds. Well-known examples of such systems are ALGORAND [54], which employs PoS to form a committee, and Ouroboros [83], which elects leaders with probability proportional to their stake. Hybrid approaches using both PoW and PoS exist, including one proposed for the Ethereum system [8], and under the name "Proof of Activity" [27]. In contrast to the above examples, PoS involves a measurement, rather than a consumption of, a resource.

Disadvantages of Proof-of-Stake. Unfortunately, PoS can only be used in systems where the "stake" of each ID is globally known. Thus, it seems likely to remain relevant primarily in the domain of cryptocurrencies. Moreover, even within that community, there are concerns about proof-of-stake. To quote researcher Dahlia Malkhi: *"I think proof-of-stake is fundamentally vulnerable ... In my opinion, it's giving power to people who have lots of money"* [35].

2.4   Resource Burning Does Not Require *Waste* of the Resource

While resource burning requires verifiable *consumption* of a resource, it does not necessarily require *waste* of that resource. For example, Von Ahn et al. [137] developed the reCAPTCHA system which channeled human effort from solving CAPTCHAs into the problem of deciphering scanned words that could not be recognized by computer. Their system achieved an accuracy exceeding professional human transcribers, and was responsible for sucesssfully transcribing hundreds of millions of words from public domain books.

In 2018, Ball et al. developed proof-of-work puzzles whose hardness is based on worst-case assumptions [25]. These puzzles are based on the Orthogonal Vectors, 3SUM, and All-Pairs Shortest Path problems, and any problem that reduces to these problems, including deciding any graph property statable in first-order logic. Hence, their work enables design of PoW puzzles that can be useful for solving computational problems of practical importance.

In [126], Shoker developed proof-of-work puzzles that solve real-world matrix-based scientific computation problems. He named this technique "Proof of Exercise".

All algorithms discussed in this paper are compatible with this type of "useful" resource burning, where the consumption of the resource solves practical problems. Our only requirement of the resource burning mechanism is that the consumption of the resource be easily verifiable, which holds true for the above results.

2.5   A General Model

We discuss broad aspects of a general model for permissionless systems. This allows us to highlight commonalities between different application domains, while retaining the same terminology throughout.

The system consists of virtual *identifiers (IDs)*. An ID is *good* if it obeys protocol and belongs to a unique user; otherwise, the ID is *bad*. Good and bad IDs cannot necessarily be distinguished *a priori*.

Communication. Communication is synchronous and occurs either via point-to-point or via a broadcast primitive. The former is typical for peer-to-peer systems and the general client-server setting. The latter corresponds to permissionless blockchains, where it is a standard assumption that a good ID may send a value to all other good IDs within a known and bounded amount of time, despite an adversary; for examples, see [30,52,54,92] and see [97] for empirical justification.

Adversary. A single adversary controls all bad IDs; this pessimistically represents perfect collusion and coordination by malicious users. Bad IDs may arbitrarily deviate from our protocol, including sending incorrect or spurious messages. The adversary can send messages to any ID at will, and can view any communications sent by good IDs before sending its own. It knows when good IDs join and depart, but it does not know in advance the private random bits generated by any good ID.

Often, the adversary is assumed to control only an α-fraction of the network resources, for $\alpha > 0$. Generally, in settings where correctness is threatened, α must be a small constant; for example, often bounded below $1/3$ or $1/4$. Alternatively, there are settings where α can be any constant bounded away from 1; typically, this corresponds to problems of performance (rather than correctness).

Tunable Costs. We measure *cost* as the amount of resource consumed. Our model is agnostic with respect to the particular resource used. However, we assume that it is possible to arbitrarily tune the cost. In particular, we assume that, for any value x, an ID can be issued a *challenge* of difficulty x that will require consumption of x units of whatever resource is used.

Resources such as computation, computer memory, and bandwidth have inherently tunable costs. For CAPTCHAs, cost could be adjusted in two possible ways. First, by adjusting the difficulty of the puzzle, by either (1) adjusting the number of alphanumeric digits or the number of images to be classified; or (2) adjusting the difficulty of an individual recognition task as described in the ScatterType CAPTCHA system [24]. Second, by adjusting the *expected* difficulty by adjusting a probability of being required to solve a CAPTCHA.

Joins and Departures. Often, the system is dynamic, with IDs joining or departing over time. There is no *a priori* method for determining whether a joining ID is good or bad. Joins and departures by bad IDs may be scheduled in a worst-case fashion, and pessimistically we often assume the adversary also has a limited ability to schedule these events for good IDs. We will generally assume a lower bound on the number of IDs in the system, and that the lifetime of the system is polynomial in this lower bound.

Key Notation. Through out this work, let T denote the **adversarial spending rate**, which is the cost to the adversary over the system lifetime divided by the lifetime of the system. Let the **algorithmic spending rate**, A, be the cost to all good IDs over the system lifetime divided by the lifetime of the system.

In the blockchain and DHT problems, we let J_G denote the **good ID join rate**, which is the number of good IDs that join during the system lifetime divided by the lifetime of the system. Finally, for the review spam problem, we let P_G denote the **good posting rate**, which is the number of posts made by good IDs during the system lifetime divided by lifetime of the system.

2.6   Game Theoretic Analysis

For many of our problems, we can analyze the defense of a system as a two-player zero sum game [45] as follows. There is an adversary that can choose to attack or not, and an algorithm that can choose to defend or not. There is a system invariant, which the algorithm seeks to protect, that has some value V. There is a function f that gives the cost incurred when the algorithm chooses to defend as follows: if the adversary spends T to attack, then the algorithm will spend $f(T)$ to defend. Thus the payoff matrix for the algorithm is given below.

Adversary

		Attack	¬Attack
Algorithm	Defend	$T - f(T)$	$-f(0)$
	¬Defend	$-V$	0

Solving this game, we get that in the Nash equilibrium, the algorithm player will defend with probability $p = \frac{V}{T-f(T)+f(0)+V}$. Thus, the expected utility of the game to the algorithm player will be $\frac{-Vf(0)}{T-f(T)+f(0)+V}$. In many of our problems, $f(T) = f(0) + o(T)$, and so we obtain a value that is $\Theta\left(\frac{-Vf(0)}{T+V}\right)$. Smaller T optimizes the utility for the adversary, in which case, the expected utility of the algorithm is $\Theta(-f(0))$.

3   Blockchains and Cryptocurrencies

A *blockchain* is a distributed ledger. In particular, it is a distributed data structure that stores transactions between IDs in a network. Each transaction represents flow of a resource from one ID to another. Every transaction added must be legitimate, in the sense that the source ID owns the resource to be transferred, as indicated by the distributed ledger, at the time of the transaction. Importantly, transactions can only be added to the blockchain, and once added, can never be deleted or edited.

3.1   GENID and DEFID

Perhaps the current, most frequently-used application of resource-burning is for blockchains. Permissionless blockchains are vulnerable to Sybil attacks [89]. The next two problems use resource burning to defend against this. Recall that the adversary controls an α-fraction of the resource that is being burned.

The GENID Problem. The problem stated below, GENID, was first defined and studied by Aspnes, Jackson, and Krishnamurthy [11]. They proposed a solution with latency of 3 rounds, and $\tilde{O}(n^2)$ bits sent per good ID, at a burned resource cost of $O(1)$ per good ID.

Problem 1. **GENID**

Model: Initial set of IDs; n of which are good, with the rest are controlled by an adversary.
Goal: All good IDs decide on a set of IDs S such that: (1) all good IDs are in S; and (2) at most a $O(\alpha)$ fraction of the IDs in S are adversarial.

Several other solutions to GENID have been proposed in the literature [4, 10, 67, 81]. Andrychowicz and Dziembowski described an algorithm with a latency of $\Theta(n)$ rounds; $\tilde{O}(n^2)$ bits sent per good ID; and a burned resource cost of $\tilde{O}(1)$ per good ID [10]. Concurrent to this work, Katz, Miller and Shi [81] proposed another solution with similar costs. Hao et al. [67] improved on these results via using a randomized leader election protocol. Their algorithm has, in expectation, a latency of $\Theta\left(\frac{\ln n}{\ln \ln n}\right)$ rounds; $\tilde{O}(n)$ bits sent per good ID; and a burned resource cost of $\Theta\left(\frac{\ln n}{\ln \ln n}\right)$ per good ID.

The most recent work in this domain is by Aggarwal et al. [4], which requires in expectation: $O(1)$ latency; $O(n)$ bits sent per good ID; and a burned resource cost of $O(1)$ per good ID.

It is still not known if these costs can be reduced for the general problem, or for an "almost-everywhere" versions of the problem, where all but a $o(1)$ fraction of the IDs must learn S. To the best of our knowledge, there are no current lower-bounds on the problem.

The DEFID Problem. The following problem, called DEFID, considers the GENID problem in the presence of churn.

Problem 2. **DEFID**

Model: Stream of IDs joining and leaving a network.
Goal: At most an $O(\alpha)$-fraction of bad IDs in the network at any time.

A first algorithm to solve DEFID was proposed in by Gupta, Saia and Young in [58]. It required algorithmic spend rate of $O(J_G + T)$; recall that J_G is the join rate of good IDs per time step, and T is the spend rate of the adversary. Note that this result holds without any additional assumptions. Gupta, Saia and Young further improved this result in [59, 60] to $O(J_G + \sqrt{T J_G})$, subject to two assumptions on the join rate of good IDs, which are found to be supported by real-world data [59].

Specifically, the assumptions needed are as follows. Define an **epoch** to be the length of time it takes for the fraction of good IDs to change by $3/4$ fraction. First, the join rate for good IDs changes by at most a multiplicative factor between any two successive epochs. Second, in any epoch the actual join rate for good IDs over any "sufficiently large" period of time is within constant factors of the join rate for good IDs over the entire epoch.

An asymptotically matching lower bound was obtained for a large class of algorithms [59]. An open problem is to generalize this bound to all algorithms.

4   Distributed Hash Tables

Distributed hash tables (DHTs) are a popular P2P distributed data structure [3,80,87,96,117,129] with several implementations over the years [46,128,141]. Generally, the design entails hashing attributes of a user's machine to a key value (or ID) in a virtual space; similarly, for data items. The various DHT constructions differ in their overlay topologies, but typically IDs need only maintain state on a small number of neighbors, and routing is possible with a small number of messages, where small means at most logarithmic in the number of IDs in the system.

These systems are vulnerable to attack. A bad ID that participates in routing can drop or corrupt any message it receives. A good ID can be completely isolated from the rest of the network if all of its neighbors are bad; this is often referred to as an **eclipse attack** [63,127]. Finally, content can be compromised if bad IDs alone are responsible for storing a particular data item. Generally, the behavior of bad IDs is modeled by Byzantine faults. For almost two decades, there has been a sustained interest in the design of secure DHTs that can tolerate such attacks [135].

Byzantine Fault Tolerance in DHTs. A popular approach to tolerating bad IDs depends makes use of **groups**: these are small sets of IDs, each of which have a good majority. Intuitively, a group is used in place of an individual peer, and the group members act by using majority action or secure multiparty computation to coordinate actions. For example, routing can be performed robustly via all-to-all communication between each pair of groups along the path from source to destination. Examples of group-based DHT constructions include [21–23,48,72, 105,122,125,151].

As an alternative to using groups, bad IDs may be tolerated by employing some form of redundant routing [32,65,74,78,82,104]. Several other results do not explicitly apply to DHTs, although they may be compatible. For example, the challenge of tolerating bad IDs is exacerbated in highly-dynamic P2P systems, and there is a growing body of work in this area [14–18,57]. Self-healing networks are another approach for achieving security, where bad IDs are identified and evicted [84,120,121].

In all of these works, a critical assumption is that the fraction of bad IDs is a small constant. However, given that DHTs are often permissionless, this assumption is easily violated via a Sybil attack. Thus, while many tools have been developed for securing DHTs against Byzantine faults, additional work is required to limit the fraction of bad IDs in the permissionless setting.

Sybil Resistance. Several approaches have been proposed for mitigating the Sybil attack. The influence of bad IDs can be limited via containment schemes that leverage the network topology in structured overlays [124] and in social networks [7,86,99,143,152–154]. However, the information required—particularly social networks—may not always be available.

An alternative defense is to use measurements of communication latency or wireless signal strength to verify the uniqueness of IDs [26,38,53,91,140]. However, these techniques are sensitive to measurement accuracy.

For DHTs, an early result by Danezis et al. [37] gives a heuristic to limit the impact of bad IDs using bootstrapping information, but unfortunately provides no formal guarantees. Results that employ resource burning are scarce. The use of computational puzzles in decentralized systems is explored by Borisov [31] and Tegeler and Fu [132] as a means for identifying and excluding bad IDs from the system. Computational puzzles are also used by Rowaihy et al. [116] to throttle the rate of bad IDs added to a structured P2P system; however, this does not limit their number. Arguably the best-known result is the SybilControl scheme by Li et al. [88], which provides for a DHT construction that limits the number of bad IDs through the use of computational puzzles. Good IDs periodically challenge their neighbors under the Chord DHT topology [129,130], and blacklist those who do not respond with a solution in time. Experimental results indicate that this approach, in conjunction with limited data replication, allows for almost all searches to succeed.

4.1 Why DEFID is Not Enough

The DEFID problem (Sect. 3.1) captures many of the challenges required for secure DHTs. However, current solutions to DEFID depend heavily on a means to coordinate resource burning. The main approach is to use a committee—a small set of IDs with a good majority—which issue resource-burning challenges. To apply results on DEFID to DHTs requires decentralizing the functionality provided by the committee.

Additionally, while DEFID always guarantees a minority of bad IDs, this is not enough. In particular, to ensure reliable routing and protection from eclipse attacks, group-based approaches demand that all groups have a minority of bad IDs. Fortunately, there are already clever techniques to spread the bad IDs uniformly across the groups. Informally, when a new ID joins a group, some IDs in the group are evicted and resettled in random locations, and their replacements are selected uniformly at random [21–23,57].

Unfortunately, performing such shuffling for every joining ID, even when there are no bad IDs in the system, incurs large bandwidth costs. A major open problem is to devise an algorithm that minimizes both bandwidth and resource-burning costs, as a function of adversarial spend rate.

4.2 The Permissionless DHT Problem

Problem 3 gives our formal problem in this domain. It assumes that the adversary controls an $\alpha < 1/3$ fraction of the burnable resource. We now describe some ideas about how to solve it.

Problem 3. **A Secure DHT in the Permissionless Setting**

Model: The adversary has complete control over the scheduling of joins and departures for bad IDs and limited control for good IDs. There is no explicit assumption that the good IDs are in the majority at all times.
Goal: A DHT that enables secure and efficient routing between any two good IDs in the system.

Recall from Sect. 3.1 that DEFID imposes a cost of $O(J_G + \sqrt{TJ_G})$ on the good IDs. Informally, a plausible extension to this result is for each group in the DHT to act as a committee that runs an algorithm to solve DEFID. In many group-based constructions, a good ID belongs to a number of groups that is logarithmic in the system size. Consequently, the algorithmic spend rate is likely to increase by a logarithmic factor. This yields our conjectured bound of $\tilde{O}(\sqrt{TJ_G} + J_G)$. Note that this aligns with Position 2 since costs to the good IDs are low when the adversary expends little effort (or does not attack at all), and grows slowly relative to the adversary's cost when a significant attack occurs. In the absence of a single committee that can track global information (such as the join rate of IDs), setting the hardness of challenges is tricky, and new ideas are needed to obtain the conjectured upper bound.

Finally, while we have focused on DHTs, new defenses for them might generalize to providing security in permissionless settings for other structured P2P systems [12,20,47,51,62,71,158].

5 Application-Layer DDoS Attacks

A denial-of-service (DoS) attack prevents good IDs from accessing resources of a system. A distributed denial-of-service (DDoS) attack occurs when multiple bad IDs carry out a coordinated DoS attack. In an application-layer DDoS attacks, an adversary attacks by issuing many requests for system resources, as opposed to say swamping the network bandwidth. Here, we discuss defenses against application-layer DDoS attacks based on resource burning.

Filtering Methods. Many DDoS defenses rely on techniques for filtering out malicious traffic, including IP profiling [94,155]; CAPTCHAs [109,136]; capability-based schemes [9,149][4]; and anomaly detection [70]. An extensive survey of defenses can be found in [157]. Unfortunately, these techniques are imperfect, and an adversary may bypass them by issuing traffic that appears legit-

---

[4] Informally, this refers to a scheme where the source makes a "capability" request and, if approved by the receiver, will then obtain prioritized service from those routers along the path between the source and the receiver.

imate. This has led to resource-burning defenses against DDoS attacks, which are sometimes referred to in the literature as currency-based or resource-based schemes [139].

Resource-Burning Approaches. A number of proposed defenses require IDs to solve puzzles before their requests for service are honored [19, 76, 77, 112]. A challenging aspect of these proposals is the lack of a theoretically-backed method to tune the puzzle difficulty. To address this issue, Mankins et al. [95] propose a pricing mechanism to set the difficulty based on the service-request type; however, the pricing functions are set by the server *a priori*, and may fail as the incentives or capabilities of the attacker change over time. A dynamic strategy to determine puzzle difficulty is given by Wang and Reiter [142]. A client requesting service chooses the puzzle difficulty based on the effort it is willing to expend, while the server prioritizes service according to the difficulty of the puzzles solved. However, this approach may starve IDs with limited resources, and requires the server to maintain state on the difficulty of the puzzles solved. Finally, Noureddine et al. [108] employ a game-theoretic model to pre-compute the difficulty of puzzles assuming all IDs (good and bad) are rational.

An alternative resource—communication capacity—is consumed by the *speak-up* defense of Walfish et al. [138]. During an attack, it is common for bad IDs to bombard the server with requests, using much (or all) of the data rate available to the adversary. Speak-up encourages good IDs to respond in kind by increasing their respective request rates. A front-end server known as a "thinner" randomly drops requests in order to impose a manageable service load. If the aggregate capacity of the good IDs is comparable to that of the bad IDs, then this resource-burning scheme can allow good IDs to obtain a commensurate amount of service.

5.1   The Application-Layer DDoS Problem

There are many similarities between the application-layer DDoS attack and the Sybil attack. The DDoS model is not purely permissionless, since the server is a trusted authority. However, the attacks involve IDs whose distinctness cannot be ascertained, and where an adversary may create many bad IDs to facilitate attacks. In this sense, the DDoS model is a hybrid of permissionless and permissioned systems. Thus, it is not surprising that resource burning would be useful to defend against DDoS attacks.

In this vein, we propose the open problem below.

Problem 4. **Application-Layer DDoS Attacks**

Model: There are n good client IDs and a good server. An adversary controls an α-fraction of the consumable resource, and can generate any number of bad client IDs. Client IDs can request service from the server at any time. The server must decide which requests to service based on its own limited resources.

Goal: The good clients obtain a $1 - O(\alpha)$ fraction of the service provided by the server.

Problem 4 shares much in common with DEFID (Sect. 3.1). Requests from client IDs correspond to join events; satisfying requests corresponds to departures. Here, α need not be bounded, since we are not making a correctness guarantee analogous to maintaining a good majority in DEFID. Rather, our new requirement concerns performance: good IDs receive a $1 - O(\alpha)$ fraction of service. In this sense, Problem 3 seems strictly easier than DEFID.

However, a new difficulty is heterogeneity: requests may differ in the amount of effort required to service them. Thus, enforcing a bound on the fraction of bad requests serviced does not ensure that the goal of Problem 4 will be met. In light of this issue, it may be helpful to consider a weighted version of DEFID, and whether existing solutions can be extended to this more general setting. While we are optimistic that for large T, $o(T)$ is possible for Problem 4, a tight upper bound is an interesting direction for future work.

6   Review Spam

Online user-generated reviews play an important role in influencing the purchasing decisions of consumers. These systems are subject to manipulation where an adversary employs multiple accounts to create fake reviews that falsely promote or disparage a product [50]; this malicious behavior is often referred to as **review spam**, but also goes by other labels such as *astroturfing* [133] and *opinion spam* [69].

Review spam threatens online retailers—such as Amazon or Walmart [33,50]—and merchants who depend on income from online sales. While online review systems typically have some form of admission control, such as requiring credentials for the creation of an account, this can be bypassed. For example, an attacker can hire users that possess a sufficient online presence in order to engage in review spam [36,66], and social-media credentials can be automatically generated [133]; examples of these attacks are described in [69,93].

In response to this threat, the research community has proposed various strategies for detecting fraudulent reviews; these employ a range of techniques including machine learning [34,73], anomaly detection [123,145,146], linguistic evaluation [79,115], graph analysis [5,28,66], and many others. A comprehensive overview of these techniques is given in [64,144,150].

Progress in this area offers the ability to classify a review as either spam or legitimate, with some small error probability; for example, the work in [110] achieve an accuracy of almost 90%. This classification functionality is a promising ingredient for designing more general tools for mitigating review spam.

6.1 The Review Spam Problem

The problem of review spam largely aligns with our general model in Sect. 2.5. While online systems often require some credentials for creating an account, this admission control can be circumvented, and the system is effectively permissionless. However, the review spam model has some novel features. IDs join the system, but they may never formally depart. Even IDs that are regularly in use may have periods where the corresponding user is offline. Thus, any attempt to simultaneously challenge all IDs, in order to reveal some as bad, will fail.

On the positive side, as noted above, machine learning can now help. In particular, we may assume a classifier that correctly classifies reviews as spam or not with some fixed probability of error. Over a sufficiently large number of reviews, this classifier can be used to obtain a good approximation of the current fraction of spam reviews, and this information can be used to set the amount of resource burning required to post a review. Our conjecture of $O(T^{2/3} + P_G)$ in Table 1 follows from a preliminary analysis that leverages a classifier in this way. Informally, we increase the cost for posting a review when a significant attack is ongoing—that is, many reviews are diagnosed as spam by the classifier. Otherwise, we reset the cost to the lowest level.

We formalize the challenge of review spam as Problem 5.

Problem 5. **Review Spam**

Model: IDs post reviews online. A classifier labels each post as legitimate or as spam, with some fixed error probability. Each spam post has unit cost, reflecting its negative impact on system usability. The algorithm can also set an arbitrary resource-burning cost for each new post, based on the classification of past posts.

Goal: Minimize costs due to spam posts plus resource-burning costs incurred from legitimate posts.

---

7   Conclusion

In this paper, we surveyed the literature on resource burning and established it as critical a tool for securing permissionless systems. We described results from four domains: blockchains, DHTs, application-layer distributed DDoS attacks, and review spam. We noted shared security vulnerabilities in both permissionless and hybrid systems, and how resource burning is well-suited for addressing common threats.

We observed that resource burning costs are prohibitively high for most current systems. Thus, a high-priority area for theoretical research is the design of resource-burning defenses that reduce these costs. In particular, whenever possible, good IDs should spend at a rate which is asymptotically less than the adversary when the system is under attack. To encourage research efforts, we defined several open problems, along with conjectured upper bounds for these problems.

Acknowledgements. We are grateful to the organizers of SIROCCO 2020 for inviting this paper, and we thank Valerie King for helpful feedback on our manuscript.

References

1. Abadi, M., Burrows, M., Manasse, M., Wobber, T.: Moderately hard, memory-bound functions. ACM Trans. Internet Technol. (TOIT) **5**(2), 299–327 (2005)
2. Abraham, I., Malkhi, D.: The Blockchain Consensus Layer and BFT. Bull. EATCS: Distrib. Comput. Column (2017)
3. Abraham, I., Malkhi, D., Dobzinski, O.: Land: stretch $(1 + \epsilon)$ locality-aware networks for DHTs. In: Proceedings of the 15th Annual ACM-SIAM Symposium on Discrete algorithms (SODA), pp. 550–559 (2004)
4. Aggarwal, A., Movahedi, M., Saia, J., Zamani, M.: Bootstrapping public blockchains without a trusted setup. In: Proceedings of the 2019 ACM Symposium on Principles of Distributed Computing, pp. 366–368. ACM (2019)
5. Akoglu, L., Chandy, R., Faloutsos, C.: Opinion fraud detection in online reviews by network effects. In: Seventh International AAAI Conference on Weblogs and Social Media (2013)
6. Ali, I.M., Caprolu, M., Pietro, R.D.: Foundations, properties, and security applications of puzzles: a survey. CoRR abs/1904.10164 (2019). http://arxiv.org/abs/1904.10164
7. Alvisi, L., Clement, A., Epasto, A., Lattanzi, S., Panconesi, A.: SoK: the evolution of sybil defense via social networks. In: Proceedings of the IEEE Symposium on Security and Privacy, pp. 382–396 (2013)
8. Hertig, A.: Ethereum's big switch: the new roadmap to proof-of-stake (2017). urlwww.coindesk.com/ethereums-big-switch-the-new-roadmap-to-proof-of-stake/. Accessed 28 Nov 2019
9. Anderson, T., Roscoe, T., Wetherall, D.: Preventing internet denial-of-service with capabilities. ACM SIGCOMM Comput. Commun. Rev. **34**(1), 39–44 (2004)
10. Andrychowicz, M., Dziembowski, S.: PoW-based distributed cryptography with no trusted setup. In: Gennaro, R., Robshaw, M. (eds.) CRYPTO 2015. LNCS, vol. 9216, pp. 379–399. Springer, Heidelberg (2015). https://doi.org/10.1007/978-3-662-48000-7_19
11. Aspnes, J., Jackson, C., Krishnamurthy, A.: Exposing computationally-challenged byzantine impostors. Technical report, Technical Report YALEU/DCS/TR-1332, Yale University (2005). http://www.cs.yale.edu/homes/aspnes/papers/tr1332.pdf
12. Aspnes, J., Shah, G.: Skip graphs. In: Fourteenth Annual ACM-SIAM Symposium on Discrete Algorithms (SODA), pp. 384–393 (2003)
13. Ateniese, G., Bonacina, I., Faonio, A., Galesi, N.: Proofs of space: when space is of the essence. In: Abdalla, M., De Prisco, R. (eds.) SCN 2014. LNCS, vol. 8642, pp. 538–557. Springer, Cham (2014). https://doi.org/10.1007/978-3-319-10879-7_31
14. Augustine, J., Molla, A.R., Morsy, E., Pandurangan, G., Robinson, P., Upfal, E.: Storage and search in dynamic peer-to-peer networks. In: Proceedings of the Twenty-fifth Annual ACM Symposium on Parallelism in Algorithms and Architectures (SPAA), pp. 53–62 (2013)

15. Augustine, J., Pandurangan, G., Robinson, P.: Fast byzantine agreement in dynamic networks. In: Proceedings of the ACM Symposium on Principles of Distributed Computing (PODC), pp. 74–83 (2013)
16. Augustine, J., Pandurangan, G., Robinson, P.: Fast byzantine leader election in dynamic networks. In: Moses, Y. (ed.) DISC 2015. LNCS, vol. 9363, pp. 276–291. Springer, Heidelberg (2015). https://doi.org/10.1007/978-3-662-48653-5_19
17. Augustine, J., Pandurangan, G., Robinson, P., Roche, S., Upfal, E.: Enabling robust and efficient distributed computation in dynamic peer-to-peer networks. In: Proceedings of the IEEE 56th Annual Symposium on Foundations of Computer Science (FOCS), pp. 350–369 (2015)
18. Augustine, J., Pandurangan, G., Robinson, P., Upfal, E.: Towards robust and efficient computation in dynamic peer-to-peer networks. In: Proceedings of the Twenty-third Annual ACM-SIAM Symposium on Discrete Algorithms (SODA), pp. 551–569 (2012)
19. Aura, T., Nikander, P., Leiwo, J.: DOS-resistant authentication with client puzzles. In: Christianson, B., Malcolm, J.A., Crispo, B., Roe, M. (eds.) Security Protocols 2000. LNCS, vol. 2133, pp. 170–177. Springer, Heidelberg (2001). https://doi.org/10.1007/3-540-44810-1_22
20. Awerbuch, B., Scheideler, C.: The hyperring: a low-congestion deterministic data structure for distributed environments. In: Proceedings of the 15th Annual ACM-SIAM Symposium on Discrete Algorithms (SODA), pp. 318–327 (2004)
21. Awerbuch, B., Scheideler, C.: Robust random number generation for peer-to-peer systems. In: Shvartsman, M.M.A.A. (ed.) OPODIS 2006. LNCS, vol. 4305, pp. 275–289. Springer, Heidelberg (2006). https://doi.org/10.1007/11945529_20
22. Awerbuch, B., Scheideler, C.: Towards a scalable and robust DHT. In: Proceedings of the 18th ACM Symposium on Parallelism in Algorithms and Architectures (SPAA), pp. 318–327 (2006)
23. Awerbuch, B., Scheideler, C.: Towards scalable and robust overlay networks. In: Proceedings of the 6th International Workshop on Peer-to-Peer Systems (IPTPS), p. n. pag. (2007)
24. Baird, H.S., Moll, M.A., Wang, S.Y.: ScatterType: a legible but hard-to-segment CAPTCHA. In: Proceedings of the Eighth International Conference on Document Analysis and Recognition (ICDAR), pp. 935–939 (2005)
25. Ball, M., Rosen, A., Sabin, M., Vasudevan, P.N.: Proofs of work from worst-case assumptions. In: Shacham, H., Boldyreva, A. (eds.) CRYPTO 2018. LNCS, vol. 10991, pp. 789–819. Springer, Cham (2018). https://doi.org/10.1007/978-3-319-96884-1_26
26. Bazzi, R.A., Konjevod, G.: On the establishment of distinct identities in overlay networks. In: Proceedings 24th Annual ACM Symposium on Principles of Distributed Computing (PODC), pp. 312–320 (2005)
27. Bentov, I., Lee, C., Mizrahi, A., Rosenfeld, M.: Proof of activity: extending bitcoin's proof of work via proof of stake [extended abstract] y. ACM SIGMETRICS Perform. Eval. Rev. **42**(3), 34–37 (2014)
28. Beutel, A., Xu, W., Guruswami, V., Palow, C., Faloutsos, C.: CopyCatch: stopping group attacks by spotting lockstep behavior in social networks. In: Proceedings of the 22nd International Conference on World Wide Web (WWW), pp. 119–130 (2013)
29. Binmore, K., et al.: Playing for Real: A Text on Game Theory. Oxford University Press, Oxford (2007)
30. BitcoinWiki: Bitcoinwiki network (2019). https://en.bitcoin.it/wiki/Network. Accessed 28 Nov 2019

31. Borisov, N.: Computational puzzles as sybil defenses. In: Proceedings of the Sixth IEEE International Conference on Peer-to-Peer Computing (P2P), pp. 171–176 (2006)
32. Castro, M., Druschel, P., Ganesh, A., Rowstron, A., Wallach, D.S.: Secure routing for structured peer-to-peer overlay networks. In: Proceedings of the 5th Usenix Symposium on Operating Systems Design and Implementation (OSDI), pp. 299–314 (2002)
33. CBS News, A.P.: Buyer beware: scourge of fake reviews hitting Amazon, Walmart and other major retailers (2019). https://www.cbsnews.com/news/buyer-beware-a-scourge-of-fake-online-reviews-is-hitting-amazon-walmart-and-other-major-retailers/
34. Chau, D.H., Pandit, S., Faloutsos, C.: Detecting fraudulent personalities in networks of online auctioneers. In: Fürnkranz, J., Scheffer, T., Spiliopoulou, M. (eds.) PKDD 2006. LNCS (LNAI), vol. 4213, pp. 103–114. Springer, Heidelberg (2006). https://doi.org/10.1007/11871637_14
35. CoinDesk: Vulnerable? Ethereum's Casper Tech Takes Criticism at Curacao Event (2018). https://www.coindesk.com/fundamentally-vulnerable-ethereums-casper-tech-takes-criticism-curacao
36. Cracked: I get paid to write fake reviews for amazon (2016). https://www.cracked.com/personal-experiences-2376-i-get-paid-to-write-fake-reviews-amazon.html
37. Danezis, G., Lesniewski-Laas, C., Kaashoek, M.F., Anderson, R.: Sybil-resistant DHT routing. In: di Vimercati, S.C., Syverson, P., Gollmann, D. (eds.) ESORICS 2005. LNCS, vol. 3679, pp. 305–318. Springer, Heidelberg (2005). https://doi.org/10.1007/11555827_18
38. Demirbas, M., Song, Y.: An RSSI-based scheme for sybil attack detection in wireless sensor networks. In: Proceedings of the 2006 International Symposium on on World of Wireless, Mobile and Multimedia Networks (WOWMOM), pp. 564–570 (2006)
39. Digiconomist: Bitcoin energy consumption index (2020). https://digiconomist.net/bitcoin-energy-consumption
40. Dinger, J., Hartenstein, H.: Defending the sybil attack in P2P networks: taxonomy, challenges, and a proposal for self-registration. In: Proceedings of the First International Conference on Availability, Reliability and Security (ARES), pp. 756–763 (2006)
41. Douceur, J.R.: The sybil attack. In: Druschel, P., Kaashoek, F., Rowstron, A. (eds.) IPTPS 2002. LNCS, vol. 2429, pp. 251–260. Springer, Heidelberg (2002). https://doi.org/10.1007/3-540-45748-8_24
42. Dwork, C., Goldberg, A., Naor, M.: On memory-bound functions for fighting spam. In: Boneh, D. (ed.) CRYPTO 2003. LNCS, vol. 2729, pp. 426–444. Springer, Heidelberg (2003). https://doi.org/10.1007/978-3-540-45146-4_25
43. Dwork, C., Naor, M.: Pricing via Processing or combatting junk mail. In: Brickell, E.F. (ed.) CRYPTO 1992. LNCS, vol. 740, pp. 139–147. Springer, Heidelberg (1993). https://doi.org/10.1007/3-540-48071-4_10
44. Dziembowski, S., Faust, S., Kolmogorov, V., Pietrzak, K.: Proofs of space. In: Gennaro, R., Robshaw, M. (eds.) CRYPTO 2015. LNCS, vol. 9216, pp. 585–605. Springer, Heidelberg (2015). https://doi.org/10.1007/978-3-662-48000-7_29
45. Easley, D., Kleinberg, J., et al.: Networks, Crowds, and Markets, vol. 8. Cambridge University Press, Cambridge (2010)
46. Falkner, J., Piatek, M., John, J.P., Krishnamurthy, A., Anderson, T.: Profiling a million user DHT. In: Proceedings of the 7th ACM SIGCOMM Conference on Internet Measurement, pp. 129–134 (2007)

47. Fiat, A., Saia, J.: Censorship resistant peer-to-peer content addressable networks. In: Proceedings of the Thirteenth ACM Symposium on Discrete Algorithms (SODA), pp. 94–103 (2002)
48. Fiat, A., Saia, J., Young, M.: Making chord robust to byzantine attacks. In: Brodal, G.S., Leonardi, S. (eds.) ESA 2005. LNCS, vol. 3669, pp. 803–814. Springer, Heidelberg (2005). https://doi.org/10.1007/11561071_71
49. FitzGibbon, C.D., Fanshawe, J.H.: Stotting in Thomson's gazelles: an honest signal of condition. Behav. Ecol. Sociobiol. **23**(2), 69–74 (1988)
50. Forbes: Amazon's fake review problem is getting worse (2019). https://www.forbes.com/sites/emmawoollacott/2019/04/16/amazons-fake-review-problem-is-getting-worse/#f6988195f525
51. Fraigniaud, P., Gauron, P.: D2B: a de bruijn based content-addressable network. Theoret. Comput. Sci. **355**(1), 65–79 (2006)
52. Garay, J., Kiayias, A., Leonardos, N.: The bitcoin backbone protocol: analysis and applications. In: Oswald, E., Fischlin, M. (eds.) EUROCRYPT 2015. LNCS, vol. 9057, pp. 281–310. Springer, Heidelberg (2015). https://doi.org/10.1007/978-3-662-46803-6_10
53. Gil, S., Kumar, S., Mazumder, M., Katabi, D., Rus, D.: Guaranteeing spoof-resilient multi-robot networks. In: Proceedings of Robotics: Science and Systems, Rome, Italy, July 2015
54. Gilad, Y., Hemo, R., Micali, S., Vlachos, G., Zeldovich, N.: Algorand: scaling byzantine agreements for cryptocurrencies. In: Proceedings of the 26th Symposium on Operating Systems Principles (SOSP), pp. 51–68 (2017)
55. Gilbert, S., Newport, C., Zheng, C.: Who are you? Secure identities in ad hoc networks. In: Kuhn, F. (ed.) DISC 2014. LNCS, vol. 8784, pp. 227–242. Springer, Heidelberg (2014). https://doi.org/10.1007/978-3-662-45174-8_16
56. Gilbert, S., Zheng, C.: SybilCast: broadcast on the open airwaves. In: Proceedings of the 25th Annual ACM Symposium on Parallelism in Algorithms and Architectures (SPAA), pp. 130–139 (2013)
57. Guerraoui, R., Huc, F., Kermarrec, A.M.: Highly dynamic distributed computing with byzantine failures. In: Proceedings of the 2013 ACM Symposium on Principles of Distributed Computing (PODC), pp. 176–183 (2013)
58. Gupta, D., Saia, J., Young, M.: Proof of work without all the work. In: Proceedings of the 19th International Conference on Distributed Computing and Networking (ICDCN) (2018)
59. Gupta, D., Saia, J., Young, M.: Peace through superior puzzling: an asymmetric sybil defense. In: Proceedings of the 33rd IEEE International Parallel and Distributed Processing Symposium (IPDPS), pp. 1083–1094 (2019)
60. Gupta, D., Saia, J., Young, M.: ToGCom: an asymmetric sybil defense. arXiv preprint arXiv:2006.02893 (2020)
61. Hartline, J.D., Roughgarden, T.: Optimal mechanism design and money burning. In: Proceedings of the 40th Annual ACM Symposium on Theory of Computing, pp. 75–84 (2008)
62. Harvey, N.J.A., Jones, M.B., Saroiu, S., Theimer, M., Wolman, A.: Skipnet: a scalable overlay network with practical locality properties. In: USENIX Symposium on Internet Technologies and Systems (2003)
63. Heilman, E., Kendler, A., Zohar, A., Goldberg, S.: Eclipse attacks on bitcoin's peer-to-peer network. In: Proceedings of the 24th USENIX Conference on Security Symposium, pp. 129–144 (2015)

64. Heydari, A., AliTavakoli, M., Salim, N., Heydari, Z.: Detection of review spam: a survey. Expert Syst. Appl. **42**(7), 3634–3642 (2015). https://doi.org/10.1016/j.eswa.2014.12.029. http://www.sciencedirect.com/science/article/pii/S0957417414008082

65. Hildrum, K., Kubiatowicz, J.: Asymptotically efficient approaches to fault-tolerance in peer-to-peer networks. In: Fich, F.E. (ed.) DISC 2003. LNCS, vol. 2848, pp. 321–336. Springer, Heidelberg (2003). https://doi.org/10.1007/978-3-540-39989-6_23

66. Hooi, B., Song, H.A., Beutel, A., Shah, N., Shin, K., Faloutsos, C.: FRAUDAR: bounding graph fraud in the face of camouflage. In: Proceedings of the 22nd ACM SIGKDD International Conference on Knowledge Discovery and Data Mining (KDD), pp. 895–904. Association for Computing Machinery, New York (2016). https://doi.org/10.1145/2939672.2939747

67. Hou, R., Jahja, I., Luu, L., Saxena, P., Yu, H.: Randomized view reconciliation in permissionless distributed systems, pp. 2528–2536 (2018)

68. Huck, S., Müller, W.: Burning money and (pseudo) first-mover advantages: an experimental study on forward induction. Games Econ. Behav. **51**(1), 109–127 (2005)

69. Hunt, K.M.: Gaming the system: fake online reviews v. consumer law. Comput. Law Secur. Rev. **31**(1), 3–25 (2015). http://www.sciencedirect.com/science/article/pii/S0267364914001824

70. Hussain, A., Heidemann, J., Papadopoulos, C.: A framework for classifying denial of service attacks. In: Proceedings of the Conference on Applications, Technologies, Architectures, and Protocols for Computer Communications (SIGCOMM), pp. 99–110 (2003)

71. Jagadish, H., Ooi, B.C., Vu, Q.H.: BATON: a balanced tree structure for peer-to-peer networks. In: Proceedings of the 31st International conference on Very Large Data Bases (VLDB), pp. 661–672 (2005)

72. Jaiyeola, M.O., Patron, K., Saia, J., Young, M., Zhou, Q.M.: Tiny groups tackle byzantine adversaries. In: Proceedings of the IEEE International Parallel and Distributed Processing Symposium, IPDPS, pp. 1030–1039 (2018)

73. Jindal, N., Liu, B.: Opinion spam and analysis. In: Proceedings of the 2008 International Conference on Web Search and Data Mining, pp. 219–230 (2008)

74. Johansen, H., Allavena, A., van Renesse, R.: Fireflies: scalable support for intrusion-tolerant network overlays. In: ACM SIGOPS Operating Systems Review, pp. 3–13 (2006)

75. John, R., Cherian, J.P., Kizhakkethottam, J.J.: A survey of techniques to prevent sybil attacks. In: Proceedings of the International Conference on Soft-Computing and Networks Security (ICSNS), pp. 1–6 (2015)

76. Juels, A., Brainard, J.: Client puzzles: a cryptographic countermeasure against connection depletion attacks. In: Proceedings of the Network and Distributed System Security Symposium (NDSS), pp. 151–165 (1999)

77. Kaiser, E., Feng, W.C.: Mod_kaPoW: mitigating DoS with transparent proof-of-work. In: Proceedings of the 2007 ACM CoNEXT Conference, pp. 74:1–74:2 (2007)

78. Kapadia, A., Triandopoulos, N.: Halo: High-assurance locate for distributed hash tables. In: Proceedings of the Network and Distributed System Security Symposium (NDSS) (2008)

79. Karami, A., Zhou, B.: Online review spam detection by new linguistic features. In: iConference 2015 Proceedings (2015)

80. Kaashoek, M.F., Karger, D.R.: Koorde: a simple degree-optimal distributed hash table. In: Kaashoek, M.F., Stoica, I. (eds.) IPTPS 2003. LNCS, vol. 2735, pp. 98–107. Springer, Heidelberg (2003). https://doi.org/10.1007/978-3-540-45172-3_9

81. Katz, J., Miller, A., Shi, E.: Pseudonymous secure computation from time-lock puzzles. IACR Cryptol. ePrint Arch. 2014, 857 (2014). http://eprint.iacr.org/2014/857

82. Khan, S.M., Mallesh, N., Nambiar, A., Wright, M.K.: The dynamics of salsa: a robust structured P2P system. Netw. Protocols Algorithms **2**, 40–60 (2010)

83. Kiayias, A., Russell, A., David, B., Oliynykov, R.: Ouroboros: a provably secure proof-of-stake blockchain protocol. In: Katz, J., Shacham, H. (eds.) CRYPTO 2017. LNCS, vol. 10401, pp. 357–388. Springer, Cham (2017). https://doi.org/10.1007/978-3-319-63688-7_12

84. Knockel, J., Saad, G., Saia, J.: Self-healing of byzantine faults. In: Higashino, T., Katayama, Y., Masuzawa, T., Potop-Butucaru, M., Yamashita, M. (eds.) SSS 2013. LNCS, vol. 8255, pp. 98–112. Springer, Cham (2013). https://doi.org/10.1007/978-3-319-03089-0_8

85. Laurie, B., Clayton, R.: "Proof-of-work" proves not to work. In: Proceedings of the 3rd Annual Workshop on Economics and Information Security (WEIS) (2004)

86. Lesniewski-Laas, C., Kaashoek, M.F.: Whanau: a sybil-proof distributed hash table. In: Proceedings of the 7th USENIX Conference on Networked Systems Design and Implementation, NSDI 2010 , p. 8 (2010)

87. Li, D., Lu, X., Wu, J.: FISSIONE: a scalable constant degree and low congestion DHT scheme based on Kautz graphs. In: Proceedings IEEE 24th Annual Joint Conference of the IEEE Computer and Communications Societies, vol. 3, pp. 1677–1688 (2005)

88. Li, F., Mittal, P., Caesar, M., Borisov, N.: SybilControl: practical sybil defense with computational puzzles. In: Proceedings of the Seventh ACM Workshop on Scalable Trusted Computing, pp. 67–78 (2012)

89. Lin, I.C., Liao, T.C.: A survey of blockchain security issues and challenges. IJ Netw. Secur. **19**(5), 653–659 (2017)

90. Liu, D., Camp, L.J.: Proof of work can work. In: Proceedings of the 5th Workshop on the Economics of Information Security (WEIS) (2006)

91. Liu, Y., Bild, D.R., Dick, R.P., Mao, Z.M., Wallach, D.S.: The Mason test: a defense against sybil attacks in wireless networks without trusted authorities. IEEE Trans. Mob. Comput. **14**(11), 2376–2391 (2015)

92. Luu, L., Narayanan, V., Zheng, C., Baweja, K., Gilbert, S., Saxena, P.: A secure sharding protocol for open blockchains. In: Proceedings of the 2016 ACM SIGSAC Conference on Computer and Communications Security (CCS), pp. 17–30 (2016)

93. Malbon, J.: Taking fake online consumer reviews seriously. J. Consum. Policy **36**(2), 139–157 (2013)

94. Malliga, S., Tamilarasi, A., Janani, M.: Filtering spoofed traffic at source end for defending against DoS/DDoS attacks. In: Proceedings of the International Conference on Computing, Communication and Networking, pp. 1–5. IEEE (2008)

95. Mankins, D., Krishnan, R., Boyd, C., Zao, J., Frentz, M.: Mitigating distributed denial of service attacks with dynamic resource pricing. In: Proceedings of the Seventeenth Annual Computer Security Applications Conference, pp. 411–421. IEEE (2001)

96. Maymounkov, P., Mazières, D.: Kademlia: a peer-to-peer information system based on the XOR metric. In: Druschel, P., Kaashoek, F., Rowstron, A. (eds.) IPTPS 2002. LNCS, vol. 2429, pp. 53–65. Springer, Heidelberg (2002). https://doi.org/10.1007/3-540-45748-8_5

97. Miller, A., et al.: Discovering bitcoin's public topology and influential nodes (2015). http://cs.umd.edu/projects/coinscope/coinscope.pdf
98. Miller, G.: Spent: Sex, Evolution, and Consumer Behavior. Penguin, New York (2009)
99. Mohaisen, A., Hollenbeck, S.: Improving social network-based sybil defenses by rewiring and augmenting social graphs. In: Kim, Y., Lee, H., Perrig, A. (eds.) WISA 2013. LNCS, vol. 8267, pp. 65–80. Springer, Cham (2014). https://doi.org/10.1007/978-3-319-05149-9_5
100. Mohaisen, A., Kim, J.: The sybil attacks and defenses: a survey. Smart Comput. Rev. **3**(6), 480–489 (2013)
101. Mónica, D., Leitao, L., Rodrigues, L., Ribeiro, C.: On the use of radio resource tests in wireless ad-hoc networks. In: Proceedings of the 3rd Workshop on Recent Advances on Intrusion-Tolerant Systems, pp. 21–26 (2009)
102. Moran, T., Orlov, I.: Simple proofs of space-time and rational proofs of storage. In: Boldyreva, A., Micciancio, D. (eds.) CRYPTO 2019. LNCS, vol. 11692, pp. 381–409. Springer, Cham (2019). https://doi.org/10.1007/978-3-030-26948-7_14
103. Nakamoto, S.: Bitcoin: a peer-to-peer electronic cash system (2008). http://bitcoin.org/bitcoin.pdf
104. Nambiar, A., Wright, M.: Salsa: a structured approach to large-scale anonymity. In: Proceedings of the 13th ACM Conference on Computer and Communications Security, pp. 17–26 (2006)
105. Naor, M., Wieder, U.: Novel architectures for P2P applications: the continuous-discrete approach. In: Proceedings of the 15th ACM Symposium on Parallelism in Algorithms and Architectures (SPAA) (2003)
106. Newsome, J., Shi, E., Song, D., Perrig, A.: The sybil attack in sensor networks: analysis & defenses. In: Proceedings of the 3rd International Symposium on Information Processing in Sensor Networks (IPSN), pp. 259–268 (2004)
107. Nolan, C.: The Dark Knight. Quote from the scene where the Joker sets a large pile of money ablaze (2008)
108. Noureddine, M.A., et al.: Revisiting client puzzles for state exhaustion attacks resilience. In: Proceedings of the 49th Annual IEEE/IFIP International Conference on Dependable Systems and Networks (DSN), pp. 617–629 (2019)
109. Oikonomou, G., Mirkovic, J.: Modeling human behavior for defense against flash-crowd attacks. In: Proceedings of the IEEE International Conference on Communications, pp. 1–6. IEEE (2009)
110. Ott, M., Choi, Y., Cardie, C., Hancock, J.T.: Finding deceptive opinion spam by any stretch of the imagination. In: Proceedings of the 49th Annual Meeting of the Association for Computational Linguistics: Human Language Technologies, pp. 309–319. Association for Computational Linguistics, USA (2011)
111. Park, S., Kwon, A., Fuchsbauer, G., Gaži, P., Alwen, J., Pietrzak, K.: SpaceMint: a cryptocurrency based on proofs of space. In: Meiklejohn, S., Sako, K. (eds.) FC 2018. LNCS, vol. 10957, pp. 480–499. Springer, Heidelberg (2018). https://doi.org/10.1007/978-3-662-58387-6_26
112. Parno, B., Wendlandt, D., Shi, E., Perrig, A., Maggs, B., Hu, Y.C.: Portcullis: protecting connection setup from denial-of-capability attacks. ACM SIGCOMM Comput. Commun. Rev. **37**(4), 289–300 (2007)
113. Penn, D.J.: The evolutionary roots of our environmental problems: toward a darwinian ecology. Q. Rev. Biol. **78**(3), 275–301 (2003)
114. Pogue, D.: Time to kill off captchas. Sci. Am. **306**(3), 23–23 (2012)

115. Rayana, S., Akoglu, L.: Collective opinion spam detection: bridging review networks and metadata. In: Proceedings of the 21th ACM SIGKDD International Conference on Knowledge Discovery and Data Mining, KDD 2015, pp. 985–994. Association for Computing Machinery, New York (2015). https://doi.org/10.1145/2783258.2783370

116. Rowaihy, H., Enck, W., McDaniel, P., La Porta, T.: Limiting Sybil attacks in structured P2P networks. In: Proceedings of the 26th IEEE International Conference on Computer Communications (INFOCOM), pp. 2596–2600 (2007)

117. Rowstron, A.I.T., Druschel, P.: Pastry: scalable, decentralized object location, and routing for large-scale peer-to-peer systems. In: Proceedings of the IFIP/ACM International Conference on Distributed Systems Platforms, pp. 329–350 (2001)

118. Saad, G.: The Evolutionary Bases of Consumption. Psychology Press (2007)

119. Saad, G., Vongas, J.G.: The effect of conspicuous consumption on men's testosterone levels. Organ. Behav. Hum. Decis. Process. **110**(2), 80–92 (2009)

120. Saad, G., Saia, J.: Self-healing computation. In: Proceedings of the International Symposium on Stabilization, Safety, and Security of Distributed Systems (SSS), pp. 195–210 (2014)

121. Saad, G., Saia, J.: A theoretical and empirical evaluation of an algorithm for self-healing computation. Distrib. Comput. **30**(6), 391–412 (2017)

122. Saia, J., Young, M.: Reducing communication costs in robust peer-to-peer networks. Inform. Process. Lett. **106**(4), 152–158 (2008)

123. Savage, D., Zhang, X., Yu, X., Chou, P., Wang, Q.: Detection of opinion spam based on anomalous rating deviation. Expert Syst. Appl. **42**(22), 8650–8657 (2015). https://doi.org/10.1016/j.eswa.2015.07.019. http://www.sciencedirect.com/science/article/pii/S0957417415004790

124. Scheideler, C., Schmid, S.: A distributed and oblivious heap. In: Albers, S., Marchetti-Spaccamela, A., Matias, Y., Nikoletseas, S., Thomas, W. (eds.) ICALP 2009, Part II. LNCS, vol. 5556, pp. 571–582. Springer, Heidelberg (2009). https://doi.org/10.1007/978-3-642-02930-1_47

125. Sen, S., Freedman, M.J.: Commensal cuckoo: secure group partitioning for large-scale services. ACM SIGOPS Oper. Syst. **46**(1), 33–39 (2012)

126. Shoker, A.: Sustainable blockchain through proof of exercise. In: 2017 IEEE 16th International Symposium on Network Computing and Applications (NCA), pp. 1–9. IEEE (2017)

127. Singh, A., Ngan, T.W., Druschel, P., Wallach, D.S.: Eclipse attacks on overlay networks: threats and defenses. In: Proceedings IEEE International Conference on Computer Communications (INFOCOM), pp. 1–12 (2006)

128. Steiner, M., En-Najjary, T., Biersack, E.W.: A global view of KAD. In: Proceedings of the 7th ACM SIGCOMM Conference on Internet Measurement, pp. 117–122 (2007)

129. Stoica, I., Morris, R., Karger, D., Kaashoek, M.F., Balakrishnan, H.: Chord: a scalable peer-to-peer lookup service for internet applications. In: Proceedings of the Conference on Applications, Technologies, Architectures, and Protocols for Computer Communications (SIGCOMM), pp. 149–160 (2001)

130. Stoica, I., et al.: Chord: a scalable peer-to-peer lookup protocol for internet applications. IEEE/ACM Trans. Netw. **11**(1), 17–32 (2003). https://doi.org/10.1109/TNET.2002.808407

131. Sundie, J.M., Kenrick, D.T., Griskevicius, V., Tybur, J.M., Vohs, K.D., Beal, D.J.: Peacocks, porsches, and thorstein veblen: Conspicuous consumption as a sexual signaling system. J. Pers. Soc. Psychol. **100**(4), 664 (2011)

132. Tegeler, F., Fu, X.: SybilConf: computational puzzles for confining sybil attacks. In: Proceedings of the IEEE Conference on Computer Communications Workshops (INFOCOM), pp. 1–2 (2010)
133. The Guardian, G.M.: The need to protect the internet from 'astroturfing' grows ever more urgent (2011). https://www.theguardian.com/environment/georgemonbiot/2011/feb/23/need-to-protect-internet-from-astroturfing
134. Thorstein, V.: The Theory of the Leisure Class: An Economic Study of Institutions. BW Huebsch, New York (1912)
135. Urdaneta, G., Pierre, G., van Steen, M.: A survey of DHT security techniques. ACM Comput. Surv. **43**(2), 1–53 (2011)
136. von Ahn, L., Blum, M., Hopper, N.J., Langford, J.: CAPTCHA: using hard AI problems for security. In: Biham, E. (ed.) EUROCRYPT 2003. LNCS, vol. 2656, pp. 294–311. Springer, Heidelberg (2003). https://doi.org/10.1007/3-540-39200-9_18
137. Von Ahn, L., Maurer, B., McMillen, C., Abraham, D., Blum, M.: reCAPTCHA: human-based character recognition via web security measures. Science **321**(5895), 1465–1468 (2008)
138. Walfish, M., Vutukuru, M., Balakrishnan, H., Karger, D., Shenker, S.: DDoS defense by offense. In: Proceedings of the 2006 Conference on Applications, Technologies, Architectures, and Protocols for Computer Communications (SIGCOMM), pp. 303–314 (2006)
139. Walfish, M., Vutukuru, M., Balakrishnan, H., Karger, D., Shenker, S.: DDoS defense by offense. ACM Trans. Comput. Syst. (TOCS) **28**(1), 3 (2010)
140. Wang, H., Zhu, Y., Hu, Y.: An efficient and secure peer-to-peer overlay network. In: Proceedings of the IEEE Conference on Local Computer Networks, pp. 764–771 (2005)
141. Wang, L., Kangasharju, J.: Measuring large-scale distributed systems: case of BitTorrent mainline DHT. In: IEEE 13th International Conference on Peer-to-Peer Computing (P2P), pp. 1–10 (2013)
142. Wang, X., Reiter, M.K.: Defending against denial-of-service attacks with puzzle auctions. In: Proceedings of the 2003 IEEE Symposium on Security and Privacy, p. 78 (2003)
143. Wei, W., Xu, F., Tan, C.C., Li, Q.: SybilDefender: a defense mechanism for sybil attacks in large social networks. IEEE Trans. Parallel Distrib. Syst. **24**(12), 2492–2502 (2013)
144. Wu, Y., Ngai, E.W., Wu, P., Wu, C.: Fake online reviews: literature review, synthesis, and directions for future research. Decis. Support Syst. **132**, 113280 (2020). https://doi.org/10.1016/j.dss.2020.113280. http://www.sciencedirect.com/science/article/pii/S016792362030035X
145. Xie, S., Wang, G., Lin, S., Yu, P.S.: Review spam detection via temporal pattern discovery. In: Proceedings of the 18th ACM SIGKDD International Conference on Knowledge Discovery and Data Mining, pp. 823–831. Association for Computing Machinery, New York (2012). https://doi.org/10.1145/2339530.2339662
146. Xie, S., Wang, G., Lin, S., Yu, P.S.: Review spam detection via time series pattern discovery. In: Proceedings of the 21st International Conference on World Wide Web, WWW 2012 Companion, pp. 635–636. Association for Computing Machinery, New York (2012). https://doi.org/10.1145/2187980.2188164
147. Yan, J., El Ahmad, A.S.: Captcha robustness: a security engineering perspective. Computer **44**(2), 54–60 (2011)

148. Yan, J., El Ahmad, A.S.: Usability of CAPTCHAs or usability issues in CAPTCHA design. In: Proceedings of the 4th Symposium on Usable Privacy and Security, SOUPS 2008, pp. 44–52. Association for Computing Machinery, New York (2008). https://doi.org/10.1145/1408664.1408671

149. Yang, X., Wetherall, D., Anderson, T.: TVA: A DoS-limiting network architecture. IEEE/ACM Trans. Netw. **16**(6), 1267–1280 (2008)

150. Ma, Y., Li, F.: Detecting review spam: challenges and opportunities. In: 8th International Conference on Collaborative Computing: Networking, Applications and Worksharing (CollaborateCom), pp. 651–654 (2012)

151. Young, M., Kate, A., Goldberg, I., Karsten, M.: Towards practical communication in Byzantine-resistant DHTs. IEEE/ACM Trans. Netw. **21**(1), 190–203 (2013)

152. Yu, H.: Sybil defenses via social networks: a tutorial and survey. SIGACT News **42**(3), 80–101 (2011)

153. Yu, H., Gibbons, P.B., Kaminsky, M., Xiao, F.: SybilLimit: a near-optimal social network defense against sybil attacks. IEEE/ACM Trans. Netw. **18**(3), 885–898 (2010)

154. Yu, H., Kaminsky, M., Gibbons, P.B., Flaxman, A.: SybilGuard: defending against sybil attacks via social networks. In: Proceedings of the 2006 Conference on Applications, Technologies, Architectures, and Protocols for Computer Communications (SIGCOMM) vol. 36, pp. 267–278, August 2006

155. Yu, S., Thapngam, T., Liu, J., Wei, S., Zhou, W.: Discriminating DDoS flows from flash crowds using information distance. In: Proceedings of the Third International Conference on Network and System Security, pp. 351–356. IEEE (2009)

156. Zahavi, A.: Mate selection - a selection for a handicap. J. Theor. Biol. **53**(1), 205–214 (1975)

157. Zargar, S.T., Joshi, J., Tipper, D.: A survey of defense mechanisms against distributed denial of service (DDoS) flooding attacks. IEEE Commun. Surv. Tutor. **15**(4), 2046–2069 (2013)

158. Zatloukal, K.C., Harvey, N.J.A.: Family trees: an ordered dictionary with optimal congestion, locality, degree, and search time. In: Proceedings of the 15th Annual ACM-SIAM Symposium on Discrete Algorithms (SODA), pp. 308–317 (2004)

Mobile Robots

ANTS on a Plane

Abhinav Aggarwal and Jared Saia[✉]

Department of Computer Science, University of New Mexico,
Albuquerque, NM, USA
{abhiag,saia.cs}@unm.edu

Abstract. In the ANTS (Ants Nearby Treasure Search) problem, multiple searchers, starting from a central location, search for a treasure. The searchers cannot communicate and have few bits of initial knowledge, called *advice*, when they begin the search. In this paper, we initiate the study of ANTS in the geometric plane.

Our main result is an algorithm, GOLDENFA, that tolerates arbitrarily many crash failures caused by an adaptive adversary, and requires no bits of advice. GOLDENFA takes $O\left(\left(L + \frac{L^2(t+1)}{ND}\right)\log L\right)$ expected time to find the shape, for a shape of diameter D, at distance L from the central location, with N searchers, $t < N$ of which suffer adversarial crash-failures.

We complement our algorithm with a lower bound, showing that it is within logarithmic factors of optimal. Additionally, we empirically test GOLDENFA, and a related heuristic, and find that the heuristic is consistently faster than the state-of-the-art. Our algorithms and analysis make critical use of the Golden Ratio.

Keywords: Golden Ratio · Reliability · Computational geometry · Natural algorithms

1 Introduction

How can multiple simple searchers best find a target? Feinerman, Korman and others formalized this question by defining the ANTS (Ants Nearby Treasure Search) problem, where many searchers, all starting at a central location, seek a hidden target [9–11]. In this paper, we extend results on the ANTS problem in two key directions. Our first extension is to consider search on a 2-dimensional plane, rather than on a grid graph. This has two advantages for applications involving geometric search.[1] First, it allows us to more easily design search algorithms for targets of different sizes and shapes. Second, it avoids the problem of choosing the correct granularity for the grid graph. In particular, if the granularity is too low, then the target may not overlap any node. But if the granularity is too high, it places a high computational burden on the searchers.

---

[1] Our own motivating application is drones searching for gas plumes [1,30].

With apologies to the cast and crew of the Hollywood classic *Snakes on a Plane*. This work is supported by the National Science Foundation grant CNS 1816250.

A. W. Richa and C. Scheideler (Eds.): SIROCCO 2020, LNCS 12156, pp. 47–62, 2020.
https://doi.org/10.1007/978-3-030-54921-3_3

Our second extension is ensuring provable robustness to adversarial failures, without requiring communication among searchers. Importantly, our algorithm can tolerate all but 1 searcher crashing, and the efficiency of our algorithm decreases only linearly with the actual number of faults, even when that number is not known in advance.

Our Model. N searchers start at a central location, called the *nest*. We define a *treasure* to be a convex shape with ratio of diameter to width equal to a fixed constant. Recall that the *diameter* of a convex shape is the largest distance between two parallel lines that are both tangent to the boundary of the shape, and that the *width* is the smallest such distance.[2]

A treasure of diameter D is placed adversarially at a distance L from the nest, where L is measured from the nest to the geometric center of the treasure. The searchers are synchronous in the sense that they all move at the same speed, and that local computation is instantaneous. The searchers cannot communicate with each other, and have zero bits of initial knowledge, including no knowledge of L or D. We measure the time it takes for some searcher to first locate the treasure. We refer to this as the *search time* of our algorithm. Our failure model is based on an adaptive adversary considered in [23]. In particular, an omniscient adversary chooses $t < N$ searchers that suffer crash failures at times chosen by the adversary.

We assume that every searcher has the ability to turn at angles of both π and $2\pi/\phi$, where $\phi = \frac{1+\sqrt{5}}{2}$ is the Golden Ratio. When turning at an angle of α, let $\beta = 2\pi - \alpha$, be the remaining angle in the circle. Then to turn at an angle of π, requires that the searcher has the ability to turn until $\alpha = \beta$. To turn at an angle of $2\pi/\phi$, requires that the searcher have the ability to turn until $\frac{2\pi}{\alpha} = \frac{\alpha}{\beta}$.

1.1 Our Results

Our upper bound, summarized in the theorem below, considers N searchers looking for a treasure of diameter D at distance L, with $t < N$ crash failures (See Table 1).

Theorem 1. *There exists an algorithm,* GOLDENFA, *that in the presence of up to $t < N$ crash failures, is able to locate a treasure of unknown diameter D, placed adversarially at an unknown distance L from the nest, in expected search time $O\left(\left(L + \frac{L^2(t+1)}{ND}\right)\log L\right)$.*

Additionally, GOLDENFA requires zero bits of initial knowledge, called *advice*; it is uniform in that the searchers know nothing about N, and have no unique identifiers.

We prove lower-bounds, showing that the expected run time of GOLDENFA is within logarithmic factors of optimal among a class of spoke-based algorithms. A *spoke-based* algorithm is one where the searchers only search along line segments, where each line-segment has an end-point in the nest, the central location where the searchers all start. See Sect. 6 for details.

---

[2] For example, a treasure can be a circle, regular polygon, or rectangle with constant aspect ratio.

Table 1. A comparison of GOLDENFA and the algorithms by Feinerman and Korman [10] (abbreviated as F&K). While the latter are not provably robust against adversarial crash failures, GOLDENFA can efficiently handle all but one searchers to fail, even when these failures are scheduled by an adaptive adversary.

Algorithm	Advice (bits)	Robustness	Runtime
F& K (advice)	$O(\log \log N)$	Not robust	$O\left(L + \frac{L^2}{N}\right)$ for $D = \Theta(1)$
F& K (no advice)	0	Not robust	$O\left(\left(L + \frac{L^2}{N}\right)\log^{1+\varepsilon} N\right)$ for fixed $\varepsilon > 0$ and $D = \Theta(1)$
GOLDENFA	0	$t < N$	$O\left(\left(L + \frac{L^2(t+1)}{ND}\right)\log L\right)$

Our algorithm makes use of the Golden Ratio, both to ensure robustness and to ensure good coverage during the search. To the best of our knowledge, our algorithm is the first for the ANTS problem that makes use of this value.

1.2 Novelty and Technical Challenges

Our upper bound makes critical use of the Golden Ratio, and the difficultly to approximate it rationally. In particular, we can write any number as a (possibly infinite) continued fraction [18] of the form $x_1 + \cfrac{1}{x_2 + \cfrac{1}{x_3 + \dots}}$, where the x_i values are all integers for $i \geq 1$. The degree to which the original number is well-approximated by a finite continued fraction depends on how large the x_i values are. For example, if x_2 is large, then the absolute difference between x_1 and the original number is small; if x_3 is large, then the absolute difference between $x_1 + 1/x_2$ and the original number is small, and so forth.

When $x_i = 1$ for all $i \geq 1$, we obtain an irrational number that is most difficult to approximate. To find this most difficult to approximate irrational number, we set $y = 1 + \frac{1}{y}$, and solve the resulting quadratic equation to obtain a solution $y = \frac{1+\sqrt{5}}{2}$, which is the celebrated Golden Ratio ϕ.

Using ϕ to Spread-Out Spokes. In our algorithm, searchers proceed from the nest in line segments that we call *spokes*. Each new spoke is oriented at arc length ϕ, along the unit circle, from the previous one. The fact that ϕ is difficult to approximate with a rational number has useful implications in ensuring the angles between spokes are "well-spread". For example, if we start at the point 0 on a unit circle, and iteratively add points by moving clockwise by arc distance ϕ, then we will end up with near uniform distance between points (See Lemma 3 and [20,29]). In particular, if x spokes are added this way, then the maximum arc length on a unit circle between neighboring spokes is $O(1/x)$ by the Three Gap Theorem (see Lemma 3). This allows us to locate the treasure efficiently, when D is unknown. Interestingly, this has connections to how plants add leaves as they grow. In particular, if the next leaf is added by moving arc length ϕ along a unit circle, this ensures that leaves are well-spread, which increases their exposure to sunlight [27].

Using ϕ to Handle Failures. In our algorithm, each searcher creates the first spoke at a random heading and then iteratively proceeds to the next spoke by moving an arc distance ϕ along the unit circle (see Fig. 1). Thus, even in the presence of $t = N - 1$ failures, the gaps between the spokes generated by the single remaining searcher decrease linearly and the treasure is found. This way, our algorithm eventually succeeds even when all but one searcher crashes.

Unknown L. Since L is unknown, we must carefully balance increasing spoke lengths and decreasing arc lengths between spokes over time. Simple doubling of spoke lengths over time is inefficient. Instead, our algorithm proceeds in epochs, where in epoch i, we search along spokes of length $2^0, 2^1, \ldots, 2^{i-1}$. In each epoch, we ensure that the amount of time spent searching along spokes of length 2^j is the same for all $0 \leq j < i$. We do this by having 2^{i-1} spokes of length 2^0, 2^{i-2} spokes of length 2^1, and so on up to 2^0 spokes of length 2^{i-1} (see Fig. 2). Additionally, the angles between these spokes is determined using the Golden Ratio so that the angular gaps decrease linearly with the number of spokes.

1.3 Paper Organization

The rest of the paper is organized as follows. We discuss related work in Sect. 2 and some technical preliminaries in Sect. 3. We describe GOLDENFA in Sect. 4 and analyze it in Sect. 5. We then give our lower bounds in Sect. 6. We provide empirical results on GOLDENFA, comparing it with existing work, in Sect. 7. Finally, we conclude and discuss areas for future work in Sect. 8.

2   Related Work

Search is a fundamental problem in biology, where survival depends on search for mates, prey and other resources. It is also a common problem in robotics and mobile computing. Collective search, where multiple searchers must coordinate, is a key problem in computer science, robotics and in social insects. Ant- and bee-inspired algorithms have been particularly influential in swarm robotics research [16,22,28].

ANTS. Feinerman, Korman et al. [9–11] introduced the ANTS problem where multiple searchers starting from the same central location search for a treasure. Searchers are simple in that they cannot communicate and have few bits of initial knowledge, called advice, when they first leave the nest. Research on this problem now extends in multiple directions including: tradeoffs between computational resources and knowledge of searchers and the search time [7,9,24]; tradeoffs between communication and search time [3,23,25]; fault-tolerance [23]; handling asynchronous searchers [8,25]; and game theoretic analysis of rational searchers [4]. As stated previously, our model is equivalent to that of [9–11], except that we search for a convex treasure in the infinite plane, rather than a single vertex on an infinite grid. We note that the paper [11], while alluding to search on the plane, actually performs search on a two dimensional grid, by assuming each agent has a "bounded field of view of say ε" (Sect. 2 of [11]).

There are two potential benefits to avoiding this type of discretization. First, in some search applications, such as gas plume detection [30], there may be no clear analogue to a bounded "field of view". In this case, choosing ε too large risks missing the treasure, but choosing ε too small increases computational load on the searchers, since coordinate storage space seems to grow as $\Omega(\log(1/\varepsilon))$. Second, searching in the geometric plane more naturally allows for consideration of different shapes and sizes for the treasure. In many search applications, this seems important since targets are likely to be large or to be co-located, in both biological [2,13] and engineering systems [14,30].

Golden Ratio. Our algorithms make critical use of the celebrated Golden Ratio. This ratio is the limit of the ratio of consecutive numbers in the Fibonacci sequence. Fibonacci generated the sequence as an idealized model of a reproducing rabbit population assuming overlapping generations [6]. It was documented in India many centuries earlier, and has been observed in numerous biological systems including the arrangement of pine cones, unfurling of fern leaves, and the arrangement of sunflower seeds that optimally fills the circular area of the flower [27]. The Golden Ratio and Fibonacci numbers have been used in computer science for various applications like obtaining optimal schedules for security games [17], Fibonacci hashing [20], bandwidth sharing [15], data structures [12] and game theoretic models for blocking-resistant communication [19]. See [26] for a fascinating discussion of the history and applications of the Golden Ratio.

Crash Faults. To the best of our knowledge, work by Langner et al. [23] is the only other result that tolerates adversarial crash failures for a problem similar to ANTS. However, their model significantly deviates from ANTS in that they allow communication. In particular, constant-sized messages can be exchanged between searchers when they are both at the same location. Additionally, their searchers are much more restricted than ours in that they are modeled by finite-state automata. They describe an algorithm that locates a single target in $O\left(L + L^2/N + Lt\right)$ time, while tolerating $t \leq cN$ crash failures for some constant $c < 1$. In contrast, our algorithm can handle any $t < N - 1$, and does not require communication.

3   Technical Preliminaries

Let $\phi = (1 + \sqrt{5})/2$ denote the Golden ratio. For $m \geq 1$, let F_m denote the mth Fibonacci number, so that $F_1 = F_2 = 1$ and $F_m = F_{m-1} + F_{m-2}$ for all $m \geq 3$. Given integer n, let $m(n)$ denote the index of the largest Fibonacci number not greater than n.

Lemma 1. *For all $x \geq 1$, the following properties hold:*

1. $\lfloor \log_\phi x + 1 \rfloor \leq m(x) \leq \lceil \log_\phi x + 2 \rceil$.
2. $\frac{1}{\phi^3 x} \leq \phi^{-m(x)} \leq \frac{1}{x}$.

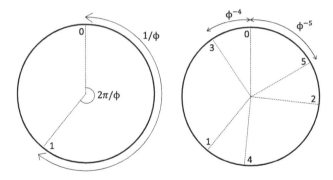

Fig. 1. A schematic of the gaps induced by points placed on the unit circle, following Lemma 3. When the arc distance between successive points is $\phi \equiv \phi^{-1} \mod 1$, then the gaps decrease as the number of points increase.

Proof. For (1), using the fact that $F_r \leq \phi^{r-1}$ for all r, it holds that $F_{\lfloor \log_\phi x+1 \rfloor} \leq \phi^{\lfloor \log_\phi x+1 \rfloor - 1} \leq x$. Similarly, since $F_r \geq \phi^{r-2}$ for all r, it holds that $F_{\lceil \log_\phi x+2 \rceil} \geq \phi^{\lceil \log_\phi x+2 \rceil - 2} \geq x$. For (2), using the result obtained in (1), we obtain $\phi^{-m(x)} \leq \phi^{-\lfloor \log_\phi x+1 \rfloor} \leq \frac{1}{x}$. Similarly, $\phi^{-m(x)} \geq \phi^{-\lceil \log_\phi x+2 \rceil} \geq \frac{1}{\phi^3 x}$. □

We define a *unit circle* as a circle around the nest with circumference one.

Lemma 2. *Let the treasure be oriented so that its diameter is perpendicular to the spoke ending at the diameter midpoint. Let α be the arc length on the unit circle made by the two spokes that are tangent to the diameter. Then,*

1. $\alpha = \frac{1}{\pi} \sin^{-1} \left(\frac{D}{2L} \right)$; and
2. $\alpha \geq \frac{D}{2\pi L}$.

Proof. Part (1) follows by definition. Part (2) follows from the Maclaurin expansion [21] of $\sin^{-1} x$, from which it follows that $\sin^{-1} x \geq x$. □

Our analysis makes use of the following lemma regarding the Three Gap Theorem by Swierczkowski [29] (also known as the Steinhaus Conjecture) for Golden-ratio based gaps between successive points on the circumference of a unit circle (see Fig. 1). In this lemma, the set of points on the unit circle is equivalent to the set of points generated by our algorithm. This holds since $\phi^{-1} \equiv \phi \mod 1$, because $\phi^{-1} = \phi - 1$. The last sentence of the lemma follows immediately from Lemma 1(2).

Lemma 3 (Restatement of Corollary 2 from [29]). *Let C be a circle of circumference 1 and p_0 be a fixed starting point on C. For $k \geq 0$, let p_k be the point which makes an arc of length $k\phi$ from p_0, measured clockwise. Let $n \geq 1$ and F_m be the largest Fibonacci number no more than n. Then, the set of points $P_n = \{p_0, p_1, \ldots, p_n\}$ partition C into disjoint arcs, each of which has length ϕ^{-m}, ϕ^{-m+1} or ϕ^{-m+2}. In particular, this implies that every disjoint arc has length between $\frac{1}{\phi^3 n}$ and $\frac{\phi^2}{n}$.*

Algorithm 1: The GOLDENFA Algorithm.

/* Each searcher independently performs the following steps.   */

1    $i \leftarrow 1$;

2 **while** *treasure not found* **do**

3   *direction* $\leftarrow$ uniformly random heading on the unit circle;

4   **for** $j \in \{0, \ldots, i-1\}$ **do**

5   Traverse 2^{i-j} spokes of length 2^j. The first spoke is at heading *direction*. Each subsequent spoke has heading that increases clockwise by arc distance of ϕ, along the unit circle, from the heading of the previous spoke.;

6   **end**

7      $i \leftarrow i+1$;

8 **end**

In the rest of the paper, we will assume circular treasures but our results hold for all convex shapes where the ratio of the diameter to the width is a fixed constant. We assume that L is the distance from the nest to the center of the circular treasure.

Our algorithm is designed to search in the real plane, $\mathbb{R}^2$. We note, however, that it can be adapted to search in the infinite two-dimensional grid as follows. For every spoke generated by our algorithm, create a walk on the grid that visits every edge incident to every face in the grid that is intersected by the spoke. This ensures that we will find any treasure that overlaps a grid vertex. Additionally, it increases total search time by at most a constant factor.

4   GoldenFA

Algorithm 1 describes our main algorithm, GOLDENFA. The algorithm proceeds in *epochs* numbered iteratively starting at $i = 1$. In epoch number, i, each searcher initially chooses a random initial heading *direction*. Then for all j, $1 \leq j \leq (i-1)$, the searcher traverses along 2^{i-j} spokes of length 2^j. Each spoke starts and ends at the nest. For each value of j, the first of these spokes is at heading *direction*, and each subsequent spoke has heading that increases clockwise along the unit circle at arc length of ϕ from the previous spoke. Thus in epoch i, a total of $\sum_{j=0}^{i-1} 2^{i-j} = 2^{i+1} - 1$ spokes are traversed. If the treasure is not found after these traversals, epoch i ends and epoch $i + 1$ begins.

Figure 2 illustrates two epochs of GOLDENFA when $N = 2$.

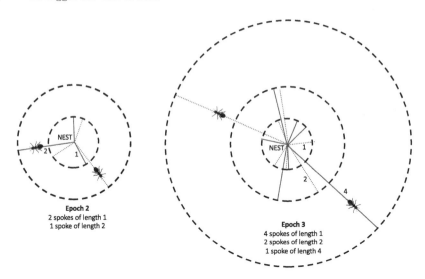

Fig. 2. A schematic of GOLDENFA is shown for the spokes made by two searchers in epochs 2 and 3 (red solid lines for searcher 1 and green dotted lines for searcher 2). Both searchers choose a random initial heading at the beginning of every epoch. (Color figure online)

5   Analysis

We next analyze GOLDENFA and compute its runtime as a function of the unknown parameters: treasure diameter, its distance from the nest, the number of searchers and the number of crash failures. In the following, all log terms are base 2.

Lemma 4. *In epoch $i \geq \log L$, the probability that a single searcher finds the treasure is at least $\alpha 2^{i-(\log L)-5}$, where $\alpha \geq \frac{D}{2\pi L}$. In any epoch $i \geq \log L + \log \frac{\phi^2}{\alpha} + 1$, a searcher finds the treasure with probability 1.*

Proof. When $i \geq \log L$, there will be $2^{i-\lceil \log L \rceil}$ spokes of length at least L, where the first of these spokes has a uniformly random orientation, and the remainder are spread out at successive clockwise arc distances of ϕ. By Lemma 3, the maximum arc length between any neighboring pair of these $2^{i-\lceil \log L \rceil}$ spokes is $\phi^2 2^{-i+\lceil \log L \rceil}$. By Lemma 2, if any of these spokes intersect an arc of length α, where $\alpha \geq \frac{D}{2\pi L}$, then the searcher will find the treasure. Thus, for $i \geq \log L + \log \frac{\phi^2}{\alpha} + 1$, a searcher is guaranteed to find the treasure.

By Lemma 3, the minimum arc length between any neighboring pair of x spokes is $\frac{1}{\phi^3 x}$. Thus, when $x \leq \frac{1}{\alpha \phi^3}$, all spokes are arc distance at least α apart. If there are x such spokes of length at least L, the probability that one of these spokes intersects the treasure is $x\alpha$. To see this, imagine fixing the x spokes, and then letting the α length arc associated with the treasure move uniformly at random on the circumference of the unit circle. The total measure of locations where the treasure may fall so that it intersects a spoke is then $x\alpha$. Hence, when $\log L \leq i \leq \lceil \log L \rceil + \log \frac{1}{\phi^3 \alpha}$, the probability that a single searcher finds the treasure is at least $\alpha 2^{i-\lceil \log L \rceil} \geq \alpha 2^{i-(\log L)-1}$.

Finally, note that for $\lceil \log L \rceil + \log \frac{1}{\phi^3 \alpha} < i < \log L + \log \frac{\phi^2}{\alpha} + 1$, the probability that a single searcher finds the treasure is at least $1/\phi^3$. Thus, in this range, the probability of finding the treasure is at least $\alpha \frac{1}{\phi^5} 2^{i-\lceil \log L \rceil} \geq \alpha 2^{i-(\lceil \log L \rceil + 4)} \geq \alpha 2^{i-(\log L)-5}$. □

Theorem 2. *In the presence of up to $t < N$ crash failures, GoldenFA takes an expected number of times steps that is $O\left(\left(L + \frac{L^2(t+1)}{ND}\right) \log L\right)$.*

Proof. First, we consider the case where $2(t+1) > N$. By Lemma 4, when $i \geq \log L + \log \frac{\phi^2}{\alpha} + 1$, all searchers will find the treasure. Thus, in this case, the total time of GoldenFA is no more than

$$\sum_{i=1}^{\log L + \log \frac{\phi^2}{\alpha} + 1} i 2^i = O\left(\frac{L\phi^2}{\alpha} \log \frac{L\phi^2}{\alpha}\right).$$

This is the claimed number of time steps when $t = \Theta(N)$, since $\alpha = \Theta(D/L)$.

Next, assume that $2(t+1) \leq N$. We first compute the expected number of searchers that find the treasure in each epoch, and then use this expectation to bound, for each epoch, the probability that the total number of searchers that find the treasure is no more than the total number of faults.

For any epoch i, let S_i be a random variable giving the number of searchers that find the treasure in epoch i. By Lemma 4, and linearity of expectation, we have that for $\log L \leq i \leq \log L + \log \frac{\phi^2}{\alpha} + 1$,

$$E(S_i) \geq N\alpha 2^{i-(\log L)-5},$$

where $\alpha \geq \frac{D}{2\pi L}$. Since each searcher finds the treasure independently, we can use Chernoff bounds on S_i (See [5], Exercise 1.1). These show that $Pr(S_i < (1/2)\mu_L) \leq e^{-\mu_L/8}$, where $\mu_L = N\alpha 2^{i-(\log L)-5}$ is a lower bound on the expected value. Let

$$i^* = (\log L) + 5 + \max\left(0, \log \frac{2(t+1)}{N\alpha}\right).$$

Then $E(S_i) \geq 2(t+1)$, when $i \geq i^*$ and $2(t+1) \leq N$. Thus, for $i \geq i^*$,

$$Pr(S_i < t+1) \leq e^{-N\alpha 2^{i-(\log L)-8}}.$$

This bound holds even for $i \geq \log L + \log \frac{\phi^2}{\alpha} + 1$, since for i in that range, $Pr(S_i < t+1) = 0$, since each searcher finds the treasure with probability 1.

[**Jared:** *Page 9, 2nd to last displayed part Why do you need t+1 here? The conclusion that the adversary cannot prevent the searchers from finding the treasure would have been the same with t rather than t+1, would it not?*]

Let X be a random variable giving the number of epochs until more than t searchers find the treasure. Note that $Pr(X \geq i) \leq Pr(S_{i-1} < t + 1)$, where $Pr(S_0 < t + 1) = 1$. Then, we can bound the expected search time of our algorithm as follows.

$$\sum_{i \geq 1} i2^i Pr(X \geq i) = \sum_{i \geq 1} i2^i Pr(S_{i-1} < t + 1)$$
$$\leq \sum_{1 \leq i < i^*} i2^i + \sum_{i \geq i^*} i2^i e^{-N\alpha 2^{i-(\log L)-8}}$$

Let S_1 be the value of the first sum. Note that:

$$S_1 = \sum_{1 \leq i < i^*} i2^i$$
$$= O\left(L \log L + \left(\frac{L(t+1)}{\alpha N}\right) \log\left\lceil\frac{L(t+1)}{\alpha N}\right\rceil\right)$$

Let S_2 be the value of the second sum. Note that:

$$S_2 = \sum_{i \geq i^*} i2^i e^{-N\alpha 2^{i-(\log L)-8}}$$
$$\leq 2i^* 2^{i^*} \sum_{j \geq 1} j2^j e^{-N\alpha 2^{j+i^*-(\log L)-9}}$$
$$\leq 2i^* 2^{i^*} \sum_{j \geq 1} \exp\left(\ln j + j \ln 2 - 2^{j+i^*-\left(\log \frac{L}{N\alpha}\right)-9}\right)$$

In the above, the second line holds by noting that for all $j \geq 1$ and $x \geq 1$, $(x+j)2^{x+j} \leq 2(x2^x)(j2^j)$, and letting $x = i^*$. Next, we bound the exponent:

$$\ln j + j \ln 2 - 2^{j+i^*-\left(\log \frac{L}{N\alpha}\right)-9} \leq \ln j + j \ln 2 - 2^{j-3}$$
$$\leq -j$$

In the above, the first line holds since $i^* = (\log L) + 5 + \max\left(0, \log \frac{2(t+1)}{N\alpha}\right) \geq 5 + \log \frac{2L(t+1)}{N\alpha} \geq 6 + \log \frac{L}{N\alpha}$. The second line holds when $j \geq 7$, since then $\ln j + j \ln 2 - 2^{j-3} \leq -j$. Hence, the infinite summation is $O(1)$. Thus, we have that

$$S_2 \leq 2i^* 2^{i^*} \sum_{j \geq 1} \exp\left(\ln j + j \ln 2 - 2^{j+i^*-\left(\log \frac{L}{N\alpha}\right)-9}\right)$$
$$= O\left(L \log L + \left(\frac{L(t+1)}{\alpha N}\right) \log\left\lceil\frac{L(t+1)}{\alpha N}\right\rceil\right)$$

By Lemma 2, $\alpha \geq \frac{D}{2\pi L}$, so the total expected cost of GOLDENFA is:

$$O\left(L \log L + \left(\frac{L^2(t+1)}{ND}\right) \log\left\lceil\frac{L^2(t+1)}{ND}\right\rceil\right).$$

Finally, note that $\log \left\lceil \frac{L^2(t+1)}{ND} \right\rceil = O(\log L + \log(t+1) - \log N - \log D) = O(\log L)$, since $N \geq t+1$.

Thus, we can simplify the above to:

$$O\left(\left(L + \frac{L^2(t+1)}{ND}\right)\log L\right).$$

Note that this is tight since if the log term on the right is less than $\log L$, it means that $\frac{L^2(t+1)}{ND} \leq L$, in which case the first summand dominates.   □

6   Lower Bound for Spoke-Based Algorithms

We now prove a lower bound on the number of time steps that *any* spoke-based algorithm must take to locate the treasure in the presence of adversarial crash failures.

Theorem 3. *In the presence of up to $t < N$ crash failures, any spoke-based algorithm requires $\Omega\left(L + \left(\frac{L^2(t+1)}{ND}\right)\log L\right)$ time steps to locate the treasure.*

Proof. By Yao's lemma [31], the search time of the best randomized algorithm equals the search time of the best deterministic algorithm against a known randomized adversary. Thus, we compute the search time for the best deterministic algorithm against a known but randomized adversarial placement of the treasure.

Adversarial Strategy. Let x be a positive integer and y be an integer chosen uniformly at random in $[0, x]$. Let $L = 2^y$ and $L/D = 2^{x-y}$. Choose an integer z uniformly at random in $[0, L/D]$. Let the treasure be an ellipse with diameter D and an arbitrary small width. Place this ellipse so that its center is at distance L from the nest, in the direction from the nest that is oriented at arc distance $\frac{zD}{L}$ along the unit circle. Rotate the ellipse so that its diameter is perpendicular to the ray connecting the nest and the center of the ellipse.

Lower Bound Against This Strategy. Assume the algorithm knows the value x; the randomized adversarial strategy above; and t, the number of faults that will occur. The algorithm can be represented as a sequence, σ, of tuples. Each tuple corresponds to some searcher's first visit to a region that is a possible treasure location. In particular, tuple (ℓ, a) corresponds to a visit to any point in the ellipse with center at distance ℓ from the nest, and orientation that is arc length a along the unit circle centered at the nest. The tuples in σ are all sorted by time of visit to first point in the ellipse, with ties broken arbitrarily.

First, note that there is 1 unique tuple of length 2^x: $(2^x, 0)$; 2 unique tuples of length 2^{x-1}: $(2^{x-1}, 0)$ and $(2^{x-1}, 1/2)$; 4 unique tuples of length 2^{x-2}: $(2^{x-2}, 0)$, $(2^{x-2}, 1/4)$, $(2^{x-2}, 1/2)$, $(2^{x-2}, 3/4)$; and so forth. Next, observe that each unique tuple, (ℓ, a), appears in σ at least $t+1$ times. This is necessary since each possible ellipse must be visited by $t+1$ searchers in order for the algorithm to be robust to t adversarial faults. Finally, note that visiting any point on the ellipse corresponding to (ℓ, a) requires movement of $\Omega(\ell)$, no matter at what tuple, (ℓ', a') the searcher visiting (ℓ, a) was previously at. To see this, first note that if $\ell = \ell'$, then the distance travelled between these two tuples is $\Omega(\ell)$ since there must be a trip to the nest between the tuples, because the algorithm is spoke-based. Second, if $\ell \neq \ell'$, draw two squares centered at the nest, one enclosing (ℓ, a), and the other enclosing (ℓ', a'). Then, note that the minimum distance between the squares is $\Omega(\ell)$.

The expected total distance travelled by all searchers in the algorithm is then given as follows. Select a tuple in σ uniformly at random, and sum up the lengths of all tuples preceding, and including, the selected tuple in σ. Let X be the random variable giving this sum. Note that X stochastically dominates the following random variable, X': Let σ' be a sequence where each tuple in σ of the type (ℓ, a) is expanded to ℓ copies. Select a tuple uniformly at random in σ' and let X' be the index of the selected tuple.

Note that $E(X')$ is half the length of σ', and that the length of $\sigma' = (t+1)x2^x$. Thus, $E(X) \geq E(X') \geq (1/2)(t + 1)x2^x$. Finally, note that, since there are N searchers, the expected search time is at least the total distance travelled by all searchers divided by N. By linearity of expectation, the expected search time is thus at least $\frac{(t+1)x2^x}{2N}$. Since $2^x = L^2/D$, then $\frac{(t+1)x2^x}{2N} = \Omega\left(\frac{(t+1)L^2 \log L}{ND}\right)$. The lower bound is this value plus L, since no matter the values of D, N and t, the total search time is always at least L. □

7 Empirical Evaluation

We implement GOLDENFA and algorithms from [10] to empirically evaluate how search time changes as we increase: the ratio of the diameter of treasure to distance to treasure (D/L); the number of searchers (N); and the fraction of random crash failures (t/N). We compare GOLDENFA to algorithms from Feinerman and Korman in [10].

7.1 Setup

We implemented four algorithms. GOLDENFA is our algorithm from Sect. 4. F&K-ADVICE is Algorithm 1 from [10]; it requires $O(\log \log N)$ bits of advice. F&K-NOADVICE is Algorithm 2 from [10] with $\epsilon = .01$; it requires zero bits of advice. Since the value for ϵ in Algorithm 2 is not specified in [10], we conducted experiments to determine that the setting $\epsilon = .01$ performs well empirically.

GOLDENFA-HEURISTIC is the last algorithm. In this algorithm, for epoch $i \geq 1$, there are $\lceil c(1 + \alpha)^i \rceil$ spokes of length $(1 + \alpha)^i$, for parameters $c, \alpha > 0$. Similar to the GOLDENFA, each spoke in this is at arc distance equal to the Golden Ratio from the previous. We set $c = 1.9$ and $\alpha = 7$, since they perform well empirically.

In all of our experiments the treasure is a circle with diameter D. For each data point plotted, 150 trials were run and the average search time was plotted. The location of the treasure was kept fixed throughout all trials. The search

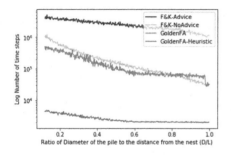 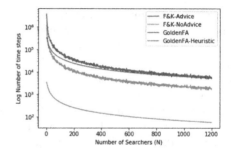

Fig. 3. Search time versus D/L; $L = 500$, $N = 1$, and D is varied.

Fig. 4. Search time versus N; $L = 500$ and $D = 4$.

time reported is time steps, where one time step is the amount of time it takes a searcher to travel a distance of 1. All algorithms were implemented in Python 3.6, and all experiments were run on a Macbook Pro with 2.6 GHz Intel Core i7 processor and 16GB RAM.

7.2 Results

Our results show how search time changes as we vary three different values. In particular, we include three plots giving results of experiments based on varying (1) the ratio D/L, where D is the treasure diameter and L the distance from the nest to the center of the treasure; (2) the number of searchers N; (3) and the fraction of faults t/N. In each plot, search time is the independent variable, and it is plotted on a logarithmic scale.

Search Time versus D/L. Our first experiment tracks search time as the ratio D/L increases. The value of L is fixed at 500, and D increases from 1 to 500.

Figure 3 shows how search time decreases as D/L increases from .1 to 1. As the plot shows, search time decreases for all algorithms. GOLDENFA-HEURISTIC consistently has the best search time across values tested, with performance that is always between 1 and 2 orders of magnitude better than all other algorithms, when D/L is greater than about .15. Next, in performance, are GOLDENFA and F&K-NOADVICE. Initially F&K-NOADVICE has worse search time than GOLDENFA, but as D/L increases, they both trend towards roughly similar performance. Last, in the plot is F&K-ADVICE, which does not decrease nearly as much as the other algorithms as D/L increases.

It is surprising that F&K-NOADVICE performs better than F&K-ADVICE as D/L increases. We conjecture this holds because (1) F&K-NOADVICE has an algorithmic parameter (ϵ), while F&K-ADVICE has none; and (2) we optimized this parameter based on empirical feedback.

Search Time versus N**.** Our second experiment tracks search time versus the number of searchers, N. Figure 4 shows the outcome when $L = 500$, $D = 4$, and N varies from 1 to 200.

In this plot, search time of all algorithms decreases with N. GOLDENFA-HEURISTIC performs about 2 orders of magnitude better than any other algorithm, for all values of N tested. Next comes F&K-NOADVICE, which performs up to a factor of about 5 better than the remaining algorithms.

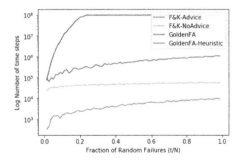

Fig. 5. Search time versus the fraction of failures (t/N); $L = 500, D = 4$ and $N = 100$.

Finally, GOLDENFA and F&K-ADVICE are last, with performance roughly equal for large N.

Search Time versus t/N**.** Our last experiment tracks search time as the ratio t/N increases, where t is the number of crash failures and N is the number of searchers. In these experiments, we hold the following values fixed: $L = 500$, $D = 4$, $N = 100$; and we vary t from 0 to 99. For each value of t, a random subset of t searchers are removed after the first 100 time steps of the algorithm. To prevent any algorithm from running forever, a hard timeout was set at 10^8 time steps.

The results are given in Fig. 5. Again GOLDENFA-HEURISTIC has fastest search time over the entire range of values tested, with performance a bit less than an order of magnitude better than then next fastest algorithm, F&K-NOADVICE. F&K-NOADVICE has search times which increase slowly as t/N increases. GOLDENFA comes next with a search time that increases more rapidly with t/N. Finally, F&K-ADVICE comes last, with search time increasing rapidly with t/N until it times out when t/N is roughly about .20. Our theoretical analysis suggests that search time for GOLDENFA would increase roughly linearly with t/N. Results from this experiment suggest this is the case with slope of approximately 10 for search time as a function of t/N.

8   Conclusion and Future Work

We have described an algorithm, GOLDENFA that solves the ANTS problem by finding a treasure that is a convex shape with any diameter D, even in the presence of $t < N$ crash failures. We have proven that our algorithm takes $O\left(\left(L + \frac{L^2(t+1)}{ND}\right)\log L\right)$ expected search time, where L is the distance from the nest to the treasure and N is the number of searchers. Additionally, we have proven a near-matching lower bound on search time for a class of "spoke-algorithms"s, which search only via line segments emanating from the nest. Our algorithm uses the Golden Ratio to spread out search spokes uniformly, even in the presence of many crash failures.

Several interesting problems remain including the following. Can we develop a non-spoke-based algorithm that removes the logarithmic terms in our search time, but is still robust to failures and does not require advice? Fibonacci spirals are quite commonly used in nature for space-filling applications, so they may be useful for this open problem.

Another interesting open problem is to extend our results for multiple treasures with different shapes and orientations. It is possible for a treasure to have a large L but its orientation is such that the nearest point to the nest is only $\Theta(1)$ units away. This treasure can be located in $\Theta(1)$ time steps by searching along a spiral around the nest. However, when rotated, this treasure can be oriented in a way that now requires $O(L^2/D)$ time steps.

References

1. Barchyn, T.E., Hugenholtz, C.H., Fox, T.A.: Plume detection modeling of a drone-based natural gas leak detection system. Elem. Sci. Anthropocene **7**(1) (2019)
2. Beverly, B.D., McLendon, H., Nacu, S., Holmes, S., Gordon, D.M.: How site fidelity leads to individual differences in the foraging activity of harvester ants. Behav. Ecol. **20**(3), 633–638 (2009)
3. Boczkowski, L., Natale, E., Feinerman, O., Korman, A.: Limits on reliable information flows through stochastic populations. PLoS Comput. Biol. **14**(6), e1006195 (2018)
4. Collet, S., Korman, A.: Intense competition can drive selfish explorers to optimize coverage. In: Symposium on Parallelism in Algorithms and Architectures (SPAA) (2018)
5. Dubhashi, D.P., Panconesi, A.: Concentration of Measure for the Analysis of Randomized Algorithms. Cambridge University Press, Cambridge (2009)
6. Dunlap, R.A.: The Golden Ratio and Fibonacci Numbers. World Scientific, Singapore (1997)
7. Emek, Y., Langner, T., Stolz, D., Uitto, J., Wattenhofer, R.: How many ants does it take to find the food? Theoret. Comput. Sci. **608**, 255–267 (2015)
8. Emek, Y., Langner, T., Uitto, J., Wattenhofer, R.: Solving the ANTS problem with asynchronous finite state machines. In: Esparza, J., Fraigniaud, P., Husfeldt, T., Koutsoupias, E. (eds.) ICALP 2014. LNCS, vol. 8573, pp. 471–482. Springer, Heidelberg (2014). https://doi.org/10.1007/978-3-662-43951-7_40
9. Feinerman, O., Korman, A.: Memory lower bounds for randomized collaborative search and implications for biology. In: Aguilera, M.K. (ed.) DISC 2012. LNCS, vol. 7611, pp. 61–75. Springer, Heidelberg (2012). https://doi.org/10.1007/978-3-642-33651-5_5
10. Feinerman, O., Korman, A.: The ANTS problem. Distrib. Comput. **30**(3), 149–168 (2017)
11. Feinerman, O., Korman, A., Lotker, Z., Sereni, J.S.: Collaborative search on the plane without communication. In: Proceedings of the 2012 ACM Symposium on Principles of Distributed Computing (PODC), pp. 77–86. ACM (2012)
12. Fredman, M.L., Tarjan, R.E.: Fibonacci heaps and their uses in improved network optimization algorithms. J. ACM (JACM) **34**(3), 596–615 (1987)
13. Gordon, D.M.: Ants at Work: How an Insect Society is Organized. Simon and Schuster, New York (1999)

14. Hecker, J.P., Carmichael, J.C., Moses, M.E.: Exploiting clusters for complete resource collection in biologically-inspired robot swarms. In: International Conference on Intelligent Robots and Systems IROS, pp. 434–440 (2015)
15. Itai, A., Rosberg, Z.: A golden ratio control policy for a multiple-access channel. IEEE Trans. Autom. Control **29**(8), 712–718 (1984)
16. Karaboga, D., Akay, B.: A survey: algorithms simulating bee swarm intelligence. Artif. Intell. Rev. **31**(1–4), 61–85 (2009)
17. Kempe, D., Schulman, L.J., Tamuz, O.: Quasi-regular sequences and optimal schedules for security games. In: ACM-SIAM Symposium on Discrete Algorithms (SODA), pp. 1625–1644. Society for Industrial and Applied Mathematics (2018)
18. Khinchin, A.I.: Continued Fractions, vol. 525. P. Noordhoff (1963)
19. King, V., Saia, J., Young, M.: Conflict on a communication channel. In: Proceedings of the 30th Annual ACM SIGACT-SIGOPS Symposium on Principles of Distributed Computing, pp. 277–286. ACM (2011)
20. Knuth, D.E.: The Art of Computer Programming, Volume 3: Searching and Sorting. Addison-Wisley, Reading (1973)
21. Kreyszig, E.: Advanced Engineering Mathematics, 9th edn. Wiley, Hoboken (2008)
22. Krieger, M.J., Billeter, J.B., Keller, L.: Ant-like task allocation and recruitment in cooperative robots. Nature **406**(6799), 992 (2000)
23. Langner, T., Uitto, J., Stolz, D., Wattenhofer, R.: Fault-tolerant ANTS. In: Kuhn, F. (ed.) DISC 2014. LNCS, vol. 8784, pp. 31–45. Springer, Heidelberg (2014). https://doi.org/10.1007/978-3-662-45174-8_3
24. Lenzen, C., Lynch, N., Newport, C., Radeva, T.: Trade-offs between selection complexity and performance when searching the plane without communication. In: Proceedings of the 2014 ACM Symposium on Principles of Distributed Computing (PODC), pp. 252–261. ACM (2014)
25. Lenzen, C., Radeva, T.: The power of pheromones in ant foraging. In: Workshop on Biological Distributed Algorithms (BDA) (2013)
26. Livio, M.: The Golden Ratio: The Story of Phi, the World's Most Astonishing Number. Broadway Books, New York (2008)
27. Naylor, M.: Golden, $\sqrt{2}$, and π flowers: a spiral story. Math. Mag. **75**(3), 163–172 (2002)
28. Şahin, E.: Swarm robotics: from sources of inspiration to domains of application. In: Şahin, E., Spears, W.M. (eds.) SR 2004. LNCS, vol. 3342, pp. 10–20. Springer, Heidelberg (2005). https://doi.org/10.1007/978-3-540-30552-1_2
29. Świerczkowski, S.: On successive settings of an arc on the circumference of a circle. Fundam. Math. **46**, 187–189 (1958)
30. Williams, S.C.P.: Studying volcanic eruptions with aerial drones. Proc. Natl. Acad. Sci. **110**(27), 10881–10881 (2013)
31. Yao, A.C.C.: Probabilistic computations: toward a unified measure of complexity. In: Symposium on Foundations of Computer Science (FOCS), pp. 222–227. IEEE (1977)

Local Gathering of Mobile Robots in Three Dimensions

Michael Braun, Jannik Castenow$^{(\boxtimes)}$, and Friedhelm Meyer auf der Heide

Heinz Nixdorf Institute and Computer Science Department, Paderborn University,
Paderborn, Germany
{braunm,janniksu,fmadh}@mail.upb.de

Abstract. In this work, we initiate the research about the GATHERING problem for robots with limited viewing range in the three-dimensional Euclidean space. In the GATHERING problem, a set of initially scattered robots is required to gather at the same position. The robots' capabilities are very restricted – they do not agree on any coordinate system or compass, have a limited viewing range, have no memory of the past and cannot communicate.

We study the problem in two different time models, in $\mathcal{F}$SYNC (fully synchronized discrete rounds) and the continuous time model. For $\mathcal{F}$SYNC, we introduce the 3D-GO-TO-THE-CENTER-strategy and prove a runtime of $\Theta\left(n^2\right)$ that matches the currently best runtime bound for the same model in the Euclidean plane [SPAA'11] .

Our main result is the generalization of contracting strategies (continuous time model) from [Algosensors'17] to the three-dimensional case. In contracting strategies, every robot that is located on the global convex hull of all robots' positions moves with full speed towards the inside of the convex hull. We prove a runtime bound of $\mathcal{O}\left(\Delta \cdot n^{3/2}\right)$ for *any* three-dimensional contracting strategy, where Δ denotes the diameter of the initial configuration. This comes up to a factor of $\sqrt{n}$ close to the lower bound of $\Omega\left(\Delta \cdot n\right)$ which is already true in two dimensions.

In general, it might be hard for robots with limited viewing range to decide whether they are located on the global convex hull and which movement maintains the connectivity of the swarm, rendering the design of *concrete* contracting strategies a challenging task. We prove that the continuous variant of 3D-GO-TO-THE-CENTER is contracting and keeps the swarm connected. Moreover, we give a simple design criterion for three-dimensional contracting strategies that maintains the connectivity of the swarm and introduce an exemplary strategy based on this criterion.

Keywords: Mobile robots · Local strategies · Gathering · Continuous time

1 Introduction

We study a scenario where a distributed system of mobile entities (called *robots*) is supposed to establish a certain formation, also denoted as a *pattern*. The robots

© Springer Nature Switzerland AG 2020
A. W. Richa and C. Scheideler (Eds.): SIROCCO 2020, LNCS 12156, pp. 63–79, 2020.
https://doi.org/10.1007/978-3-030-54921-3_4

are scattered in a d-dimensional Euclidean space (usually the Euclidean plane) and have to coordinate their movements in a distributed manner to reach the desired formation. The robots' capabilities depend on the exact model and formation problem but are typically very restricted. Usually, the robots do not agree on a common coordinate system or compass, cannot communicate with each other and have only limited sensing capabilities. One extensively studied coordination problem is the PATTERN FORMATION problem, dealing with questions such as: Which patterns are generally formable by a set of robots? Which capabilities do the robots need? Given a specific pattern, for which initial configurations is this pattern formable? Interestingly, it has been proven that there are only *two* patterns that might be formable starting in an arbitrary input configuration. These are the patterns POINT and UNIFORM CIRCLE. Forming the pattern POINT is known under a more common name – the GATHERING problem, which studies the task of gathering a set of robots on the same position. Both of these problems have been extensively studied under several different assumptions, involving the viewing range (local or global), the synchronization (synchronous or asynchronous activation), the extent (robots can or cannot occupy the same position) or the opacity of robots, to name only a few. However, most of these models have in common that the robots operate in the two-dimensional Euclidean plane. A natural extension would be to consider the three-dimensional Euclidean space, where the robots have the ability to fly, such as drones, or to move underwater. Existing results about robots in the three-dimensional Euclidean space are very scarce, rely on strong assumptions (such as axis agreement) and do not consider any runtime analyses of the proposed strategies. Our work initiates the study of GATHERING of robots in three-dimensions, in one of the weakest possible models – robots do not agree on any coordinate system or compass, are *oblivious* (have no memory of the past) and have only a local view.

1.1   Model and Time Notions

We consider a set $\mathcal{R}$ of n robots $r_1, \ldots, r_n$, each of which occupies a single point in $\mathbb{R}^3$ at each time. As such, robots can neither block each other's views nor paths, and multiple robots are allowed to occupy the same position at the same time. The position of robot r_i at time t is denoted by $p_i(t)$. The positions of all robots at time t, $\mathcal{P}_t = (p_1(t), \ldots, p_n(t))$ are collectively called the *configuration* at time t. The Euclidean distance between points $x, y \in \mathbb{R}^3$ is denoted as $d(x, y)$. For a subset of the three-dimensional Euclidean space $\mathcal{P} \subseteq \mathbb{R}^3, d(x, \mathcal{P})$ is used as a shorthand for $\min_{y \in \mathcal{P}} d(x, y)$.

The overall abilities of the robots are rather limited: They are not allowed to communicate with each other, they are *identical* (they cannot be distinguished) and are *oblivious*, meaning they have no memory of the past. Furthermore they do not share a common coordinate system or orientation. Robots are only able to observe the space around them within a limited viewing range of 1, i.e. a robot r_i can see the position of another robot r_j if and only if $d(p_i(t), p_j(t)) \leq 1$. Two robots r_i and r_j with $d(p_i(t), p_j(t)) \leq 1$ are also called *neighbors*. The set of all neighbors of r_i at time t is called the *neighborhood* of r_i and is denoted as $\mathcal{R}_i(t)$.

This limited viewing range can also be considered to induce a unit ball graph $\text{UBG}_t = (\mathcal{R}, E_t)$ at time t, whose nodes consist of the robots and where the set of edges E_t contains an edge $\{r_i, r_j\}$ if and only if $d(p_i(t), p_j(t)) \leq 1$. This graph is also called the *visibility graph* at time t. Note that the UBG is a generalization of the two-dimensional *unit disk graph* (UDG) to three dimensions.

Starting from a configuration of n robots in the three-dimensional Euclidean space that is connected at time 0, i.e. UBG_0 is connected, the goal is to gather all robots in one point. This problem will be referred to as the (three-dimensional) GATHERING problem. Note that the eventual gathering point is not predefined and can instead be chosen by the robots at runtime. This also imposes a subgoal during the execution of any algorithm that solves this problem: It has to be ensured that UBG_t remains connected. Otherwise, the limited viewing range of the robots, combined with the fact that they do not share coordinate systems, makes it impossible for any deterministic algorithm to restore connectivity and the robots can no longer converge to the same point [1].

Throughout this work, we consider two different notions of time: The fully synchronous $\mathcal{F}$SYNC model and the continuous time model.

$\mathcal{F}$**sync:** In $\mathcal{F}$SYNC, all robots operate in fully synchronous `Look-Compute-Move` (LCM) cycles. In the `Look` phase, a robot r_i observes its environment, detects the set of all visible robots $\mathcal{R}_i(t)$ and stores a snapshot in its local memory. Based on this snapshot, r_i computes a target point in the `Compute` phase. Finally, in the `Move` phase, r_i moves to that target point. The execution of a single LCM cycle is also denoted as one *round*.

Continuous Time Model: Generally, the continuous time model can be seen as a continuous variant of $\mathcal{F}$SYNC, in which robots only move an infinitesimal small distance towards their target points [8]. At every point in time, the movement of each robot r_i can be expressed by a *velocity vector* $\boldsymbol{v}_i(t)$ with $0 \leq \|\boldsymbol{v}_i(t)\| \leq 1$, i.e. the maximal speed of a robot is bounded by 1. In contrast to $\mathcal{F}$SYNC, the function $p_i \colon \mathbb{R}_{>0} \to \mathbb{R}^3$, representing the position of r_i at time t, is a continuous function and also called the *trajectory* of r_i. Although the trajectories are continuous, they are not necessarily differentiable because robots are able to change their speed and direction non-continuously. However, natural movement strategies have (right) differentiable trajectories. Thus, the velocity vector of a robot $\boldsymbol{v}_i \colon \mathbb{R}_{>0} \to \mathbb{R}^3$ can be seen as the (right) derivative of p_i.

1.2   Our Contribution

The contribution of this paper is twofold. We consider the fully synchronous $\mathcal{F}$SYNC model and the continuous time model. For $\mathcal{F}$SYNC, we introduce the strategy 3D-GO-TO-THE-CENTER (3D-GTC), which is the three-dimensional generalization of GO-TO-THE-CENTER (GTC), invented for robots operating in the Euclidean plane [1]. The main idea of 3D-GTC is that robots move towards the center of the smallest enclosing sphere of all robots within their viewing radius, while ensuring that the configuration stays connected. We prove a runtime bound of $\Theta\left(n^2\right)$ for 3D-GTC which matches the runtime of the two-dimensional GTC strategy.

For the continuous time model, we generalize the class of contracting strategies [10] to three dimensions. In contracting strategies, every robot that lies on the convex hull of all robots' positions moves always with speed 1 into a direction that points inside or on the boundary of the convex hull. We prove that every (three-dimensional) contracting gathering strategy gathers all robots on a single point in time at most $\mathcal{O}\left(\Delta \cdot n^{3/2}\right)$, where Δ denotes the (geometric) diameter of the initial configuration, i.e. the maximum Euclidean distance between any pair of robots. This runtime bound differs from the runtime bound for two-dimensional contracting strategies by a factor of $\sqrt{n}$. The lower bound is $\Omega\left(\Delta \cdot n\right)$ and already holds for the two-dimensional case [10]. The main open question is whether $\mathcal{O}\left(\Delta \cdot n^{3/2}\right)$ is tight or can be improved to $\mathcal{O}\left(\Delta \cdot n\right)$.

Note that a contracting strategy is not necessarily local. Therefore, we finally present two local, contracting strategies. Our first example is the continuous variant of 3D-GTC, called CONT-3D-GTC. We prove that the strategy is contracting and thus gathers the robots in time $\mathcal{O}\left(\Delta \cdot n^{3/2}\right)$. In addition, we present the class of *tangential-normal* strategies. These strategies are local and maintain connectivity. As an example for a strategy that is both tangential-normal and contracting, we introduce the MOVE-ON-ANGLE-MINIMIZER strategy.

1.3   Related Work

In this overview over related work, we focus on the GATHERING problem for synchronized robots with local visibility in the Euclidean plane. Beyond that, we give a summary about research concerning robot coordination problems in the three-dimensional Euclidean space. For other models and coordination problems, which involve, among others, less synchronized schedulers or robots with a global view, we refer the reader to the recent survey [7].

Ando, Suzuki and Yamashita introduced the GTC-strategy for fully synchronous robots with local view [1]. In GTC, every robot moves in every round towards the center of the smallest enclosing circle of all robots within its viewing range while ensuring that the swarm remains connected. Ando et al. could prove that GTC solves the GATHERING-problem in finite time. Later on, Degener et al. could prove a tight runtime bound of $\Theta\left(n^2\right)$ for GTC [5]. By now, this is the best known runtime bound for a strategy that solves GATHERING of robots with local visibility and without agreement on any coordinate system or compass in $\mathcal{F}$SYNC.

Faster runtimes could so far only be obtained under different assumptions – for example by introducing one-axis agreement or changing the time model. Poudel and Sharma proved that it is possible to gather a swarm of robots with local view in time $\mathcal{O}\left(\Delta\right)$, where Δ denotes the diameter of the initial configuration [12]. The main assumption for their strategy is that the robots agree on one axis of their coordinate systems.

The second time model we consider in this paper is the continuous time model, introduced by Gordon et al. [8]. In this time model robots do not operate in synchronized rounds but continuously observe their environment and move while having a bounded maximal speed. Gordon et al. propose a gathering strategy for the continuous time model. In their strategy, all robots that locally assume that they are located on the global convex hull move with maximal speed along the bisector formed by vectors to their neighbors along the global convex hull. This strategy has later been called MOVE-ON-BISECTOR by Degener et al. They could also prove runtime of $\Theta(n)$ [4].

The main result of this paper is based on a more general view on continuous GATHERING strategies in the Euclidean plane – the class of *contracting* strategies in which all robots that are located on the global convex hull of all robots move with maximal speed into a direction that points inside of the convex hull [10]. Li et al. could prove a runtime of $\mathcal{O}(\Delta \cdot n)$ for *any* contracting strategy. Note that MOVE-ON-BISECTOR is also a contracting strategy but has a significantly faster runtime than $\mathcal{O}(\Delta \cdot n)$. However, there are contracting strategies with a runtime of $\Omega(\Delta \cdot n)$ [10].

In the three-dimensional Euclidean space there is so far, to the best of our knowledge, no strategy known that solves GATHERING of robots with limited viewing range. More generally, literature about robots operating in three-dimensional spaces is very scarce. We summarize the literature briefly. In [2] the authors show that gathering of robots in the three-dimensional Euclidean space is possible – under the assumptions that robot have a global view but are not transparent and that the robots agree on one axis of their coordinate systems. Tomita et al. study a different problem – the plane formation problem [13]. In the plane formation problem, the goal is that eventually all robots are located on the same plane, while ensuring that no two robots occupy the same position. The authors show that this problem is not solvable for every initial configuration, give a characterization of all start configurations for which the problem is solvable and introduce an algorithm that solves the problem for the latter set of configurations. Yamauchi, Uehara and Yamashita generalize this result further and study the more general PATTERN FORMATION for synchronized robots in the three-dimensional Euclidean space [14]. They characterize the set of all patterns that might be formable depending on symmetries of the initial configuration.

2 Gathering in $\mathcal{F}$sync

In this section, the three-dimensional GATHERING problem will be studied under the $\mathcal{F}$SYNC model. The results can be considered as a generalization of those obtained by Degener et al. [5] for the two-dimensional setting. It will be shown that a generalization of GTC by Ando et al. [1] solves the gathering problem in three dimensions in $\Theta(n^2)$ rounds.

2.1 3D-Go-To-The-Center

The strategy 3D-GO-TO-THE-CENTER (3D-GTC) is a generalization of GO-TO-THE-CENTER to the three-dimensional Euclidean space and is summarized

---

Algorithm 1. 3D-Go-To-The-Center (3D-GTC)

1: $\mathcal{R}_i(t) := \{$positions of robots visible from r_i, including r_i at time $t\}$
2: $\mathcal{S}_i(t) :=$ smallest enclosing sphere of $\mathcal{R}_i(t)$
3: $c_i(t) :=$ center of $\mathcal{S}_i(t)$ ▷ target point
4: **for all** $r_j \in \mathcal{R}_i(t)$ **do**   ▷ Maintain connectivity
5:     $m_j :=$ midpoint between $p_i(t)$ and $p_j(t)$
6:     $\mathcal{B}_j(t) :=$ ball with radius $\frac{1}{2}$ and center m_j
7:     $\ell_j :=$ maximum distance r_i can move towards $c_i(t)$ without leaving $\mathcal{B}_j(t)$
8: $L_i := \min_{r_j \in \mathcal{R}_i(t)} \ell_j$
9: Move towards $c_i(t)$ for a distance of L_i

---

in Algorithm 1. A key component is the computation of a smallest enclosing sphere (SES) of a set of points $\mathcal{P}$. This is a sphere of minimal radius that contains all points in $\mathcal{P}$ with the following properties:

Proposition 1 [6]. *Let $\mathcal{S}$ be the smallest enclosing d-sphere (SES) of a point set $\mathcal{P} \subset \mathbb{R}^d$. Then the center c of $\mathcal{S}$ is a convex combination of at most $d + 1$ points in $\mathcal{P}$ that lie on the surface of $\mathcal{S}$. Especially,*

1. c lies in S
2. c minimizes the maximum distance to the points in $\mathcal{P}$.

Intuitively, 3D-GTC works by attempting to locally move robots closer together. This is achieved by letting each robot r_i compute the SES of its neighborhood $\mathcal{R}_i(t)$ and then moving towards its center $c_i(t)$. Additionally, the strategy follows the subgoal of maintaining connectivity of UBG_{t+1}. This is achieved by limiting the distance a robot r_i moves towards its target $c_i(t)$, such that for any of its neighbors r_j, it stays within a distance of $\frac{1}{2}$ of the midpoint between the positions of r_i and r_j at time t. Thus, if both r_i and r_j perform this strategy, the distance between their positions at the start of the next round $t + 1$ is at most 1, maintaining visibility. By the argumentation above, the following Lemma holds.

Lemma 1. *If UBG_0 is connected, UBG_t remains connected for all $t \geq 0$.*

Overall, the only difference to the original GTC strategy for two dimensions lies in the computation of a smallest enclosing sphere in the 3D case over a smallest enclosing circle in the 2D case. In fact, if the three-dimensional version is applied to a configuration of robots that is coplanar with respect to some plane h, it acts just as if the robots' positions were projected to h and the two-dimensional version was applied to the resulting two-dimensional subspace. This is a result of the fact that computing a SES of a set of coplanar points is equivalent to computing a smallest enclosing circle instead.

From this observation, we can immediately conclude that the lower bound on the runtime of the two-dimensional version of the strategy shown by Degener et al. [5] also applies to the three-dimensional case by simply embedding the two-dimensional worst-case start configuration within three-dimensional space: In the configuration, n robots are positioned on a circle such that the distance between two neighbors is 1. This causes the robots to only take small steps of size $\mathcal{O}(1/n)$ towards the center of the circle, leading to a gathering time of $\Omega(n^2)$.

Theorem 1. *There is a start configuration such that* 3D-GTC *takes* $\Omega(n^2)$ *rounds to gather the robots in one point.*

With a generalization of the analysis of [5], we can also prove an upper runtime bound of $\mathcal{O}\left(n^2\right)$. For the proof, we refer the reader to the full version of this paper [3].

Theorem 2. *Given n robots in a connected starting configuration $\mathcal{P} \in \mathbb{R}^3$ in the Euclidean space,* 3D-GTC *gathers the robots in $\mathcal{O}(n^2)$ rounds.*

The combination of both theorems yields a tight runtime of $\Theta(n^2)$.

3   Continuous Gathering

Now, we consider the GATHERING problem within the continuous time model. For the Euclidean plane, Li et al. [10] introduced the class of *contracting strategies*. This definition can also be applied to three dimensions: Let CH_t denote the closed convex hull of the robots' configuration $\mathcal{P}_t$ at time t and let Corn_t denote the vertices of CH_t. The class of contracting strategies can be defined as follows:

Definition 1. *In the continuous time model, a movement strategy for n robots is called* contracting *if for every time t such that the cardinality of Corn_t is strictly greater than 1, every robot in Corn_t moves with speed 1 in a direction that points to CH_t.*

The main idea of our analysis is to project the three-dimensional configuration (including the velocity vectors) to a two-dimensional plane. The projected robots then perform something similar to a contracting strategy where they move towards the inside of the projected convex hull with varying speeds. However, when looking at only a single projection plane, some velocity vectors might even have a length of 0 in the projection at some points in time (in case the projection plane is chosen orthogonal to the velocity vector). Thus, the analysis of Li et al. cannot be directly applied to the projection as this analysis assumes that all robots on the convex hull move with speed 1 towards the inside. Instead, we analyze not only one but all possible (meaningfully different) projections, since – intuitively – for a majority of all possible projection planes, the projected length of a velocity vector must be larger than a constant ε.

3.1   Preliminaries

The following lemma is a useful tool for the analysis of continuous strategies stating how the distance between two robots changes over time.

Lemma 2 ([9]). *Consider two robots r_i and r_j with differentiable trajectories at time t. Their distance $d(p_i(t), p_j(t))$ at time t changes with speed*

$$d'(p_i(t), p_j(t)) = -(\|v_i(t)\| \cdot \cos \beta_{i,j}(t) + \|v_j(t)\| \cdot \cos \beta_{j,i}(t)),$$

where $\beta_{i,j}(t)$ is the angle between $v_i(t)$ and the line segment $\overline{p_i(t)p_j(t)}$.

The main tool for the analysis of contracting strategies in the three-dimensional Euclidean space are projections of the robots' configuration onto a two-dimensional plane. Let $h(x)$ be the plane through the origin with normal vector x and let Π_x denote the orthogonal projection onto $h(x)$. Now, given a configuration $\mathcal{P}$ of n robots, consider their projection $\hat{\mathcal{P}}^{(x)} = \{\Pi_x p_i(t) \mid p_i(t) \in \mathcal{P}\}$ onto $h(x)$ along with the projections of their movement vectors $\hat{v}_i^{(x)}(t) = \Pi_x v_i(t)$. Furthermore, denote the convex hull of $\hat{\mathcal{P}}$ as $PCH_t(x)$. See also Fig. 1.

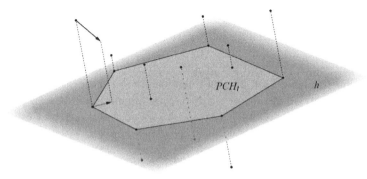

Fig. 1. A configuration of robots being projected onto a plane $h(x)$. The mapping of the orthogonal projection Π_x is illustrated by dashed lines and the projected convex hull $PCH_t(x)$ is shown in light grey. One of the robots' movement vectors as well as its projection are represented by arrows.

If the robots perform a contracting strategy in the three-dimensional space, their projections also move towards the inside of the projected convex hull $PCH_t(x)$ since Π_x is a linear transformation and therefore preserves convexity. However, the lengths of the projected movement vectors $\hat{v}_i(t)$ are going be smaller than 1 in general. For a given projection onto a plane $h(x)$, the minimum length of the $\hat{v}_i^{(x)}(t)$ will be called the *projected speed* and is denoted by $\varepsilon_x = \min_{r_i \in \mathcal{R}} \|\Pi_x v_i(t)\|$. Note that ε_x can even be 0 in case $h(x)$ is orthogonal to any velocity vector. The following notion of the *length* of $PCH_t(x)$ will be used as a part of a progress measure for three-dimensional contracting strategies:

Definition 2. *(Length) Let $m_1(t), m_2(t), ..., m_{k(t)}(t)$ be the vertices of $PCH_t(x)$ (ordered counter-clockwise), where $k(t)$ is the number of vertices at time t. The length $\ell(t, x)$ of $PCH_t(x)$ is defined as the sum of its edge lengths: $\ell(t, x) = \sum_{\iota=1}^{k(t)} d(m_\iota(t), m_{\iota-1}(t))$, where $m_0 := m_{k(t)}(t)$.*

Note that if the diameter of the starting configuration was Δ, the length of a given projection can be at most $\pi\Delta$ (if it approximates a circle). Furthermore, if $\ell(t, \boldsymbol{x}) = 0$, then the robots have either gathered in the original three-dimensional space or have formed a line that is parallel to $\boldsymbol{x}$. In the latter case it only takes further time of at most $O(\Delta)$ for the robots to gather, as those robots that form the endpoints of the line have no choice but to move towards each other. The following Lemma provides a statement about how the length changes over time.

Lemma 3. *For time t, let $h(\boldsymbol{x})$ be a plane with projected speed $\varepsilon_{\boldsymbol{x}}$, such that $\ell(t, \boldsymbol{x}) > 0$ and no two robots with different positions in $\mathbb{R}^3$ get projected onto the same point on $h(\boldsymbol{x})$. Then $\ell'(t, \boldsymbol{x}) \le -\frac{8\varepsilon_{\boldsymbol{x}}}{n}$.*

Proof. Because $\Pi_{\boldsymbol{x}}$ is a linear transformation, each of the $m_\iota(t)$ (corners of $\mathrm{PCH}_t(\boldsymbol{x})$) must also be the projection of one of the vertices of the original, three-dimensional convex hull CH_t. Therefore, they possess velocity vectors that point towards the inside of CH_t by the definition of a contracting strategy. Now consider the projections of these velocity vectors onto $h(\boldsymbol{x})$: Let $\hat{\boldsymbol{v}}_i^{(\boldsymbol{x})}(t) := \Pi_{\boldsymbol{x}}\boldsymbol{v}_i(t)$. By assumption, we have $||\hat{\boldsymbol{v}}_i^{(\boldsymbol{x})}(t)|| \ge \varepsilon_{\boldsymbol{x}}$. Using this, it is now possible to bound $\ell'(t, \boldsymbol{x})$: Let $\alpha_\iota(t)$ be the internal angle of $\mathrm{PCH}_t(\boldsymbol{x})$ at $m_\iota(t)$.

Note that in general, it may happen that two corner robots of CH_t got projected onto the same point on $h(\boldsymbol{x})$ for some $\boldsymbol{x}$. By one of the assumptions of the lemma, this is not true. Therefore, we know that each corner $m_\iota(t)$ of $\mathrm{PCH}_t(\boldsymbol{x})$ contains only a single robot. This means that each $\alpha_\iota(t)$ is split into two parts, $\hat{\beta}_{\iota,\iota-1}(t)$ and $\hat{\beta}_{\iota-1,\iota}(t)$ by $m_\iota(t)$'s velocity vector $\hat{\boldsymbol{v}}_\iota^{(\boldsymbol{x})}(t)$, such that $\alpha_\iota(t) = \hat{\beta}_{\iota,\iota-1}(t) + \hat{\beta}_{\iota-1,\iota}(t)$. Using Lemma 2, the derivative of $\ell(t)$ can now be bounded as follows: Recall that $\ell'(t, \boldsymbol{x}) = \sum_{\iota=1}^{k(t)} d'(m_\iota(t), m_{\iota-1}(t))$:

$$\ell'(t, \boldsymbol{x}) = \sum_{\iota=1}^{k(t)} d'(m_\iota(t), m_{\iota-1}(t)) \tag{1}$$

$$= \sum_{\iota=1}^{k(t)} -\left(||\hat{\boldsymbol{v}}_\iota^{(\boldsymbol{x})}(t)|| \cos \hat{\beta}_{\iota,\iota-1}(t) + ||\hat{\boldsymbol{v}}_{\iota-1}^{(\boldsymbol{x})}(t)|| \cos \hat{\beta}_{\iota-1,\iota}(t)\right) \tag{2}$$

$$\le -\varepsilon_{\boldsymbol{x}} \sum_{\iota=1}^{k(t)} \cos \hat{\beta}_{\iota,\iota-1}(t) + \cos \hat{\beta}_{\iota-1,\iota}(t) \tag{3}$$

$$= -\varepsilon_{\boldsymbol{x}} \sum_{\iota=1}^{k(t)} \frac{2(\alpha_\iota(t) - \pi)^2}{\pi^2} \tag{4}$$

$$= -\frac{2\varepsilon_{\boldsymbol{x}}}{\pi^2} \sum_{\iota=1}^{k(t)} (\alpha_\iota(t) - \pi)^2 \tag{5}$$

For Eq. (4) observe that for $\vartheta \in [0, 1]$ and $\alpha \in [0, \pi]$, it holds that $\cos(\alpha\vartheta) + \cos(\alpha(1-\vartheta)) \ge \frac{2(\alpha-\pi)^2}{\pi^2}$ [11]. Now, the Cauchy-Schwarz inequality can be applied along with the fact that the sum of the inner angles of a convex polygon with k corners is $(k - 2) \cdot \pi$.

$$\ell'(t, \boldsymbol{x}) \leq -\frac{2\varepsilon_x}{k(t) \cdot \pi^2} \cdot \left(\sum_{\iota=1}^{k(t)} (\alpha_\iota(t) - \pi)\right)^2 = -\frac{2\varepsilon_x}{k(t)\pi^2} \cdot \left((k(t) - 2) \cdot \pi - k(t)\pi\right)^2$$

$$= -\frac{8\varepsilon_x}{k(t)} \leq -\frac{8\varepsilon_x}{n}$$

This concludes the proof.   □

Note that this also means that $\ell(t, \boldsymbol{x})$ is monotonically decreasing over time.

3.2 Proof of the Upper Bound

The main idea of the analysis is to track the lengths $\ell(t, \boldsymbol{x})$ for all (meaningfully different) projection planes $h(\boldsymbol{x})$. Since the length of the normal vector does not matter, it is enough to consider only vectors $\boldsymbol{x}$ of length 1. Additionally, a vector $\boldsymbol{x}$ and its reflection about the origin $-\boldsymbol{x}$ describe the same plane. Therefore it is enough to consider those vectors that lie on the surface of a unit hemisphere U centered around the origin (w.l.o.g. the one above the XY-plane).

The integral of the lengths $\ell(t, \boldsymbol{x})$ with respect to $\boldsymbol{x}$ on the surface of U at time t can now be used as a measure to track the progress of a three-dimensional gathering strategy:

$$L(t) = \iint_U \ell(t, \boldsymbol{x}) dA$$

If $L(t) = 0$, the robots have gathered. If one of the $\ell(t, \boldsymbol{x})$ prematurely becomes 0, then the robots are collinear and gather in further time $O(\Delta)$.

Lemma 4. $L(0) \leq 2\pi^2 \Delta$.

Proof. Since $\ell(t, \boldsymbol{x}) \leq \pi\Delta$ (if $\mathrm{PCH}_t(\boldsymbol{x})$ approximates a circle), we conclude

$$L(0) \leq \iint_U \pi\Delta dA = \pi\Delta \iint_U dA$$

The remaining integral part is a surface integral over a hemisphere. By observing that the surface area of a unit hemisphere is 2π, the lemma follows.   □

The goal of the proof is to show that there is at least a constant $(1 - \alpha)$-fraction of projection planes $h(\boldsymbol{x})$ with projected speed at least ε for some constants α and ε. This can then be used to show that $L(t)$ decreases by a constant amount at each point in time using Lemma 3.

Now consider a projection plane $h(\boldsymbol{x})$. If this plane has projected speed smaller than ε at time t, then there is a movement vector $\boldsymbol{v}_i(t)$, such that $\angle(\boldsymbol{x}, \boldsymbol{v}_i(t)) < \sin^{-1}\varepsilon$. We say that $\boldsymbol{v}_i(t)$ *blocks* $h(\boldsymbol{x})$. Conversely, given a $\boldsymbol{v}_i(t)$, we can determine the set of all the $h(\boldsymbol{x})$ that are blocked by this $\boldsymbol{v}_i(t)$:

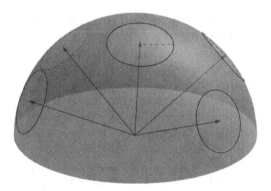

Fig. 2. A figure illustrating how movement vectors block areas of the unit hemisphere U. Around each movement vector $v_i(t)$, there is a spherical cap of radius ε. Each plane corresponding to a normal vector x lying in one of those spherical caps is blocked.

Lemma 5. *At time t, the movement vector $v_i(t)$ blocks vectors from an area of $2\pi\left(1 - \sqrt{1 - \varepsilon^2}\right)$ on U from reaching projected speed ε.*

Proof. W.l.o.g. it can be assumed that $v_i(t)$ has a positive z-component, i.e. lies on U. Otherwise it can be reflected about the origin and it will still affect the exact same planes.

Now consider the spherical cap of U with base radius ε and apex $v_i(t)$ and let C be its curved surface (see Fig. 2 for an illustration). For all vectors $x \in C$, $h(x)$ is blocked from reaching projected speed ε. The area of C can be computed by $A_C = 2\pi r^2(1 - \cos\theta) = 2\pi(1 - \cos(\sin^{-1}\varepsilon)) = 2\pi(1 - \sqrt{1 - \varepsilon^2})$ □

Since there are n robots, the area blocked by their movement vectors is at most $n \cdot 2\pi(1 - \sqrt{1 - \varepsilon^2})$, whereas the total surface of U is 2π. If we want the movement vectors to block only an α-fraction of U's surface, the ε can be chosen accordingly:

Lemma 6. *Let $0 \leq \alpha \leq 1$. Then for a minimum speed of $\varepsilon = \frac{\sqrt{2n\alpha - \alpha^2}}{n}$, there is at most an α-fraction of the surface of U that is blocked with respect to ε.*

Proof. U has a surface of 2π and the robots' movement vectors block an area of at most $n \cdot 2\pi(1 - \sqrt{1 - \varepsilon^2})$. We want to choose ε such that the following holds:

$$\alpha 2\pi = n \cdot 2\pi(1 - \sqrt{1 - \varepsilon^2}) \iff \varepsilon = \frac{\sqrt{2n\alpha - \alpha^2}}{n}$$

□

Using this lemma, it is now possible to bound the decrease of the progress measure $L(t)$ for a given α:

Lemma 7. *For a time $t \geq 0$ such that $\ell(t, x) > 0$ for all $x \in U$ and $0 \leq \alpha \leq 1$, then $L'(t) \leq -16\pi \cdot (1 - \alpha) \cdot \frac{\sqrt{2n\alpha - \alpha^2}}{n^2}$.*

Proof. Choose $\varepsilon = \frac{\sqrt{2n\alpha - \alpha^2}}{n}$ according to Lemma 6, i.e. there is only at most an α-fraction of the surface of U that is blocked. Since Lemma 3 only applies to those $\boldsymbol{x}$ for which no two robots get projected onto the same point, the $\boldsymbol{x}$ for which this is the case still have to be considered. However, there is only a finite number $\binom{n}{2}$ of such vectors out of the uncountably many that form U and they are only singular points on U. Therefore, they can be ignored when considering the integral $L(t)$. By Lemma 3, there is an $(1-\alpha)$-fraction of vectors $\boldsymbol{x}$ from the surface of U (which has size 2π) with $\ell'(t, \boldsymbol{x}) \le -\frac{8\varepsilon}{n} = -8\frac{\sqrt{2n\alpha - \alpha^2}}{n^2}$. Using this, we can bound $L'(t)$:

$$L'(t) = \frac{d}{dt}\left(\iint_U \ell(t, \boldsymbol{x})dA\right) = \iint_U \ell'(t, \boldsymbol{x})dA$$

$$\le (1-\alpha) \cdot 2\pi \cdot -8\frac{\sqrt{2n\alpha - \alpha^2}}{n^2} = -16\pi \cdot (1-\alpha) \cdot \frac{\sqrt{2n\alpha - \alpha^2}}{n^2}$$

$\square$

By choosing the α appropriately, the main result can now be obtained:

Theorem 3. *A set of n robots controlled by a contracting strategy gathers in time $\mathcal{O}\left(\Delta \cdot n^{3/2}\right)$ from an initial configuration with diameter Δ.*

Proof. By Lemma 4, we have $L(0) \le 2\pi^2 \Delta$. By Lemma 7, $L(t)$ decreases by at least $16\pi \cdot (1-\alpha) \cdot \frac{\sqrt{2n\alpha - \alpha^2}}{n^2}$ for a given α as long as $\ell(t, \boldsymbol{x}) > 0$ for all $\boldsymbol{x} \in U$. However, if there is an $\boldsymbol{x} \in U$ with $\ell(t, \boldsymbol{x}) = 0$, then the robots are collinear along some line that is parallel to $\boldsymbol{x}$ and take further time $O(\Delta)$ to gather.

Now choose $\alpha = \frac{1}{2}$ and consider an arbitrary time t such that $\ell(t, \boldsymbol{x}) > 0$ for all $\boldsymbol{x} \in U$. Then $L'(t) \le -8\pi\frac{1}{n^{3/2}}$. Therefore it takes time at most $(2\pi^2 \Delta)/(8\pi\frac{1}{n^{3/2}}) = \frac{\pi}{4}\Delta n^{3/2}$ until $L(t)$ is zero. This leads to a gathering time of $\mathcal{O}\left(\Delta \cdot n^{3/2}\right) + \mathcal{O}(\Delta) \in \mathcal{O}\left(\Delta \cdot n^{3/2}\right)$. $\square$

3.3 Continuous-3D-Go-To-The-Center

Next, a continuous version of 3D-GTC which was already presented for the discrete time setting, will be considered as a concrete example of a contracting strategy. The two-dimensional version of this strategy was adapted for continuous time by Li et al. [11]. Compared to the discrete time version, no additional measures have to be taken to preserve connectivity, as it can be shown that this happens naturally in the continuous case. The strategy is summarized in Algorithm 2.

To show that CONT-3D-GTC is contracting, it must first be verified that connectivity of the visibility graph $UBG_t = (\mathcal{R}, E_t)$ is maintained at all times. The same reasoning that was used in the two-dimensional case by Li et al. [11] can also be applied here:

Lemma 8. *Let $\mathcal{R}$ be a set of robots in the three-dimensional Euclidean space that follows the CONT-3D-GTC strategy. If $\{r_i, r_j\}$ is an edge in UBG_t at time t, then $\{r_i, r_j\}$ is an edge in $UBG_{t'}$ at $t' \ge t$. Thus, CONT-3D-GTC maintains the connectivity of UBG_t.*

Algorithm 2. CONTINUOUS-3D-GO-TO-THE-CENTER (CONT-3D-GTC)

1: $\mathcal{R}_i(t) :=$ {positions of robots visible from r_i, including r_i at time t}
2: $\mathcal{S}_i(t) :=$ smallest enclosing sphere of $\mathcal{R}_i(t)$
3: $c_i(t) :=$ center of $\mathcal{S}_i(t)$
4: Move towards $c_i(t)$ with speed 1, or stay on $c_i(t)$ if r_i is already positioned on it.

Proof. Consider a robot r_i with neighborhood $\mathcal{R}_i(t)$ at time t. Let $Q_i(t)$ be the intersection of the unit balls of all robots in $\mathcal{R}_i(t)$. Since the SES of $\mathcal{R}_i(t)$ can have a radius of at most 1 and contains all robots in $\mathcal{R}_i(t)$, its center $c_i(t)$ must lie in $Q_i(t)$.

Consider some neighbor $r_j \in \mathcal{R}_i(t)$ of r_i and assume that there is some future point in time $t' > t$, such that $d(p_i(t'), p_j(t')) > 1$, i.e. r_i and r_j are no longer neighbors. Since the movement of robots is continuous, there must be some time $t^* \in [t, t']$, for which $d(p_i(t^*), p_j(t^*)) = 1$.

Now let L denote the intersection of the unit balls of r_i and r_j at time t^*. Any point in L is within distance at most 1 of both r_i and r_j. Furthermore L is a superset of both $Q_i(t^*)$ and $Q_j(t^*)$, meaning the target points $c_i(t^*)$ and $c_j(t^*)$ of both r_i and r_j also lie in L. Therefore, r_i and r_j can only move in the direction of points that are in distance at most 1 from both of them, meaning their distance can never exceed 1, creating a contradiction to the assumption that their distance is greater than 1 at time t'.   □

It remains to show that CONT-3D-GTC is a contracting strategy. This follows directly from Lemma 8 and Proposition 1, which states that the center of a SES is a convex combination of the points it encloses, meaning any target point computed by the strategy lies within the convex hull of the current configuration.

Theorem 4. CONT-3D-GTC *is a contracting, local strategy and thus gathers the robots in time* $\mathcal{O}\left(\Delta \cdot n^{3/2}\right)$.

3.4   Tangential-Normal Strategies

Previously, we showed a runtime bound for a relatively general class of (not necessarily local) gathering strategies and introduced a concrete example in CONT-3D-GTC. However, when designing a local strategy, additional care has to be taken to maintain the visibility graph UBG_t to successfully solve the GATHER-ING problem. It would be useful to also have a relatively simple design criterion that ensures this property. For this purpose, we will focus on robots' *local convex hulls* and introduce the notion of tangential-normal strategies. Let $\mathrm{CH}(\mathcal{R}_i(t))$ denote the local convex hull of robot r_i, i.e. the convex hull of r_i's neighborhood. Furthermore, let $\mathrm{Adj}_t(i)$ denote the set of robots that are adjacent to r_i on $\mathrm{CH}(\mathcal{R}_i(t))$ if r_i lies on $\mathrm{CH}(\mathcal{R}_i(t))$ itself. The main idea is to identify those velocity vectors that lead to a decrease in distance to all neighboring robots. These vectors are the normal vectors of *tangential planes*:

Definition 3. *Given a convex polyhedron $P \subset \mathbb{R}^3$ and a vertex $p \in P$. A tangential plane h_p w.r.t. P through p is a plane that only intersects P at the vertex p.*

Note that as long as P is actually convex, such a plane always exists and can – for example – be obtained by taking the plane through one of the faces adjacent to p and slightly rotating it. Based on this notion, we define the class of *tangential-normal* strategies in which the corner robots of local convex hulls move along the normal vectors of tangential planes:

Definition 4. *In the continuous time model, a gathering strategy for n robots is called* tangential-normal *if for every time t in which the robots have not yet gathered, each robot $r_i \in \mathcal{R}$ that is on a corner of its own local convex hull $CH(\mathcal{R}_i(t))$ moves with speed 1 along the normal vector of a tangential plane w.r.t. $CH(\mathcal{R}_i(t))$ through p_i while other robots do not move.*

The following lemma characterizes the normal vectors of tangential planes and will be used to show the desired properties of tangential-normal strategies.

Lemma 9. *Let p_i be a corner of a convex polyhedron P and let E_i be the set of edges of P adjacent to p_i. Then a plane h through p_i with normal vector $\boldsymbol{n}$ is a tangential plane w.r.t. P if and only if for each edge $e \in E_i$, $\angle(\boldsymbol{n}, e) < \frac{\pi}{2}$*

Proof. First, note that by the convexity of P and since h only intersects with it in p_i, the entire rest of P lies on one side of h. However, if there was an edge e with $\angle(\boldsymbol{n}, e) \geq \frac{\pi}{2}$, this would mean that e lies on the opposite side of or directly on h, both of which are contradictions to h being a tangential plane.

For the other direction of the statement, let $\boldsymbol{n}$ be a vector such that for each edge $e \in E_i$, $\angle(\boldsymbol{n}, e) < \frac{\pi}{2}$. This property now immediately yields that all edges $e \in E_i$ lie on the same side of the plane $h : \boldsymbol{n} \cdot (x - p_i) = 0$ defined by $\boldsymbol{n}$ and the point p_i, making h a tangential plane w.r.t. P through p_i. $\qquad\square$

Theorem 5. *Let $\mathcal{R}$ be a set of robots controlled by a tangential-normal strategy. Then, for each pair of robots $r_i, r_j \in \mathcal{R}$ and time t such that $\{r_i, r_j\}$ is an edge in UBG_t, $\{r_i, r_j\}$ is an edge in $\mathrm{UBG}_{t'}$ for all $t' \geq t$. Thus, tangential-normal strategies maintain the connectivity of UBG_t.*

Proof. Let $\mathcal{R}$ be a set of robots that follows a tangential-normal strategy. Consider a time t and a robot $r_i \in \mathcal{R}$ that lies on the corner of its own local convex hull $CH(\mathcal{R}_i(t))$ and let $v_i(t)$ be the normal vector of a tangential plane h_{p_i} w.r.t. $CH(\mathcal{R}_i(t))$. By Lemma 9, for each adjacent robot $r_j \in \mathrm{Adj}_t(i)$, it holds that $\beta_{i,j}(t) < \frac{\pi}{2}$. Since the cosine is positive on the interval $[0, \frac{\pi}{2}]$, Lemma 2 yields that r_i contributes a strict decrease in distance to all of its neighbors. A neighbor r_j now either does not move or also contributes a decrease in distance to r_i. This also means that for each pair of robots r_i and r_j that can see each other, the distance between r_i and r_j cannot increase, guaranteeing that the visibility graph remains connected. $\qquad\square$

Note however that while the tangential-normal property is a sufficient condition for ensuring connectivity, it is not a necessary condition. In particular, CONT-3D-GTC is not tangential-normal but is still able to maintain the connectivity of the visibility graph.

3.5   Move-on-Angle-Minimizer

Next, we introduce a strategy based on the tangential-normal criterion. It is based around the idea to find a movement vector that somehow causes a large decrease in distance to all neighbors. Since a smaller angle causes a greater decrease in distance (according to Lemma 2), one intuitive approach might be to find a movement vector that minimizes the maximal angle to all neighbors on the convex hull. To this end, the notion of an *angle minimizer* will be introduced. Let $V = \{v_1, v_2, ..., v_k\} \subset \mathbb{R}^3$ be a set of vectors that lie on one side of a plane through the origin. Then the vector $x^* = \mathrm{argmin}_{x \in \mathbb{R}^3} \max_{v_i \in V} \angle(x, v_i)$ is called an angle minimizer of V.

Now, we define a strategy in which each robot that is a corner of its local convex hull r_i moves along the angle minimizer of the edges between itself and the robots in $\mathrm{Adj}_t(i)$. This strategy will be called MOVE-ON-ANGLE-MINIMIZER and is summarized in Algorithm 3.

Algorithm 3. MOVE-ON-ANGLE-MINIMIZER

1: $\mathcal{R}_i(t) := \{$positions of robots visible from r_i, including r_i at time $t\}$
2: $CH(\mathcal{R}_i(t)) := $ Convex hull of r_i's neighborhood
3: **if** r_i is on a corner of $CH(\mathcal{R}_i(t))$ **then**
4:     $x^* = \mathrm{argmin}_{x \in \mathbb{R}^3} \max_{r_j \in \mathrm{Adj}_t(i)} \angle(x, p_j(t) - p_i(t))$
5:     r_i moves along x^* with speed 1
6: **else**
7:     r_i does not move

Note that if a robot r_i's local convex hull is two-dimensional, the angle minimizer is identical to the angle bisector of the inner angle at r_i. Therefore, this strategy can also be viewed as a generalization of MOVE-ON-BISECTOR for two-dimensional continuous gathering [8], for which Kempkes et al. [4] could show an optimal gathering time of $\Theta(n)$.

It will now be shown that the presented strategy is both a tangential-normal and a contracting strategy. By Lemma 9 and the existence of a tangential plane, we already know that there is a possible movement vector that has an angle of less than $\pi/2$ to all neighbors on the local convex hull. Therefore, the same must hold for x^*, immediately showing that MOVE-ON-ANGLE-MINIMIZER is a tangential-normal strategy.

Lemma 10. MOVE-ON-ANGLE-MINIMIZER *is a tangential-normal strategy.*

Computing x^*. In order to see that MOVE-ON-ANGLE-MINIMIZER is also a contracting strategy, we look at a method to compute the angle minimizer.

Let $\hat{v} = v/\|v\|$ denote the respective normalized vector of v and let $\hat{V} = \{\hat{v} \mid v \in V\}$ for a set V of vectors. Then the following holds:

Lemma 11. *Let $V \subset \mathbb{R}^3$ be a set of vectors that all lie on one side of a plane through the origin. The center c of the smallest enclosing sphere of $\hat{V}$ is an angle minimizer of V.*

Proof. Let x be a vector such that $\angle(x, v_i) \leq \pi/2$ for all $v_i \in V$. Such a vector exists, since there is a plane such that all v_i lie on one side of this plane.

Now consider the normalized vectors $\hat{v}_i$. They lie on the surface of the unit sphere centred on the origin. Let C_x be the minimal spherical cap centred on the vector x such that all the $\hat{v}_i$ lie on its surface. The vector $\hat{v}_j \in \hat{V}$ with the maximal angle to x lies on the edge of the base of C_x. The maximal angle can now be computed using the radius r of C_x as $\angle(x, \hat{v}_j) = \sin^{-1} r$.

Since $\sin^{-1}$ is monotonically increasing on the interval $[0, 1]$, finding the angle minimizer x^* now amounts to finding the center c of a spherical cap with minimal radius, which can be achieved by computing the smallest enclosing sphere of $\hat{V}$. □

By applying the fact that the SES of $\hat{V}$ is a convex combination of $\hat{V}$ (Proposition 1), this lemma together with Lemma 10 immediately yields that MOVE-ON-ANGLE-MINIMIZER is also a contracting strategy.

Theorem 6. MOVE-ON-ANGLE-MINIMIZER *is a tangential-normal and a contracting strategy. Thus, it gathers the robots in time* $\mathcal{O}\left(\Delta \cdot n^{3/2}\right)$.

References

1. Ando, H., Suzuki, Y., Yamashita, M.: Formation and agreement problems for synchronous mobile robots with limited visibility. In: Proceedings of the 1995 IEEE International Symposium on Intelligent Control, ISIC 1995, pp. 453–460. IEEE, August 1995. https://doi.org/10.1109/ISIC.1995.525098
2. Bhagat, S., Chaudhuri, S.G., Mukhopadhyaya, K.: Gathering of opaque robots in 3D space. In: Proceedings of the 19th International Conference on Distributed Computing and Networking, ICDCN 2018, Varanasi, India, 4–7 January 2018, pp. 2:1–2:10 (2018). https://doi.org/10.1145/3154273.3154322
3. Braun, M., Castenow, J., Meyer auf der Heide, F.: Local gathering of mobile robots in three dimensions (2020). https://arxiv.org/abs/2005.07495
4. Degener, B., Kempkes, B., Kling, P., Meyer auf der Heide, F.: Linear and competitive strategies for continuous robot formation problems. TOPC **2**(1), 2:1–2:18 (2015). https://doi.org/10.1145/2742341
5. Degener, B., Kempkes, B., Langner, T., Meyer auf der Heide, F., Pietrzyk, P., Wattenhofer, R.: A tight runtime bound for synchronous gathering of autonomous robots with limited visibility. In: Rajaraman, R., Meyer auf der Heide, F. (eds.) SPAA 2011: Proceedings of the 23rd Annual ACM Symposium on Parallelism in Algorithms and Architectures, San Jose, CA, USA, 4–6 June 2011 (Co-located with FCRC 2011), pp. 139–148. ACM (2011). https://doi.org/10.1145/1989493.1989515

6. Elzinga, D.J., Hearn, D.W.: The minimum covering sphere problem. Manage. Sci. **19**(1), 96–104 (1972). https://doi.org/10.1287/mnsc.19.1.96
7. Flocchini, P., Prencipe, G., Santoro, N. (eds.): Distributed Computing by Mobile Entities, Current Research in Moving and Computing. Lecture Notes in Computer Science, vol. 11340. Springer, Heidelberg (2019). https://doi.org/10.1007/978-3-030-11072-7
8. Gordon, N., Wagner, I.A., Bruckstein, A.M.: Gathering multiple robotic a(ge)nts with limited sensing capabilities. In: Dorigo, M., Birattari, M., Blum, C., Gambardella, L.M., Mondada, F., Stützle, T. (eds.) ANTS 2004. LNCS, vol. 3172, pp. 142–153. Springer, Heidelberg (2004). https://doi.org/10.1007/978-3-540-28646-2_13
9. Kling, P., Meyer auf der Heide, F.: Continuous protocols for swarm robotics. In: Flocchini, P., Prencipe, G., Santoro, N. (eds.) Distributed Computing by Mobile Entities, Current Research in Moving and Computing. LNCS, vol. 11340, pp. 317–334. Springer, Cham (2019). https://doi.org/10.1007/978-3-030-11072-7_13
10. Li, S., Markarian, C., Meyer auf der Heide, F., Podlipyan, P.: A continuous strategy for collisionless gathering. In: Fernández Anta, A., Jurdzinski, T., Mosteiro, M.A., Zhang, Y. (eds.) ALGOSENSORS 2017. LNCS, vol. 10718, pp. 182–197. Springer, Cham (2017). https://doi.org/10.1007/978-3-319-72751-6_14
11. Li, S., Meyer auf der Heide, F., Podlipyan, P.: The impact of the gabriel subgraph of the visibility graph on the gathering of mobile autonomous robots. In: Chrobak, M., Fernández Anta, A., Gąsieniec, L., Klasing, R. (eds.) ALGOSENSORS 2016. LNCS, vol. 10050, pp. 62–79. Springer, Cham (2017). https://doi.org/10.1007/978-3-319-53058-1_5
12. Poudel, P., Sharma, G.: Universally optimal gathering under limited visibility. In: Spirakis, P., Tsigas, P. (eds.) SSS 2017. LNCS, vol. 10616, pp. 323–340. Springer, Cham (2017). https://doi.org/10.1007/978-3-319-69084-1_23
13. Tomita, Y., Yamauchi, Y., Kijima, S., Yamashita, M.: Plane formation by synchronous mobile robots without chirality. In: 21st International Conference on Principles of Distributed Systems, OPODIS 2017, Lisbon, Portugal, 18–20 December 2017, pp. 13:1–13:17 (2017). https://doi.org/10.4230/LIPIcs.OPODIS.2017.13
14. Yamauchi, Y., Uehara, T., Yamashita, M.: Brief announcement: Pattern formation problem for synchronous mobile robots in the three dimensional euclidean space. In: Proceedings of the 2016 ACM Symposium on Principles of Distributed Computing, PODC 2016, Chicago, IL, USA, 25–28 July 2016, pp. 447–449 (2016). https://doi.org/10.1145/2933057.2933063

Improved Lower Bounds for Shoreline Search

Stefan Dobrev[1], Rastislav Královič[2(✉)], and Dana Pardubská[2]

[1] Slovak Academy of Sciences, Bratislava, Slovakia
Stefan.Dobrev@savba.sk
[2] Comenius University in Bratislava, Bratislava, Slovakia
{kralovic,pardubska}@dcs.fmph.uniba.sk

Abstract. Shoreline search is a natural and well-studied generalisation of the classical cow-path problem: k initially co-located unit speed agents are searching for a line (called *shoreline*) in 2 dimensional Euclidean space. The shoreline is at (a possibly unknown) distance δ from the starting point O of the agents. The goal is to minimize the *competitive ratio* $\frac{T_\delta}{\delta}$, where T_δ is the worst case (over all possible locations of the shoreline at distance δ) time until the shoreline is found.

Upper bounds conjectured to be optimal have been established for all $k \geq 1$[4], however lower bounds have been severely lacking. Recent paper [1] showed an improved lower bound for $k = 2$ and gave the first non-trivial lower bounds for $k \geq 3$. While for $k \geq 4$ the lower bounds match the best known upper bounds, that is not the case for $k < 4$.

In this paper we improve the lower bound for $k = 2$ from 3 to $(1 + \sqrt{3} + \pi/6) \approx 3.2556$, and for $k = 3$ from $\sqrt{3}$ to 2. These lower bounds apply for known δ, matching the corresponding upper bounds. In fact, for $k = 3$ our lower bound matches the upper bound for unknown δ as well.

We achieve these results by employing a novel simple virtual colouring technique, allowing us to transform the problem of covering the (uncountably many) points of the circle of radius δ (whose tangents represent all possible shorelines at distance δ) to a combinatorially much simpler problem of finding the shortest path from the centre to three specific tangents of this circle.

1 Introduction

Searching is one of the fundamental problems of computer science, with numerous real life applications. A wide variety of models have been investigated, considering all kinds of search space, searchers, their power and multiplicity, as well as different cost functions. One of the very natural variants (motivated e.g. by motion planning in operations research) is a search in 2 dimensional space by one or more search agents: There is an object (finite shape, circle, line, ..., we will call it the *target*) to be found at an unknown location in the 2D space. The

Research supported by VEGA 1/0601/20.

search agents are initially collocated at the origin O. We are looking for a set of fixed trajectories, one for each agent, each starting at O, such that the length of the trajectory hitting the target is minimized in the worst case (w.r.t. target's location). A natural way to evaluate the quality of the search strategy is *competitive analysis*, i.e. dividing the search time by the best possible search time if the location of the target was known, i.e. by the distance δ of the target from the origin.

In this paper, we limit ourselves to one of the simpler targets, namely to lines. A simplified version of the problem (searching for a line of unknown slope but at a known distance away) has it roots in 50's: It was first proposed by Bellman [7] and solved by Isbell [14]. The general case of unknown δ was introduced by Baeza-Yates [4] in late 80's. Upper bounds conjectured to be optimal for all $k \geq 1$ soon followed. However, similarly as for its 1D variant (the notorious *cow-path* problem), establishing lower bounds has proved to be much more difficult. In fact, until recently, there have been no lower bounds for $k > 2$ and only very weak lower bounds for $k = 1$ and $k = 2$.

This paper has been inspired by [1], where the authors present strong lower bounds for $k \geq 2$, in fact matching the upper bounds for $k \geq 4$. Of particular interest was the hint that there might exist possibility to improve upon the (old, and previously conjectured optimal) upper bound for $k = 3$.

1.1 Our Contribution

The case of $k = 3$: Improving the lower bound from $\sqrt{3}$ [1] to 2, matching the upper bound from [4] and disproving the hypothesis that this upper bound can be improved. In addition to the bound being stronger, our proof is also quite a bit simpler than the proof in [1].
The case of $k = 2$: We start by improving the lower bound from 3 to $\sqrt{10}$ using a very simple proof. We subsequently improve the lower bound to $(1+\sqrt{3}+\pi/6) \approx 3.2556$, matching the upper bound from [4] for the case of known δ.

In fact, all our lower bounds apply for known δ. As the tight lower bounds have been known for $k = 1$ and $k \geq 4$, this closes the case of known δ.

We achieve these results by employing a novel simple virtual colouring technique, allowing us to transform the problem of covering the (uncountably many) points of the circle (whose tangents represent all possible shorelines at distance δ) to a combinatorially much simpler problem of finding the shortest path from the centre to three specific tangents of this circle.

1.2 Related Work

Search theory has a long and rich history; just skimming the vast search literature is way beyond of scope of this paper. As such, we will focus on result most relevant to the topic of this paper. Even the much narrower field of continuous searching in 2D space has its roots in 50's, motivated by motion planning and operations research. Even the simplest continuous search problem – finding a

point on an unlimited line by a unit-speed agent, when the location and direc-
tionet of the sought point is unknown (the so-called *cow-path* problem) and the
search time should be optimized w.r.t. to the point's distance from the agent's
origin – has inspired spirited research over the decades. The fruits of this work
have been summarized in numerous surveys, e.q. [8,10,11]; see also [9] for a
newest one with focus on group search and evacuation. In fact, several books
have been devoted to the underlying mathematical theory, with [2,3] being the
most relevant and influential.

The problem of searching for a line of arbitrary slope in 2D plane at a known
(i.e. unit) distance away was first posed by Bellman [7]. The problem was solved
by Isbell [14], showing that the optimal worst case distance walked by a single
robot is $U_1 = 1 + 7\pi/6 + \sqrt{3} \approx 6.3972$. Gluss [12] investigated randomized
algorithms minimizing the statistical expectation of the distance travelled, as
well as the problem of finding a circle of known radius and known distance
[13]. The latter can be seen as a generalization of the shoreline problem, which
corresponds to the case when the radius of the circle goes to infinity. The problem
of finding a circle of known radius r is motivated by a problem of finding a
point/object visible from a distance r.

The generalized version for unknown distance to the line was first considered
by Baeza-Yates [5]. The authors assume one searcher, but deal with numerous
variants of the model, starting with the 1D case (cow-path problem), progress via
multi-ray cow-path problem and then investigate various 2D models (e.g. limiting
the searcher to move only in axis-parallel directions, considering the shoreline to
be axis-parallel, as well as the general shoreline problem we consider here). For
the general shoreline problem, an upper bound based on a logarithmic spiral is
presented, with a competitive ratio of 13.81.

The results for shoreline search were generalized to multiple searchers in [4].
The authors were interested not only in optimal time, but also in optimal total
distance traveled. They made an interesting observation that for two agents, the
total distance traveled is optimized by an asymmetric algorithm. Regarding the
optimal time, they showed an upper bound (let us call it U_2) of $1 + \sqrt{3} + \pi/6 \approx$
3.2556 for two searchers searching for a line at a known distance δ away, as
well as an upper bound of ≈ 5.2644 (based on two equally-spaced logarithmic
spirals) for unknown δ. For $k \geq 3$ searchers, they proposed a simpler algorithm:
each agent moves away from the origin along the equally separated rays. The
competitive ratio of this algorithm is $1/\cos\frac{\pi}{k}$.

These upper bounds have stood the test of time and have been conjectured
optimal. However, for a long time, only few (and rather weak) lower bounds have
been proven. For $k = 1$, Isbell [14] proved the optimality of U_1 for known δ, while
for $k = 2$ a rather weak lower bound of 1.5593 was shown by Baeza-Yates [5].
Decades later, Langetepe [17] proved a lower bound of 12.5385 for the case of
axis-parallel shoreline, however his proof is limited to spiral-based trajectories.
The paper also provides a matching upper bound, improving upon the previous
upper bound of 12.5406 by Jez [15]. In [16], the optimality of spiral search is
shown for a related problem of searching for a point in plane, in which a point is

found if it lies on the line segment connecting the agent's current position with the O.

The lack of lower bounds has been recently ameliorated by Georgeu et al. [1], where they showed a lower bound of $\sqrt{3}$ for $k = 3$, a lower bound of 3 for $k = 2$, as well as lower bound of $1\cos\frac{\pi}{k}$ for $k \geq 4$. This closed the gap between the upper and lower bounds for $k \geq 4$, as well as significantly reduced them for $k = 2$ and $k = 3$.

2   Lower Bounds

Consider a ring R of radius 1 centered at O. Assume each agent i has a unique colour c_i. Let T be the time when the shoreline has been found. Let π_i be the path/trail of the agent i from time 0 to time T inclusively, and let $p_i(t)$ denote the position of agent i at time $t \in [0, T]$.

Let us assign the colours to points of R as follows:

- whenever $p_i(t)$ is on R, colour it c_i
- whenever $p_i(t)$ is outside R, draw the tangents from $p_i(t)$ to R. Let x and y be the two points of R where these tangents touch R. Assign colour c_i to x and y (Fig. 1).

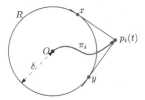

Fig. 1. Colouring R. (Color figure online)

Note that some points of R might not have any colour, while others might have several. Furthermore, as the agent's trajectory is continuous and the time domain is a closed set, the points of colour c_i also form a closed set.

Lemma 1. *If there is an uncoloured point r of R then there exists a line at distance 1 which has not been located by the agents.*

Proof. The tangent of R at r is the shoreline that the agents did not find.   □

Let $S \subseteq R$. We say that a trajectory (or a set of trajectories) *covers* S iff every point of S is assigned a colour using the above technique.

Hence, in order to show the lower bound on the time needed to find the shoreline, it is sufficient to show a lower bound on time needed to cover the whole circle R by the multiple agents. The key to our lower bounds is a technique to further reduce this problem to the much simpler problem of covering only constant number of specific points by a single agent.

2.1   Three Agents

Lemma 2. *If* 3 *agents locate every line at distance* 1 *from* O, *then there exist two points* x *and* y *on* R *holding an angle of* $2\pi/3$ *with* O *and having the same colour.*

Proof. First, because the sets of the same colour are closed and (by Lemma 1) no point is uncoloured, either all points of R are of the same colour (and the lemma holds), or there exists a point x having two colours (w.l.o.g. c_1 and c_2). Consider now the points y and z at an angle $2\pi/3$ from x. Either one of them has colour c_1 or c_2, in which case it forms an equal-coloured pair with x, or both of them have colour c_3, forming equal-coloured pair themselves.   □

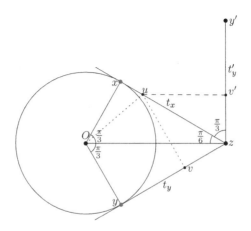

Fig. 2. $|Ouv| = |Ouv'| > |Oz|$ (Color figure online)

Now we are ready to prove the lower bound for $k = 3$.

Theorem 1. *Any three agent algorithm for locating the shoreline at known distance* 1 *from the origin needs at least* 2 *time units.*

Proof. By Lemma 2 there are two points x and y on R forming an angle $2\pi/3$ and having the same colour c_i. Let t_x and t_y be the tangents touching R at x and y, respectively. Then the agent i must have, during its travel, touched both t_x and t_y, otherwise x and y would not have received their colour.

What is the shortest path agent i could have taken to touch both t_x and t_y? W.l.o.g. assume that i touched t_x not later than it touched t_y. Let t'_y be the reflection of t_y over t_x. Then the length of this shortest path is the same as the length of the shortest path from O to t'_y while not crossing t'_y before crossing t_x (refer to Fig. 2). Since $\angle xOy = 2\pi/3$, $\angle y'zO = \pi/2$ and therefore the shortest such path is a straight line from O to z. As

$$|Oz| = \frac{1}{\cos \angle xOz} = \frac{1}{\cos \pi/3} = 2$$

the theorem follows.   □

2.2 Two Agents

Lemma 3. *If 2 agents locate every line at distance 1 from O, then there are three equal-color points x, y, z on R with $\angle xOy = \angle yOz = \pi/2$.*

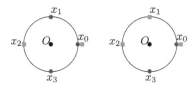

Fig. 3. Two colours for four points. (Color figure online)

Proof. For the same reason as in Lemma 2, either all points of R are of the same colour (and the lemma holds), or there must exist a point x_0 which has both colours. Consider the other three points, equally separated by $\pi/2$. Let's name them x_1, x_2 and x_3 in counterclockwise order (see Fig. 3). If x_1 and x_3 are of the same colour, then x_3, x_0, x_1 are the sought set. If they are different, one of them (say x_i) must have the same colour as x_2. Then x_2, x_i and x_0 (not necessarily in that order) form the sought set.   □

Theorem 2. *Any two agent algorithm A for locating the shoreline at known distance 1 takes time at least $\sqrt{10} \approx 3.1623$.*

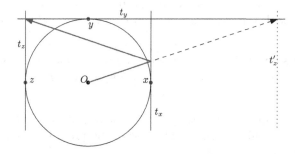

Fig. 4. Optimal trajectory to touch t_x, t_y and t_z. t_z' is a reflection of t_z by t_x. (Color figure online)

Proof. Using the same arguments as before, there must be an agent that starts at O and touches t_x, t_y and t_z. t_y cannot be the first tangent touched by π_b, as the shortest such path would be of length more than $2 + \sqrt{2} > U_2$, violating the optimality of π_b. Due to symmetry, w.l.o.g. we may assume that π_b touches t_x before touching t_z. It is easy to verify that the shortest such trajectory is depicted in Fig. 4, and that its length is $\sqrt{1^2 + 3^2} = \sqrt{10}$.   □

The lower bound of Theorem 2 does not match the best known upper bound $U_2 = (1 + \sqrt{3} + \pi/6) \approx 3.2556$ [4]. This is so because Theorem 2 aims to be as simple as possible, at the cost of the strength of the bound.

In the rest of this subsection we present a tighter analysis that matches the upper bound (Fig. 5).

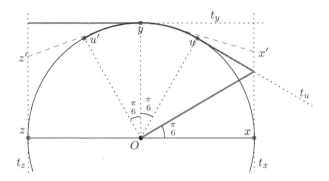

Fig. 5. The upper bound for $k = 2$ shown as solid path. The dashed path is the shortest trajectory from x' to z' that covers the arc uu'. (Color figure online)

In order to simplify presentation, we assume the agents have colours blue and red (their paths are π_b and π_r), and x, y and z are blue.

Let u and u' be the two points on R at an angle $\frac{\pi}{6}$ from y. Note that u is the point where the path of the upper bound by Baeza-Yates starts to follow R.

We first deal with the case that all points on the arc between u and u' are blue:

Lemma 4. *The shortest trajectory π_b covering the arc uu' while touching t_x, t_y and t_z is the blue line corresponding to the upper bound by Baeza-Yates [6] and is of length $U_2 = (1 + \sqrt{3} + \pi/6) \approx 3.2556$.*

Proof. Using the same arguments as in Theorem 2 we may assume that π_b touches t_x before touching t_y or t_z. The rest of the proof is based on the ideas from [6] and [4]:

First, note that π_b is convex, as any shortcut via the non-convex part would shorten it, violating the assumption of its optimality [6]. Let x' and z' be the lowest points where π_b touches t_x and t_z, respectively. Applying Lemma 3.1 from [6] (to a rotated and shifted setting where x' and z' lie on the horizontal axis with x' to the right of z', so that preconditions of this Lemma are satisfied) directly yields that the shortest path from x' to z' while covering the arc uu' is the upper convex envelope π' of x', z' and the arc uu'. Let $d(\pi')$ be the length of this path. In order to find the shortest π_b, it suffices to find x' and z' such that $|Ox'| + d(\pi')$ is minimized.

As π_b touches t_x first, x' lies below y, therefore $y \in \pi_b$. Hence, the positions of z' and x' can be optimized independently. z' is obviously optimized by going perpendicularly from y to t_z, while the optimal position of x' has been determined in [4] to be given by $\angle xOx' = \pi/6$, yielding $d(\pi_b) = U_2$.   □

The remaining case is that of arc uu' containing a point that is not blue.

Definition 1. *Let a, b and c be three distinct points on R. We say that the triangle abc is* BalancedRight *(or, in short, BRT) iff*

- $\angle aOb = \frac{\pi}{2}$
- $\min(\angle bOc, \angle aOc) \geq \frac{2\pi}{3}$

We will need the following technical Lemma:

Lemma 5. *If $\triangle abc$ is a BRT then the shortest path starting at O and touching t_a, t_b and t_c (not necessarily in that order) is longer than U_2.*

Proof. Let π_{xyz} denote the shortest path starting at O and touching t_x, t_y and t_z, in that order. We have to show that $|\pi_{xyz}| > U_2$ for every ordering of t_a, t_b and t_c.

Let t_{xy} denote a reflection of t_y over t_x and t_{xyz} denote a reflection of t_{xz} over t_{xy} (refer to Fig. 6). Then π_{xyz} corresponds to the shortest path from O to t_{xyz}, while crossing t_x before crossing t_y or t_z, then crossing t_{xy} before crossing t_{xz} and finally touching t_{xyz}. In some cases (π_{abc} and π_{bca}), π_{xyz} is the shortest line from O to t_{xyz}, while in other cases (π_{cab} and π_{cba}, π_{acb}) it has to go to the intersection of t_{xy} and t_{xz} as the shortest line from O to t_{xyz} would cross t_{xz} before t_{xy}. Finally, in the case of π_{bac} it has to make a detour to avoid crossing t_y before t_x.

Note that thanks to a rather small range of positions c can be in the BRT, the shortest paths in Fig. 6 are representative of what happens in any BRT (observe that as the BRT becomes more symmetric, π_{abc} and π_{bac} are getting closer, while π_{acb} starts to resemble π_{bca}).

Let $h(c)$ denote the distance from c' to $a'b'$ (the *height* of the $\triangle a'b'c'$). We will show that for each of these π_{xyz} it holds that $|\pi_{xyz}| > 1 + h(c)$: As at least 1 is needed to exit the triangle abc, it is sufficient to show that in each case the length of the path outside $\triangle a'b'c'$ (lets call this *external length*) is at least $h(c)$.

- for π_{bac}, the external length is exactly $h(c)$
- for π_{cab} and π_{cba} this is straightforward from the definition of $h(c)$
- for π_{abc} and π_{acb} the external length is longer than the external length of $\pi_{bac} = h(c)$
- for π_{bca} the external length is longer than $|a'c''| > h(c)$.

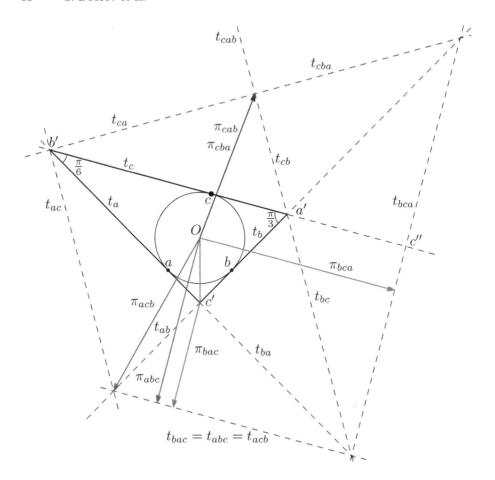

Fig. 6. All the shortest paths starting at O and touching t_a, t_b and t_c (Color figure online)

Hence, it is sufficient to show that $h(c) > U_2 - 1$. $h(c)$ is minimized when c is in the extreme position for BRT, i.e. when $\angle bOc = \frac{2\pi}{3}$. In such case,

$$\frac{h(c)}{|a'c'|} = \sin\frac{\pi}{3} = \frac{\sqrt{3}}{2}$$

hence

$$h(c) = \frac{\sqrt{3}|a'c'|}{2}$$

$$|a'c'| = |a'b| + |bc'| = 1 + \cot\frac{\pi}{6} = 1 + \sqrt{3}$$

Combining this yields $h(c) = \frac{\sqrt{3}(1+\sqrt{3})}{2} \approx 2.336$. As $U_2 \approx 3.2556$, the Lemma holds. $\qquad\square$

Note that the sharp inequality in Lemma 5 does not contradict the upper bound of U_2: The upper bound is achieved when the arc uu' is fully coloured blue, while Lemma 5 deals with the case that there is a red point in this arc.

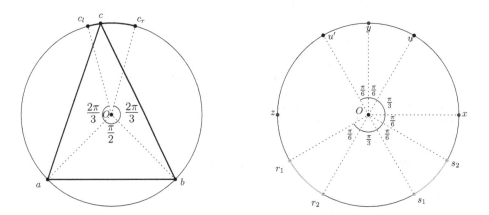

Fig. 7. Left: BRT, right: existence of a monochromatic BRT. (Color figure online)

Lemma 6. *Let x,y and z be three blue points given by Lemma 3 and let u and u' be two points at angle $\frac{\pi}{6}$ from y. If there exists a red point in the arch uu' then there exists a monochromatic BRT.*

Proof. Let points r_1, r_2, s_1 and s_2 be located according to Fig. 7, right. If any of the points of r_1, r_2, s_1, s_2 are blue, then they form a blue BRT either with xy or with yz. Hence, r_1, r_2, s_1 and s_2 are red. Therefore, $r_1 s_1$ form a red BRT with any red point on the arch uy while $r_2 s_2$ form a red BRT with any red point on the arch yu'. As, by assumption, there is a red point in the arch uu', the lemma follows. □

Combining Lemma 4 with Lemmas 5 and 6 directly yields.

Theorem 3. *Any two agent algorithm A for locating the shoreline at known distance 1 takes time at least U_2, i.e. the symmetric algorithm from Fig. 3 of [4] is time optimal.*

3 Conclusions

In this paper, we have closed the last gaps for the case of known δ, as well as for $k = 3$ and unknown δ. However, the gaps for $k = 1$ and $k = 2$ and unknown δ remain open.

We are hopeful that the techniques and insight from this paper will prove useful in closing or at least narrowing those gaps as well.

More generally, many variants of the shoreline problem are wide open: One can consider different costs (average case, the sum of traveled distances, need for every agent to reach the shoreline – the *evacuation* problem), as well as different models of agents (e.g. of different speed, possibly faulty, perhaps memory-limited) or the search space (discrete grids, graphs).

References

1. Acharjee, S., Georgiou, K., Kundu, S., Srinivasan, A.: Lower bounds for shoreline searching with 2 or more robots. In: OPODIS (2019)
2. Alpern, S., Fokkink, R., Gasieniec, L., Lindelauf, R., Subrahmanian, V.: Search Theory: A Game Theoretic Perspective. Springer, New York (2013). https://doi.org/10.1007/978-1-4614-6825-7
3. Alpern, S., Gal, S.: The Theory of Search Games and Rendezvous, vol. 55. Springer, Boston (2006). https://doi.org/10.1007/b100809
4. Baeza-Yates, R., Schott, R.: Parallel searching in the plane. Comput. Geom. **5**(3), 143–154 (1995)
5. Baeza-Yates, R.A., Culberson, J.C., Rawlins, G.J.E.: Searching with uncertainty extended abstract. In: Karlsson, R., Lingas, A. (eds.) SWAT 1988. LNCS, vol. 318, pp. 176–189. Springer, Heidelberg (1988). https://doi.org/10.1007/3-540-19487-8_20
6. Baeza-Yates, R.A., Culberson, J.C., Rawlins, G.J.E.: Searching with uncertainty (technical report 239). Indiana University, Computer Science Department (1988)
7. Bellman, R.: A minimization problem. Bull. AMS **62**(3), 270 (1956)
8. Benkoski, S.J., Monticino, M.G., Weisinger, J.R.: A survey of the search theory literature. Naval Res. Logist. (NRL) **38**(4), 469–494 (1991)
9. Czyzowicz, J., Georgiou, K., Kranakis, E.: Group search and evacuation. In: Flocchini, P., Prencipe, G., Santoro, N. (eds.) Distributed Computing by Mobile Entities. LNCS, vol. 11340, pp. 335–370. Springer, Cham (2019)
10. Dobbie, J.M.: A survey of search theory. Oper. Res. **16**(3), 525–537 (1968)
11. Gal, S.: Search games. In: Wiley Encyclopedia of Operations Research and Management Science (2010)
12. Gluss, B.: An alternative solution to the "lost at sea" problem. Naval Res. Logist. Q. **8**(1), 117–122 (1961)
13. Gluss, B.: The minimax path in a search for a circle in a plane. Naval Res. Logist. Q. **8**(4), 357–360 (1961)
14. Isbell, J.R.: An optimal search pattern. Naval Res. Logist. Q. **4**(4), 357–359 (1957)
15. Jeż, A., Łopuszański, J.: On the two-dimensional cow search problem. Inf. Proc. Lett. **109**(11), 543–547 (2009)
16. Langetepe, E.: On the optimality of spiral search, pp. 1–12, January 2010
17. Langetepe, E.: Searching for an axis-parallel shoreline. Theor. Comput. Sci. **447**, 85–99 (2012)

Guarding a Polygon Without Losing Touch

Barath Ashok[1], John Augustine[1(✉)], Aditya Mehekare[2], Sridhar Ragupathi[2], Srikkanth Ramachandran[2], and Suman Sourav[3]

[1] Indian Institute of Technology Madras, Chennai, India
augustine@iitm.ac.in
[2] National Institute of Technology, Tiruchirappalli, India
[3] Advanced Digital Sciences Center, Singapore, Singapore

Abstract. We study the classical *Art Gallery Problem* first proposed by Klee in 1973 from a mobile multi-agents perspective. Specifically, we require an optimally small number of agents (also called guards) to navigate and position themselves in the interior of an unknown simple polygon with n vertices such that the collective view of all the agents covers the polygon.

We consider the *visibly connected setting* wherein agents must remain connected through line of sight links – a requirement particularly relevant to multi-agent systems. We first provide a centralized algorithm for the visibly connected setting that runs in time $O(n)$, which is of course optimal. We then provide algorithms for two different distributed settings. In the first setting, agents can only perceive relative proximity (i.e., can tell which of a pair of objects is closer) whereas they can perceive exact distances in the second setting. Our distributed algorithms work despite agents having no prior knowledge of the polygon. Furthermore, we provide lower bounds to show that our distributed algorithms are near optimal.

Our visibly connected guarding ensures that (i) the guards form a connected network and (ii) the polygon is fully guarded. Consequently, this guarding provides the distributed infrastructure to execute any geometric algorithm for this polygon.

Keywords: Art gallery problem · Mobile agents · Swarm robotics · Visibility · Line of sight communication

1 Introduction

The *Art Gallery Problem* is a classical computational geometry problem that seeks to minimize the number of guards (or agents in our context) required to guard an art gallery (represented by a simple polygon P comprising vertices $\{p_1, p_2, \ldots, p_n\}$). To successfully guard the gallery, every point inside the polygon

---

The authors are listed in alphabetical order.

© Springer Nature Switzerland AG 2020
A. W. Richa and C. Scheideler (Eds.): SIROCCO 2020, LNCS 12156, pp. 91–108, 2020.
https://doi.org/10.1007/978-3-030-54921-3_6

must be visible to at least one guard, i.e., for every point in the polygon, there must exist at least one guard such that the segment joining the point to the guard does not intersect the exterior of the polygon. This computational geometry problem was first posed by Klee in 1973, and thereafter has been widely studied over the years (see [16, 26, 30, 31]).

In this paper, we investigate a variation of the problem called the *visibly connected art gallery problem* from a distributed multi-agents perspective. We require an optimally small number of agents (also called guards) with omni-directional vision to navigate an unknown simple polygon with n vertices in a coordinated manner and position themselves in its interior such that the collective view of all the agents covers the polygon. Additionally, for the connected art-gallery problem (as in [25]), it is required that the agents maintain line-of-sight connectivity. More precisely, the visibility graph [22, 25] comprising agents as nodes and edges between agents that are within line of sight of each other (unobstructed by polygon edges) must be a connected graph.

The visibly connected art gallery problem was studied as early as 1993 by Liaw *et al.* [22] in the centralized setting, but only for the special case of spiral polygons. Hernandez-Penalver [19] considered simple polygons and showed that $\lfloor n/2 \rfloor - 1$ guards are sufficient and sometimes necessary. Pinciu [28] presented a centralized algorithm based on iteratively processing the dual graph of the polygon's triangulation. Although [28] lacks the analysis, one can infer that it runs in time linear in n, the number of vertices of the polygon. However, the algorithm is somewhat complicated and not amenable for parallel or distributed computing. Obermeyer et al. [25] provided a distributed algorithm that is capable of handling polygons with holes, but unfortunately requires $O(n^2)$ rounds.

Our work is motivated by recent advancements in unmanned aerial vehicles (UAVs), especially those capable of automated sensing (either through photogrammetry or LiDAR) and communication (typically through line-of-sight electromagnetic radio waves). Such UAVs are typically deployed into unknown territories from which they are required to navigate, learn, and perform useful tasks in a coordinated manner. We model these UAVs as point agents that start from a common starting point assumed without loss of generality (w.l.o.g.) to be p_1. Agents operate in synchronous rounds during which they can look, compute, communicate and move. They are required to coordinate with each other and achieve full visibility coverage of the polygon while maintaining line-of-sight connectivity with each other.

A connected visibility graph ensures that there exists a path between every pair of guards. So visibly connected guards can simultaneously maintain coverage of the polygon and execute distributed computing protocols through line of sight communication.

Our Contributions. We begin with a description of a centralized algorithm in Sect. 3 that takes a simple polygon P with n vertices as input and produces a placement of at most $\lfloor n/2 \rfloor - 1$ guards that satisfy the requirements of the visibly connected art gallery problem. This algorithm only requires $O(n)$ time. Here, we introduce a notion of triplets (three connected nodes) in the weak dual graph $\mathcal{D}$ (defined formally in Sect. 2). Informally, $\mathcal{D}$ is the graph whose nodes are triangles of a triangulation of P and arcs connect pairs of triangles that share an edge. We show that $\mathcal{D}$ can be decomposed into $O(n)$ triplets that are connected and cover $\mathcal{D}$, and then we compute a set of visibly connected guards by placing guards – one per triplet – positioned strategically within the triangles pertaining to each triplet. Although Pinciu [28] has already presented an $O(n)$ algorithm, we believe that our algorithm is simpler and, more importantly, amenable to parallel computing. In particular, we show that our algorithm can be adapted to run in the PRAM model in time linear in the diameter of $\mathcal{D}$.

Next, we turn our attention to distributed computing models in which the agents are independent mobile computing entities that must interact with each other to solve the visibly connected art gallery problem. We define two model variants based on agents' perception capabilities. The *depth perception* variant wherein the agents can accurately perceive depth (i.e., distances) is inspired by UAVs with LiDAR technology [23]. On the other hand, the *proximity percep-tion* variant only provides the agents with relative proximity. For concreteness, we limit the proximity perception variant to being able to sense which of any two objects (i.e., edges or vertices of the polygon) is closer. It is inspired by photogrammetry [1], which is cheaper and only guarantees coarser perception.

We present algorithms for both cases. The algorithm for the proximity perception variant, presented in Sect. 4, operates by exploring visible territories within the polygon (formally defined in Sect. 2). We describe two forms of exploration, one in a breadth-first manner and the other in a depth-first manner. Since the polygon structure is completely unknown to the agents, for each level of the breadth-first exploration, the nodes must communicate to the root in order to ensure that a sufficient number of agents are provisioned to explore that level. Our algorithm – taking the best out of both explorations –runs in $O(\min(\tilde{d}^2, n))$ rounds. The term $\tilde{d}$ is a natural notion of diameter of P called *minimal v-diameter* that we define formally in Sect. 4. Informally, it is the largest diameter among all visibility graphs pertaining to minimal placement of connected guards that cover all vertices in P. The candidate placements are minimal in the sense that the removal of guards either leads to lack of coverage or loss of connectivity.

When the agents can perceive depth, we exploit this capability to place agents based on the medial axis of the polygon P (defined formally in Sect. 2), which is a well-known tree-like structure that captures the "shape" of the polygon. Since depth perception is more powerful than proximity perception, we can take the best of all options to ensure a running time of $O(\min(\tilde{d}^2, D^2, n))$ time, where D is the (unweighted) diameter of the medial axis tree. Each of our algorithms require at most $O(n)$ agents for an initial placement, but a subsequent post-processing ensures that at most $\lfloor n/2 \rfloor - 1$ agents are placed in the polygon, which is optimal.

To complement our complexity claims, we consider the weaker problem wherein the robots are not required to be placed in a visibly connected guarding position, but rather just that the entire polygon must be explored by the agents.

The exploration problem only requires that for every point in P, some agent must have been within line of sight of that point at some time instant during the course of the algorithm. Clearly, any solution to the visibly connected guard placement problem will also be a solution for the exploration problem. Thus, we focus on showing a lower bound for the exploration problem. Specifically, we show that, for any deterministic algorithm operating on a polygon (with $\tilde{d}, D \in o(\log n)$) to even centrally coordinate $\Theta(n)$ agents to explore an initially unknown polygon, we can construct a polygon that requires $\Omega(D^2)$ (or $\Omega(\tilde{d}^2)$) communication rounds even with depth perception.

Unfortunately, due to space limitation, we have deferred some of our proofs to the full version [2].

Related Work. The classical art gallery problem was first introduced by Klee in 1973. Chvátal [7] showed by an induction argument that $\lfloor \frac{n}{3} \rfloor$ guards are always sufficient and occasionally necessary for any simple polygon with n vertices. Fisk [12] proved the same result via an elegant coloring argument. Lee and Lin in [20] proved that determining a set of minimum number of guards that can guard a given polygon is NP-hard. Consequently, researchers have focused on approximate solutions starting from an $O(\log n)$ approximation provided by Ghosh [17] in 1987, along with a conjecture that the problem admitted a polynomial time constant approximation algorithm. However, Eidenbenz et al. [9] showed that the problem was APX-hard, thereby precluding the possibility of a PTAS unless P = NP. After several improvements over the years, in 2017, Bhattacharya et al. [4] have reported constant factor approximation algorithms for the classical art gallery problem as well as for several well-studied variants.

In literature, based on the different restrictions placed on the shape of the galleries or the powers of the guards, several variations of "art gallery problems" have been studied. See [26,30], and [31] for details.

The *connected art-gallery problem* was first introduced by Liaw et al. in 1993 [22], where they refer to the problem as *minimum cooperative guards problem* and study it on k-spiral polygons (polygons with a maximal chain of k consecutive reflex vertices, i.e., vertices with internal angle $>180°$). It was also shown [22] that this problem is NP-Hard for simple polygons but can be solved in linear time in spiral and 2-spiral polygons (also see [29] for results on k-spiral graphs). For simple polygons with n vertices, Hernández-Penalver [19] proved by induction that $\lfloor \frac{n}{2} \rfloor - 1$ guards are always sufficient to obtain a connected guarding. Moreover, they also show that $\lfloor \frac{n}{2} \rfloor - 1$ guards are necessary for some polygons. The same result was also shown by Pinciu via an elegant coloring argument in [27].

In [21], Liaw et al. relax the strong connectivity condition from [22] and consider the case where there are no isolated vertices in the guards visibility graph. This problem of guarded guards where the overall connectivity of the guards visibility graph is non-essential has also been studied in [24,28].

In the distributed setting, Obermeyer et al. [25] study this problem in polygonal environment with holes. They first design a centralized incremental partition algorithm (defined therein) and from that obtain the distributed deployment algorithm by a distributed emulation of the centralized algorithm. The authors give a deployment of agents that is guaranteed to achieve full visibility coverage of the polygon with n vertices and h holes in $O(n^2 + nh)$ time, given that there are at least $\lfloor \frac{n+2h-1}{2} \rfloor$ agents. This work closely relates to our work. While the scope of [25] includes polygons with holes, their algorithm is not optimized for time and their ideas lead to algorithms that require $O(n^2)$ communication rounds even for simple polygons without holes. Notable prior works with ideas leading to [25] can be found in [10, 13–15].

Organization of the Paper. In Sect. 2, we present some preliminary definitions including several geometric definitions pertaining to polygons as well as formal definitions of the distributed computing models. In Sect. 3, we provide a centralized algorithm to solve the visibly connected art gallery problem and then show how to parallelize it. In Sect. 4 and Sect. 5, we present distributed algorithms under proximity perception and depth perception, respectively. We complement our upper bounds with a lower bound that is proved in Sect. 6. We then conclude with some remarks and future works in Sect. 7.

2   Preliminaries

Let P be a simple polygon with $n \geqslant 4$ vertices; we use P to refer to the polygonal region including both the interior and the boundary. In general, for a polygonal region $K \subseteq P$, we use ∂K for the set of vertices of P that lie on the boundary of this polygonal region. The ordered list of vertices of P are denoted $p_1, p_2, \ldots, p_n$, and thus, $\partial P \triangleq \{p_1, p_2, \ldots, p_n\}$. Each open line segment connecting p_i to $p_{i(\bmod\ n)+1}$, $1 \leqslant i \leqslant n$, is denoted e_i. We assume that the vertices are in general position, i.e., (i) no three vertices are collinear, and (ii) no four vertices are co-circular. We use the term *object* to refer to either a vertex or an edge. Thus, the objects of P are $\{p_i\}_i \cup \{e_i\}_i$.

We use the notation $g \in P$ for some point g to indicate that g can either be a vertex, lie on an edge, or lie in the interior of P. Two points $g_1 \in P$ and $g_2 \in P$ are said to be in line of sight of each other if the open line segment $\overline{g_1 g_2}$ lies entirely within P. We use V_g^P (or just V_g when P is clear from context) to denote the visibility polygon of a point $g \in P$, which is defined as the subset of P that contains all points that are in line of sight from g. See Fig. 1 for an illustration.

We also borrow a useful definition from Obermeyer *et al.* [25] for *vertex-limited visibility polygon* $\bar{V}_g^P$ for a point $g \in P$ w.r.t. P, which is a modified form of V_g^P. Notice that V_g^P could have vertices that are not vertices in P; call such vertices *spurious vertices*. To get $\bar{V}_g^P$, perform the following operation repeatedly until there are no more spurious vertices: pick a spurious vertex v with predecessor p and successor s and crop the visibility polygon by cutting along the line segment $\overline{ps}$ and removing the portion that lacks the point g from further consideration. Note that either the predecessor or the successor may themselves be spurious. In Fig. 1, note the first vertex v that is clipped has successor s that is itself a spurious vertex. An edge that is in $\bar{V}_g^P$ but not in V_g^P is called a *gap edge.* We also define a way to crop a polygon (cf. Fig. 1). Formally, for any pair of vertices p_i and p_j such that $\overline{p_i p_j}$ lies entirely within the interior of P and $c \in P \setminus \overline{p_i p_j}$, we define $\mathsf{crop}(P, c, p_i, p_j)$ to be the subset of P obtained by cutting along $\overline{p_i p_j}$ and discarding the part that contains c.

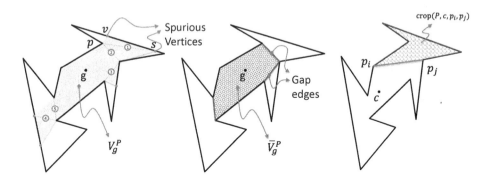

Fig. 1. Visibility polygon (left) and vertex-limited visibility polygon (at the center). The numbered line segments refer to one possible repeated sequence of cuts to arrive at vertex-limited visibility polygon. The polygon on the right depicts a cropped polygon.

Definition 1. *Let* $G = \{g_1, g_2, \ldots\}$ *be a set of points in* P. *We say that the points in* G *guard polygon* P *if* $\cup_{g \in G} V_g = P$. *In this context, we call the points in* G *as guards of* P.

The classical art gallery problem seeks to find a smallest possible set G of points that guard P, with variants including vertex and edge guarding.

In this paper, we consider a variant called the *connected art gallery problem* that, to the best of our knowledge, was introduced first by Liaw et al. [22]. In this variant, guards are connected in a suitable way, which we now formalize. We define the *visibility graph* of a set of points G within (and with respect to) P, denoted $\mathcal{G}_G^P$ (or just $\mathcal{G}$ when clear from context) as the graph with vertex set G. Two points in G are connected by an edge in $\mathcal{G}$ iff they are visible to each other within P. A set G of points in P is said to be connected (w.r.t. P) if $\mathcal{G}_G^P$ is connected. In the connected art gallery problem, we are required to compute a set G of points that guards P *and* the additional requirement that $\mathcal{G}_G^P$ is connected. It is well-known [3] that at most $\lfloor n/3 \rfloor$ guards are always sufficient to guard any polygon with n vertices. However, this bound does not hold under connected guarding.

Claim (consolidated from [27] and [28]). There exist orthogonal polygons with n vertices that require at least $n/2 - 2$ number of connected guards even if we only require them to guard the vertices of the polygon [28]. This bound increases mildly to $\lfloor n/2 \rfloor - 1$ for simple polygons with non-orthogonal edges [27].

We now define several structures associated with any polygon P. A line segment joining two vertices is said to be a *diagonal* if its interior lies entirely within the interior of P. It is easy to see that a maximal set of diagonals that do not intersect each other decomposes the polygons into a set of $n-2$ triangles called a *triangulation* of P [3]. A famous result by Chazelle [6] shows us how to find such a triangulation in $O(n)$ time. Given a triangulation T for a polygon P, the *weak dual graph* $\mathcal{D}_T^P$ (or just $\mathcal{D}$ when clear from context) is the graph (or more informatively, a tree) whose nodes are the triangles in T with edges in $\mathcal{D}_T^P$ between pairs of triangles that share a common triangle edge. Note that the weak dual graph is a tree where each tree node has a degree of at most 3.

Recall that, the term *object* refers to either a vertex or an edge of the polygon P. We define the medial axis M of the polygon P to be the (infinite) collection of points within P that are equidistant from at least two distinct objects of P (see Fig. 2(iii) for an illustration).

Claim. For any simple polygon P, the medial axis is a tree whose leaves are convex vertices of P, i.e., vertices with internal angle being less than $180°$.

Note that reflex vertices (i.e., vertices with internal angles greater than $180°$) cannot be nodes in the medial axis tree. An edge in M is a non-empty maximal set of points that are equidistant between the same set of objects. When the two objects are of the same type (either both polygonal edges or both vertices), the corresponding medial axis edge will be a straight line segment. On the other hand, if one of the objects is a vertex and the other is a polygonal edge, the corresponding medial axis edge will be a parabolic arc. The endpoints of medial axis edges are the *medial axis nodes* or just nodes. Since the number of leaves is at most n, we get:

Claim. The number of nodes in the medial axis of P will be $O(n)$.

We use D^P (or just D when clear from context) to refer to the unweighted diameter of the medial axis M. More precisely, D^P is the maximum number of edges over all paths in the medial axis tree of P.

2.1 Computational Models

We focus on the connected art gallery problem in the classical sequential setting. In this case, we assume that the sequence of vertex points $(p_1, p_2, \ldots, p_n)$ are given in order, say, as an array of points. However, for the connected art gallery problem as inspired by mobile agents that operate in a spatially distributed setting, we employ a distributed computing model based on the work by Obermeyer *et al.* [25]. For clarity, we assume a synchronous model with time discretized into a sequence of rounds and the agents executing a look-communicate-move cycle in each synchronous round; local computation is interspersed between the look-communicate-move cycles.

We assume that there are $\Theta(n)$ agents. Agents (modeling mobile robots) can be represented as points in the plane and as a result multiple agents can be co-located at the same point. Without loss of generality, assume all agents start at the same vertex somewhere in P. Furthermore, agents can only move from one vertex p_i to another vertex p_j provided $\overline{p_i p_j}$ is a diagonal in P (i.e., p_i and p_j have direct line of sight to each other). We assume that agents have unique IDs from $\{1, 2, \ldots, n\}$. Each agent g performs the following tasks in each round.

Look. The agent g first orients itself to start from a particular direction (in the direction of another vertex called its *orientation vertex*) and perform a $360°$ clockwise sweep during which it creates a view of V_g^P. The level of information that the agent can gather depends on whether the agents have depth perception or not. With depth perception, the view is simply the full visibility polygon V_g^P. Without depth perception, however, the view is limited to a sequence of alternating vertices and edges (possibly gap edges) starting from its orientation vertex. In both cases, g can also see other agents that are inside V_g.

Communicate. Two agents can communicate as long as they are visible to each other (which includes co-located agents). Communication is via message passing. Each agent can send at most one message to each agent that it can see.

Move. This step again differs based on whether agents can perceive depth or not. Let us first consider the case when agents can perceive depth. Based on the outcome of the communication and computation, each agent g chooses to move from its current location to a new location within its current visibility polygon. We assume that the agent – once it reaches its destination location – can "remember" its source position in the sense that it can spot the source location in its view after it reaches its destination. The only restriction when agents cannot perceive depth is that they are limited to moving to vertices of the polygon. For this reason, we always assume that agents will be on polygonal vertices (even at the start of time) when they cannot perceive depth.

3   Centralized Sequential and Parallel Algorithms

We first present a centralized sequential algorithm for the connected art gallery problem and then briefly show how it can be parallelized. Our approach is to decompose the weak dual graph into a suitable set of at most $\lfloor n/2 \rfloor - 1$ connected triplets and then assign a guard for every triplet. The high-level steps are outlined in Algorithm 1. Consider the weak dual graph $\mathcal{T}$, which is of course a tree with maximum degree 3. Root the tree at some node r that is of degree 1. A *triplet* is any set of three nodes in the tree that are connected. We now show a simple procedure (cf. Algorithm 2 and Fig. 2(ii) for an illustration) to decompose $\mathcal{T}$ into triplets. Subsequently, we will prove some properties of these triplets that will immediately lead us to the required centralized algorithm for the connected art gallery problem.

Lemma 1. *When two triplets share at least one common node, their associated guards can see each other.*

Algorithm 1. Centralized algorithm for the connected art gallery problem.

1: Compute a triangulation T of the polygon P [6] and then compute the weak dual graph.
2: Decompose the weak dual graph into at most $\lfloor n/2 \rfloor - 1$ triplets, i.e., groups of three connected nodes in T, as described in Algorithm 2.
3: Each triplet corresponds to three triangles arranged in such a way that there is a middle triangle that shares two edges, say a and b, with the other two triangles. Placing a guard at the common vertex between a and b for every triplet is the required solution. (See Figure 2(i) for an illustration.)

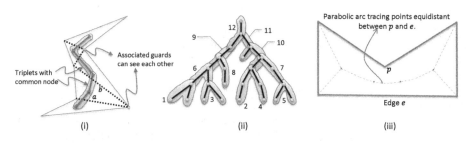

Fig. 2. (i) Illustrated placement of guards associated with triplets. Notice that the guards are associated with the two triplets sharing a common node. Consequently, they can see each other. (ii) Sequence of triplets computed at the end of each iteration of the **for** loop in Algorithm 2. (iii) Medial axis of a polygon.

Algorithm 2. Algorithm to decompose T into triplets.

Require: A tree T rooted at a node r of degree 1, with max degree three, and depth L. Note that r is assumed to be at level 0, so there are L levels from r to the farthest leaf (inclusive).
Ensure: A collection of triplets.

1: Color all internal nodes red and all leaves orange.
2: **for** $\ell \leftarrow L$ down to 2 (decrementing by 1 every iteration) **do**
3:     **while** $\exists$ orange node v at level ℓ **do**
4:         **if** v has a sibling v' that is also colored orange **then**
5:             Form a new triplet comprising v, v', and their common parent node p.
6:             Color parent p orange.
7:             Color v and v' green.
8:         **else**
9:             Form a triplet comprising v, parent p of v, and the grandparent p' of v.
10:            Color p' orange.
11:            Color v and p green.
12:         **end if**
13:     **end while**
14: **end for**
15: **if** the root is not part of some triplet **then**
16:     Form a triplet comprising the root, its child, and an arbitrarily chosen grandchild. Color all three nodes green.
17: **end if**

Having presented the algorithm to solve the connected art gallery problem in this centralized setting, we move on to analyze the algorithm. Our main focus will be on analyzing Algorithm 2. We make a series of observations formalized as lemmas and then derive the result as a consequence. For a given set of triplets, we define the *triplets graph* to be the graph with the triplets as vertices and edges between pairs of triplets that share at least one edge. We say that the set of triplets covers the tree $\mathcal{T}$ if every node is part of at least one triplet. The proof of the following lemma is deferred to the full version.

Lemma 2. *Given* t *is the number of nodes in* $\mathcal{T}$, *we claim that Algorithm 2*

1. *forms a set of triplets that covers* $\mathcal{T}$,
2. *guarantees that the triplets graph is connected,*
3. *guarantees that the number of triplets formed is at most* $\lfloor \mathsf{t}/2 \rfloor$, *and*
4. *runs in* $O(\mathsf{t})$ *time.*

Recalling that $\mathsf{t} = n - 2$, we can conclude that the sequential algorithm runs in $O(n)$ time.

Finally, we remark that the algorithm described above can be implemented in parallel specifically in the Parallel Random Access Machine (PRAM) model. In PRAM, we have several processors that operate on a shared addressable memory space. Typically, concurrent reading (i.e., multiple processors reading a word simultaneously) is acceptable but writing requires exclusivity (i.e., ensuring that at most one processor can write to a word in one time step); this is the concurrent read exclusive write (CREW) version of PRAM. Goodrich [18] has already shown how to triangulate P in $O(\log n)$ time under CREW PRAM. We logically assign one processor per triangle and ensure the parent-child relationship between triangles is extended to the processors. Then, each iteration of the **for** loop in Algorithm 2 (comprising several **while** loop iterations) can be executed in parallel. Redundant triplets that can occur when we form triplets connecting two orange siblings (see line number 4 in Algorithm 2). This can be avoided in such situations by only allowing processors corresponding to the left children to form the triplet.

Theorem 1. *Supported by Algorithm 2, Algorithm 1 solves the connected art gallery problem with at most* $\lfloor n/2 \rfloor - 1$ *guards in time that is linear in* n. *Moreover, can be solved in the CREW PRAM model with at most* $\lfloor n/2 \rfloor - 1$ *guards in time that is linear in the diameter of the weak dual graph associated with the triangulation of* P.

4   Distributed Guarding with Proximity Perception

In this section, we consider the case where each agent is able to distinguish the proximity or relative distances (without knowing the actual distances) between the various objects associated with the polygon as well as with other agents, etc. which are in its visibility polygon at any specified moment. Specifically, the agents' "look" paradigm is reflective of real-world sensing techniques where absolute distances to objects in the scene are unavailable whilst their relative distances can be inferred such as with photogrammetric vision in drones.

We give a distributed solution that solves the connected art-gallery problem and runs in $O(\min(\tilde{d}^2, n))$ rounds, where $\tilde{d}$ is the minimal v-diameter that we formally define later in this section. Our solution comprises two algorithms that are executed in parallel, one in a breadth first manner and the other in a depth first manner. Our final solution is to take the best out of both explorations. Due to space limitation, we will describe the breadth first algorithm that runs in $O(\tilde{d}^2)$ rounds in detail. We subsequently present a brief overview of the depth first exploration and defer further details to the full version [2]. Finally, we conclude with some remarks on how our algorithm can form the basis for solving other polygon problems on P.

Assuming that the vertices and the edges defining the polygon P are in general position, the agents start at some vertex in P, which we can assume w.l.o.g. to be p_1. The algorithm operates in phases. At the end of a particular phase ℓ, a subset S_ℓ of the agents have "settle" into their final positions while establishing a connected guarding of the subset P_ℓ of the polygon. For any agent i, its settled position is a vertex in P and is denoted s_i.

Moreover, the settled agents are arranged in the following hierarchical manner. W.l.o.g., let the root be agent 1, settled at p_1. We define the territory of the root, i.e., agent 1, to be $\mathsf{territory}(1) \triangleq \bar{V}_{p_1}^P$. Every other settled agent j has a parent agent $\mathsf{parent}(j)$. If agent $i = \mathsf{parent}(j)$, then we say that j is the child of i denoted as $\mathsf{child}(i)$. Each parent agent i has one child agent j per gap edge in its $\mathsf{territory}(i)$ and the child is located at one of the end points, say p_a, of a the gap edge (p_a, p_b). Thus, $s_j = p_a$. The other end of the gap edge p_b is denoted $\mathsf{orient}(j)$; intuitively, j settles at s_j and orients itself towards $\mathsf{orient}(j)$ for performing "look" operations.

Furthermore, $\mathsf{territory}(j) \triangleq \bar{V}_{s_j}^P \cap \mathsf{crop}(P, s_{\mathsf{parent}(j)}, s_j, \mathsf{orient}(j))$ i.e., $\mathsf{territory}(j)$ is the portion of s_j's vertex limited visibility polygon not containing s_j's parent and truncated by the gap edge that originated it. (See Fig. 3.) Intuitively, each agent j is only responsible for guarding its $\mathsf{territory}(j)$. Notice that by definition, territories of a parent and its child do not overlap (they share a bordering edge that is a gap edge seen originally by the parent). Moreover, territories of children of a given parent do not overlap as well (at most they share a vertex) as this would imply that the polygon contains holes, i.e., it is non-simple. For correctness, we need $\cup_j \mathsf{territory}(j) = P$, which is immediate from our algorithm.

Next, we define a *territory tree* to be a tree in which nodes are territories and edges are pairs of territories that share a common diagonal (gap) edge. Let T^* be the set of all possible territory trees that can be achieved given all possible options for the starting vertex p_1 and all possible choice of placement of child agents. Then, we define d as the maximum diameter of all such territory trees, i.e., $d \triangleq \max_{T \in T^*} \mathsf{diameter}(T)$.

Initially, $S_0 = \{1\}$ (w.l.o.g.), $s_1 = p_1$, and P_0 is simply $\bar{V}_{s_1}^P$. We are now ready to present the steps to be performed within each phase ℓ; notice that there cannot be more than d phases to the algorithm hence, $1 \leqslant \ell < d$. Intuitively, in each phase ℓ (see Algorithm 3), we incrementally construct the territories at level ℓ of the territory tree.

Algorithm 3. Phase $\ell \geqslant 1$ of the distributed algorithm for the connected art gallery problem that may use more than $\lfloor n/2 \rfloor - 1$ guards. This description assumes phases 0 to $\ell - 1$ have completed and each agent $i \in S_{\ell-1}, \ell > 1$, remembers one marked vertex. (This marking scheme ensures that a child and grandparent are not visible to one another.)

1: Every settled agent $i \in S_{\ell-1}$ performs a *look* operation into its territory(i) and counts the number of gap edges in its territory(i). Call this count b_i. Each settled agent i now up-casts b_i to the root with intermediate settled agents aggregating the quantities by adding up the numbers sent by their children.

2: Notice that at the end of the up-casting, the root will know the total number b of gap edges. The root apportions b new agents and sends them to its children according to the numbers sent by each child. Subsequently, whenever a settled agent notices new agents reaching its position, it will apportion the agents according to numbers sent by its children and the new agents will move to their assigned child of the current settled agent.

3: Each settled agent $i \in S_{\ell-1}$ whose territory(i) has some $b_i > 0$ gap edges gets exactly b_i new agents. Agent i assigns each of those new agents j to an unmarked vertex of each such gap edge and consequently, agent j marks the other vertex of that gap edge.

Lemma 3. *Repeating Algorithm 3 until all levels of the territory tree are explored, we get a distributed algorithm that, with no more than n agents, ensures that the agents position themselves in a manner that solves the connected art gallery problem. The round complexity is $O(d^2)$.*

Now, we introduce the notion of *minimal visibility connected vertex guarding* (henceforth referred to as *minimal v-guarding*) which is pivotal in this case for developing algorithmic bounds on the running time. Let P be a simple polygon with n vertices. Recall that, given a set of labelled guards $G = \{g_1, g_2, \ldots, g_k\}$ of P, we associate with G a unique graph $\mathcal{G}$ with a vertex set of size k such that when two guards are visible to each other, then they are connected by an edge in this graph $\mathcal{G}$ i.e., $e = \{i, j\} \in E[\mathcal{G}] \iff g_i$ is visible to g_j. We say that G is a *minimal v-guarding* or a *minimal v-configuration* of P whenever the following holds, $\forall v \in \mathcal{G}$, *at least* one of the following two conditions applies:

1. $\mathcal{G} - v$ has more than a single component.
2. The vertex guarding $G_{-v} := G \backslash \{g_v\}$ of P is incomplete, i.e., $\cup_{g \in G_{-v}} \partial V_g \subsetneq \partial P$ where ∂P is the set of vertices of polygon P and ∂V_g refers to the set of vertices of P visible from g.

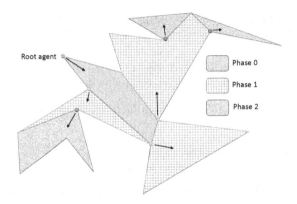

Fig. 3. Depicts the placement of agents and their respective territories at three levels. The arrows point each agent into its territory.

We then define the notion of *minimal v-diameter* $\tilde{d} := \max_{\tau \in \mathcal{M}} \text{diameter}(G_\tau)$ where $\mathcal{M}$ is the set of all possible minimal v-configurations of P and G_τ denotes the associated visibility graph of such a guarding τ from $\mathcal{M}$.

In order that we may successfully compare the efficiency of our algorithms with one another we use $\tilde{d}$ which is a polygonal parameter pertinent to our guarding problem.

Lemma 4. $d = O(\tilde{d})$, *i.e., the diameter of any territory tree on P is asymptotically bounded by the minimal v-diameter of the Polygon P.*

Using Lemma 3 along with Lemma 4 yields the following theorem:

Theorem 2. *There exist a distributed algorithm that solves the connected art gallery problem in $\mathcal{O}(\tilde{d}^2)$ time using no more than n agents limited to proximity sensing capability, where $\tilde{d}$ refers to the minimal v-diameter of P and n is the number of polygon vertices.*

An $O(n)$ Round Algorithm. While most real-world polygons may have small diameters, it is nevertheless conceivable that $\tilde{d} \in \Omega(n)$ in some cases like spirals. The algorithm that we have presented above will unfortunately require quadratic in n number of rounds for such situations, which is undesirable. In the full version [2], we describe a depth first procedure that only requires $O(n)$ rounds.

Remark 1. Both the $O(\tilde{d}^2)$ algorithm and the depth first $O(n)$ algorithm will maintain agents connected by line of sight. So we can start both algorithms simultaneously. When one of them – the winner – finishes, the other can be terminated by a terminate message that will reach all agents in at most $O(\min(\tilde{d}, n))$.

General Problem Solving Given a Visibly Connected Guard Placement. In the breadth first scenario, since the agents are settled into a visibly connected guarding position and are aware of their respective territories in the territory tree, they can gather at the root's position in a bottom-up fashion. Thus, the root, in $O(\tilde{d})$ rounds, can collect the views of all the agents and perform computation using the collective views of the agents. Similarly, taking into account the depth first scenario, we get the following generalized theorem.

Theorem 3. *Suppose $\mathcal{P}$ is a computationally tractable problem that takes a polygon P as input and either*

- *outputs information in the form of bits*
- *or requires placing available agents in positions within P.*

Then, $\mathcal{P}$ can be solved in our distributed context in $O(\min(\tilde{d}^2, n))$ rounds.

5   Distributed Guarding with Depth Perception

In this section, we give a distributed algorithm that solves the connected art gallery problem in $O(D^2)$ rounds, where D is the (unweighted) diameter of the medial axis. This algorithm's advantage is that its running time depends on D, which is more well-known than $\tilde{d}$. However, as opposed to the previous section, in this case, agents require the ability to perceive depth. The key idea of the algorithm here is if agents were placed on all internal nodes of the medial axis and some specially chosen vertices, they cover the entire graph as well as remain visibly connected.

Computing Adjacent Nodes in the Medial Axis. Imagine an agent at any point x on the medial axis. The agent can simulate the creation of a maximal disc at x to find the objects (vertices or edges) that determine x, i.e., the objects due to which x is a part of the medial axis. There would be at least two such objects that determine x. If x is determined by multiple objects (>2), it implies that x itself is a node on the medial axis, and we can consider any two consecutive objects determined by the **look** operation. For example, if a, b, c, d are 4 objects, that determine x and are ordered in accordance with the look operation, we consider the pairs ab, bc, cd and da only. Note that, only the consecutive object pairs determine the medial axis edges incident at x, and hence only those are considered.

For each pair of objects ob_1 and ob_2, there can only be three possible cases; either both are vertices, both are edges, or one of them is a vertex while the other is an edge. For all the cases, the agent at x is aware of the structure of the medial axis from x. Thus, if both ob_1 and ob_2 are vertices, then the next node of the medial axis lies on the perpendicular bisector of the line segment (ob_1, ob_2). If both ob_1 and ob_2 are edges, then the next node of the medial axis lies on the angle bisector of ob_1 and ob_2. Lastly, w.l.o.g. if ob_1 is a vertex and ob_2 is an edge, then the next node of the medial axis lies on the parabola determined by ob_1 and ob_2. Since agents have infinite computing power and depth sensing ability, they can progressively simulate maximal discs along the medial axis structure (perpendicular bisector, angle bisector or parabola) until the maximal disc encounters a new object (say ob_3). The center of the maximal disc at this instance determines the next adjacent node in the medial axis. We define the set of new adjacent nodes obtained in phase i as A_i.

Agent Placement. As in Algorithm 3, when an already placed agent a determines its adjacent set of positions on which new agents are to be placed, then a upcasts the request of the required number of agents up to the root (the spot initially containing all the agents) with intermediate agents aggregating the quantities by adding up the numbers sent by their children. The root serves the request by assigning the required number of agents. The assigned agents trace back the path to a and thereafter get placed in their determined spot.

Algorithm 4. An $O(D^2)$ time algorithm for the connected art gallery problem.

1: Starting from the initial given vertex v where all the agents are placed, a medial axis point m is determined. If v is convex, then m is given by v's adjacent node in the medial axis. Alternatively, if v is a reflex vertex, pick the nearest visible new object ob_3 (determined by the depth sensing ability of the agents) not including v and choose the center of the maximal disc determined v and ob_3 as the point m on the medial axis.
2: Consider m as the root. All agents are moved here.
3: Determine all the adjacent medial axis nodes from m (i.e., the set A_1). * *This marks the end of the first phase. Each iteration of the loop represents a subsequent phase. The algorithm continues until the entire medial axis is uncovered.* *\
4: **for** each new adjacent node $x \in A_{i-1}$ determined in the previous phase, that is not a leaf node of the medial axis tree **do**
5:     Compute set A_i (current set of new adjacent medial axis nodes of x) in parallel.
6:     Place an agent at each node $y \in A_i$ except when y corresponds to a convex vertex of the polygon (leaf node of the medial axis tree).
7:     **if** y is a part of a parabola determined by a polygon vertex and an edge and the polygon vertex does have an agent on it **then**
8:         Place an agent on the reflex vertex determining the parabola.
9:     **end if**
10: **end for**

Lemma 5. *Algorithm 4 gives a visibly connected guard placement while ensuring that the entire polygon is guarded/covered.*

Theorem 4. *There exists a distributed algorithm that solves the connected art gallery problem in $\mathcal{O}(D^2)$ time using no more than n agents, where D refers to the medial axis diameter and n is the number of polygon vertices.*

To reduce the final number of guards placed, we use similar procedure as described in Sect. 4. This gives us the following theorem.

Theorem 5. *There exists a distributed algorithm that uses no more than n agents to compute the placement of at most $\lfloor n/2 \rfloor - 1$ guard agents in a visibly connected manner, when the agents have depth sensing ability. Moreover, this algorithm takes at most $O(D^2)$ communication rounds.*

6   Lower Bound

In this section, we give lower bounds for a slightly weaker polygon exploration problem that requires for every point in P, that some agent must have been within line of sight of that point at some time instant during the course of the algorithm. Clearly, any solution to the visibly connected guard placement problem will also be a solution for the exploration problem. The lower bounds highlight the criticality of parameters like the medial axis diameter D and the minimal v-diameter $\tilde{d}$ for solving the connected art-gallery problem. The main result is summarized by the following Theorem.

Theorem 6. *For every deterministic distributed guard placement algorithm A with depth perception (resp., proximity perception) there exists a polygon P with medial axis diameter $D \in o(\log n)$ (resp., with minimal v-diameter $\tilde{d} \in o(\log n)$) such that A requires $\Omega(D^2)$ time (resp., $\Omega(\tilde{d}^2)$ time) to place the guards even when A is provisioned with a number of guards that is $\Theta(n)$.*

We show a reduction from any instance of the well-studied tree exploration game [8] to the problem of exploring a polygon, which our guarding problem subsumes. We sketch our approach here and defer details to the full version [2].

1. Firstly, we embed the tree in the Euclidean plane such that no two adjacent edges form an angle of $180°$.
2. We *thicken* the edges of the embedded tree and form a simple polygon.
3. With the embedding and thickening transformations, we reduce the problem of collaborative exploration of the underlying tree to the guard placement in the obtained polygon.

7   Conclusion and Future Works

In this paper, we have presented centralized and distributed algorithms for computing a visibly connected guard placement. Crucially, our algorithms take time that is quadratic in a couple of different notions of diameters of P, i.e., $\tilde{d}$ and D. We believe that $\tilde{d} \in O(D)$, thereby obviating the need for precise depth perception, but the proof has eluded us. It would be nice to establish this formally as this would lend to understanding the trade off between LiDAR and photogrammetry (see [5,11] for example). We also remark that our algorithms have been explained in the synchronous setting for the purpose of clarity, but they can be easily extended to the asynchronous setting. Additionally, it will be interesting to extend our works to polygons with holes or polyhedra in higher dimensions.

Acknowledgements. Barath Ashok and John Augustine were supported in part by DST/SERB Extra Mural Grant (file number EMR/2016/00301) and DST/SERB MATRICS Grant (file number MTR/2018/001198). Suman Sourav was supported in part by the National Research Foundation, Prime Minister's Office, Singapore under the Energy Programme and administrated by the Energy Market Authority (EP Award No. NRF2017EWT-EP003-047). Part of this work was done when Aditya Mehekare, Sridhar Ragupathi, Srikkanth Ramachandran and Suman Sourav visited IIT Madras. We thank Rajsekar Manokaran for pointing out LiDAR and Photogrammetry.

References

1. Aber, J.S., Marzolff, I., Ries, J.B., Aber, S.E.: Principles of photogrammetry. In: Aber, J.S., Marzolff, I., Ries, J.B., Aber, S.E. (eds.) Small-Format Aerial Photography and UAS Imagery, pp. 19–38. Academic Press, Cambridge (2019). (Chapter 3)
2. Ashok, B., Augustine, J., Mehekare, A., Ragupathi, S., Ramachandran, S., Sourav, S.: Guarding a polygon without losing touch (2020). https://arxiv.org/abs/2005.05601
3. de Berg, M., Cheong, O., van Kreveld, M., Overmars, M.: Computational Geometry: Algorithms and Applications, 3rd edn. Springer, Heidelberg (2008). https://doi.org/10.1007/978-3-540-77974-2
4. Bhattacharya, P., Ghosh, S.K., Pal, S.: Constant approximation algorithms for guarding simple polygons using vertex guards (2017)
5. Buczkowski, A.: Drone lidar or photogrammetry? Everything you need to know. https://geoawesomeness.com/drone-lidar-or-photogrammetry-everything-your-need-to-know
6. Chazelle, B.: Triangulating a simple polygon in linear time. Discrete Comput. Geom. **6**(3), 485–524 (1991). https://doi.org/10.1007/BF02574703
7. Chvátal, V.: A combinatorial theorem in plane geometry. J. Comb. Theory, Ser. B **18**(1), 39–41 (1975)
8. Disser, Y., Mousset, F., Noever, A., Skoric, N., Steger, A.: A general lower bound for collaborative tree exploration. Theor. Comput. Sci. **811**, 70–78 (2018)
9. Eidenbenz, S., Stamm, C., Widmayer, P.: Inapproximability results for guarding polygons and terrains. Algorithmica **31**(1), 79–113 (2001). https://doi.org/10.1007/s00453-001-0040-8
10. Fekete, S.P., Kamphans, T., Kröller, A., Mitchell, J.S.B., Schmidt, C.: Exploring and triangulating a region by a swarm of robots. In: Goldberg, L.A., Jansen, K., Ravi, R., Rolim, J.D.P. (eds.) APPROX/RANDOM 2011. LNCS, vol. 6845, pp. 206–217. Springer, Heidelberg (2011). https://doi.org/10.1007/978-3-642-22935-0_18
11. Filippelli, S.K., Lefsky, M.A., Rocca, M.E.: Comparison and integration of lidar and photogrammetric point clouds for mapping pre-fire forest structure. Remote Sens. Environ. **224**, 154–166 (2019)
12. Fisk, S.: A short proof of Chvátal's watchman theorem. J. Comb. Theory, Ser. B **24**(3), 374 (1978)
13. Ganguli, A.: Motion coordination for mobile robotic networks with visibility sensors. Ph.D. thesis, University of Illinois at Urbana-Champaign (2007)
14. Ganguli, A., Cortes, J., Bullo, F.: Distributed deployment of asynchronous guards in art galleries. In: 2006 American Control Conference, p. 6, June 2006
15. Ganguli, A., Cortes, J., Bullo, F.: Visibility-based multi-agent deployment in orthogonal environments. In: American Control Conference, July 2007

16. Ghosh, S.: Visibility Algorithms in the Plane. Cambridge University Press, Cambridge (2007)
17. Ghosh, S.K.: Approximation algorithms for art gallery problems. In: Proceedings of Canadian Information Processing Society Congress, pp. 429–434 (1987)
18. Goodrich, M.T.: Triangulating a polygon in parallel. J. Algorithms **10**(3), 327–351 (1989)
19. Hernández-Penalver, G.: Controlling guards. In: CCCG, pp. 387–392 (1994)
20. Lee, D., Lin, A.: Computational complexity of art gallery problems. IEEE Trans. Inf. Theory **32**(2), 276–282 (1986). https://doi.org/10.1109/TIT.1986.1057165
21. Liaw, B., Lee, R.C.T.: An optimal algorithm to solve minimum weakly cooperative guards problem for 1-spiral polygons. Inf. Process. Lett. **52**(2), 69–75 (1994)
22. Liaw, B.C., Huang, N.F., Lee, R.C.T.: The minimum cooperative guards problem on k-spiral polygons. In: CCCG (1993)
23. McManamon, P.: LiDAR Technologies and Systems. SPIE Press, Bellingham (2019)
24. Michael, T.S., Pinciu, V.: Art gallery theorems for guarded guards. Comput. Geom. Theory Appl. **26**(3), 247–258 (2003)
25. Obermeyer, K.J., Ganguli, A., Bullo, F.: Multi-agent deployment for visibility coverage in polygonal environments with holes. Int. J. Robust Nonlinear Control **21**(12), 1467–1492 (2011)
26. O'Rourke, J.: Art Gallery Theorems and Algorithms. Oxford University Press, Oxford (1987)
27. Pinciu, V.: A coloring algorithm for finding connected guards in art galleries. In: Calude, C.S., Dinneen, M.J., Vajnovszki, V. (eds.) DMTCS 2003. LNCS, vol. 2731, pp. 257–264. Springer, Heidelberg (2003). https://doi.org/10.1007/3-540-45066-1_20
28. Pinciu, V.: Connected guards in orthogonal art galleries. In: Kumar, V., Gavrilova, M.L., Tan, C.J.K., L'Ecuyer, P. (eds.) ICCSA 2003. LNCS, vol. 2669, pp. 886–893. Springer, Heidelberg (2003). https://doi.org/10.1007/3-540-44842-X_90
29. Deogun, J.S., Sarasamma, S.: On the minimum co-operative guard problem. J. Comb. Math. Comb. Comput. (JCMCC) **22**, 161–182 (1996)
30. Shermer, T.C.: Recent results in art galleries (geometry). Proc. IEEE **80**(9), 1384–1399 (1992). https://doi.org/10.1109/5.163407
31. Urrutia, J.: Art gallery and illumination problems. In: Handbook of Computational Geometry, North-Holland, pp. 973–1027 (2000). (Chapter 22)

Dynamic Graphs

Random Walks on Randomly Evolving Graphs

Leran Cai, Thomas Sauerwald, and Luca Zanetti[(✉)]

University of Cambridge, Cambridge, UK
{lc647,tms41,lz381}@cam.ac.uk

Abstract. A random walk is a basic stochastic process on graphs and a key primitive in the design of distributed algorithms. One of the most important features of random walks is that, under mild conditions, they converge to a stationary distribution in time that is at most polynomial in the size of the graph. This fundamental property, however, only holds if the graph does not change over time; on the other hand, many distributed networks are inherently dynamic, and their topology is subjected to potentially drastic changes.

In this work we study the mixing (i.e., convergence) properties of random walks on graphs subjected to random changes over time. Specifically, we consider the edge-Markovian random graph model: for each edge slot, there is a two-state Markov chain with transition probabilities p (add a non-existing edge) and q (remove an existing edge). We derive several positive and negative results that depend on both the density of the graph and the speed by which the graph changes.

Keywords: Random walks · Evolving graphs · Mixing times

1 Introduction

A random walk on a network is a simple stochastic process, defined as follows. Given an undirected graph $G = (V, E)$, the walk starts at a fixed vertex. Then, at each step, the random walk moves to a randomly chosen neighbor[1]. Due to their simplicity and locality, random walks are very useful algorithmic primitive, especially in the design of distributed algorithms. In contrast to topology-driven algorithms, algorithms based on random walks usually benefit from a strong robustness against structural changes in the network.

Random walks and related works have found various applications such as routing, information spreading, opinion dynamics, and graph exploration [3,9]. One key property of random walks is that, under mild assumptions on the underlying network, they converge to a stationary distribution – an equilibrium state

---

[1] In case of a *lazy* random walk, the walk would remain at the current location with probability $1/2$, and otherwise move to a neighbor chosen uniformly at random.

The second and third author acknowledge support by the ERC Starting Grant "Dynamic March".

A. W. Richa and C. Scheideler (Eds.): SIROCCO 2020, LNCS 12156, pp. 111–128, 2020.
https://doi.org/10.1007/978-3-030-54921-3_7

in which every vertex is visited proportionally to its degree. The time for this convergence to happen is called *mixing time*, and understanding this time is crucial for many sampling and exploration related tasks. In particular, whenever a graph has a small *mixing time*, also its *cover time* (the expected time to visit all vertices of the graph) is small as well.

While most of the classical work devoted to understanding random walks has focused on static graphs, many networks today are subject to dramatic changes over time. Hence understanding the theoretical power and limitations of dynamic graphs has become one of the key challenges in computer science [17]. Several recent works have indeed considered this problem and studied the behavior of random walks [2,3,10,15,21,22,24] or similar processes [4,5,8,11,14] on such dynamic graphs, and their applications to distributed computing [2,14,24].

In this work, we study the popular *evolving graph model*. That is, we consider sequences of graphs $G_1, G_2, \ldots$ over the same set of vertices but with a varying set of edges. This model has been studied in, for example, [3,15,25]. Both [3] and later [25] proved a collection of positive and negative results about the mixing time (and related parameters), and they assume a worst-case scenario where the changes to the graph are dictated by an oblivious, non-adaptive adversary. For example, [3] proved the following remarkable dichotomy. First, even if all graphs $G_1, G_2, \ldots$ are connected, small (but adversarial) changes to the stationary distribution can cause exponential mixing (and hitting) times. Secondly, if the sequence of connected graphs share the same stationary distribution, i.e., the degrees (or relative degrees) of vertices are time-invariant, then mixing and hitting times are polynomial. This assumption about a time-invariant stationary distribution is crucial in the majority of the positive results in [3,25].

In contrast to [3,25], we do not impose such assumptions, but instead study a model with incremental changes. Specifically, we consider a setting where the evolving graph model changes *randomly* and study the so-called *edge-Markovian random graph* $\mathcal{G}(n, p, q)$, which is defined as follows (see Definition 2.5 for a more formal description). For each edge slot, there is a two-state Markov chain that switches from off to on with probability p and from on to off with probability q. This model can be seen as a dynamic version of the Erdős-Rényi random graph, and has been studied in the context of information spreading and flooding [5–7]. While these results demonstrate that information disseminates very quickly on these dynamic graphs, analysing the convergence properties of a random walk seems to require new techniques, since degree fluctuations make the use of any "inductive" argument very difficult – from one step to another, the distribution of the walk could become "worse", whereas the set of informed (or reachable) nodes can never decrease.

In this work, we will investigate the mixing time of a random walk on such evolving graphs. It turns out that, as our results demonstrate, the mixing time depends crucially on the density as well as on the speed by which the graph changes. We remark that deriving bounds on the mixing time on $\mathcal{G}(n, p, q)$ poses some unique challenges, which are not present in the positive results of [3,25]. The main difficulty is that in $\mathcal{G}(n, p, q)$, due to the changing degrees of the vertices, there is no time-invariant stationary distribution, and the traditional notion of mixing time must be adapted to our dynamic setting. Informally, what we ask, then, is how many steps the walk needs to take before the distance to a *time-dependent* stationary distribution becomes *small enough*. Furthermore, in contrast to static graphs, where the distance between the distribution of the walk and the stationary distribution can only decrease, in dynamic graphs the distance to the time-dependent stationary distribution might increase with time. For this reason, we also ask that the distribution of the walk remains close to a time-dependent stationary distribution for a *long enough* interval of time (for a precise definition of our notion of mixing time, see Definition 2.7). We believe this requirement is necessary for our definition of mixing time to be useful in potential applications.

Further Related Work. Recently, [15] analysed the cover time of so-called "Edge-Uniform Stochastically-Evolving Graphs", that include our model as a special case (i.e., the history is $k = 1$). Their focus is on a process called "Random Walk with a Delay", where at each step the walk picks a (possible) neighbor and then waits until the edge becomes present. In [15, Theorem 4], the authors also relate this process to the standard random walk, and prove a worst-case upper bound on the cover time. However, one of the key differences to [15] is that we will study the *mixing time* instead of the *cover time*.

In [26], the authors analysed a continuous-time version of the edge-Markovian random graph. However, unlike the standard random walk, they consider a slightly different process: when the random walk tries to make a transition from a vertex u, it picks one of the $n - 1$ other vertices and moves there *only if* the edge is present; otherwise it remains in place. For this process, they were able to derive very tight bounds on the mixing time and establish a cutoff phenomenon. The same random walk was also analysed on a dynamic graph model of the d-dimensional grid in [19, 20] and, more generally, in [12].

1.1   Main Results

We study the mixing properties of random walks on edge-Markovian random graphs $\mathcal{G}(n, p, q)$. In particular, we consider six different settings of parameters p and q, which separates edge-Markovian models based on how fast graphs change over time (slowly vs. fast changing), and how dense graphs in the dynamic sequence are (sparse vs. semi-sparse vs dense).

As noted in previous works (see, e.g., [5]), a dynamic sequence sampled from $\mathcal{G}(n, p, q)$ will eventually converge to an Erdős-Rényi random graph $\mathcal{G}(n, \tilde{p})$ where $\tilde{p} = \frac{p}{p+q}$ (for the sake of completeness, we give a proof of this fact in Appendix A). We use the expected degree in such random graph, which is equal to $d = (n-1)\tilde{p}$, to separate edge-Markovian models according to their density as follows:

1. **Sparse** $d = o(\log n)$
2. **Semi-sparse** $d = \Theta(\log n)$
3. **Dense** $d = \omega(\log n)$.

Notice that the sparse regime corresponds to random graphs with density below the connectivity threshold of Erdős-Rényi random graphs.

We further separate edge-Markovian models based on how fast they change over time. Let $\delta = \binom{n}{2}\tilde{p}q + \binom{n}{2}(1-\tilde{p})p$ be the expected number of changes at each step, when starting from a stationary initial graph $G_0 \sim \mathcal{G}(n,\tilde{p})$. We consider the following two opposite regimes.

1. **Fast-changing** $\delta = \Theta(dn)$.
2. **Slowly-changing** $\delta = O(\log n)$.

Notice that the fast-changing regime corresponds to graphs for which a constant fraction of edges change at each step in expectation.

Table 1. Summary of our main results (informal). See referenced theorems for the precise and complete statements.

	Fast-changing $\delta = \Theta(dn)$	Slowly-changing $\delta = O(\log n)$
Sparse $d \in [1, o(\log n)]$	$t_{\mathrm{mix}} = \infty$ Theorem 1.1	$t_{\mathrm{mix}} = \Omega(n)$ Proposition 1.4
Semi-sparse $d = \Theta(\log n)$	Coarse mixing[a] in $O(\log n)$ Proposition 3.2	$t_{\mathrm{mix}} = O(\log n)$, Theorem 1.3
Dense $d \in [\omega(\log n), n/2]$	$t_{\mathrm{mix}} = O(\log n)$ Theorem 1.2	

[a]In this regime we are not able to prove finite mixing time. However, we show that the distribution of the walk will "flatten out" after $O(\log n)$ steps. We refer to this behavior as *coarse* mixing.

The main results of our work are presented in Table 1. Here, we assume G_0 is sampled from the stationary graph distribution $\mathcal{G}(n,\tilde{p})$. In the fast-changing regime, as highlighted in Remark A.1, this is without loss of generality. For slowly-changing models, instead, different choices of G_0 can result in drastically different outcomes with regard to the mixing time. For ease of presentation, we assume in Table 1 that $G_0 \sim \mathcal{G}(n,\tilde{p})$, but this assumption can usually be relaxed, and we refer to the full statement of the corresponding results for our actual assumptions on G_0.

Next, we formally state the four main results of our work. The formal definitions of mixing time for random walk on dynamic graphs will be presented in Sect. 2.1 (see in particular Definition 2.7 and Definition 2.8). The first theorem is a negative result that tells us that, for fast-changing and sparse edge-Markovian graphs, random walks don't have finite mixing time. Its proof will be presented in Sect. 3.1.

Theorem 1.1 (Fast-changing and sparse, no mixing). *Let* $p = \Theta(1/n)$ *and* $q = \Omega(1)$. *Then,* $t_{\mathrm{mix}}(\mathcal{G}(n, p, q)) = \infty$.

The following theorem is a positive result that establishes fast mixing time in the dense and fast-changing regime. Its proof is presented in Sect. 3.2.

Theorem 1.2 (Fast-changing and dense, fast mixing). *Let* $p = \omega\,(\log n/n)$ *and* $q = \Omega(1)$. *Then,* $t_{\mathrm{mix}}(\mathcal{G}(n, p, q)) = O(\log n)$.

The only case missing in the fast-changing regime is the semi-sparse case, where nodes have average degree $d = \Theta(\log n)$. We do not have a definitive answer on the mixing time of random walks in such case, however, we do have a partial result that guarantees at least that random walk distributions will be "well spread" over a large support after $O(\log n)$ steps (we call this behavior *coarse mixing*). This statement can be made formal by considering the ℓ_2-norm of the distribution of the walk. Because of its technical nature, we defer the formal statement to Sect. 3.2 and Proposition 3.2.

We now turn our attention to the slowly-changing regime, where at most $\delta = O(\log n)$ edges are created and removed at each step. Unlike the results for the fast-changing regime, where the choice of the starting graph G_0 does not really affect the mixing time of a random walk (see Appendix A and Remark A.1 for a discussion), in the slowly-changing regime the choice of G_0 will affect the properties of G_t for a large number of steps t.

The following theorem shows that in the slowly-changing and dense regime, under mild conditions on the starting graph $G_0 = (V, E_0)$ (which are satisfied for G_0 drawn from the limiting distribution of dense $\mathcal{G}(n, p, q)$), random walks will mix relatively fast. We use $E_0(S, V \setminus S)$ to indicate the set of edges in G_0 between a subset of vertices $S \subset V$ and its complement, and Φ_{G_0} to indicate the minimum conductance of G_0 (see Definition 2.2).

Theorem 1.3 (Slowly-changing and dense, fast mixing). *Let* $d = \Omega(\log n)$, $p = O(\log n/n^2)$, *and* $q = O(\log n/(dn))$. *Let the following assumptions on the starting graph* $G_0 = (V, E_0)$ *be satisfied for large enough constants* $c_1, c_2, c_3 > 0$.

(1) $\deg_0(x) = \Theta(d)$ *for any* $x \in V$;
(2) $|E_0(S, V \setminus S)| \geq c_2 \log n |S|$, *for any* $S \subset V$ *with* $|S| \leq c_1 \log n$;
(3) $\Phi_{G_0} \geq c_3 \log d/d$.

Then, $t_{\mathrm{mix}}(\mathcal{G}(n, p, q)) = O(\log n/\Phi_{G_0}^2)$.

Let us briefly discuss the assumptions and results of Theorem 1.3. First of all notice that the parameters p and q are defined so that the average degree is $d = \Omega(\log(n))$ and the number of changes in the graph at each step is $\delta = O(\log(n))$. Assumption (1) just requires the degree of the vertices in G_0 to be of the same order as the degree of the vertices in the limiting graph $\mathcal{G}(n, \tilde{p})$. Assumption (2) guarantees that for any small set S there are enough edges going from S to the rest of the graph. Assumption (3) is a mild condition on the conductance of G_0. The last two assumptions ensure that the conductance of G_t will not be much lower than the conductance of G_0 for a large number of steps t. Finally, notice that $O(\log n/\Phi_{G_0}^2)$ is a classic bound for the mixing time of a *static* random walk on G_0. Theorem 1.3 essentially states that, if the three assumptions are satisfied, the mixing time of a random walk on $\mathcal{G}(n, p, q)$ will not be much larger. In particular, all the three assumptions are satisfied for a starting graph $G_0 \sim \mathcal{G}(n, \tilde{p})$ with $\tilde{p} = p/(p + q)$. Furthermore, in such case $t_{\mathrm{mix}}(\mathcal{G}(n, p, q)) = O(\log n)$. The proof of this theorem can be found in Sect. 4.1.

We conclude this section by stating our result in the slowly-changing and dense regime. We prove a negative result: we show that the mixing time of $\mathcal{G}(n, p, q)$ is at least linear in n.

Proposition 1.4 (Slowly-changing and sparse, slow mixing). *Let $p = O(1/n^2)$ and $q = \omega(1/(n \log n))$. Consider a random walk on $\mathcal{G}(n, p, q)$ with starting graph $G_0 \sim \mathcal{G}(n, \tilde{p})$ with $\tilde{p} = p/(p + q)$. Then, $t_{\mathrm{mix}}(\mathcal{G}(n, p, q)) = \Omega(n)$.*

2   Notation and Definitions

2.1   Random Walk and Conductance

In this section we introduce the relevant notation and basic results about Markov chains that we will use throughout the paper. For more background on Markov chains and random walks we defer the reader to [16].

Let $\mathscr{G} = (G_t)_{t \in \mathbb{N}}$ be a sequence of undirected and unweighted graphs defined on the same vertex set V, with $|V| = n$, but with potentially different edge-sets E_t ($t \in \mathbb{N}$). We study (lazy) random walks on $\mathscr{G}$: suppose that at a time $t \geq 0$ a particle occupies a vertex $u \in V$. At step $t + 1$ the particle will remain at the same vertex u with probability $1/2$, or will move to a random neighbor of u in G_t. In other words, it will perform a single random walk step according to a transition matrix P_t, which is the transition matrix of a lazy random walk on G_t: $P_t(u, u) = 1/2$, $P_t(u, v) = 1/(2 \deg_t(u))$ (where $\deg_t(u)$ is the degree of u in G_t) if there is an edge between u and v in G_t, or $P_t(u, v) = 0$ otherwise.

Given an initial probability distribution $\mu_0 \colon V \to [0, 1]$, which is the distribution of the initial position of the walk, the t-step distribution of a random walk on $\mathscr{G}$ is equal to $\mu_t = \mu_0 P_1 \cdot P_2 \cdot \ldots \cdot P_t$. In particular, we use μ_t^x to denote the t-step distribution of the random walk starting at a vertex $x \in V$. Hence $\mu_0^x(x) = 1$ and $\mu_0^x(y) = 0$ for $x \neq y \in V$. Furthermore, we use π_t to denote the probability distribution with entries equal to $\pi_t(x) = \deg_t(x)/(2|E_t|)$ for any $x \in V$. This distribution is stationary for P_t (i.e, it satisfies $\pi_t P_t = \pi_t$) and, if G_t is connected, it is the unique stationary distribution of P_t. If G_t is disconnected, P_t will have multiple stationary distribution. However, unless stated otherwise, we will consider only the "canonical" stationary distribution π_t. Finally, while any individual P_t is *time-reversible* (it satisfies $\pi_t(x) P_t(x, y) = \pi_t(y) P_t(y, x)$ for any $x, y \in V$), a random walk on $\mathcal{G}$ may not.[2]

---

[2] For example, it might happen that $P_1 \cdots P_t(x, y) > 0$ while $P_1 \cdots P_t(y, x) = 0$. This cannot happen in the "static" case where $P_1 = \cdots = P_t = P$ with P reversible.

Recall that if P is a transition matrix of a reversible Markov chain, it has n real eigenvalues, which we denote with $-1 \le \lambda_n(P) \le \cdots \le \lambda_1(P) = 1$. If P is the transition matrix of a lazy random walk on a graph G, it holds that $\lambda_n(P) \ge 0$. Moreover, $\lambda_1(P) < 1$ if and only if G is connected.

For two probability distributions $f, g \colon V \to [0,1]$, the *total variation distance* between f and g is defined as $\|f - g\|_{TV} := \frac{1}{2} \sum_{x \in V} |f(x) - g(x)|$. We denote with $\|f\|_2 = \left(\sum_{x \in V} f^2(x) \right)^{1/2}$ and $\|f\|_\infty = \max_{x \in V} |f(x)|$ the standard ℓ_2 and ℓ_∞ norms of f. Given a probability distribution $\pi \colon V \to \mathbb{R}_+$, we also define the $\ell_2(\pi)$-norm as $\|f\|_{2,\pi} := \sqrt{\sum_{x \in V} f^2(x) \pi(x)}$. By Jensen's inequality, it holds for any f, g that $2 \cdot \|f - g\|_{TV} \le \|f - g\|_{2,\pi}$. The lemma below relates the decrease in the distance to stationarity after one random walk step to the spectral properties of its transition matrix.

Lemma 2.1 (Lemma 1.13 in [18], rephrased). *Let P be the transition matrix of a lazy random walk on a graph $G = (V, E)$ with stationary distribution π. Then, for any $f \colon V \to \mathbb{R}$, we have that*

$$\left\| \frac{fP}{\pi} - 1 \right\|_{2,\pi}^2 \le \lambda_2(P)^2 \left\| \frac{f}{\pi} - 1 \right\|_{2,\pi}^2.$$

In the lemma above and throughout the paper, a division between two functions is to be understood entry-wise, while 1 refers to a function always equal to one. An important quantity which can be used to obtain bounds on $\lambda_2(P)$ is the *conductance* of G, which is defined as follows.

Definition 2.2 *The conductance of a non-empty set $S \subseteq V$ in a graph G is defined as:*

$$\Phi_G(S) := \frac{|E(S, V \setminus S)|}{\mathrm{vol}(S)},$$

where $\mathrm{vol}(S) := \sum_{x \in V} \deg(x)$ *and $E(S, V \setminus S)$ is the set of edges between S and $V \setminus S$. The conductance of the entire graph G is defined as*

$$\Phi_G := \min_{\substack{S \subset V: \\ 1 \le \mathrm{vol}(S) \le \mathrm{vol}(V)/2}} \frac{|E(S, V \setminus S)|}{\mathrm{vol}(S)}.$$

The conductance of G and the second largest eigenvalue of the transition matrix P of a lazy random walk in G are related by the so-called discrete Cheeger inequality [1], which we state below.

Theorem 2.3 (Cheeger inequality). *Let P be the transition matrix of a lazy random walk on a graph G. Then, it holds that*

$$1 - \lambda_2(P) \le \Phi_G \le 2\sqrt{1 - \lambda_2(P)}.$$

Finally, we use the notation $o_n(1)$ to denote any function $f : \mathbb{N} \to \mathbb{R}$ such that $\lim_{n \to +\infty} f(n) = 0$. We often drop the subscript n.

2.2   Dynamic Graph Models

In this section we formally introduce the random models of (dynamic) graphs that are the focus of this work. We start by recalling the definition of the Erdős-Rényi model of (static) random graphs.

Definition 2.4 (Erdős-Rényi model). $G = (V, E) \sim \mathcal{G}(n, p)$ *is a random graph such that* $|V| = \{1, \ldots, n\}$ *and the* $\binom{n}{2}$ *possible edges appear independently, each with probability* p.

We now introduce the *edge-Markovian* model of dynamic random graphs, which has been studied both in the context of information spreading in networks [5,6] and random walks [15]. This model is the focus of our work.

Definition 2.5 (Edge-Markovian model). *Given a starting graph* G_0, *we denote with* $(G_t)_{t \in \mathbb{N}} \sim \mathcal{G}(n, p, q)$ *a sequence of graphs such that* $G_t = (V, E_t)$, *where* $V = \{1, \ldots, n\}$ *and, for each* $t \in \mathbb{N}$, *any pair of distinct vertices* $u, v \in V$ *will be connected by an edge in* G_t *independently at random with the following probability:*

$$\mathbb{P}\left[\{u, v\} \in E_{t+1} \mid G_t\right] = \begin{cases} 1 - q & \text{if } \{u, v\} \in E_t \\ p & \text{if } \{u, v\} \notin E_t. \end{cases}$$

Notice that different choices of a starting graph G_0 will induce different probability distributions over $(G_t)_{t \in \mathbb{N}}$. In general, we try to study $\mathcal{G}(n, p, q)$ by making the fewest possible assumptions on our choice of G_0. Moreover, as pointed out for example in [15], $(G_t)_{t \in \mathbb{N}} \sim \mathcal{G}(n, p, q)$ converges to $\mathcal{G}(n, \tilde{p})$ with $\tilde{p} = p/(p + q)$. We leave considerations about the speed of this convergence and how this affects our choice of G_0 to Appendix A and, in particular, Remark A.1.

2.3   Mixing Time of Random Walks on Dynamic Graphs

One of the most studied quantities in the literature about time-homogeneous (i.e., static) Markov chains (random walks included) is the mixing time, i.e., the time it takes for the distribution of the chain to become close to stationarity. Formally, it is defined as follows.

Definition 2.6 (Mixing time for time-homogeneous Markov chains). *Let* μ_t^x *be the* t-*step distribution of a Markov chain with state space* V *starting from* $x \in V$. *Let* π *be its stationary distribution. For any* $\epsilon > 0$, *the* ϵ-*mixing time is defined as*

$$t_{\mathrm{mix}}(\epsilon) := \min\{t \in \mathbb{N} : \max_{x \in V} \|\mu_t^x - \pi\|_{TV} \leq \epsilon\}.$$

A basic fact in random walk theory states that a lazy random walk on a connected undirected graph $G = (V, E)$ has always a finite mixing time. In particular, if $|V| = n$, $t_{\mathrm{mix}}(1/4) = O(n^3)$. Moreover, considering a different ϵ does not significantly change the mixing time: for any $\epsilon > 0$,

$t_{\mathrm{mix}}(\epsilon) = O(t_{\mathrm{mix}}(1/4)\log(1/\epsilon))$ (see, e.g., [16]). Also, it is a well-known fact that $\|\mu_t^x - \pi\|_{TV}$ is non-increasing.

However, in the case of random walks on dynamic graphs, convergence to a time-invariant stationary distribution does not, in general, happen. For this reason, other works have studied alternative notions of mixing for dynamic graphs, such as merging [23], which happens when a random walk "forgets" the vertex where it started. In this work, instead, we focus on a different approach that we believe best translates the classical notion of mixing from the static to the dynamic case. More precisely, let us consider a dynamic sequence of graphs $(G_t)_{t \in \mathbb{N}}$ with corresponding stationary distributions $(\pi_t)_{t \in \mathbb{N}}$. Our goal is to establish if there exists a time t such that the distribution μ_t of the walk at time t is close to π_t. Moreover, to make this notion of mixing useful in possible applications, we require that μ_s remains close to π_s for a reasonably large number of steps $s \geq t$. Formally, we introduce the following definition of mixing time for dynamic graph sequences.

Definition 2.7 (Mixing time for dynamic graph sequences). *Let $\mathcal{G} = (G_t)_{t \in \mathbb{N}}$ be a dynamic graph sequence on a vertex set V, $|V| = n$. The mixing time of a random walk in $\mathcal{G}$ is defined as*

$$t_{\mathrm{mix}}(\mathcal{G}) = \min\left\{ t \in \mathbb{N} \colon \forall s \in [t, t + \sqrt{n}), \ \forall x \in V, \ \|\mu_s^x - \pi_s\|_{TV} = o_n(1) \right\},$$

where π_s is the stationary distribution of a random walk in G_s, and μ_s^x is the s-step distribution of a random walk in $\mathcal{G}$ that started from $x \in V$.

First observe we require that the total variation distance between μ_s and π_s goes to zero as the number of vertices increases.[3] This is motivated by the fact that the distance to stationarity, unlike in the static case, might not tend to zero as the number of steps t goes to infinity. However, we ask that the distance to stationarity is smaller than a threshold which decreases for larger sized graphs. Secondly, we require that such distance remains small for $\sqrt{n}$ steps (recall n is the number of vertices in the graph). This is due to the fact that, for all dynamic graph models we consider, we cannot hope for such distance to stay small arbitrarily long. However, we believe that $\sqrt{n}$ steps is a long enough period of time for mixing properties to be useful in applications.

Since our goal is to study the mixing property of $\mathcal{G}(n, p, q)$, we now introduce a definition of mixing time for edge-Markovian models that takes into account the probabilistic nature of such graph sequences. Essentially, we say that the mixing time of $\mathcal{G}(n, p, q)$ is t if a random walk on a dynamic sequence of graphs sampled from $\mathcal{G}(n, p, q)$ mixes (according to the previous definition) in t steps with high probability over the sampled dynamic graph sequence.

Definition 2.8 (Mixing time for edge-Markovian models). *Given an edge-Markovian model $\mathcal{G}(n, p, q)$, its mixing time is defined as*

$$t_{\mathrm{mix}}(\mathcal{G}(n, p, q)) = \min\left\{ t \in \mathbb{N} \colon \mathbb{P}_{\mathscr{G} \sim \mathcal{G}(n,p,q)} \left[t_{\mathrm{mix}}(\mathscr{G}) \leq t \right] \geq 1 - o_n(1) \right\}.$$

---

[3] We are implicitly assuming there is an infinite family of dynamic graph sequences with increasing n.

Finally, we remark that, while in static graphs connectivity is a necessary prerequisite to mixing, random walks on sequences of disconnected dynamic graphs might nonetheless exhibit mixing properties. Examples of this behavior were studied in [25].

3   Results for the Fast-Changing Case

3.1   Negative Result for Mixing in the Sparse and Fast-Changing Case

In this section we consider random walks on sparse and fast-changing edge-Markovian graphs. In particular, we study $\mathcal{G}(n, p, q)$ with $0 < q = \Omega(1)$ and $p = \frac{1}{n}$. Since $\Omega(1)$, by Remark A.1, we can restrict ourselves to consider the case where $G_0 \sim \mathcal{G}(n, \tilde{p})$ with $\tilde{p} = p/(p+q)$. We prove the following theorem.

Theorem 1.1 (Fast-changing and sparse, no mixing). *Let* $p = \Theta(1/n)$ *and* $q = \Omega(1)$. *Then,* $t_{\min}(\mathcal{G}(n, p, q)) = \infty$.

The key idea behind this result is that, due to the fast-changing nature of graphs in this model, the degrees of the nodes also change rapidly. In particular, for a linear number of nodes such as u, there is at least one neighbor $v_{\min}$ in the neighbors of u whose degree may change from one constant in round t to basically any other constant (this also makes use of the assumption on p, ensuring that the graph is sparse). The proof then exploits that, due to the "unpredictable" nature of this change, the probability mass received by $v_{\min}$ in round $t + 1$ is likely to cause a significant difference between $\mu_{t+1}(u)$ and $\pi_{t+1}(u)$. Since this holds for a linear number of nodes u, we obtain a sufficiently large lower bound on the total variation distance, and the theorem is established. The complete proof will appear in the full version of the paper.

3.2   Positive Result for Mixing in the Dense and Fast-Changing Case

In this section we analyse the mixing properties of $\mathcal{G}(n, p, q)$ for $p = \Omega(\log n/n)$ and $q = \Omega(1)$. Since q is large, for simplicity we will assume throughout this section that $G_0 \sim \mathcal{G}(n, \tilde{p})$, where $\tilde{p} = \frac{p}{p+q}$ (see Remark A.1 for an explanation of why this is not a restriction). The following theorem is the main result.

Theorem 1.2 (Fast-changing and dense, fast mixing). *Let* $p = \omega(\log n/n)$ *and* $q = \Omega(1)$. *Then,* $t_{\min}(\mathcal{G}(n, p, q)) = O(\log n)$.

While in this paper we study for simplicity only lazy random walks on graphs, to prove Theorem 1.2, however, we need to introduce *simple* random walks on graphs: given a graph $G = (V, E)$, a simple random walk on G has transition matrix Q such that, for any $x, y \in V$, $Q(x, y) = 1/\deg(x)$ if $\{x, y\} \in E$, $Q(x, y) = 0$ otherwise. The following lemma, whose proof is the main technical part of the section, shows that if the *simple* random walk on a sequence of graphs $\mathcal{G} = (G_t)_{t \in \mathbb{N}}$ exhibits strong expansion properties, and the time-varying stationary distribution is always close to uniform, then a *lazy* random walk on $\mathcal{G}$ will be close to the stationary distribution of G_t for any t large enough. Note that a strong expansion condition on lazy random walks can never be satisfied; luckily, we just need this strong expansion condition to hold for their simple counterpart.

Lemma 3.1. *Let $(G_t)_{t\in\mathbb{N}}$ be a sequence of graphs, and $(P_t)_{t\in\mathbb{N}}$ (resp. $(Q_t)_{t\in\mathbb{N}}$) the corresponding sequence of transition matrices for a lazy (resp. non-lazy) random walk. Assume there exists $1 < C = O(1)$ such that, for any $t \geq 1$ and any $x \in V$, $1/(C \cdot n) \leq \pi_t(x) \leq C/n$. Moreover, also assume that, for any $t \in \mathbb{N}$, $\max\{|\lambda_2(Q_t)|, |\lambda_n(Q_t)|\} \leq \lambda = o(1)$. Then, there exists an absolute constant C' such that, w.h.p., for any $t \geq C' \log n$ and any starting distribution μ_0,*

$$\left\| \frac{\mu_t}{\pi_t} - 1 \right\|_{2,\pi_t}^2 \leq 10C^2(C-1)^2,$$

where $\mu_t = \mu_0 P_1 \cdots P_t$.

We now show how it can be used to derive Theorem 1.2. First recall that since we are assuming $G_0 \sim \mathcal{G}(n, \tilde{p})$, all graphs in the sequence $(G_t)_{t\in\mathbb{N}}$ are sampled (non-independently) from $\mathcal{G}(n, \tilde{p})$ (see Appendix A). Furthermore, for any $t \in \mathbb{N}$, the assumptions of Theorem 1.2 on $\lambda_2(Q_t)$ and $\lambda_n(Q_t)$ are satisfied with probability $1 - o(1/n^2)$ for any graph sampled from $\mathcal{G}(n, \tilde{p})$ with $\tilde{p} > 2 \log n/n$ by [13, Theorem 1.1]. Moreover, for $\tilde{p} = \omega(\log n/n)$, by a standard Chernoff bound argument we can show that, with probability $1 - o(1/n^2)$, all vertices of a graph sampled from $\mathcal{G}(n, \tilde{p})$ have degree $(1 + o_n(1))n\tilde{p}$. This implies that, for any t, w.h.p, the stationary distribution of G_t satisfies the assumptions of Lemma 3.1 with $C = 1 + o(1)$, which yields Theorem 1.2.

It is natural to ask if we can relax the condition on p. Assume for example that p, q are such that $\tilde{p} = p/(p+q) > 2 \log n$. By [13, Theorem 1.1], the conditions on λ are still satisfied. However, it only holds that $C = \Theta(1)$. Therefore, Lemma 3.1 can only establish that the $\ell_2(\pi_t)$-distance to stationarity is a constant (potentially larger than 1). This, unfortunately, does not give us any meaningful bound on the total variation distance. However, if the ℓ_2-distance between two distributions μ and π is small, $\mu(x)$ cannot be much larger than $\pi(x)$. In a sense, this result can be interpreted as a *coarse* mixing property. This is summarised in the following proposition.

Proposition 3.2. *Let $(G_t)_{t\in\mathbb{N}} \sim \mathcal{G}(n, p, q)$ with $p/(p+q) > 2 \log n/n$ and $q = \Omega(1)$. Let π_t be the stationary distribution of G_t. Then, there exists absolute constants $c_1, c_2 > 0$ such that, for any starting distribution μ_0 and any $c_1 \log n \leq t \leq \sqrt{n} + c_1 \log n$, it holds that*

$$\mathbb{P}\left[\left\| \frac{\mu_t}{\pi_t} - 1 \right\|_{2,\pi_t}^2 \leq c_2 \right] \geq 1 - o_n(1).$$

4   Results for the Slowly-Changing Case

4.1   Positive Result for Mixing in the Dense and Slowly-Changing Case

The aim of this section is to prove the following theorem.

Theorem 1.3 (Slowly-changing and dense, fast mixing). *Let $d = \Omega(\log n)$, $p = O(\log n/n^2)$, and $q = O(\log n/(dn))$. Let the following assumptions on the starting graph $G_0 = (V, E_0)$ be satisfied for large enough constants $c_1, c_2, c_3 > 0$.*

(1) $\deg_0(x) = \Theta(d)$ for any $x \in V$;
(2) $|E_0(S, V \setminus S)| \geq c_2 \log n|S|$, for any $S \subset V$ with $|S| \leq c_1 \log n$;
(3) $\Phi_{G_0} \geq c_3 \log d/d$.

Then, $t_{\mathrm{mix}}(\mathcal{G}(n, p, q)) = O(\log n/\Phi_{G_0}^2)$.

We start by stating that, if the three assumptions of Theorem 1.3 are satisfied, then, for any $t = O(nd \log n)$, the conductance of G_t is not much worse than the conductance of G_0 (with high probability).

Lemma 4.1 (Conductance lower bound). *Let $d = \Omega(\log n)$, $p = O(\log n/n^2)$, and $q = O(\log n/(dn))$. Assume that G_0 satisfies assumptions (1), (2), (3) of Theorem 1.3. Then, there exists a constant $c > 0$ such that, for any $t = O(nd \log n)$ and any vertex $v \in V$,*

$$\mathbb{P}\left[\deg_t(v) \leq \frac{1}{2}\deg_0(v)\right] = O(n^{-4})$$

and

$$\mathbb{P}[\Phi_{G_t} \geq c \cdot \Phi_{G_0}] = 1 - O(n^{-4}).$$

The proof of this lemma proceeds as follows: for any $S \subset V$, when an edge is randomly added or removed from the graph, we show that the probability that $|E_t(S, V \setminus S)|$ increases is usually larger than the probability it decreases. Therefore, we model $|E_t(S, V \setminus S)|$ as a random walk on $\mathbb{N}$ with a bias towards large values of $|E_t(S, V \setminus S)|$, i.e., a *birth-and-death* chain. Using standard arguments about birth-and-death chains, we show it is very unlikely that $|E_t(S, V \setminus S)|$ becomes much smaller than $|E_0(S, V \setminus S)|$. By a similar argument we also show that the degrees of all nodes in S are approximately the same as their original degrees in G_0. This ensures that the conductance of a single set S is preserved after $t = O(dn \log n)$ steps. We then use a union bound argument to show that, with high probability, the conductance of the entire graph is preserved. For certain values of d, however, we cannot afford to use a union bound on *all* possible sets of vertices. To overcome this, we show that only applying the union bound for connected sets S would suffice. By bounding the number of such sets with respect to the maximum degree in G_0, we establish the lemma.

We can now give an outline of the proof of Theorem 1.3. The idea is to show that $\left\|\frac{\mu_{t+1}}{\pi_{t+1}} - 1\right\|_{2,\pi_{t+1}}$ is smaller than $\left\|\frac{\mu_t}{\pi_t} - 1\right\|_{2,\pi_t}$ (unless the latter is already very small). We do this by first relating $\left\|\frac{\mu_t}{\pi_t} - 1\right\|_{2,\pi_t}$ with $\left\|\frac{\mu_{t+1}}{\pi_t} - 1\right\|_{2,\pi_t}$. More precisely, we can use Lemma 2.1 and Lemma 4.1 to show that the latter is smaller than the former by a multiplicative factor that depends on Φ_{G_0}. Then, we bound the difference between $\left\|\frac{\mu_{t+1}}{\pi_t} - 1\right\|_{2,\pi_t}$ and $\left\|\frac{\mu_{t+1}}{\pi_{t+1}} - 1\right\|_{2,\pi_{t+1}}$. In particular, by exploiting the fact that at each step only $O(\log n)$ random edges can be deleted with high probability, we are able to show that $\left\|\frac{\mu_{t+1}}{\pi_{t+1}} - 1\right\|_{2,\pi_{t+1}}$ is not much larger than $\left\|\frac{\mu_{t+1}}{\pi_t} - 1\right\|_{2,\pi_t}$. Finally, by putting together all these argument, we show that $\left\|\frac{\mu_t}{\pi_t} - 1\right\|_{2,\pi_t}$ is monotonically decreasing in t, at least until the walk is mixed. This establishes the theorem.

Proof of Theorem 1.3. We establish the theorem by showing that, unless $\left\|\frac{\mu_t}{\pi_t} - 1\right\|_{2,\pi_t}$ is already small, $\left\|\frac{\mu_t}{\pi_t} - 1\right\|_{2,\pi_t}$ will significantly decrease at each step. In particular we relate $\left\|\frac{\mu_t}{\pi_t} - 1\right\|_{2,\pi_t}$ to $\left\|\frac{\mu_{t+1}}{\pi_{t+1}} - 1\right\|_{2,\pi_{t+1}}$ in two steps:

(1) We lower bound the change between $\left\|\frac{\mu_t}{\pi_t} - 1\right\|_{2,\pi_t}$ and $\left\|\frac{\mu_{t+1}}{\pi_t} - 1\right\|_{2,\pi_t}$;
(2) We upper bound the difference between $\left\|\frac{\mu_{t+1}}{\pi_t} - 1\right\|_{2,\pi_t}$ and $\left\|\frac{\mu_{t+1}}{\pi_{t+1}} - 1\right\|_{2,\pi_{t+1}}$.

Step 1: The first step follows from a simple spectral argument. Indeed, by Lemma 2.1, we have that

$$\left\|\frac{\mu_{t+1}}{\pi_t} - 1\right\|_{2,\pi_t}^2 \leq \lambda_2^2(P_t) \left\|\frac{\mu_t}{\pi_t} - 1\right\|_{2,\pi_t}^2,$$

where $\lambda_2(P_t)$ is the second largest eigenvalue of P_t, the transition matrix of G_t.

Step 2: We now upper bound the expected difference between $\left\|\frac{\mu_{t+1}}{\pi_t} - 1\right\|_{2,\pi_t}$ and $\left\|\frac{\mu_{t+1}}{\pi_{t+1}} - 1\right\|_{2,\pi_{t+1}}$. In the following analysis we condition on the event that at any time t, $|E_t| \in [(1 - o(1))nd, (1 + o(1))nd]$ where $d = (n - 1)\tilde{p}$. This event happens with probability $1 - o(1)$ by Lemma 4.1. Recall that

$$\left\|\frac{\mu_t}{\pi_t} - 1\right\|_{2,\pi_t}^2 = \sum_{y \in V} \pi_t(y) \left(\frac{\mu(y)}{\pi_t(y)} - 1\right)^2 = \left(\sum_{y \in V} \frac{\mu_t^2(y)}{\pi_t(y)}\right) - 1.$$

Hence, we have that

$$
\mathbb{E}\left[\left\|\frac{\mu_{t+1}}{\pi_{t+1}}-1\right\|_{2,\pi_{t+1}}^2 - \left\|\frac{\mu_{t+1}}{\pi_t}-1\right\|_{2,\pi_t}^2\right]
$$

$$
= \sum_{y\in V}\mathbb{E}\left[\mu_{t+1}^2(y)\left(\frac{1}{\pi_{t+1}(y)}-\frac{1}{\pi_t(y)}\right)\right]
$$

$$
= \sum_{y\in V}\mathbb{E}\left[\mu_{t+1}^2(y)\left(\frac{2|E_{t+1}|}{\deg_{t+1}(y)}-\frac{2|E_t|}{\deg_t(y)}\right)\right]
$$

$$
\le 2(1+o(1))|E|\sum_{y\in V}\mu_{t+1}^2(y)\mathbb{E}\left[\left(\frac{1}{\deg_{t+1}(y)}-\frac{1}{\deg_t(y)}\right)\right] \tag{4.1}
$$

$$
\le 2(1+o(1))|E|\sum_{y\in V}\mu_{t+1}^2(y)\frac{(1-\frac{1}{2})\deg_t(y)}{\frac{1}{2}\deg_t(y)\cdot\deg_t(y)}(1-(1-q)^{\deg_t(y)}) \tag{4.2}
$$

$$
\le \frac{2(1+o(1))}{1-o(1)}\sum_{y\in V}\cdot\frac{\mu_{t+1}^2(y)}{\deg_t(y)/((1-o(1))|E|)}(1-(1-q)^{\deg_t(y)})
$$

$$
\le \frac{2(1+o(1))}{1-o(1)}\cdot(1-(1-q)^{\deg_t(y)})\sum_{y\in V}\frac{\mu_{t+1}^2(y)}{\pi_t(y)}
$$

$$
\le O\left(\frac{\log n}{n}\right)\left(\left\|\frac{\mu_{t+1}}{\pi_t}-1\right\|_{2,\pi_t}^2+1\right) \tag{4.3}
$$

where $|E|=nd$ and $d=(n-1)\tilde{p}$. From line (4.1) to line (4.2) we upper bound the expectation by only considering the cases where the difference is positive, i.e., $\deg_t(y)\ge\deg_{t+1}(y)$. In line (4.2), by Lemma 4.1 we know $\deg_{t+1}(y)$ will not be smaller than $\frac{1}{2}\cdot\deg_t(y)$ with probability $1-O(n^{-4})$. Moreover, the probability $1-(1-q)^{\deg_t(y)}$ is the probability that at least one of the edges connected to y at time t changes at $t+1$. In line (4.3), we hide unimportant constants in the O-notation and we use the inequality $(1-q)^{\deg_t(y)}\ge 1-q\cdot\deg_t(y)$. Since $q=O(\log n/(dn))$ by assumption, we get $O(\log n/n)$ in line (4.3).

By combining the two steps above we have

$$
\mathbb{E}\left[\left\|\frac{\mu_{t+1}}{\pi_{t+1}}-1\right\|_{2,\pi_{t+1}}^2 - \left\|\frac{\mu_t}{\pi_t}-1\right\|_{2,\pi_t}^2\right]
$$

$$
\le O\left(\frac{\log n}{n}\right)\left(\left\|\frac{\mu_{t+1}}{\pi_t}-1\right\|_{2,\pi_t}^2+1\right)-(1-\lambda_2^2(P_t))\left\|\frac{\mu_t}{\pi_t}-1\right\|_{2,\pi_t}^2
$$

$$
\le O\left(\frac{\log n}{n}\right)\left(\lambda_2^2(P_t)\left\|\frac{\mu_t}{\pi_t}-1\right\|_{2,\pi_t}^2+1\right)-(1-\lambda_2^2(P_t))\left\|\frac{\mu_t}{\pi_t}-1\right\|_{2,\pi_t}^2
$$

$$
\le \left(\frac{n+\log n}{n}\cdot\lambda_2^2(P_t)-1\right)\left\|\frac{\mu_t}{\pi_t}-1\right\|_{2,\pi_t}^2+O\left(\frac{\log n}{n}\right)
$$

Therefore, it holds that

$$\mathbb{E}\left[\left\|\frac{\mu_{t+1}}{\pi_{t+1}} - \mathbf{1}\right\|_{2,\pi_{t+1}}^2\right] \leq \left(\frac{n + \log n}{n}\right) \lambda_2^2(P_t) \cdot \left\|\frac{\mu_t}{\pi_t} - \mathbf{1}\right\|_{2,\pi_t}^2 + O\left(\frac{\log n}{n}\right).$$

By Theorem 2.3 and the laziness of the walk,

$$\frac{\Phi_{G_t}^2}{2} \leq 1 - \lambda_2(P_t) \leq 2\Phi_{G_t}.$$

Since we assume the conductance is lower bounded by $O(\log d/d)$, we have $\lambda_2(P_t) \leq 1 - O(\log^2 d/d^2)$ and hence $((n + \log n)/n)\lambda_2^2(P_t) \leq 1$. Therefore, in expectation, the ℓ_2 distance shrinks by a constant factor (unless it's already small in the first place). Therefore, by standard arguments, after $O(\log n/\Phi_{G_0}^2)$ rounds the expected distance to π_t is at most $O(\sqrt{\log n/n})$. By Lemma 4.1, we know this holds for $poly(n)$ time steps. Finally, it suffices to apply Markov's inequality and a union bound to show the expected distance is small with probability $1 - O(n^{-4})$ on a polynomially long time interval as required by Definition 2.7. $\square$

4.2 Negative Result for Mixing in the Sparse and Slowly Changing Case

Proposition 1.4 (Slowly-changing and sparse, slow mixing). *Let* $p = O(1/n^2)$ *and* $q = \omega(1/(n\log n))$. *Consider a random walk on* $\mathcal{G}(n, p, q)$ *with starting graph* $G_0 \sim \mathcal{G}(n, \tilde{p})$ *with* $\tilde{p} = p/(p + q)$. *Then,* $t_{\mathrm{mix}}(\mathcal{G}(n, p, q)) = \Omega(n)$.

Proof. Consider the graph $G_0 \sim \mathcal{G}(n, \tilde{p})$. Notice that $\tilde{p} = o(\log n/n)$ is well below the connectivity threshold of Erdős-Rényi random graphs. Therefore, with high probability, there is at least one isolated vertex in G_0; call this vertex u and assume the random walk starts from that vertex. The probability that u remains isolated in the steps $1, 2, \ldots, t$ is at least

$$(1 - p)^{(n-1)\cdot t} \geq (1 - O(1/n^2))^{(n-1)\cdot t} \geq 1 - O(t/n).$$

Therefore, with at least constant nonzero probability, there exists a constant $c > 0$ such that, for any $t \leq c \cdot n$, $\mu_t^u(u) = 1$. Since $\pi_t(u) = 0$, this implies that $\|\mu_t^u - \pi^t\|_{TV} = 1$. $\square$

Actually the proof reveals a stronger "non-mixing" property; if the random walk starts from a vertex that is isolated in G_0, then this vertex will remain isolated for $\Theta(1/(np))$ rounds in expectation, and in this case the random walk did not move at all!

5 Conclusion

In this work we investigated the mixing time of random walks on the edge-Markovian random graph model. Our results cover a wide range of different densities and speeds by which the graph changes. On a high level, these findings provide some evidence to the intuition that both "high density" and "slow change" correlate with fast mixing.

For further work, one interesting setting that is not fully understood is the semi-sparse ($d = \Theta(\log n)$) and fast-changing ($q = \Omega(1) > 0$) case. While we proved that the random walk achieves some coarse mixing in $O(\log n)$, we conjecture that strong mixing is not possible. Another possible direction for future work is, given the bounds on the mixing time at hand, to derive tight bounds on the cover time. Finally, it would be also interesting to study the mixing time in a dynamic random graph model where not all edge slots are present (similar to the models studied in [12,15], where the graph at each step is a random subgraph of a fixed, possibly sparse, network).

A   Mixing Times for the Graph Chain of Edge-Markovian Models

It is well known that the edge-Markovian graph model $\mathcal{G}(n, p, q)$ converges to an Erdős-Rényi model $\mathcal{G}(n, \tilde{p})$ where $\tilde{p} = \frac{p}{p+q}$, which is the stationary distribution of the original edge-Markovian model. The mixing time of the graph chain has not been proven formally in previous works. Hence, we provide a proof for the sake of completeness. We remark that since an edge-Markovian model is a time-homogeneous (i.e., static) Markov chain, the classical definition of mixing time (Definition 2.6) applies.

Theorem A.1 (Graph chain mixing time). *For an edge-Markovian model* $\mathcal{G}(n, p, q)$*, the graph distribution converges to the graph distribution of the random graph model* $\mathcal{G}(n, \tilde{p})$ *where* $\tilde{p} = \frac{p}{p+q}$*. For any* $\epsilon \in (0, 1)$*, the mixing time of the graph chain* $\mathcal{G}(n, p, q)$ *is* $t_{\mathrm{mix}}(\epsilon) = O\left(\frac{\log(n/\epsilon)}{\log(1/|1-p-q|)}\right)$ *for* $p + q \neq 1$*, and* $t_{\mathrm{mix}}(\epsilon) = 1$ *if* $p + q = 1$*.*

Proof. Every edge slot can be represented by a two-state (close/open) Markov chain with transition matrix

$$P = \begin{pmatrix} 1 - p & p \\ q & 1 - q \end{pmatrix}$$

and stationary distribution $\left(\frac{q}{p+q}, \frac{p}{p+q}\right)$. By using standard Markov chain arguments (see, e.g., [16, Chapter 1]), the distance to the stationary distribution shrinks at each step by a factor of $|1 - p - q|$, i.e.,

$$\|\mu_{t+1} - \pi\|_{TV} \leq |1 - p - q| \, \|\mu_t - \pi\|_{TV} \,.$$

Therefore, when $p + q \neq 1$, the mixing time $t_{\mathrm{mix}}(\epsilon)$ of this two-state Markov chain is $O\left(\frac{\log(1/\epsilon)}{\log(|1-p-q|)}\right)$ where $\epsilon < 1$. For all the $\binom{n}{2}$ edge slots, the time that all of them mix is $O\left(\frac{\log\binom{n}{2}+\log(1/\epsilon)}{\log(|1-p-q|)}\right)$. When $p+q=1$, instead, the graph mixes immediately, which confirms the fact that in this regime the graph model is equivalent to a sequence of independent graphs from $\mathcal{G}(n,\tilde{p})$. $\qquad\square$

Remark A.1. Theorem A.1 essentially tells us that, whenever at least one between p and q is large (e.g., $\Omega(1)$), the graph chain quickly converges to $\mathcal{G}(n,\tilde{p})$ with $\tilde{p}=\frac{p}{p+q}$. This suggests that for a fast-changing edge-Markovian model $\mathcal{G}(n,p,q)$ with $q=\Omega(1)$, we can consider w.l.o.g. the starting graph G_0 as sampled from $\mathcal{G}(n,\tilde{p})$.

References

1. Alon, N., Milman, V.D.: λ_1, isoperimetric inequalities for graphs, and superconcentrators. J. Combin. Theory Ser. B **38**(1), 73–88 (1985)
2. Augustine, J., Pandurangan, G., Robinson, P.: Distributed algorithmic foundations of dynamic networks. SIGACT News **47**(1), 69–98 (2016)
3. Avin, C., Koucký, M., Lotker, Z.: Cover time and mixing time of random walks on dynamic graphs. Random Struct. Algorithms **52**(4), 576–596 (2018)
4. Berenbrink, P., Giakkoupis, G., Kermarrec, A., Mallmann-Trenn, F.: Bounds on the voter model in dynamic networks. In: 43rd International Colloquium on Automata, Languages, and Programming (ICALP 2016). LIPIcs, vol. 55, pp. 146:1–146:15 (2016)
5. Clementi, A., Crescenzi, P., Doerr, C., Fraigniaud, P., Pasquale, F., Silvestri, R.: Rumor spreading in random evolving graphs. Random Struct. Algorithms **48**(2), 290–312 (2016)
6. Clementi, A., Monti, A., Pasquale, F., Silvestri, R.: Information spreading in stationary Markovian evolving graphs. IEEE Trans. Parallel Distrib. Syst. **22**(9), 1425–1432 (2011)
7. Clementi, A.E.F., Macci, C., Monti, A., Pasquale, F., Silvestri, R.: Flooding time of edge-markovian evolving graphs. SIAM J. Discrete Math. **24**(4), 1694–1712 (2010)
8. Clementi, A., Silvestri, R., Trevisan, L.: Information spreading in dynamic graphs. Distrib. Comput. **28**(1), 55–73 (2014). https://doi.org/10.1007/s00446-014-0219-2
9. Cooper, C.: Random walks, interacting particles, dynamic networks: randomness can be helpful. In: Kosowski, A., Yamashita, M. (eds.) SIROCCO 2011. LNCS, vol. 6796, pp. 1–14. Springer, Heidelberg (2011). https://doi.org/10.1007/978-3-642-22212-2_1
10. Denysyuk, O., Rodrigues, L.: Random walks on evolving graphs with recurring topologies. In: Kuhn, F. (ed.) DISC 2014. LNCS, vol. 8784, pp. 333–345. Springer, Heidelberg (2014). https://doi.org/10.1007/978-3-662-45174-8_23
11. Giakkoupis, G., Sauerwald, T., Stauffer, A.: Randomized Rumor Spreading in Dynamic Graphs. In: Esparza, J., Fraigniaud, P., Husfeldt, T., Koutsoupias, E. (eds.) ICALP 2014. LNCS, vol. 8573, pp. 495–507. Springer, Heidelberg (2014). https://doi.org/10.1007/978-3-662-43951-7_42
12. Hermon, J., Sousi, P.: Random walk on dynamical percolation. arXiv preprint arXiv:1902.02770 (2019)
13. Hoffman, C., Kahle, M., Paquette, E.: Spectral gaps of random graphs and applications. International Mathematics Research Notices, May 2019

14. Kuhn, F., Oshman, R.: Dynamic networks: models and algorithms. SIGACT News **42**(1), 82–96 (2011)
15. Lamprou, I., Martin, R., Spirakis, P.: Cover time in edge-uniform stochastically-evolving graphs. Algorithms **11**(10), 15 (2018). (Paper No. 149)
16. Levin, D.A., Peres, Y.: Markov Chains and Mixing Times. American Mathematical Society, Providence (2017)
17. Michail, O., Spirakis, P.G.: Elements of the theory of dynamic networks. Commun. ACM **61**(2), 72 (2018)
18. Montenegro, R., Tetali, P.: Mathematical aspects of mixing times in Markov chains. Found. Trends Theor. Comput. Sci. **1**(3), x+121 (2006)
19. Peres, Y., Sousi, P., Steif, J.: Mixing time for random walk on supercritical dynamical percolation. Probab. Theory Relat. Fields **176**, 809–849 (2020). https://doi.org/10.1007/s00440-019-00927-z
20. Peres, Y., Stauffer, A., Steif, J.E.: Random walks on dynamical percolation: mixing times, mean squared displacement and hitting times. Probab. Theory Relat. Fields **162**(3–4), 487–530 (2015). https://doi.org/10.1007/s00440-014-0578-4
21. Saloff-Coste, L., Zúñiga, J.: Merging for time inhomogeneous finite Markov chains. I. Singular values and stability. Electron. J. Probab. **14**, 1456–1494 (2009)
22. Saloff-Coste, L., Zúñiga, J.: Merging for inhomogeneous finite Markov chains, Part II: Nash and log-Sobolev inequalities. Ann. Probab. **39**(3), 1161–1203 (2011)
23. Saloff-Coste, L., Zúñiga, J.: Merging and stability for time inhomogeneous finite Markov chains. In: Surveys in Stochastic Processes, pp. 127–151. EMS Series of Congress Reports, European Mathematical Society, Zürich (2011)
24. Sarma, A.D., Molla, A.R., Pandurangan, G.: Distributed computation in dynamic networks via random walks. Theor. Comput. Sci. **581**, 45–66 (2015)
25. Sauerwald, T., Zanetti, L.: Random walks on dynamic graphs: Mixing times, hitting times, and return probabilities. In: 46th International Colloquium on Automata, Languages, and Programming (ICALP 2019). LIPIcs, vol. 132, pp. 93:1–93:15 (2019)
26. Sousi, P., Thomas, S.: Cutoff for random walk on dynamical Erdos-Renyi graph. arXiv preprint arXiv:1807.04719 (2018)

Non-strict Temporal Exploration

Thomas Erlebach[ID] and Jakob T. Spooner[(✉)][ID]

School of Informatics, University of Leicester, Leicester, UK
jts21@leicester.ac.uk

Abstract. A *temporal graph* $\mathcal{G} = \langle G_1, ..., G_L \rangle$ is a sequence of graphs $G_i \subseteq G$, for some given *underlying graph* G of order n. We consider the *non-strict* variant of the TEMPORAL EXPLORATION problem, in which we are asked to decide if $\mathcal{G}$ admits a sequence W of consecutively crossed edges $e \in G$, such that W visits all vertices at least once and that each $e \in W$ is crossed at a *timestep* $t' \in [L]$ such that $t' \geq t$, where t is the timestep during which the previous edge was crossed. This variant of the problem is shown to be NP-complete. We also consider the hardness of approximating the exploration time for yes-instances in which our order-n input graph satisfies certain assumptions that ensure exploration schedules always exist. The first is that each pair of vertices are contained in the same component at least once in every period of n steps, whilst the second is that the temporal diameter of our input graph is bounded by a constant c. For the latter of these two assumptions we show $O(n^{\frac{1}{2}-\varepsilon})$-inapproximability and $O(n^{1-\varepsilon})$-inapproximability in the $c = 2$ and $c \geq 3$ cases, respectively. For graphs with temporal diameter $c = 2$, we also prove an $O(\sqrt{n}\log n)$ upper bound on worst-case time required for exploration, as well as an $\Omega(\sqrt{n})$ lower bound.

1 Introduction

Given a connected, undirected graph G of order n, an $O(n)$ upper bound on the length of a minimal walk that *explores* G (i.e., visits in an arbitrary order, all $v \in V(G)$ at least once) can be easily obtained by considering the length of an Euler tour around a spanning tree of G. The situation is altered considerably if we allow for the edge-set of the graphs in our input space to change over the course of some discretised time period, assuming that the vertex set remains constant at each point in this period. Such graphs have in recent years been referred to as *temporal, dynamic* or *time-varying*, and indeed it is known that there exist infinitely many graphs $\mathcal{G}$ of this sort that are connected at each point in time, and such that their exploration requires $\Omega(n^2)$ moves (where a move can consist of traversing an edge, or waiting at the current vertex), where $n = |V(\mathcal{G})|$ [8]. Due in large part to the frequency at which highly dynamic networks arise in the modelling of practical, real-life situations, an effort to better understand temporal graph models, along with the various optimisation problems defined upon them (e.g., the exploration problem considered here), has been made in recent years. For a more detailed introduction to the concept of temporal graphs and related combinatorial problems the reader is referred to [17].

© Springer Nature Switzerland AG 2020
A. W. Richa and C. Scheideler (Eds.): SIROCCO 2020, LNCS 12156, pp. 129–145, 2020.
https://doi.org/10.1007/978-3-030-54921-3_8

Much of the existing work regarding temporal graph exploration sees a temporal graph defined as a length-L sequence of static graphs $\mathcal{G} = \langle G_1, ..., G_L \rangle$, where each G_i has the same vertex set as some given *underlying graph* G, but can have an edge set that is a proper subset of $E(G)$. The TEMPORAL EXPLORATION problem (TEXP for short) then asks that, given a temporal graph $\mathcal{G}$ and some prespecified start vertex $s \in V(\mathcal{G})$, we compute a *foremost exploration schedule* starting from s – a sequence of edges crossed during strictly increasing timesteps (equivalently, at most one edge can be crossed per timestep), such that all vertices are visited at least once and that the timestep in which the last unvisited vertex is reached is minimal.

In this paper, we relax the condition that edges in a feasible exploration schedule must be crossed during strictly-increasing timesteps, and allow for any number of edges to be crossed in each step. Such a scenario arises for example in delay-tolerant networks [20]. Such networks tend to be disconnected at any time, and the speed at which the network topology changes is often much slower than the speed at which messages can be transmitted. Therefore, a mobile agent could visit any network node in its current connected component before the topology changes. It is clear that allowing for an agent to make an arbitrary number of moves across edges in a single time step alters the nature of the exploration problem considerably. In particular, it no longer makes sense to restrict our input space to always-connected graphs, since a trivial bound of a single step can be obtained by employing the same Euler tour-based technique that can be used to explore any static graph. As such, it is more natural to assume that a given input graph $\mathcal{G}$ consists of a number of disjoint components in each step. This, however, means that we cannot always guarantee that, for arbitrary $\mathcal{G}$ and start vertex $s \in V(\mathcal{G})$, $\mathcal{G}$ admits an exploration schedule starting at s. Given that we relax the requirement that edges be crossed in strictly-increasing timesteps, we dub the problem of deciding, in this model, whether or not a given temporal graph $\mathcal{G}$ admits an exploration schedule NON-STRICT TEXP, showing it to be NP-complete in general. We then consider two seemingly natural assumptions regarding the connectivity of the vertices which, when satisfied by our input graph, ensure that exploration is always possible. The first of these (which we name *pairwise vertex-togetherness*) posits that every pair of vertices will be contained in the same component at least once every n steps, where n is the graph's order – we prove $O(n^{1-\epsilon})$-inapproximability in this case. The second assumption insists that every pair of vertices in our input graph are able to reach one another in at most a constant c many steps. We note that this is equivalent to insisting that our input graph have temporal diameter bounded by a constant c (using the natural adaptation of the definition of temporal diameter from [19] to the non-strict model). For the latter assumption an obvious $O(n)$ upper bound on exploration time exists, and we show that when $c \geq 3$ this bound is in fact tight. For $c = 2$, we prove upper and lower bounds of $O(\sqrt{n} \cdot \log n)$ and $\Omega(\sqrt{n})$ respectively, leaving just a $\Theta(\log n)$ factor's gap between the two. Amongst other things, we also consider the hardness of approximating optimal solutions for the cases of temporal diameter 2 and ≥ 3, and lower bounds showing

$O(n^{\frac{1}{2}-\varepsilon})$ inapproximability in the former case and $O(n^{1-\varepsilon})$ inapproximability in the latter are provided (where n is the order of the input graph).

2   Related Work

Brodén et al. [3] considered the TEMPORAL TRAVELLING SALESPERSON PROBLEM on a graph with n vertices that is complete in every timestep, but had edge-costs which differed between 1 and 2 from step to step. Even when each edge's cost can change at most k times during the lifetime of the graph, they showed that the problem is NP-complete, but were able to provide a $(2 - \frac{2}{3k})$-approximation. Michail and Spirakis [19] showed the same problem to be APX-hard and provided an improved $(1.7 + \epsilon)$-approximation. Bui-Xuan et al. proposed in [4] a number of natural objectives to consider when computing a temporal walk/path, amongst which were *fastest* (minimum difference between departure and arrival time) and *foremost* (minimum arrival time) which is considered here. Also introduced in [19] was the TEMPORAL EXPLORATION problem (TEXP), by which we are asked to decide whether a given temporal graph admits an exploration schedule. They showed that this is NP-complete when no restrictions are placed on the input graph, and that even when the graph is connected in every timestep, approximating foremost exploration schedules with ratio $(2 - \epsilon)$ is NP-hard. Erlebach et al. [8] considered the TEMPORAL EXPLORATION problem under the *always-connected* property introduced in [19], improving the previously best-known inapproximability ratio to $O(n^{1-\epsilon})$. Complementing the aforementioned $\Omega(n^2)$ lower bound on the time needed to explore general always-connected temporal graphs they proved a $O(n^2)$ upper bound, as well as a number of subquadratic/superlinear upper/lower bounds for restricted subclasses of always-connected temporal graphs. In a similar vein, Bodlaender and van der Zanden [2] considered TEXP when the input graph has pathwidth at most 2 in every step, showing the decision variant to be NP-complete under these restrictions. In [1], Akrida et al. consider a variant of TEXP in which a candidate solution must return to the vertex from which it initially departed, focusing on the case in which the input graph has an underlying star. They gave an $O(n \log n)$-time algorithm deciding whether a given *temporal star* is explorable or not, under the restriction that each edge of the star is present in at most 2 or 3 time steps. In [15] and [14], the problem of temporal exploration is considered on the classes of temporal graphs with underlying cycles and cactuses, respectively. In [9], the authors prove an $O(dn^{1.75})$ bound on the number of time steps required to explore any temporal graph with degree bounded by d in each step, a considerable improvement over the previously best known $O(\frac{n^2 \log d}{\log n})$ bound [10]. Interestingly, the same bound can also be extended to general always-connected graphs when restrictions are relaxed and a computed exploration schedule is allowed to cross two edges in any given timestep – this is owed to the fact that the square of any static graph G admits a bounded-degree spanning tree. Notions of strict/non-strict paths which respectively allow for a single/infinitely many edge(s) to be crossed in any given timestep have been considered before, notably by Kempe et al. in [16] and Fluschnik et al. in [12].

Various other studies related to variants of exploration/path problems in temporal graphs have been considered. For example, the authors in [13] and [7] consider the problem of temporal exploration from a distributed standpoint. Casteigts et al. [6] consider a variant of the problem of finding a path between a given pair of vertices s and t, in which there is an upper bound on the number of timesteps that a computed path P is allowed to wait at any vertex $v \in P$ before crossing the next edge. For a more comprehensive overview of temporal graph problems and the various temporal graph classes on which they may be defined, the reader is referred to [5,17].

3   Graph Model and Problem Definition

Throughout the following denote by $[n]$ the set of integers $\{1, 2, ..., n\}$, and by $[x, n]$ $(x < n)$ the set of integers $\{x, x + 1, ..., n\}$. A standard way of defining a temporal graph within the literature is as a length-L sequence of graphs $\mathcal{G} = \langle G_1, ..., G_L \rangle$. Here L is the *lifetime* of the graph, and we require that for every $i \in [L]$, G_i is a subgraph of the *underlying graph* G of $\mathcal{G}$. In particular, we have that $V(G_i) = V(G)$ and $E(G_i) \subseteq E(G)$ for all $i \in [L]$.

As was previously noted, in the context of the non-strict variant of TEXP, it no longer makes sense to restrict our attention to the class of always-connected graphs. Therefore, we assume that for a given temporal graph $\mathcal{G} = \langle G_1, ..., G_L \rangle$, G_i $(i \in [L])$ consists of some number ≥ 1 of distinct connected components. Moreover, since any number of edges can be crossed in a given step, the edge structure of each component no longer matters – all that is important when attempting to compute an exploration schedule W is which component $C \in G_i$ is occupied by W in timestep i, since all $v \in C$ can be explored during that step. We can therefore, without loss of generality, use the following definition of a *non-strict temporal graph*:

Definition 1 (Non-strict temporal graph, $\mathcal{G}$). *A non-strict temporal graph $\mathcal{G} = \langle G_1, ..., G_L \rangle$ with vertex set $V := V(\mathcal{G})$ and lifetime L is an indexed sequence of partitions $G_i = \{C_{i,1}, ..., C_{i,s_i}\}$, with $i \in [L]$. For all $i \in [L]$, every $v \in V(G)$ satisfies $v \in C_{i,j_i}$ for a unique $j_i \in [s_i]$.*

Definition 2 (Non-strict temporal walk, W). *A length-k non-strict temporal walk $W = C_{1,j_1}, C_{2,j_2}, ..., C_{k,j_k}$ through a non-strict temporal graph $\mathcal{G} = \langle G_1, ..., G_L \rangle$ is a sequence of components C_{i,j_i} such that, for all $i \in [k]$, $C_{i,j_i} \in G_i$ and $j_i \in [s_i]$. Additionally, $k \in [L]$, where L is the lifetime of the graph upon which W is defined. We also require that $C_{i,j_i} \cap C_{i+1,j_{i+1}} \neq \emptyset$ for all $i \in [k-1]$, so that it is ensured that the $(i + 1)$-th component visited by W can be reached from the i-th component if and only if W ends step t_i at a vertex that lies in the intersection of these two components. For all $i \in [k]$ we say W visits all $v \in C_{i,j_i}$.*

A non-strict temporal walk $W = C_{1,j_1}, ..., C_{k,j_k}$ around a given graph $\mathcal{G}$ is an *exploration schedule* if and only if, for all $v \in V(\mathcal{G})$, there exists an $i \in [k]$

such that $v \in C_{i,j_i}$. Throughout the remainder of the paper, we may refer to non-strict temporal graphs and non-strict temporal walks simply as graphs and walks, respectively. Further, we might speak in terms of a *mobile agent* (or *agent*) which we assume to be, at any timestep, following a non-strict temporal walk around a given non-strict temporal graph in an attempt to explore it. We define the decision version of the NON-STRICT TEMPORAL EXPLORATION problem as follows:

Definition 3 (Non-Strict Temporal Exploration). *An instance of the* NON-STRICT TEMPORAL EXPLORATION *(NS-TEXP) problem is given as a tuple* $(\mathcal{G}, s)$, *where* $\mathcal{G}$ *is a given non-strict temporal graph with lifetime L and underlying graph G, and $s \in V(G)$. The problem then asks that we decide whether* $\mathcal{G}$ *admits an exploration schedule* $W = C_{1,j_1}, ..., C_{k,j_k}$ *starting from s, i.e., such that* $s \in C_{1,j_1}$.

If we consider only the set of yes-instances of NS-TEXP, i.e., those instances $(\mathcal{G}, s)$ such that $\mathcal{G}$ admits an exploration schedule starting from $s \in V(\mathcal{G})$, then it also makes sense for us to consider optimisation variants of NS-TEXP. In particular, we consider a variant FOREMOST-NON-STRICT TEXP (FNS-TEXP for short) which asks that we compute a *foremost* exploration schedule $W = C_{1,j_1}, ..., C_{k,j_k}$, i.e., one for which $k \leq l$ for any other exploration schedule $W' = C'_{1,j_1}, ..., C'_{l,j_l}$.

4 Deciding Whether Exploration Is Possible

Theorem 1. NON-STRICT TEMPORAL EXPLORATION *is NP-complete.*

Proof. The problem is in NP because an exploration schedule is of polynomial size (note that the input size is $\Omega(NL)$ for a temporal graph with N vertices and lifetime L, while an exploration schedule has size $O(NL)$) and its validity may be checked in polynomial time. To prove that the problem is NP-hard we give a reduction from 3SAT. Let instance I of 3SAT be given by variables $x_1, ..., x_n$ and clauses $c_1, ..., c_m$. Without loss of generality, assume that no clause contains x_i and $\bar{x}_i$ for any i. We proceed by creating a temporal graph $\mathcal{G}$ with vertex set $V(\mathcal{G}) = \{s\} \cup \{x_i^T, x_i^F : 1 \leq i \leq n\} \cup \{c_j : 1 \leq j \leq m\}$ and lifetime $2n$. The connected components of the graph in each step t are as follows (assume that all unmentioned vertices in each step are disconnected in $\mathcal{G}$): In step 1, let $\{s, x_1^T, x_1^F\}$ form one component. In every subsequent step $2i$ with $i \in [n]$, let there be a *true component* containing x_i^T and all clause vertices c_j that correspond to a clause of I which is satisfied by setting $x_i = 1$, as well as a *false component* containing x_i^F and all clause vertices c_j corresponding to clauses satisfied by setting $x_i = 0$. In all remaining steps $2i - 1$ for $i \in [2, n]$, let there be a component $\{x_{i-1}^T, x_{i-1}^F, x_i^T, x_i^F\}$. To complete the proof we show that there exists a satisfying assignment for I if and only if there exists an exploration schedule of $\mathcal{G}$.

($\Longrightarrow$) Since I is satisfiable, there exists a satisfying assignment $\alpha : X \to \{0, 1\}$ of boolean values to all $x_i \in X$. We claim that the following produces an exploration schedule W of $\mathcal{G}$: In the $(2i-1)$-th step ($i \in [n]$), W will be positioned in the component containing the vertices $\{x_{i-1}^T, x_{i-1}^F, x_i^T, x_i^F\}$ (as well as s in case $i = 1$). Explore both x_i^T and x_i^F, finishing at x_i^T if $\alpha(x_i) = 1$ or x_i^F is $\alpha(x_i) = 0$. At the start of step $2i$ ($i \in [n]$), W can either be in the component containing x_i^T or the one containing x_i^F (depending now on the value of $\alpha(x_i)$), and can explore all vertices c_j in that component and move back to x_i^T or x_i^F.

By definition of our reduction, the c_j explored by the produced walk W are precisely those which correspond to the clauses satisfied by assignment α. Since α is satisfying, each clause is satisfied by setting $x_i = \alpha(x_i)$ for at least one x_i; as such, W explores all vertices x_i^T, x_i^F and c_j as required.

($\Longleftarrow$) Let $\mathcal{G}$ be the input graph of the NON-STRICT TEXP instance produced by our reduction from I. Assume that $\mathcal{G}$ admits an exploration schedule W. By construction, for all $j \in [m]$, moves to and from vertices c_j can only be made during steps in which they are contained within the true/false component. Since W is an exploration schedule it must visit all c_j, and so for each j there must exist some $i \in [n]$ such that W initially reaches c_j within either the true or false component of step $2i$. Since each c_j is placed in the true/false component of step $2i$ only when $x_i = 1/x_i = 0$ satisfies the corresponding clause of I, a satisfying assignment for I can be obtained by checking, for every $i \in [n]$, whether W visits the true or false component, setting $x_i = 1$ in the former case and $x_i = 0$ in the latter. (Note that in steps $2i$ during which neither the true/false component are visited, we can choose an arbitrary setting for x_i.) $\square$

5   Exploration with Pairwise Vertex-Togetherness

We next consider instances of NS-TEXP for which the input graph $\mathcal{G}$ satisfies the following assumption, which we refer to as *pairwise vertex-togetherness*:

Assumption 1 (Pairwise vertex-togetherness). *All pairs of vertices $u, v \in V(\mathcal{G})$ are contained in the same connected component at least once during every period of N steps, where $N = |V(\mathcal{G})|$.*

The following algorithm enables us to explore any graph $\mathcal{G}$, with lifetime $L \geq N$, such that $\mathcal{G}$ satisfies Assumption 1: Start at the specified start vertex s, and in any of the steps $1 \leq i \leq N$ in which s is contained in the same connected component as some currently unexplored vertices, visit those vertices and move back to s by the end of step i. To see that this in fact produces an exploration schedule, observe that by Assumption 1 s will be contained in the same connected component as each $v \in V(\mathcal{G})$ at least once during the steps $1 \leq i \leq N$. Note that this also implies an N-approximation algorithm for the NON-STRICT TEXP problem (consider the instance in which the graph in the first step consists of a single connected component). In complement to this observation, we state the following result:

Theorem 2. *Even when the input graph $\mathcal{G}$ satisfies Assumption 1, it is **NP**-hard to approximate solutions to FOREMOST-NON-STRICT TEXP with ratio $\Theta(N^{1-\varepsilon})$ for any $\varepsilon > 0$, where N is the order of the input graph.*

Proof. Let $NST = \langle \mathcal{G}, s \rangle$ be any instance of the NON-STRICT TEXP decision problem, obtained by performing the reduction of Theorem 1 on an instance of 3SAT with n literals and m clauses. By the reduction, the graph $\mathcal{G}$ consists of $2n$ literal vertices, x_i^T and x_i^F ($1 \leq i \leq n$), m clause vertices c_j ($1 \leq j \leq m$), and an additional start vertex s. To reduce to FNS-TEXP from NST, we construct an instance $FNST = \langle \mathcal{G}', s' \rangle$ as follows: Let $V(\mathcal{G}') = V(\mathcal{G}) \cup \{d_1, ..., d_{n^c}\}$, where the vertices $d_1, d_2, ..., d_{n^c}$ are n^c dummy vertices (for some constant $c > 1$). Let $N = |V(\mathcal{G}')| = 2n + m + n^c$, and let $L = N$ be the lifetime of $\mathcal{G}'$. The components of $\mathcal{G}'$ in each step of its lifetime are defined to be as follows: In step 1, the graph consists of the connected component $\{s, d_1, d_2, ..., d_{n^c}\}$, and all clause and literal vertices lie disconnected in their own components. In the steps $t \in [2, 2n + 1]$, the step t components of $\mathcal{G}'$ are the same as the step $t - 1$ components of $\mathcal{G}$, and we create an additional n^c components containing each of the d_i. During the steps $t \in [2n + 2, N - 1]$, every vertex lies in one component on its own, and then in the N-th and final step all vertices belong to one single component.

Since, during the steps $t \in [2n + 2, N - 1]$, all vertices are disconnected in $\mathcal{G}'$, it follows that no new vertices can be explored during any of these steps. We therefore distinguish between the following two cases, showing that deciding whether $O(n)$ time steps suffice to explore $\mathcal{G}'$ or whether $\Theta(n^c)$ timesteps are required also decides whether or not $\mathcal{G}$ admits an exploration schedule.

$\mathcal{G}'$ **can be explored in** $2n + 1$ **steps:** By construction, none of the vertices c_j, x_i^T or x_i^F can be reached in $\mathcal{G}'$ from s until the start of the second step. Therefore, it must be that any exploration schedule with length $\leq 2n + 1$ starting at s visits all of these vertices during the steps $t \in [2, 2n + 1]$. Observe now that, by construction, the step $t \in [2, 2n + 1]$ components of $\mathcal{G}'$ are the step $t - 1$ components of $\mathcal{G}$, and that these steps constitute the entire lifetime of $\mathcal{G}$ – from this it follows that there must exist a valid exploration schedule of $\mathcal{G}$.

Exploring $\mathcal{G}'$ **requires** N **steps:** We claim that, since $\mathcal{G}'$ requires N steps to be explored completely, it must be that $\mathcal{G}$ admits no exploration schedule. To see this, recall once more that the step $t \in [2, 2n + 1]$ components of $\mathcal{G}'$ are the step $t - 1$ components of $\mathcal{G}$ (with each $t - 1 \in [1, 2n]$) and so it must be that no temporal walk W starting at s in time step 1 visits all vertices by the end of time step $2n$. Otherwise, we would be in case (1) and it would have been possible to explore $\mathcal{G}'$ by the end of step $2n + 1$ by visiting s and all d_i in step 1, then following an exploration schedule for $\mathcal{G}$ in $\mathcal{G}'$ during the steps $t \in [2, 2n + 1]$.

Since deciding whether or not $\mathcal{G}$ can be fully explored is NP-complete, it follows from the above case analysis that it is NP-hard to approximate solutions to FOREMOST NON-STRICT TEXP instances in which $\mathcal{G}$ satisfies Assumption 1) with ratio $\frac{\Theta(n^c)}{\Theta(n)} = \Theta(N^{\frac{c-1}{c}}) = \Theta(N^{1-\varepsilon})$, where $\varepsilon = \frac{1}{c} > 0$ can be forced arbitrarily close to 0 by choosing the constant c large enough. □

6   Exploration with Bounded Temporal Diameter

One further assumption which, when satisfied, ensures that complete exploration of a given temporal graph $\mathcal{G}$ is always possible (provided that the lifetime of $\mathcal{G}$ is suitably long) is the following:

Assumption 2 (Constant-bounded temporal diameter). *For every pair of vertices* $u, v \in V(\mathcal{G})$, *u can reach v in at most c steps (for some* $c = O(1)$) *and this holds from any time step t.*

An obvious upper bound on the number of steps required to fully explore any temporal graph $\mathcal{G}$ of order n such that $\mathcal{G}$ satisfies Assumption 2 (for a constant c), and $(\mathcal{G}, s)$ is a yes-instance of NS-TEXP, is $c(n-1)$: Starting from s, an agent in $\mathcal{G}$ could repeatedly select an arbitrary unexplored vertex and move to it in at most c steps, repeating this process $n-1$ times (once for each vertex $v \in V(\mathcal{G}) - \{s\}$).

6.1   Hardness of the Decision Problem for Temporal Diameter 2

The following result is concerned with the NP-completeness of deciding instances $(\mathcal{G}, s)$ of NS-TEXP in which $\mathcal{G}$ satisfies Assumption 2 for $c = 2$:

Theorem 3. *Deciding* NON-STRICT TEMPORAL EXPLORATION *is NP-complete, even when restricted to instances in which the input graph $\mathcal{G}$ satisfies Assumption 2 for $c = 2$.*

Proof. The reduction is from the NP-complete problem of 3SAT restricted to instances in which each literal occurs in at most 4 clauses, which we dub 3SAT* [21]. Given an instance I of 3SAT* comprised of $n \geq 3$ variables and $m = O(n)$ clauses, we proceed by constructing a non-strict temporal graph $\mathcal{G}$ (with lifetime $n + 3$) that satisfies the connectivity assumption for $c = 2$, such that $\mathcal{G}$ is fully explorable if and only if I is satisfiable. To do so we create 2 *literal vertices* x_i^T and x_i^F for each variable x_i of 3SAT* instance I. We then create $n+3$ *clause copy vertices* $c_{j,k}$ ($k \in [n+3]$) for all m clauses of I. Finally, we create $2(2n+m(n+3))^2$ many *connectivity vertices* v_i ($i \in [2(2n+m(n+3))^2]$) and divide them into two groups, the *red* group and the *blue* group, each of size $(2n+m(n+3))^2$. In steps 1 and 2, arrange the red connectivity vertices as a $2n+m(n+3)$ by $2n+m(n+3)$ grid, then let all rows of this grid lie in separate components during step 1, and all columns of the grid lie in separate components during step 2. Arbitrarily set the start vertex to be $s = x_1^T$. To the first row component in step 1, add the vertices x_i^T and x_i^F (for all $i \in [n]$), along with all blue connectivity vertices. Then, add each of the remaining clause vertices to a unique component (of which there are $2n + m(n+3) - 1$ remaining). In step 2, we now arrange the blue connectivity vertices as a $2n + m(n+3)$ by $2n + m(n+3)$ grid, with each of the step 2 components initially containing one red column and one blue row. Now, add each of the non-connectivity vertices (i.e., all literal and clause-copy vertices) to an arbitrary component, ensuring that no component contains more than one non-connectivity vertex (this is possible since there are $2n + m(n+3)$ such vertices and the same number of components). In all steps $t \in [3, n+3]$, we let the blue vertices alternate between being columns and rows, so that in each step there are exactly $2n + m(n+3)$ components. From step 3 onward, all red connectivity vertices will belong to a unique but arbitrarily selected component. All literal and clause-copy vertices should be added to one of the $2n + m(n+3)$

components in step 3. For all steps $t \in [4, n+3]$, add the literal vertex x_{t-3}^T to the first component, along with all clause-copies that correspond to a clause of 3SAT* instance I satisfied by setting $x_{t-3} = 1$; to the second component, add the literal vertex x_{t-3}^F alongside all clause-copies corresponding to a clause satisfied by setting $x_{t-3} = 0$. (We will from here onward refer to the 'first' and 'second' component as the *true* and *false* component.) In any of these steps, all remaining literal and clause-copy vertices should be assigned to an arbitrary but unique component that are neither the true or false component of that step. Clearly, $\mathcal{G}$ satisfies the $c = 2$ connectivity assumption, since in step 2 we have that all $2n + m(n+3)$ components contain exactly one of the red vertices in each of the step 1 components, and for all pairs of consecutive steps i and $i+1$ for $i > 1$, the same holds for the blue vertices. Therefore, starting in step i, it is possible to be positioned in any step $i+1$ component (and therefore reach any vertex in at most a single step) by moving to the appropriate red/blue vertex and waiting until the start of the next step. To complete the proof, we show that I is satisfiable if and only if $\mathcal{G}$ is explorable:

($\Longrightarrow$) To construct an exploration schedule of $\mathcal{G}$ from a satisfying assignment for 3SAT instance I, we can use the first three steps of $\mathcal{G}$'s lifetime in order to visit all blue/red connectivity vertices, as well as the literal vertices. For the remaining steps $t \in [4, n+3]$, visit the true component in step i if $x_{t-3} = 1$ or the false component otherwise; this is possible due to the connectivity vertices. It is clear, by arguments similar to those used in the proof of Theorem 1, that $\mathcal{G}$ is an exploration schedule since it was constructed from a satisfying assignment for I.

($\Longleftarrow$) First observe that no c_j can be reached until step 2. We have $n+3$ $c_{j,k}$ associated with the j-th clause of I; since there are only $n+2$ remaining steps it is not possible to visit all of the copies associated with the j-th clause in steps in which they are not contained in either the true or false component (i.e., one per timestep). Therefore, for all $j \in [m]$ there is at least one timestep in which ≥ 2 $c_{j,k}$ are visited whilst both contained in the true or false component, and so by construction the remaining $n+2$ copies can also be visited during that same step. Moreover, since W is an exploration schedule all $c_{j,k} \in V(\mathcal{G})$ ($j \in [m], k \in [n+3]$) must be visited; hence all clauses of I can be satisfied by setting variable $x_{t-3} = 1$ if W visits the true component in step $t \in [4, n+3]$, and $x_{t-3} = 0$ otherwise (an arbitrary setting of x_{t-3} suffices when neither are visited). $\qquad\square$

Due to the fact that any graph $\mathcal{G}$ satisfying Assumption 2 for some constant c also satisfies it for every $d > c$, we obtain as a corollary of Theorem 3 the following:

Corollary 1. *Deciding* NON-STRICT EXPLORATION *is* NP-complete *when restricted to graphs satisfying Assumption 2 for any $c \geq 3$.*

6.2   Lower Bounds on Exploration Time

The following three theorems are concerned with bounds on the amount of time required to explore graphs $\mathcal{G}$ that satisfy Assumption 2 for certain values of c. Throughout them, we consider only graphs $\mathcal{G}$ of order n with lifetime $L \geq c(n-1)$ in order to ensure that exploration is always possible.

Theorem 4. *There exists an infinite family of temporal graphs such that each member satisfies Assumption 2 for $c = 3$ (and thus also for all $c \geq 3$), has order $n \in \{7, 10, 13, ...\}$, and requires $\Omega(n)$ time steps to be explored in its entirety.*

Proof. Let $n = 3m + 1$ (for some $m \geq 2$) be the order of the temporal graph. Take an arbitrary $u \in V(\mathcal{G})$, then partition $V(\mathcal{G}) - \{u\}$ into three distinct parts $X = \{x_1, ..., x_m\}$, $Y = \{y_1, ..., y_m\}$ and $Z = \{z_1, ..., z_m\}$. In odd steps, we define the graph to consist of the components $X \cup \{u\}$ and $\{y_i, z_i\}$ for all $i \in [m]$. In even steps, it should consist of components $Y \cup \{u\}$ and $\{x_i, z_i\}$ for all $i \in [m]$. Furthermore, we (arbitrarily) set $s = x_1$. (An example of the construction can be seen in Fig. 1.) One can easily check that any pair of vertices in X (Y) can reach one another in at most 2 steps, and that vertices in X and Y are able to reach one another in at most 3 via u. Reaching any $x_j \in X$ ($y_j \in Y$) from any $z_i \in Z$ can also be achieved in at most three steps by waiting until the next step in which z_i and x_i (y_i) are contained in the same component, moving to x_i (y_i) in that step, and then to x_j (y_j) in the following step. Finally, consider reaching any vertex z_j from any vertex z_i ($i, j \in [m]$), starting at some time step t. Unless $i = j$, the quickest way to reach z_j from z_i is by first moving to the vertex $v \in \{x_i, y_i\}$ that lies in the same component as z_i in the current step. By construction, v will be contained in the same component as the vertex $v' \in \{x_j, y_j\}$; move to v' from v in step $t + 1$, finally moving to z_j from v' during step $t + 2$. In total this takes exactly 3 steps. Since any exploration schedule has to visit all $m = (n - 1)/3$ z_i at some point and reaching one from the previous takes exactly 3 steps, it follows that any exploration schedule of $\mathcal{G}$ has duration $\Omega(n)$. To complete the proof, observe that any graph that satisfies Assumption 2 for a constant b also satisfies Assumption 2 for any $c > b$.   □

One direct consequence of Theorem 4 is that the aforementioned $c(n - 1)$ upper bound on the length of exploration schedules in graphs satisfying Assumption 2 is in fact tight (asymptotically speaking) when $c \geq 3$. For the case in which $c = 2$, we now present a lower bound construction that requires $\Omega(\sqrt{n})$ steps to explore.

Theorem 5. *There exists an infinite family of graphs, the members of which satisfy Assumption 2 for $c = 2$, have order $n \in \{4, 9, 16, ...\}$, and require $\Omega(\sqrt{n})$ steps to be completely explored.*

Proof. Let $n = x^2$ for any $x \geq 2$. We now construct a graph $\mathcal{G}_n = \langle G_1, ..., G_L \rangle$, with $L = n$: Partition the vertex set into x parts of size x, arbitrarily labelling the vertices in the i-th part $v_{i,j}$ for $j \in [x]$. Arrange the vertices in the form of an x by x grid, with the first row consisting of the vertices $v_{1,1}, v_{1,2}, ..., v_{1,x}$, the

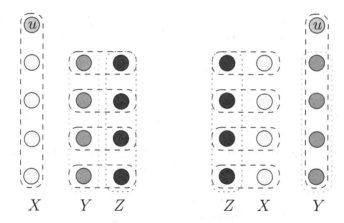

Fig. 1. The construction of Theorem 4 for $m = 4$. The left image is the graph in odd steps, with the graph in even steps displayed on the right. Black dashed lines mark the vertices contained in the components of either step.

second row of vertices $v_{2,1}, v_{2,2}, ..., v_{2,x}$, and so on and so forth (as in Fig. 2). We refer now to the collection of vertices $v_{1,j}, v_{2,j}, ..., v_{x,j}$ as the j-th column of the grid. Now, in every odd step $i = 1, 3, 5, ...$, let G_i be a partition of $V(\mathcal{G}_n)$ into the rows of the grid, and in every even step $i = 2, 4, 6, ...$, let G_i be a partition of $V(\mathcal{G}_n)$ into the columns. To see that $\mathcal{G}_n$ satisfies Assumption 2 (for $c = 2$), notice that in any pair of consecutive steps $t, t+1$ with $t \in [n-1]$, an agent can use one of those steps to change its row coordinate in the grid, and the other step to change its column coordinate.

To complete the proof, observe that in any step each component contains exactly $\sqrt{n}$ vertices. From this it follows that, during a single step, at most $\sqrt{n}$ unvisited vertices can be visited; hence at least $\Omega(\sqrt{n})$ steps are required of any exploration schedule. ☐

6.3   Upper Bounds on Exploration Time

The following result further concerns the case when $c = 2$, tightening the gap between the trivial $O(cn)$ upper bound and the $\Omega(\sqrt{n})$ lower bound of Theorem 5.

Theorem 6. *Any temporal graph $\mathcal{G}$ of order n satisfying Assumption 2 for $c = 2$ can always be explored in $O(\sqrt{n} \cdot \log n)$ time steps.*

Proof. We first show that for any consecutive pair of steps t and $t+1$, $\mathcal{G}$ consists of at most $\sqrt{n}$ components in at least one of these two steps. This is immediately obvious when each component of step t contains $\geq \sqrt{n}$ vertices (Case 1). Hence, we focus on the case in which at least one component of $\mathcal{G}$ during step t contains at most $\sqrt{n}$ vertices (Case 2). First, observe that in order for a graph to satisfy Assumption 2 for $c = 2$ it is required that, regardless of the time step t and the currently situated vertex/connected component, an agent can be positioned in any one of the step $t + 1$ components of $\mathcal{G}$ by the start of that same step. This implies that the number of connected components of $\mathcal{G}$ in step $t + 1$ is bounded from above by the size of the smallest component in step t (which by assumption of the case, is $\leq \sqrt{n}$).

Now, we show how to construct an exploration schedule W with duration $O(\sqrt{n} \cdot \log n)$. The idea is to divide the lifetime of $\mathcal{G}$ into consecutive blocks of three steps and within each, explore at least a $\frac{1}{\sqrt{n}}$ fraction of the currently unvisited vertices. More specifically, in the $(3j - 2)$-th step ($j \geq 1$), apply the above case analysis, taking $i = 3j - 1$ so that $i + 1 = 3j$. If Case 1 applies, let C^{max} be the component at time step $3j - 1$ which contains the largest number of previously unexplored vertices, resolving ties arbitrarily. Use time step $(3j - 2)$ to move to some vertex that is contained in C^{max} in step $3j - 1$ (by Assumption 2, this is always possible), exploring all unexplored vertices contained in C^{max} during step $3j - 1$. If Case 2 applies, let C^{max} be the component at time step $3j$ which contains the largest number of previously unexplored vertices, again resolving ties arbitrarily. In this case, wait at the current vertex until time step $3j - 1$, then move to some vertex contained in C^{max}, which is again possible by Assumption 2, and explore all unexplored vertices contained in C^{max} during time step $3j$. In either case, the graph consists of at most $\sqrt{n}$ components during the time step in which the agent is positioned in C^{max}. Let U be the set of previously unexplored vertices (which is initially the set $V - \{s\}$). The vertices in U are distributed amongst the $\leq \sqrt{n}$ components of step $3j - 1$ or $3j$, and since C^{max} contains the largest number of them, it follows that $|C^{max} \cap U| \geq \frac{|U|}{\sqrt{n}}$, as required.

Repeat the above process, exploring at least a $\frac{1}{\sqrt{n}}$ fraction of the previously unexplored vertices in each block of 3 consecutive steps until the number of unexplored vertices is less than 1. Since we began with $n - 1$ unexplored vertices (we consider s to be automatically explored), we get that after $k = 3x$ steps (for any $x \in \{1, 2, 3, ...\}$), the remaining number of previously unvisited vertices is at most $n \cdot (1 - 1/\sqrt{n})^k$. We require that $n \cdot (1 - \frac{1}{\sqrt{n}})^k < 1$, which can be transformed into $n < (\frac{\sqrt{n}}{\sqrt{n}-1})^k$. Taking the logarithm of both sides yields

$$\log n < k \cdot \log \left(\frac{\sqrt{n}}{\sqrt{n} - 1} \right) \iff k > \frac{\log n}{\log(1 + \frac{1}{\sqrt{n}-1})}.$$

Since $\log(1 + x) > x/(1 + x)$ for any $x > 0$, it then follows that the right-hand side of the previous inequality satisfies

$$\frac{\log n}{\log(1 + \frac{1}{\sqrt{n}-1})} < \frac{\log n}{\frac{1}{\sqrt{n}-1} / \frac{\sqrt{n}}{\sqrt{n}-1}} = \frac{\log n}{1/\sqrt{n}} = \sqrt{n} \cdot \log n.$$

Hence, as soon as the number of elapsed time steps k is greater than $\sqrt{n} \cdot \log n$, the number of remaining unexplored vertices is fewer than 1, and so the algorithm requires $O(\sqrt{n} \cdot \log n)$ steps to explore $\mathcal{G}$. $\qquad \square$

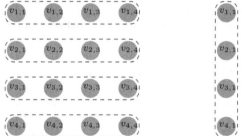 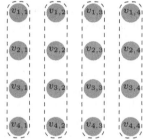

Fig. 2. The construction of Theorem 5 for $x = 4$. The left image is the graph in odd steps, with the graph in even steps displayed on the right. Blue dashed lines mark the vertices contained in the components of either step. (Color figure online)

6.4   Inapproximability Results

The constructive nature of the proof of Theorem 6 implies the existence of a $\Theta(\sqrt{N} \cdot \log N)$-approximation algorithm for instances of FOREMOST-NON-STRICT TEXP in which the given graph has order N and satisfies Assumption 2 for $c = 2$. The following result leaves only a $\Theta(\log N)$ gap between the best possible ratio achievable by any approximation algorithm for this problem and the ratio achieved by the algorithm implied by Theorem 6.

Theorem 7. *It is NP-hard to approximate solutions to instances of* FOREMOST-NON-STRICT *TEXP that satisfy Assumption 2 for $c = 2$ with approximation guarantee $O(N^{\frac{1}{2}-\varepsilon})$ for any $\varepsilon > 0$, where N is the order of the given graph.*

Proof. Take an arbitrary instance I of 3SAT* and let $\mathcal{G}$ be the corresponding instance of NON-STRICT TEXP generated via the construction of Theorem 3. Alter the construction so that there are now n^b ($b > 1$) clause-copies corresponding to each of the m clauses of 3SAT* instance I, and call the resulting graph $\mathcal{G}'$. Furthermore, let the lifetime L of $\mathcal{G}'$ be ∞, and define the components of the resulting graph $\mathcal{G}'$ (of order $N = 2n + mn^b + 2(2n + mn^b)^2 = O(n^{2b+2})$) to be the same, during the first $n + 3$ steps, as the components of $\mathcal{G}$ (but with the additional clause-copy vertices added to the appropriate components). During all subsequent steps $t \in [n + 4, \infty]$, let the blue connectivity vertices alternate between being arranged as the rows and columns of a $2n + mn^b$ by $2n + mn^b$ grid, adding exactly one of the non-connectivity vertices (i.e., the literal/clause-copy vertices) to each of the components formed by the blue rows/columns. Take the red connectivity vertices and add them all to one arbitrary component in every subsequent step.

Next, observe that, by the same reasoning used in the proof of Theorem 3, $\mathcal{G}'$ admits an exploration schedule of length at most $n + 3$ if and only if 3SAT* instance I is satisfiable. As a result, if $\mathcal{G}'$ cannot be fully explored by the end of the $(n + 3)$-th step, then it must be that there exists one clause c in 3SAT* instance I whose corresponding clause-copy vertices have not yet all been explored. Note now that at most $n+3$ of these clause-copy vertices can have been explored during the first $n+3$ steps, and so at least $n^b-(n+3)$ remain unexplored. During any step $t \in [n+4, \infty]$, at most one of these remaining clause-copies can be explored, and so it follows that $\Theta(n^b)$ steps are required to explore them all. This implies that deciding whether $\mathcal{G}'$ can be explored in $\Theta(n)$ steps or whether $\Theta(n^b)$ steps are required also decides whether or not 3SAT* instance I is satisfiable. As such, it follows that approximating solutions to NON-STRICT TEXP on graphs satisfying Assumption 2 for $c = 2$ with approximation ratio strictly better than $\Theta(n^b)/\Theta(n) = \Theta(N^{\frac{1}{2}}/n^2) = \Theta(N^{\frac{1}{2}}/N^{\frac{1}{b+1}}) = \Theta(N^{\frac{1}{2}-\varepsilon'})$ is NP-hard, where $\varepsilon' = \frac{1}{b+1}$ and can be made arbitrarily close to 0 by choosing b large enough. The theorem follows for any $\varepsilon > 0$ by forcing $\varepsilon' \geq \varepsilon$ arbitrarily close to ε.   □

Theorem 8. *It is NP-hard to approximate solutions to instances of* FOREMOST-NON-STRICT TEXP *by which Assumption 2 is satisfied for some* $c \geq 3$ *with approximation guarantee* $O(N^{1-\varepsilon})$ *for any* $\varepsilon > 0$, *where* N *is the order of the given graph.*

Proof. Let I be some instance of 3SAT* consisting of $n > 3$ variables v_i ($i \in [n]$) and $m = O(n)$ clauses. We wish to construct a non-strict temporal graph $\mathcal{G}$ (with lifetime $L = 3|V(\mathcal{G})|$) that satisfies Assumption 2 for $c = 3$ but no $d < 3$, and which admits an exploration schedule of length $O(n)$ if and only if I is satisfiable, otherwise requiring $\Omega(n^b)$ steps. To this end, we initially let $|V(\mathcal{G})| = N = 3mn^b + 1 = O(n^{b+1})$ for some $b \geq 2$. We take $2mn^b$ of the vertices in $V(\mathcal{G})$ and partition them into equisized sets $X = \{x_1, ..., x_{mn^b}\}$ and $Y = \{y_1, ..., y_{mn^b}\}$; let an additional vertex be known as u. The mn^b remaining vertices will be known as the *clause-copy* vertices $c_{j,k}$, with exactly n^b of them associated with j-th clause of I.

We now show how the components in each step of $\mathcal{G}$'s lifetime are to be arranged. In all steps $t \in [3n]$ such that $t \neq 3i$ for some $i \in [n]$, if t is odd, we place all $x \in X$ and u in the same connected component, whilst the mn^b vertices $y \in Y$ form a matching with the mn^b clause-copy vertices (this matching can be arbitrary, but will remain consistent in all considered steps). On the other hand, if $t \neq 3i$ and is even, then all $v \in Y \cup \{u\}$ form one component, whilst the vertices in X form a matching with the clause-copy vertices.

In all steps $t \in [3n]$ such that $t = 3i$ for some $i \in [n]$, if t is odd then all $v \in X \cup \{u\}$ form one connected component; let the start vertex $s = x_{mn^b}$. We create one component containing the vertex $y_1 \in Y$, along with all clause-copies corresponding to the clauses of I satisfied by setting $v_1 = 0$. To another component, we add the vertex $y_2 \in Y$, along with all clause-copies corresponding to the clauses of I satisfied by setting $v_1 = 1$. (In such steps, we will now refer to the components containing y_1 and y_2 as the 'true' and 'false' components, respectively.) All remaining clause-copies (i.e., those corresponding to clauses that are satisfied by neither a 0 nor 1 setting of v_i) will then form a matching with the remaining y_j ($j \in [mn^b] - \{1, 2\}$). (Note that there are always enough y_j to ensure this is possible, since at least one clause will be satisfied by either a 0 or 1 setting of each v_i, and so there can be at most $(m-1)n^b$ clause-copies to match with the $\leq mn^b - 2$ remaining y_j.) When t is even, the components are the same but the roles of sets X and Y are switched, so that now the components containing x_1 and x_2 are respectively the true and false components. During the steps $t \in [3n + 1, 3N]$, the components alternate between being arranged as they are in odd/even steps $t' \in [3n]$ such that $t' \mod 3 \neq 0$, depending on the parity of step t. It is straightforward to check that $\mathcal{G}$ satisfies Assumption 2 for $c = 3$. Moreover, consider any pair of clause-copies starting from any time step $t \geq 3n + 1$ and observe that 3 steps are in fact required to reach one from the other. We now demonstrate that I is satisfiable if and only if $\mathcal{G}$ is explorable in at most $3n$ steps, showing that at least $\Omega(n^b)$ steps are required otherwise.

($\implies$) By arguments similar to those used in the proof of Theorem 1 we are able to construct from a satisfying assignment α for I an exploration schedule W of $\mathcal{G}$ with length at most $3n$. To do so, we use the steps $t \in [3i - 2, 3i]$ to move to the true/false component in step $3i$ if α sets $v_i = 1/v_i = 0$, respectively.

($\impliedby$) By arguments similar to those used in the proof of Theorem 1, we construct an assignment α for I by setting $v_i = 1$ if W visits the true component during the $3i$-th step, or set $v_i = 0$ if the false component is visited (with an arbitrary setting for v_i if W visits neither). This works since, if some clause of I had all n^b of its associated copies explored separately in steps when not in the true/false component of $\mathcal{G}$, then it would take at least $n^b > 3n$ (for $b \geq 2$ and $n > 3$) steps to visit them all, a contradiction to W's length being $\leq 3n$. As such, for every $j \in [m]$ there must be an $i \in [n]$ such that >1 distinct copies associated with c_j, hence by construction all copies, are visited in the true/false component of step $3i$. It follows that α must be satisfying.

Moreover, if I has no satisfying assignment then any exploration schedule W must spend $\Theta(n^b)$ steps exploring all clause copies associated with ≥ 1 clause of I – otherwise there exists a schedule that visits all clause copies in a true/false component from which we could obtain a satisfying assignment for I. As a result, we may conclude that it is NP-hard to approximate instances of NON-STRICT TEXP which satisfy Assumption 2 for any $c \geq 3$ with ratio strictly better than $\Theta(n^b)/3n = \Theta(n^{b-1}) = O(N^{\frac{b-1}{b+1}}) = O(N^{1-\frac{2}{b+1}}) = O(N^{1-\varepsilon'})$, where $\varepsilon' = 2/(b + 1)$ can be made arbitrarily close to 0 by selecting b large enough. The theorem follows for any $\varepsilon > 0$ by forcing $\varepsilon' \geq \varepsilon$ arbitrarily close to ε. $\square$

7   Conclusion

We considered the problem of NON-STRICT TEMPORAL EXPLORATION, a variant of the TEMPORAL EXPLORATION problem in which the requirement that edges in a candidate exploration schedule are crossed at strictly increasing timesteps is weakened, so that an edge may be crossed at a timestep greater than or equal to the timestep in which the last was crossed. We showed that deciding NON-STRICT TEXP under these relaxed conditions is NP-complete.

The hardness of approximating solutions to FOREMOST-NON-STRICT TEXP when the input graphs satisfy either of two distinct vertex-connectivity assumptions was also considered. For order n graphs satisfying the *pairwise vertex-togetherness* assumption (Assumption 1), we proved that it is NP-hard to approximate solutions with ratio $O(n^{1-\varepsilon})$ for any $\varepsilon > 0$. For the second of these two assumptions, which posits that every pair of vertices can reach one another within $c = O(1)$ steps, we proved $O(n^{1-\varepsilon})$-inapproximability and $O(n^{\frac{1}{2}-\varepsilon})$-inapproximability in the $c \geq 3$ and $c = 2$ cases, respectively. Also shown was that, when $c = 2$, the graph of any yes-instance of NON-STRICT TEXP can be explored in at most $O(\sqrt{n}\log n)$ timesteps. In complement to this, a lower bound construction which requires of any exploration algorithm at least $\Omega(\sqrt{n})$ steps was described. Closing the remaining $\Theta(\log n)$ gap presents an interesting direction for future work, as does the analysis of exploration time for graphs satisfying other assumptions that ensure exploration of a graph $\mathcal{G}$ is always possible. For example, one could examine the effect of some of the connectivity/reachability-ensuring measures presented in [18] within the non-strict model; [11] and [8] also consider temporal graphs with periodically-repeating properties whose effects could be interesting to explore within the model considered here.

References

1. Akrida, E.C., Mertzios, G.B., Spirakis, P.G.: The temporal explorer who returns to the base. In: Heggernes, P. (ed.) CIAC 2019. LNCS, vol. 11485, pp. 13–24. Springer, Cham (2019). https://doi.org/10.1007/978-3-030-17402-6_2
2. Bodlaender, H.L., van der Zanden, T.C.: On exploring always-connected temporal graphs of small pathwidth. Inf. Process. Lett. **142**, 68–71 (2019). https://doi.org/10.1016/j.ipl.2018.10.016
3. Brodén, B., Hammar, M., Nilsson, B.J.: Online and offline algorithms for the time-dependent TSP with time zones. Algorithmica **39**(4), 299–319 (2004). https://doi.org/10.1007/s00453-004-1088-z
4. Bui-Xuan, B., Ferreira, A., Jarry, A.: Computing shortest, fastest, and foremost journeys in dynamic networks. Int. J. Found. Comput. Sci. **14**(2), 267–285 (2003). https://doi.org/10.1142/S0129054103001728
5. Casteigts, A., Flocchini, P., Quattrociocchi, W., Santoro, N.: Time-varying graphs and dynamic networks. IJPEDS **27**(5), 387–408 (2012)
6. Casteigts, A., Himmel, A.S., Molter, H., Zschoche, P.: The computational complexity of finding temporal paths under waiting time constraints. CoRR abs/1909.06437 (2019). https://arxiv.org/abs/1909.06437
7. Di Luna, G.A., Dobrev, S., Flocchini, P., Santoro, N.: Live exploration of dynamic rings. In: 36th IEEE International Conference on Distributed Computing Systems (ICDCS 2016), pp. 570–579. IEEE (2016). https://doi.org/10.1109/ICDCS.2016.59
8. Erlebach, T., Hoffmann, M., Kammer, F.: On temporal graph exploration. In: Halldórsson, M.M., Iwama, K., Kobayashi, N., Speckmann, B. (eds.) ICALP 2015. LNCS, vol. 9134, pp. 444–455. Springer, Heidelberg (2015). https://doi.org/10.1007/978-3-662-47672-7_36
9. Erlebach, T., Kammer, F., Luo, K., Sajenko, A., Spooner, J.T.: Two moves per time step make a difference. In: 46th International Colloquium on Automata, Languages, and Programming (ICALP 2019). LIPIcs, vol. 132, pp. 141:1–141:14. Schloss Dagstuhl-Leibniz-Zentrum für Informatik (2019). https://doi.org/10.4230/LIPIcs.ICALP.2019.141

10. Erlebach, T., Spooner, J.T.: Faster exploration of degree-bounded temporal graphs. In: 43rd International Symposium on Mathematical Foundations of Computer Science (MFCS 2018). LIPIcs, vol. 117, pp. 36:1–36:13. Schloss Dagstuhl-Leibniz-Zentrum für Informatik (2018). https://doi.org/10.4230/LIPIcs.MFCS.2018.36

11. Flocchini, P., Mans, B., Santoro, N.: On the exploration of time-varying networks. Theor. Comput. Sci. **469**, 53–68 (2013). https://doi.org/10.1016/j.tcs.2012.10.029

12. Fluschnik, T., Molter, H., Niedermeier, R., Renken, M., Zschoche, P.: Temporal graph classes: a view through temporal separators. Theor. Comput. Sci. **806**, 197–218 (2020). https://doi.org/10.1016/j.tcs.2019.03.031

13. Gotoh, T., Flocchini, P., Masuzawa, T., Santoro, N.: Tight bounds on distributed exploration of temporal graphs. In: 23rd International Conference on Principles of Distributed Systems (OPODIS 2019). LIPIcs, vol. 153, pp. 22:1–22:16. Schloss Dagstuhl - Leibniz-Zentrum für Informatik (2019). https://doi.org/10.4230/LIPIcs.OPODIS.2019.22

14. Ilcinkas, D., Klasing, R., Wade, A.M.: Exploration of constantly connected dynamic graphs based on cactuses. In: Halldórsson, M.M. (ed.) SIROCCO 2014. LNCS, vol. 8576, pp. 250–262. Springer, Cham (2014). https://doi.org/10.1007/978-3-319-09620-9_20

15. Ilcinkas, D., Wade, A.M.: Exploration of the T-interval-connected dynamic graphs: the case of the ring. In: Moscibroda, T., Rescigno, A.A. (eds.) SIROCCO 2013. LNCS, vol. 8179, pp. 13–23. Springer, Cham (2013). https://doi.org/10.1007/978-3-319-03578-9_2

16. Kempe, D., Kleinberg, J.M., Kumar, A.: Connectivity and inference problems for temporal networks. J. Comput. Syst. Sci. **64**(4), 820–842 (2002). https://doi.org/10.1006/jcss.2002.1829

17. Michail, O.: An introduction to temporal graphs: an algorithmic perspective. Internet Math. **12**(4), 239–280 (2016). https://doi.org/10.1080/15427951.2016.1177801

18. Michail, O., Chatzigiannakis, I., Spirakis, P.G.: Causality, influence, and computation in possibly disconnected synchronous dynamic networks. J. Parallel Distrib. Comput. **74**(1), 2016–2026 (2014). https://doi.org/10.1016/j.jpdc.2013.07.007

19. Michail, O., Spirakis, P.G.: Traveling salesman problems in temporal graphs. Theor. Comput. Sci. **634**, 1–23 (2016). https://doi.org/10.1016/j.tcs.2016.04.006

20. Sobin, C.C., Raychoudhury, V., Marfia, G., Singla, A.: A survey of routing and data dissemination in delay tolerant networks. J. Netw. Comput. Appl. **67**, 128–146 (2016). https://doi.org/10.1016/j.jnca.2016.01.002

21. Tovey, C.A.: A simplified NP-complete satisfiability problem. Discr. Appl. Math. **8**(1), 85–89 (1984). https://doi.org/10.1016/0166-218X(84)90081-7

Exploration of Time-Varying Connected Graphs with Silent Agents

Stefan Dobrev[1], Rastislav Královič[2(✉)], and Dana Pardubská[2]

[1] Slovak Academy of Sciences, Bratislava, Slovakia
Stefan.Dobrev@savba.sk
[2] Comenius University in Bratislava, Bratislava, Slovakia
{kralovic,pardubska}@dcs.fmph.uniba.sk

Abstract. Exploration is a fundamental task in mobile computing. We study the version where a group of cooperating agents is situated in a graph, and the task is to make sure that every vertex of the graph is visited by some agent. We consider discrete-time evolving graphs with an adaptive adversary: the adversary observes the actions of the agents, and can choose the graph for the next step arbitrarily with the only restriction that it must be connected. We are interested in solving the problem with weakest possible agents. We provide an exploration algorithm where the agents can not interact in any way among themselves or with the vertices (no messages, whiteboards, etc), and even don't sense each other. They are only aware of the others from the results of a mutual-exclusion mechanism in the vertices. We show that $2m - n + 1$ agents are sufficient, where m is the number of edges. Interestingly, $m - n + 1$ agents are needed even in an offline setting when they are controlled by a central entity.

We don't know whether the algorithm achieves polynomial exploration time. However, we provide a different algorithm that uses $O(n^4)$ agents in a slightly stronger model (the agents can observe the number of agents in a vertex, and their actions), but achieves the exploration time $O(n^9)$.

1 Introduction

The question how an agent can explore an unknown graph (i.e. visit all its vertices) is one of the older computer science problems. In 1977, Rosenkrantz et al. [21] studied graph exploration by an agent that can see the identifiers of neighboring vertices. In fact, the paper posed a question which, in spite of much effort, remains open till now. About the same time, exploration of graphs by means of various types of automata (finite, pebble, counter) has been studied, e.g. in [3]. Also, teams of cooperating automata have been considered (e.g. [4,20]). The main question in these works is what types of graphs can be explored by (teams of) automata of given type. Later works asked what are the memory requirements of an agent in order to explore any given graph (e.g. [2,12,14]). A variant of exploration where the agent has to visit all edges has also been considered ([1,7,9,19]).

The research was supported by Slovak grant VEGA 1/0601/20.

A. W. Richa and C. Scheideler (Eds.): SIROCCO 2020, LNCS 12156, pp. 146–162, 2020.
https://doi.org/10.1007/978-3-030-54921-3_9

All the mentioned results require the graph to have a distinct labelling of incident ports in every vertex. They differ in whether identifiers of vertices are available to the agents, or possibly identifiers of the neighboring vertices as well. Sometimes, the local port labelling is required to fulfill some global property (e.g. the sense of direction [10]). A series of papers [8,13,16,17,22] investigates how a properly chosen port labeling may help the exploration by a single agent.

When considering teams of agents, crucial role affecting the capabilities of the team play the means of mutual communication and synchronization. It is thus natural that the problem of exploration has been studied in various models: the agents may be synchronous or asynchronous, may communicate directly, indirectly (by means of whiteboards or tokens), or only implicitly (by means of observing positions and/or using mutual-exclusion mechanisms in vertices), etc. See [6] for a survey.

The problems connected with the changes of communication topology in time have been also in the center of the research attention for a long time (e.g. [23]). However, the various models of networks changing in time have proliferated recently. We follow the general framework of [5], where different models presented in the literature so far have been unified under the notion of *time-varying graphs* (TVG). In particular, we study what the Class 9 from [5] called *constant connected TVG*: the networks evolve in discrete time steps, and there are no constraints on the network apart from that in every time step, the snapshot must be a connected spanner of the underlying graph.

This model has been studied mostly for special topologies, such as rings in [18]. The study of the problem of exploration of unknown temporal graphs has commenced in [15], and our research is a continuation of this effort.

Contribution

We are mainly interested in the question which coordination/communication mechanisms are crucial in the exploration of TVGs. To answer this question, our goal is to study the agents with weakest possible communication abilities. First, we present an exploration algorithm that can explore an unknown temporal graph with $2m - n + 1$ agents, where m is the number of edges of the underlying graph. The main difference from the previous work is that our agents are silent, i.e. the only way an agent can be aware of the existence of other agents is when it is denied access to a link due to the mutual exclusion mechanism. Somewhat surprisingly, we also prove that increasing the communication abilities of the agents doesn't help too much, since $m - n + 1$ agents are needed even if they are controlled by a central authority that can observe the entire snapshot in every time step.

Unfortunately, we don't know much about the exploration time, since the proof of the correctness of the exploration algorithm yields only a superexponential bound on the exploration time. However, we present another exploration algorithm that achieves polynomial exploration time at the expense of using more agents ($O(n^4)$ of them), and a slightly stronger model: the agents still cannot communicate, but they can observe each other's presence. Some technical parts have been omitted due to space constraints.

2   Model

We are interested in discrete-time systems of synchronous agents operating in time-varying graphs. In order to describe our system, we have to define two components: the evolution of the network in time, and the properties of the agents.

We use the model of dynamic graphs according to [5]. A *time-varying graph (TVG) with discrete time* (also called *temporal graph* or *evolving graph*) consists of a graph $G = (V, E)$ called *underlying graph*, and a sequence of subgraphs $S_G = G_0, G_1, \ldots, G_t, \ldots$ such that for each t, the graph $G_t = (V, E_t)$, also called the *snapshot* at time t, has $E_t \subseteq E$. Moreover, we require that both G, and each of the G_t's are connected. Since we are interested in the worst-case behavior over all feasible sequences of snapshots, the evolution of the network can be visualized by an adversary that in each time step t observes the actions of the agents, and selects the spanner graph G_{t+1} for the next time step.

We assume that G is equipped with a local port labelling λ that satisfies the *sense of direction* (see [10]), i.e. the agents can, based on the sequence of labels of traversed links assign local labels to the vertices.

When specifying the properties of the agents, our aim is to employ the weakest means of communications with which we are still able to prove that exploration is feasible. In fact, our agents are *silent*, i.e. they can communicate neither directly, nor indirectly by whiteboards or tokens. The only form of communication is implicit by receiving feedback from the mutual-exclusion mechanism in vertices, and in some cases by observing the number of agents in the vertex.

An agent A is an anonymous entity with local memory [1]. The agents operate in the usual `look-compute-move` cycle (see e.g. [11]) as follows. At the beginning of time step t, A is located in a vertex v, and can see the local port numbers of incident edges. The agent is also notified whether it has freshly arrived to the vertex in the previous step, or is in the same vertex because the edge it wanted to traverse wasn't enabled. In this moment, A may also observe the number of agents currently located in v. Based on local computation, A may decide to either stay in v, or to move to some incident link. If several agents try to move to the same link, the adversary chooses one of them to succeed, and the rest remain in v with a notification about collision (they can still compute in the same step in order to process the notification, but no move action can be performed). Then, based on the actions of the agents, the adversary chooses the next spanner $G_{t+1} = (V, E_{t+1})$. Suppose that in step t, A succeeded to move along an edge $e = (v, w)$. If $e \in E_{t+1}$, A is transferred to w, and upon arrival learns about the success, and about the incoming port number. If, on the other hand, $e \notin E_{t+1}$, A stays in v, and learns about the failure.

---

[1] If the domain of λ is of cardinality $O(n)$, then $O(n \log n)$ bits of local memory will always be sufficient.

3   Minimal Number of Agents

Our first concern is the smallest number of agents needed to explore a network with an underlying graph G with n vertices and m edges. We show that $2m-n+1$ agents are sufficient to explore the whole graph. We use a variation of the known *rotor-router* algorithm. The original algorithm uses a counter in each vertex that counts up to the degree. Each time an agent visits the vertex, it chooses the outgoing link based on the counter, and increments the counter afterwards. In [24] it was shown that a single agent can explore a static undirected graph within $O(mD)$ steps using this algorithm. A modification of the algorithm was used in [15] for the exploration of time-varying graphs. However, in our case the agents do not have access to the whiteboards in the vertices, so each of them has to maintain a local counter for each vertex. Yet, the algorithm cannot be viewed as an independent execution of a number of instances of the rotor-router algorithm, since the agents located in the same vertex compete for the same set of outgoing links, and this interaction makes the algorithm more complex. The algorithm for an agent A is in the Algorithm 1: A maintains a local name of the current vertex, v, and a cyclic counter c_v for each vertex v. It tries to follow links according to the rotor-router algorithm. If the link is successfully acquired, but the corresponding edge is not present in E_{t+1}, A tries to acquire the same link in the next step, otherwise, it tries the cyclically following one.

Algorithm 1. Algorithm for agent A

1: **if** arrived from vertex w via port p **then**
2:     increment c_w (modulo $deg(w)$)
3:     update v to the local name of w's neighbor along p
4: **end if**
5: try to acquire the link c_v
6: **if** mutex failed **then**
7:     increment c_v (modulo $deg(v)$)
8: **end if**

Note that in this algorithms the agents don't need to even observe the presence of other agents, the only interaction is via the mutual exclusion mechanism.

Theorem 1. *Using Algorithm 1, $2m-n+1$ agents are sufficient to explore any graph, even if the adversary may choose their starting locations.*

Proof. First note the following: if an agent A successfully acquired some port p in a vertex v at time t, there will be some agent located at p in all subsequent steps until the corresponding edge e appears in the snapshot. This follows from the fact that if e is not in the snapshot, A tries to acquire p in the next step. Whether it succeeds or not, there will be some agent on p in the next step.

From this observation we immediately get that if there are at least $deg(v)$ agents located in a vertex v at some time step t, eventually at least one of them leaves v. This is due to the fact that if in a time step $t' \geq t$ all edges are occupied by agents, at least one of them must leave v due to the connectivity of $G_{t'}$. On the other hand, if some edge is not occupied by an agent, there is a mutex failure for some agents, and at least one of them increments its counter.

Since there are $2m - n + 1$ agents, we know that there is no deadlock, i.e. at least one agent moves every Δ steps, where Δ is the maximum degree of a vertex. Now suppose, for the sake of contradiction, that there are some vertices that are never visited. Let V_1 be the vertices that are visited infinitely often, and let V_2 be the remaining ones. Clearly, V_2 is not empty, and there are some vertices in V_2 that are never visited. Moreover, eventually all agents move in the V_1 part and no vertex from V_2 is entered anymore. Consider the cut edges between V_1 and V_2 from that moment. We claim that eventually all the ports are occupied by an agent. Indeed, consider a vertex $v \in V_1$ with a port p leading to a vertex in V_2. Since v is visited infinitely often, there must be an agent A that comes here infinitely often. Every time A leaves v, it increments its counter so eventually some agent occupies the port p.

It follows that eventually all ports leading to V_2 are occupied by an agent. However, in every time step there must be some edge $e \in E_t$ leading from V_1 to V_2 – a contradiction. □

On the other hand, even when the agents start from a common homebase, are equipped with a map of the underlying graph, and an arbitrary strong communication mechanism, the required number of agents cannot be reduced significantly, since we prove the following:

Theorem 2. *When all the agents start at the homebase, $m - n + 1$ agents are needed to explore all graphs even when scheduled by a global offline algorithm that observes the states of all agents (but does not know the adversary's choice of future E_t).*

The overall idea of the proof is as follows: Fix an arbitrary DFS spanning tree T rooted at the homebase. We describe an algorithm the adversary can use to select the snapshots. In each time step, the adversary observes the actions of the agents and *accepts* a set of edges into the next snapshot. The goal of the adversary is to block certain agents, i.e. to not accept the edges the agents want to traverse. In particular, the adversary tries to account one blocked agent for each non-tree edge of T.

Let T_v denote a subtree of T rooted at v. When the context is clear, we shall abuse the notation and refer by T_v to the whole subgraph induced by the vertices of T_v, e.g. when we speak about non-tree edges of T_v. Since T is a DFS tree, every non-tree edge connects a vertex with its (not necessarily direct) ancestor. In each vertex, we classify the incident edges as *parent* (leading to the parent in T, denoted as e_v), *child* edges, *up* edges (edges not in T, leading to vertices higher in T), and *down* edges. Let $up(v)$ denote the set of *up* edges of v. Let $d(v)$ denote the number of non-tree edges in the subtree T_v. At any point of the execution, we classify a vertex v as

- *unexplored* if v has not yet been visited by an agent,
- *done* if the whole subtree of T_v has already been visited,
- *active* otherwise.

At any given time step, let $f(e)$ denote the total balance of the flow of agents via the edge e up to this moment, where positive values are assigned for downward (w.r.t. T) flow. Let $l(v)$ be computed as $l(v) = \sum_{e \in \{e_v\} \cup up(v)} f(e) - d(v)$. If $l(v) \leq 0$, we say that v is *hungry*. If $l(v) \geq |up(v)| + 1$, we say that v is *full*. For the homebase h we add a fictitious parent edge e_h for which $f(e_h)$ is the initial number of agents. We say that h is hungry if $l(h) < 0$, and full if $l(h) \geq 0$.

Intuitively, the adversary would like to block an agent on every non-tree edge. In the definition of $l(v)$, the $d(v)$ represents the agents that entered v, but then left to explore T_v (although there may be agents that entered T_v via some non-tree edge that would not be accounted for in $d(v)$). The whole $l(v)$ is then seen as the excess of agents in v. In this intuition, a hungry vertex needs to receive some agents in order to explore T_v, and a full vertex has enough agents to cover all its up-edges. The core of the proof is to show that there are no in-between vertices: at any moment of time, every vertex is either hungry or full. The adversary then treats the hungry and full vertices differently: it tries to prevent agents to enter a hungry vertex, and tries to prevent agents to leave a full vertex. It may seem that agents leaving a hungry vertex or entering a full vertex are just a bonus for the adversary, but if they use a non-tree edge, they could cause problems in some other subtree, and so they have to be kept under control. That's why the adversary handles them in a special way: if an agent leaves a hungry vertex (or enters a full vertex) via a non-tree edge, the edge is marked as *special*. In a vertex with a special edge e, agents are permitted only over e, until the overall flow over e zeroes-out (at which time e ceases to be special). Let us now present the adversary's algorithm in a more formal way.

At each time step, we classify up-edges of a vertex v as *idle*, *outgoing* and *incoming*, according to the flow of agents across the edge at this time step: An edge is idle if no agent wants to cross the edge, or two agents want to cross the edge from different endpoints; in any case, accepting an idle edge does not change the numbers of agents in any vertex. In each vertex v, the adversary applies the following algorithm to select which edges from $\{e_v\} \cup up(v)$ are accepted to the next snapshot. The key part of the proof is the following lemma:

Lemma 1. *Let an active vertex v become done at the end of time step t. Then v must have been full at the beginning of round t. Furthermore, for every done vertex v, $f(e_v) \geq d(v)$ and $f(e) \geq 1$ for every up-edge e that is fully in T_v or exits T_v.*

Note that applying Lemma 1 to the homebase yields Theorem 2.

Algorithm 2. Adversary's algorithm at vertex v

1: accept all idle up-edges
2: accept e_v if it is idle
3: **if** an idle edge has been accepted **then**
4:     exit, you are done
5: **end if**
6: **if** v is *unexplored* **then**   ▷ so e_v and all up-edges are incoming
7:     **if** $l(v) = 0$ **then**   ▷ i.e. there are no non-tree edges in T_v
8:         accept e_v and all up edges   ▷ v becomes full and, if it is a leaf, it becomes *done*
9:     **else**
10:        accept e_v ▷ v becomes active
11:    **end if**
12: **else if** v is *done* or becomes *done* in the current step **then**
13:    **if** v has a special edge e **then**
14:        accept e ▷ if e is outgoing, it might stop being special
15:    **else if** e_v and all up-edges are outgoing **then**   ▷ v must have been full
16:        accept e_v
17:    **else**
18:        accept an incoming up-edge e ▷ e will become special
19:    **end if**
20: **else**   ▷ v is *active*
21:    **if** $l(v) \leq 0$ **then**
22:        **if** v has a special edge e **then**
23:            accept e ▷ if e is incoming, it might stop being special
24:        **else if** $l(v) < 0$ or e_v is outgoing **then**
25:            accept e_v
26:        **else**   ▷ $l(v) = 0$ and e_v is incoming
27:            **if** all up-edges are incoming **then**
28:                accept e_v and all up-edges
29:            **else**   ▷ there is an outgoing edge e
30:                accept an outgoing edge e ▷ e becomes special
31:            **end if**
32:        **end if**
33:    **else if** $l(v) \geq |up(v)| + 1$ **then**
34:        **if** v has a special edge e **then**
35:            accept e ▷ if e is outgoing, it might stop being special
36:        **else if** $l(v) > |up(v)| + 1$ or e_v is incoming **then**
37:            accept e_v
38:        **else**   ▷ $l(v) = |up(v)| + 1$ and e_v is outgoing
39:            **if** all up-edges are outgoing **then**
40:                accept e_v and all up-edges
41:            **else**   ▷ there is an incoming edge e
42:                accept an incoming edge e ▷ e becomes special
43:            **end if**
44:        **end if**
45:    **else**   ▷ this branch should never be taken!
46:    **end if**
47: **end if**

The following statements serve as preparation for the proof of Lemma 1. First we establish the important invariant that each vertex is either hungry or full.

Lemma 2. *If v is active, then either $l(v) \leq 0$ or $l(v) \geq |up(v)| + 1$. Moreover, v can switch between the two states only in time steps where no special edge is present.*

Following is a simple observation that agents may enter an unexplored vertex only over a tree-edge:

Lemma 3. *If v is unexplored, all vertices from T_v are unexplored.*

Lemma 4. *If v is active and hungry, then $f(e_v) \leq d(v)$, and $f(e) \leq 0$ for all up-edges. If v is active and full, then $f(e_v) \geq d(v) + 1$. Moreover, $f(e) \geq 1$ for all up-edges.*

Proof. Note that $0 \leq f(e) \leq 1$ for all non-special up-edges: non-special up-edges are only accepted when v switches between full and hungry, and it is always in alternating directions.

Due to Lemma 2, an active vertex v can switch from hungry to full only when there is no special edge, $l(v) = 0$, and all up edges and e_v are incoming. In this case, all up-edges will have $f(e) = 1$, and the flow on e_v is increased by 1. Since $l(v)$ was zero, the inflow must have been $d(v)$, and since all $f(e)$'s were zero, the whole inflow was due to $f(e_v)$. While v remains full, the invariant holds. Situation for switching from full to hungry is symmetrical.   □

Now we are ready to prove the crucial lemma:

Proof (of Lemma 1). Let v be any vertex that becomes done at the end of time step t. We proceed by induction on time, and the height of T_v.

If v is a leaf, the first part of the statement is trivial, since a leaf is never active: it is either unexplored or done. Let us suppose that v has children $u_1, \ldots, u_k$.

If a child u_i is unexplored at the beginning of round t, then u_i is a leaf. In order to see this, suppose, for the sake of contradiction, that T_{u_i} has more than one vertex. By Lemma 3, all vertices of T_{u_i} are unexplored at the beginning of time step t. However, from the construction if follows that an unexplored vertex is visited for the first time along the tree edge (on line 8 or 10). That means that leaves of T_{u_i} cannot be visited in time step t, and v is not done at the end of step t. Hence, if u_i is unexplored at the beginning of step t, it is a leaf, and there must be an agent traversing the edge from v to u_i in step t.

If a child u_i is done at the beginning of step t, by induction it must have $f(e_{u_i}) \geq d(u_i)$, and $f(e) \geq 1$ for all edges e that lead from v to T_{u_i}.

Let a child u_i be active at the beginning of step t. Since v becomes done in step t, u_i must become done, too, and by induction u_i is full at the beginning of step t. Then by Lemma 4, $f(e_{u_i}) = d(u_i) + 1$, and $f(e) \geq 1$ for all edges e that lead from v to T_{u_i}.

To summarize, $f(e) \geq 1$ for all non-tree down-edges from v, and $f(e_{u_i}) \geq d(u_i)$ for all tree down-edges from v. Moreover, v must have at least one child that is unexplored or active (otherwise v would have been done before t), and this child must have at least one additional agent coming from v. Overall, the down-edges from v incur outflow at least $d(v)+1$, which means that the up-edges must incur inflow to v at least $d(v) + 1$, too. However, if v was hungry, it would hold $l(v) \leq 0$, and the inflow would be at most $d(v)$. Due to Lemma 2, v must be full.

Now let us show the second part of the statement. If v is leaf, line 8 must have been performed, and so $f(e_v) = 1$, and $f(e) = 1$ for all up-edges at the end of step t. If v is active, we have just proved that v is full, and due to Lemma 4 $f(e_v) = d(v) + 1$, and $f(e) \geq 1$ for all up-edges at the beginning of step t. However, starting from step t, the code at line 12 applies: apart from special edge (which has always positive flow), v only accepts outgoing edge e_v if all the up-edges are outgoing too. Since all vertices of T_v are done, too, by induction on the height of the tree this can happen only when v is full, so $f(e_v) \geq d(v)$. □

Finally, we have to check that the adversary is correct:

Lemma 5. *In each time step, the set of accepted edges forms a connected graph.*

Proof. We will show that in each vertex v (other that the home base) some edge leading up-tree T is accepted. This is sufficient to ensure connectivity.

The adversary runs Algorithm 2 in each v. With the exception of the branch on line 45, at least one edge from $\{e_v\} \cup up(v)$ is always accepted. However, due to Lemma 2, the branch on line 45 is never taken. □

4   Polynomial Search Time

The Algorithm 1 succeeds in exploring the graph with almost optimal number of agents. However, the proof provides only an super-exponential bound on the number of steps needed to finish the exploration. We don't know any better bound, and analyzing the time complexity of Algorithm 1 (at least to answer the question whether it is always polynomial) is an interesting open question. However, we are able to provide another algorithm which uses more agents, but achieves a polynomial exploration time.

To simplify the presentation, we describe the algorithm in a *vertex-centric* model: in every step, the decisions are not made by particular agents; instead, each node has an independent entity that evaluates the states of the agents present in the vertex, and selects some action for each agent. Later we show how to modify the algorithm to the considered *agent-centric* model.

The basic idea of the algorithm is a simulated diffusion: The agents flow from vertices with high number of agents towards the vertices with fewer agents. The tricky part is to ensure such flow even when the accepted edges are decided by the adversary. We achieve that by employing Algorithm 4, and by careful accounting scheme that shows that even if there are instances when the number of agents in the receiver vertex exceeds the number of agents in the senders (this can happen due to concurrency), there is overall progress.

In order to glimpse the difficulty of the task, first let us examine the following simple algorithm:

Algorithm 3. Algorithm Threshold Flood

1: **while** needed **do**
2:      **if** the number of agents at vertex v at least equals its degree **then**
3:          send an agent via all incident edges
4:      **end if**
5: **end while**

This is essentially a modified flooding and one would expect that such an algorithm must always make progress. Unfortunately, this is not so, as Fig. 1 shows. Hence, a more sophisticated algorithm is needed.

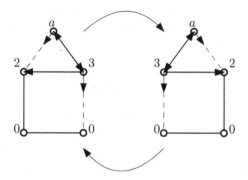

Fig. 1. Livelock in Algorithm 3. The numbers near vertices correspond to the numbers of agents. Disabled links are shown as dashed.

Let x_v^r denote the number of agents in v at the beginning of round r, and let M and T be parameters to be specified later. The algorithm proceeds in nM-step rounds described in Algorithm 4.

We first show that when a single edge is considered, the flow is always in the direction towards the vertex with fewer agents.

Lemma 6. Let $e = (u, v)$ be an edge such that $x_u^r \geq x_v^r$. Then, during round r, either

- one agent crosses from u to v and $x_u^r > x_v^r$, or
- one agent crosses from u to v and one agent crosses from v to u, or
- no agents pass between u and v

Algorithm 4. Algorithm Diffusion at vertex v

1: **for** round r from 0 to T **do**
2:     **for** time step i from 0 to $nM - n + 1$ **do** ▷ i is the step number within round r
3:         **for** each incident edge e **do**
4:             **if** $x_v^r > nM - i$ and no agent has been received or successfully sent via e in round r **then**
5:                 send an agent via e ▷ might not succeed
6:             **end if**
7:         **end for**
8:     **end for**
9: **end for**

Proof. Let $i > nM - x_u^r$ be the first time the edge e is enabled while u tries to send an agent in round r. It can happen that no such i exists (i.e. e is never enabled in round r for $i > nM - x_u^r$), then the last case applies. Otherwise, if $i > nM - x_v^r$ both u and v send an agent over e and, by construction, stop sending agents over e in round r. Hence, the middle case applies. Finally, if $i \leq nM - x_v^r$, then only u sends an agent over e and the first case applies, as no other agent is sent over e in round r.   □

Lemma 6 tells us that there is no flowback of agents from vertices with low number of agents to vertices with high number of agents. The following Lemma shows that there is a progress:

Lemma 7. *Let i be such that $n - 1 < i < nM$. Let $V_i^r = \{v \mid x_v^r \geq i\}$ and let $\overline{V_i^r}$ be the complement of V_i^r. Let E_i^r be the set of edges separating V_i^r from $\overline{V_i^r}$. If $E_i^r \neq \emptyset$, then during round r at least one agent crosses E_i^r from V_i^r to $\overline{V_i^r}$.*

While Lemma 7 indicates that there is a progress in the diffusion of the agents, it can happen that a vertex in one round receives multiple agents and it is not immediately clear whether this process really terminates, and at what time. In order to prove that it does terminate in polynomial time, we introduce the following accounting: Let the overall weight $W_r = \sum_{v \in V} w_v^r$, where the weight w_v^r of a node v at the beginning of round r is $w_v^r = \binom{x_v^r + 1}{2}$. We shall argue that for a suitable choice of M, the overall weight W_r decreases in each round. Although due to Lemma 6 agents are sent only from vertices with higher weight to vertices with lower weight, a vertex may receive multiple agents, and its weight may increase. If the parameters are not chosen properly, the overall weight may increase as can be seen in the following example.

Example 1. Let G be a complete bipartite graph with partitions of size $n/2$. Assume that at time t each vertex of one partition has k agents, while the vertices of the other partition have $k - 1$ agents each, for $M - n/2 \geq k \geq n/2$. The adversary will enable all edges at time t. Then

$$W_r = \binom{k+1}{2}\frac{n}{2} + \binom{k}{2}\frac{n}{2}$$

while

$$W_{r+1} = \left(\frac{k - n/2 + 1}{2}\right)\frac{n}{2} + \left(\frac{k + n/2}{2}\right)\frac{n}{2} = W_r + \frac{n^3}{8} - \frac{n^2}{4},$$

i.e. the overall weight increased by $\frac{n^3}{8} - \frac{n^2}{4}$.

When arguing about how W_r changes over time, we distribute the weight in a vertex at the beginning of a round among its agents: we arbitrarily order the x_v^r agents that are present in v at the beginning of round r, and assign the i-th agent weight i. For a given round, consider a directed graph D with edges E_D indicating the transfer of agents (we consider only edges that transfer an agent in one direction, since the edges that transfer agents in both directions have no effect on the resulting weights). Let out_v, and in_v be the out- and in-degree of v in D, respectively. In each vertex v, label the outgoing edges arbitrarily $1, \ldots, out_v$, and the incoming edges $1, \ldots, in_v$. This labeling defines the weight transferred by agents as follows (see Fig. 2): let $e = (u, v) \in E_D$ be an edge, and e^-, e^+ be the two labels near u, and v, respectively. Then the agent with weight $x_u^r - e^- + 1$ in u arrives to v at position e^+ (after the out_v agents already left v), and will have weight $x_v^r - out_v + e^+$.

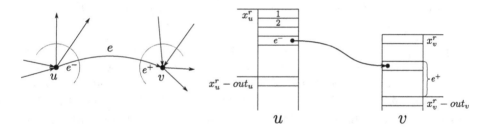

Fig. 2. An agent transferring weight from vertex u to vertex v,

Hence, the overall weight is changed as follows:

$$\begin{aligned} W_{r+1} &= W_r + \sum_{e=(u,v)\in E_D} \left(x_v^r - out_v + e^+ - (x_u^r - e^- + 1)\right) \\ &= W_r + \sum_{e=(u,v)\in E_D} (x_v^r - x_u^r) + \sum_{e=(u,v)\in E_D} \left(e^- + e^+ - 1 - out_v\right) \end{aligned} \tag{1}$$

In the next two lemmata, we argue about the two summands of (1):

Lemma 8. *Assume that at the beginning of round r there is at least one vertex with at least M agents, and at least one vertex with less than m agents for some $m \geq n - 1$. Then $\sum_{e=(u,v)\in E_D} (x_v^r - x_u^r) \leq -M + m$.*

Proof. Due to Lemma 6, $x_v^r - x_u^r < 0$ for each edge $e = (u, v) \in E_D$. We construct a specific sequence of edges to make sure the overall sum is small enough. Let

$z_0 = M$, and define the sequence of edges $e_0, e_1, \ldots$ inductively together with a sequence of numbers $z_0 > z_1 > \ldots$ as follows. For each $i \geq 0$ such that $z_i \geq m$, let $V_i = \{u \mid x_u^r \geq z_i\}$. Since $m \leq z_i \leq M$, V_i defines a non-empty cut, and due to Lemma 7 there is at least one edge in E_D going from V_i to some vertex v with $x_v^r < z_i$. Let z_{i+1} be the minimum value of x_v^r among all these vertices, and e_i be the edge leading to such vertex. Then $e_i = (u_i, v_i)$ such that $x_{u_i}^r \geq z_i$, $x_{v_i}^r = z_{i+1} < z_i$. Note that due to the minimality of each z_i, $x_{u_i}^r < z_{i-1}$ (otherwise e_i would have been chosen in the previous iterations), and so all the edges e_i are disjoint.

Let k be the length of the constructed sequence, i.e. the first index such that $z_k < m$. Then because $x_{u_i}^r \geq z_i = x_{v_{i-1}}^r$, the sum $\sum_{i=0}^{k}(x_{v_i}^r - x_{u_i}^r)$ telescopes to $\sum_{i=0}^{k}(x_{v_i}^r - x_{u_i}^r) \leq -x_{u_0}^r + x_{v_k}^r \leq -M + m$. □

Lemma 9. *For large enough* n, $\sum_{e=(u,v) \in E_D}(e^- + e^+ - 1 - out_v) \leq \frac{n^3}{3}$.

If the prerequisites of Lemma 8 and Lemma 9 hold, (1) can be rewritten

$$W_{r+1} \leq W_r + \frac{n^3}{3} - M + m \qquad (2)$$

Lemma 10. *Let* $M = \frac{2n^3}{3} + n$ *and let* nM *be the number of agents in the system. Then after* $2n^5/3 + o(n^5)$ *rounds there are no unexplored vertices left.*

Proof. As the total number of agents is nM, there will always be a vertex with at least M agents. Set $m = n$. As long as there exists an unexplored vertex, Lemma 8 applies and according to (2), W_r decreases each round by at least $M - m - n^3/3 \geq n^3/3$. As the initial weight is $\binom{nM+1}{2} < (nM+1)^2/2$ (all the agents in the home base), $\frac{(nM+1)^2}{2n^3/3} = \frac{(2n^4+3n^2+3)^2}{6n^3} = \frac{2n^5}{3} + o(n^5)$ rounds are sufficient to spread the agents so that Lemma 8 does no longer apply, and all vertices are explored. □

As each round takes nM time steps, this yields:

Theorem 3. *Let* $G = (V, E)$ *be an arbitrary base graph. Then a vertex-based algorithm that knows only* $n = |V|$, *but nothing else about* G *can explore any time-varying graph based on* G *using* $\frac{2n^4}{3} + n^2$ *agents in time* $4n^9/9 + o(n^9)$.

Improving the Time

In the initial presentation of Algorithm 4, each round lasted nM time steps. Since the overall number of agents in the system is nM, Lemma 6 ensures that agents are only transferred to vertices with fewer agents. However, we only used vertices with up to M agents in showing progress in Lemma 8, and the transfers of agents among vertices with more than M agents were not important for our argument. Hence, it is possible to modify the arguments as follows: make each round last M steps instead of nM. This means that Lemma 6 does not apply if both the vertices have more than M agents. Also, Lemma 7 will hold only for $i < M$.

We can modify the weighting scheme by capping the weights of the agents at M, i.e. $W_r = \sum_{v \in V} w_v^r$ where $w_v^r = \binom{x_v^r+1}{2}$ if $x_v^r \leq M$, and $w_v^r = \binom{M+1}{2} + M(x_v^r - M)$ if $x_v^r > M$. In the construction of the graph D we can disregard edges among vertices with $\geq M$ agents, since they don't change W_r. Split the remaining edges of E_D into E_D^s that lead from u with $x_u^r \leq M$, and E_D^b that lead from u where $x_u^r > M$. The contribution of an edge $(u, v) \in E_D^s$ into (1) is not changed, i.e. $x_u^r - out_v + e^+ - (x_u^r - e^- + 1)$, whereas the contribution of an edge from E_D^b is $x_v^r - out_v + e^+ - M$. The first summand of (1) that is dealt with in Lemma 8 then becomes $\sum_{e=(u,v) \in E_D^s} (x_v^r - x_u^r) + \sum_{e=(u,v) \in E_D^b} (x_v^r - M)$. Note that the second part is always non-positive, and the argumentation of Lemma 8 is only on edges from E_D^s anyway.

The second summand of (1) that is dealt with in Lemma 9 is changed as follows: for edges (u, v) from E_D^s the contribution is the same, i.e. $e^- + e^+ - 1 - out_v$, whereas for edges from E_D^b it is $e^+ - out_v$. Since the value only decreased in comparison to D, the upper bound from Lemma 9 is still valid.

Summarizing, we can simply shorten each round of Algorithm 4 to M steps without any adverse effects on the analysis, saving a factor of n in the time:

Theorem 4. *Let $G = (V, E)$ be an arbitrary base graph. Then a vertex-based algorithm that knows only $n = |V|$, but nothing else about G can explore any time-varying graph based on G using $\frac{2n^4}{3} + n^2$ agents in time $4n^8/9 + o(n^8)$.*

Agent-Centric Algorithm

In Algorithm 4 there is an authority in each vertex that decides which agent goes to which link. However, the decision is based only on the number of agents in the vertex, the clock (i.e. ability to count which time step currently is), and the ability to see whether some agent arrived via a given edge within the current round. Assume now that the agents start at the same time (i.e. can count the number of time steps), and are equipped with some means to distinguish the multiplicity of agents in a vertex. Also, if an agent remains in a vertex (e.g. because its chosen edge was not selected), it is notified which incident edges were selected in the previous round, and whether agents were delivered along them.

Consider a round r of Algorithm 4 in a vertex v. Note that only agents that were present in v at the beginning of r participate in transmissions (agents that arrived to v during the round just wait until next round). Moreover, for each incident edge e, the situation is as follows: either v never sends an agent along e (if some agent arrived from e), or v starts to transmit over e at certain time step, and continues trying until the transmit is successful, or the round is over.

Now we adapt the algorithm to our agent-centric model. Each step of the original algorithm will be replaced by n substeps. Let i be the first step when v starts sending agents. All agents that were present in v from the beginning of the round know which edges have already delivered agents, and hence they agree on a set of edges where the algorithm must send. In the substeps of step i, the agents try to acquire the needed links in order: At the beginning, all agents are *free*. In the j-th substep, all free agents try to acquire the j-th outgoing link. One of them succeeds, and becomes *bound* to that link; all other remain *free*, and continue to the next substep. Once an agent becomes bound to a certain link, it tries to acquire that link in all subsequent substeps of all subsequent steps, until either it succeeds in traversing the edge, or the whole round ends. Note that once there is an agent bound to some link, no other agent tries to access that link until the end of the whole round.

Recall the proof of Lemma 6, and consider an edge (u, v) with $x_u^r \geq x_v^r$. If $x_u^r > x_v^r$, then u starts sending in a sooner time step than v. During the substeps of the step when only u is sending, u's agents become bound to their respective links, and the proof follows. Hence, the only way Lemma 6 could be violated is when $x_u^r = x_v^r$, and one agent passes from u to v. The only place in the analysis where this plays role is in the proof of Lemma 8, where instead of $x_v^r - x_u^r < 0$ one has $x_v^r - x_u^r \leq 0$. This change, however, has no effect on the actual proof.

To sum up:

Theorem 5. *Let $G = (V, E)$ be an arbitrary base graph. Then an agent-based algorithm that knows only $n = |V|$, but nothing else about G can explore any time-varying graph based on G using $\frac{2n^4}{3} + n^2$ agents in time $4n^9/9 + o(n^8)$.*

Unknown n

If n is not known, the agents can't all appear at the home base, since that would reveal n to the agents (the number of agents is a function of n). Instead, we use a vertex-centric model where the algorithm can request another batch of agents at the home base until it has enough to solve the problem. Also, we suppose that the vertex can access the states of the agents, i.e. read the messages they contain.

The basic technique is to guess an upper bound n' on n, use Algorithm 4 to broadcast assuming $n' > n$, and verify whether all nodes have been reached. If not, double the guess n' and repeat the process. The verification uses two observations:

- Let S_v denote the set of edges over which v successfully sent an agent in the second last step of Algorithm 4 for n', and let R_v be the set of edges of which v received agents in the last step of this execution. The graph has been fully explored if and only if $\cup_{v \in V} S_v = \cup_{v \in V} R_v$.
- If each vertex has at least $n - 1$ agents, a simultaneous broadcast to all neighbours from each vertex does not change the number of agents in each vertex.

Note that the algorithm can be altered in a way that every explored vertex has at least $n' - 1$ agents. When Algorithm 4 terminates in time $O(n'^9)$, each vertex v with at least $n' - 1$ agents performs n' steps of broadcast, where the broadcasted information is a pair (S_v, R_v). After n' steps of broadcast, homebase can evaluate whether the current guess n' was correct. If not, it can require a new batch of agents, and start another iteration.

References

1. Albers, S., Henzinger, M.R.: Exploring unknown environments. SIAM J. Comput. **29**(4), 1164–1188 (2000)
2. Ambühl, C., Gasieniec, L., Pelc, A., Radzik, T., Zhang, X.: Tree exploration with logarithmic memory. ACM Trans. Algorithms **7**(2), 17 (2011)
3. Blum, M., Kozen, D.: On the power of the compass (or, why mazes are easier to search than graphs). In: 19th Annual Symposium on Foundations of Computer Science, Ann Arbor, Michigan, USA, 16–18 October 1978, pp. 132–142. IEEE Computer Society (1978)
4. Blum, M., Sakoda, W.J.: On the capability of finite automata in 2 and 3 dimensional space. In: 18th Annual Symposium on Foundations of Computer Science, Providence, Rhode Island, USA, 31 October–1 November 1977, pp. 147–161. IEEE Computer Society (1977)
5. Casteigts, A., Flocchini, P., Quattrociocchi, W., Santoro, N.: Time-varying graphs and dynamic networks. IJPEDS **27**(5), 387–408 (2012)
6. Das, S.: Graph explorations with mobile agents. In: Flocchini, P., Prencipe, G., Santoro, N. (eds.) Distributed Computing by Mobile Entities, Current Research in Moving and Computing. LNCS, vol. 11340, pp. 403–422. Springer, Heidelberg (2019). https://doi.org/10.1007/978-3-030-11072-7_16
7. Deng, X., Papadimitriou, C.H.: Exploring an unknown graph. J. Graph Theory **32**(3), 265–297 (1999)
8. Dobrev, S., Jansson, J., Sadakane, K., Sung, W.-K.: Finding short right-hand-on-the-wall walks in graphs. In: Pelc, A., Raynal, M. (eds.) SIROCCO 2005. LNCS, vol. 3499, pp. 127–139. Springer, Heidelberg (2005). https://doi.org/10.1007/11429647_12
9. Fleischer, R., Trippen, G.: Exploring an unknown graph efficiently. In: Brodal, G.S., Leonardi, S. (eds.) ESA 2005. LNCS, vol. 3669, pp. 11–22. Springer, Heidelberg (2005). https://doi.org/10.1007/11561071_4
10. Flocchini, P., Mans, B., Santoro, N.: Sense of direction in distributed computing. Theor. Comput. Sci. **291**(1), 29–53 (2003)
11. Flocchini, P., Prencipe, G., Santoro, N. (eds.): Distributed Computing by Mobile Entities, Current Research in Moving and Computing. LNCS, vol. 11340. Springer, Heidelberg (2019). https://doi.org/10.1007/978-3-030-11072-7
12. Fraigniaud, P., Ilcinkas, D., Peer, G., Pelc, A., Peleg, D.: Graph exploration by a finite automaton. Theor. Comput. Sci. **345**(2–3), 331–344 (2005)
13. Gasieniec, L., Klasing, R., Martin, R.A., Navarra, A., Zhang, X.: Fast periodic graph exploration with constant memory. J. Comput. Syst. Sci. **74**(5), 808–822 (2008)
14. Gasieniec, L., Radzik, T.: Memory efficient anonymous graph exploration. In: Broersma, H., Erlebach, T., Friedetzky, T., Paulusma, D. (eds.) WG 2008. LNCS, vol. 5344, pp. 14–29. Springer, Heidelberg (2008). https://doi.org/10.1007/978-3-540-92248-3_2
15. Gotoh, T., Flocchini, P., Masuzawa, T., Santoro, N.: Tight bounds on distributed exploration of temporal graphs. In: Felber, P., Friedman, R., Gilbert, S., Miller, A., (eds.) 23rd International Conference on Principles of Distributed Systems (OPODIS 2019), 17–19 December 2019, Neuchâtel, Switzerland. LIPIcs, vol. 153, pp. 22:1–22:16. Schloss Dagstuhl - Leibniz-Zentrum für Informatik (2019)
16. Ilcinkas, D.: Setting port numbers for fast graph exploration. Theor. Comput. Sci. **401**(1–3), 236–242 (2008)

17. Kosowski, A., Navarra, A.: Graph decomposition for memoryless periodic exploration. Algorithmica **63**(1–2), 26–38 (2012). https://doi.org/10.1007/s00453-011-9518-1

18. Di Luna, G., Dobrev, S., Flocchini, P., Santoro, N.: Distributed exploration of dynamic rings. Distrib. Comput. **33**(1), 41–67 (2018). https://doi.org/10.1007/s00446-018-0339-1

19. Panaite, P., Pelc, A.: Exploring unknown undirected graphs. J. Algorithms **33**(2), 281–295 (1999)

20. Rollik, H.: Automaten in planaren graphen. Acta Inf. **13**, 287–298 (1980). https://doi.org/10.1007/BF00288647

21. Rosenkrantz, D.J., Stearns, R.E., Lewis II, P.M.: An analysis of several heuristics for the traveling salesman problem. SIAM J. Comput. **6**(3), 563–581 (1977)

22. Steinová, M.: On the power of local orientations. In: Shvartsman, A.A., Felber, P. (eds.) SIROCCO 2008. LNCS, vol. 5058, pp. 156–169. Springer, Heidelberg (2008). https://doi.org/10.1007/978-3-540-69355-0_14

23. Tajibnapis, W.D.: A correctness proof of a topology information maintenance protocol for a distributed computer network. Commun. ACM **20**(7), 477–485 (1977)

24. Yanovski, V., Wagner, I.A., Bruckstein, A.M.: A distributed ant algorithm for efficiently patrolling a network. Algorithmica **37**(3), 165–186 (2003). https://doi.org/10.1007/s00453-003-1030-9

Network Communication

Optimal Packet-Oblivious Stable Routing in Multi-hop Wireless Networks

Vicent Cholvi[1]([✉]), Paweł Garncarek[2], Tomasz Jurdziński[2], and Dariusz R. Kowalski[3]

[1] Department of Computer Science, Universitat Jaume I,
Castelló de la Plana, Castelló, Spain
vcholvi@uji.es
[2] Instytut Informatyki, Uniwersytet Wrocławski, Wrocław, Poland
[3] School of Computer and Cyber Sciences, Augusta University,
Augusta, GA, USA

Abstract. Stability is an important issue in order to characterize the performance of a network, and it has become a major topic of study in the last decade. Roughly speaking, a communication network system is said to be *stable* if the number of packets waiting to be delivered (back-log) is finitely bounded at any one time.

In this paper, we introduce a new family of combinatorial structures, which we call *universally strong selectors*, that are used to provide a set of transmission schedules. Making use of these structures, combined with some known queuing policies, we propose a packet-oblivious routing algorithm which is working without using any global topological information, and guarantees stability for certain injection rates. We show that this protocol is asymptotically optimal regarding the injection rate for which stability is guaranteed.

Furthermore, we also introduce a packet-oblivious routing algorithm that guarantees stability for higher traffic. This algorithm is optimal regarding the injection rate for which stability is guaranteed. However, it needs to use some global information of the system topology.

Keywords: Wireless network · Routing · Adversarial queuing · Interference · Stability · Packet latency

1 Introduction

Stability is an important issue in order to characterize the performance of a network, and it has become a major topic of study in the last decade. Roughly speaking, a communication network system is said to be *stable* if the number of packets waiting to be delivered (backlog) is finitely bounded at any one time. The importance of such an issue is obvious, since if one cannot guarantee stability,

V. Cholvi—This work was partially supported by the Ministerio de Ciencia, Innovación y Universidades grant PRX18/000163.

A. W. Richa and C. Scheideler (Eds.): SIROCCO 2020, LNCS 12156, pp. 165–182, 2020.
https://doi.org/10.1007/978-3-030-54921-3_10

then one cannot hope to be able to ensure deterministic guarantees for most of the network performance metrics.

For many years, the common belief was that only overloaded queues (i.e., when the total arrival rate is greater than the service rate) could generate instability, while underloaded ones could only induce delays that are longer than desired, but always remain stable. However, this belief was shown to be wrong when it was observed that, in some networks, the backlogs in specific queues could grow indefinitely even when such queues were not overloaded [4,8]. These later results aroused an interest in understanding the stability properties of packet-switched networks, so that a substantial effort has been invested in that area. Among the obtained results, stability of specific scheduling policies was considered for example in [7,14,19,20]. The impact of network topologies on injection rates that guarantee stability was considered in [18,21,22]. A systematic account of issues related to universal stability was given in [2].

Whereas in wireline networks a node can transmit data over any outgoing link and simultaneously receive data over any incoming link, the situation is different in wireless networks. Indeed, nearby wireless signal transmissions that overlap in time can interfere with one another, to the effect that none can be transmitted successfully. This makes the study of stability in wireless networks more complex. As in the wireline case, a substantial effort has been invested in investigating stability in that setting. Stability in wireless networks without explicit interferences was first studied by Andrews and Zhang [5,6] and Cholvi and Kowalski [16]. Chlebus et al. [13] and Anantharamu et al. [3] studied adversarial broadcasting with interferences in the case of using single-hop radio networks. In multi-hop networks with interferences, Chlebus et al. [11] considered interactions among components of routing in wireless networks, which included transmission policies, scheduling policies and control mechanisms to coordinate transmissions with scheduling. In [10], the authors demonstrated that there is no routing algorithm guaranteeing stability for an injection rate greater than $1/L$, where the parameter L is the largest number of nodes which a packet needs to traverse while routed to its destination. They also provided a routing algorithm that guarantees stability for injection rates smaller than $1/L$. Their approach, however, is not accurate for studying stability of longer-distance packets; therefore, in this work we study how the stability of routing depends of the *conflict graph* (which we will formally define later) of the underlying wireless networks, which is independent of the lengths of the packets' routes.

Our results. In this paper, we study the stability of dynamic routing in multihop radio networks with a specific methodology of adversarial traffic that reflects interferences. We focus on packet-oblivious routing protocols; that is, algorithms that do not take into account any historical information about packets or carried out by packets. Such protocols are well motivated in practice, as real forwarding protocols and corresponding data-link layer architectures are typically packet-oblivious.

First, we give a new family of combinatorial structures, which we call *universally strong selectors*, that are used to provide a set of transmission schedules.

Table 1. Summary of the paper's results.

Routing algorithm	Scheduling policies (ALG)	Required knowledge	Maximum injection rate
USS-PLUS-ALG	LIS, SIS NFS, FTG	Bounds on the number of links and on the network's degree	$O(1/(e \cdot \Delta^H))$
COLORING-PLUS-ALG	LIS, SIS NFS, FTG	Full topology	$O(1/\Delta^H)$

Making use of these structures, combined with some known queuing policies such as Longest In System (LIS), Shortest In System (SIS), Nearest From Source (NFS) and Furthest To Go (FTG), we propose a *local-knowledge* packet-oblivious routing algorithm (i.e., which is working without using any global topological information) that guarantees stability for certain injection rates. We show that this protocol is asymptotically optimal regarding the injection rate for which stability is guaranteed, mainly, for $\Theta(1/\Delta^H)$, where Δ^H is the maximum vertex degree of the conflict graph of the wireless network.

Later, we introduce a *packet-oblivious* routing algorithm that, by using the same queueing policies, guarantees stability for higher injection rates. This algorithm is optimal regarding the injection rate for which stability is guaranteed. However, it needs to use some global information of the system topology (so called *global-knowledge*).

Table 1 summarizes the main results of this paper.

The rest of the paper is structured as follows. Section 2 contains the technical preliminaries. In Sect. 3, we introduce and study universally strong selectors, which are the core components of the deterministic local-knowledge routing algorithm that is developed in Sect. 4, where we show that it is asymptotically optimal. In Sect. 5, we present a global-knowledge routing algorithm that guarantees universal stability for higher traffic, we also show that it is optimal. In Sect. 6, we extend the results obtained for the Longest-In-System scheduling policy in Sect. 4 to other policies, mainly, SIS, NFS and FTG. This extension is based on different technical tools, mainly, on reduction to the wired model with failures studied in [1], in which SIS, NFS and FTG are stable. We end with some conclusions and future work in Sect. 7. Some technical details and proofs can be found in [15].

2 Model and Problem Definition

Wireless Radio Network. We consider a *wireless radio network* represented by a directed symmetric network graph $G = (V_G, E_G)$. It consists of *nodes* in V_G representing devices, and directed edges, called *links*, representing the fact that a transmission from the starting node of the link could be directly delivered to the ending node. The graph is symmetric in the sense that if some $(i, j) \in E_G$ then $(j, i) \in E_G$ too. Each node has a unique ID number and it knows some upper

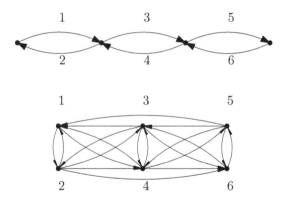

Fig. 1. Radio network G with 4 nodes and links labeled 1–6 (up). Conflict graph $H(G)$ obtained from network G (down). Observe that each link i in network G corresponds to one node i in $H(G)$.

bounds on the number m of edges in the network and the network in-degree (i.e., the largest number of links incoming to a network node).[1]

Nodes communicate via the underlying wireless network G. Communication is in synchronous rounds. In each round a node could be either transmitting or listening. Node i receives a message from a node $j \neq i$ in a round if j is the only transmitting in-neighbor of i in this round and node i does not transmit in this round; we say that the message was successfully sent/transmitted from j to i.

Conflict Graphs. We define the *conflict graph* $H(G) = (V_{H(G)}, E_{H(G)})$ of a network G as follows: (1) its vertices are links of the network (i.e., $V_{H(G)} = E_G$) and, (2) a directed edge $(u, v) \in E_{H(G)}$ if and only if a message across link $v \in E_G$ cannot be successfully transmitted while link $u \in E_G$ transmits. Note that, accordingly with the radio model, a conflict occurs if and only if the transmitter in u is also a receiver in v *or* the transmitter in u is a neighbor of the receiver in v (see Fig. 1 for an illustrative example). If network G is clear from the context, we skip the parameter G in $H(G)$ (i.e., we will use H). Note that, the links in our definition are directed in order to distinguish which transmission is blocked by which.

Routing Protocols and Transmission Schedules. We consider *packet-oblivious* routing protocols, that is, protocols which only use their hardwired memory and basic parameters of the stored packets assigned to them at injection time (such as source, destination, injection time, route) in order to decide which packet to send and when.

---

[1] In which case the performance will depend on these known estimates, instead of the actual values.

We distinguish between *global-knowledge* protocols which can use topological information given as input, and *local-knowledge* protocols that are given only basic system parameters such as the number of links or the network's in-degree.

All our protocols will be based on pre-defined transmission schedules, which will be circularly repeated—the properties of these schedules will guarantee stability for certain injection rates. These schedules will be different for different types of protocols, due to the available information based on which these schedules could be created.

Adversaries. We model dynamic injection of packets by way of an adversarial model, in the spirit of similar approaches used in [4,8,10,11,13,16,22]. An adversary represents the users that generate packets to be routed in a given radio network. The constraints imposed on packet generation by the adversary allow considering worst-case performance of deterministic routing algorithms handling dynamic traffic.

Over time, an adversary injects packets to some nodes. The adversary decides on a path a packet has to traverse upon its injection. Our task is to develop a packet-oblivious routing protocol such that the network remains *stable*; that is, the number of packets simultaneously queued is bounded by a constant in all rounds. Since an unbounded adversary can exceed the capability of a network to transmit messages, we limit its power in the following way: for any time window of any length T, the adversary can inject packets (with their paths) in such a way that each link is traversed by at most $\rho \cdot T + b$ packets, for some $0 \leq \rho \leq 1$ and $b \in \mathbb{N}^+$. We call such an adversary a (ρ, b)-*adversary*.

3 Selectors as Transmission Schedulers

In this section, we introduce a family of combinatorial structures, widely called *selectors* [12,17], that are the core of the deterministic routing algorithm presented in Sect. 4. In short, we will use specific type of selectors to provide a set of transmission schedules that assure stability when combined with suitable queuing policies.

There are many different types of selectors, with the more general one being described below:

Definition 3.1. *Given integers k, m and n, with $1 \leq m \leq k \leq n$, we say that a boolean matrix M with t rows and n columns is a (n, k, m)-selector if any submatrix of M obtained by choosing k out of n arbitrary columns of M contains at least m distinct rows of the identity matrix I_k. The integer t is referred as the size of the (n, k, m)-selector.*

In order to use selectors as transmission schedules, the parameter n is intended to refer to the number of nodes in the network, k refers to the maximum number of nodes that can compete to transmit (i.e., $k = \Delta + 1$, where Δ is the maximum degree of the network), and m refers to the number of nodes that are guaranteed to successfully transmit during the t-round schedule. Therefore, each column of the matrix M is used to define the whole transmission schedule of each node. Rows are used to decide which nodes should transmit at each time slot: In the i-th time slot, node v will transmit iff $M_{i,v} = 1$ (and v has a packet queued); the schedule is repeated after each t time slots.

Taking into account the above-mentioned approach, selectors may be used to guarantee that during the schedule, every node will successfully receive some messages.

A $(n, k, 1)$-selector guarantees that, for each node, one of its neighbors will successfully transmit during at least 1 round per schedule cycle (that is, that node will successfully receive at least one message). However, whereas the above use of selectors is helpful in broadcasting (since there is progress every time any node receives a message from a neighbor), it happens that many neighbors may have something to send, but only one of them has something for that node. Therefore, the above presented selector guarantees that each node will receive at least one message, but not necessarily will receive the one addressed to it.

A (n, k, k)-selector (which is known as *strong selector* [17]) guarantees that every node that has exactly k neighbors will receive a message from each one of them. However, it has been shown that its size $t = \Omega(\min\{n, (k^2/\log k)\log n\})$. This means that k packets will be received, but during a long amount of time.

In order to solve the above mentioned problems with known selectors, now we introduce a new type of selectors, which we call *universally strong*. Namely, a (n, k, ϵ)-*universally-strong* selector of length t guarantees that every node will receive $\epsilon \cdot t/k$ successful messages from every neighbor during t rounds. More formally:

Definition 3.2. A (n, k, ϵ)-universally-strong selector $\mathcal{S}$ is a family of t sets $T_1, \ldots, T_t \subseteq [n]$ such that for every set $A \subseteq [n]$ of at most k elements and for every element $a \in A$ there exist at least $\epsilon \cdot t/k$ sets $T_i \in \mathcal{S}$ such that $T_i \cap A = \{a\}$.

3.1   Universally Strong Selectors of Polynomial Size

Clearly, universally strong selectors make sense provided they exist and their size is moderate. In the next theorem, we prove that, for any $\epsilon \leq 1/e$, there exists a (n, k, ϵ)-universally-strong selector of polynomial size.

Theorem 3.1. *For any $\epsilon \leq 1/e$, there exists a (n, k, ϵ)-universally-strong selector of size $O(k^2 \ln n)$.*

Proof. The proof relies on the probabilistic method.

Consider a random matrix M with t rows and n columns, where $M_{i,j} = 1$ with probability p and $M_{i,j} = 0$ otherwise. Given a row i and columns $j_1, \ldots, j_k$, the probability that $M_{i,j_1} = 1$ and $M_{i,j_2} = \cdots = M_{i,j_k} = 0$ (i.e., that node j_1's transmission is not interrupted by nodes $j_2, \ldots, j_k$ in round i) is $P = p(1-p)^{k-1}$ and is maximized with $p = 1/k$. In further considerations we use matrix M generated with $p = 1/k$.

Given columns $C = \{j_1, \ldots, j_k\}$, let $X(C)$ be the number of "good" rows i such that $M_{i,j_1} = 1$ and $M_{i,j_2} = \cdots = M_{i,j_k} = 0$.

We will use the following Chernoff bound:

$$Pr[X(C) \le (1 - \delta)E[X(C)]] \le exp(-E[X(C)]\delta^2/2)$$

for $0 \le \delta \le 1$.

Using $E[X(C)] = Pt$ and $\delta = (kP - \epsilon)/(kP)$, we obtain:

$$Pr[X(C) \le \epsilon t/k] \le exp(-Pt\delta^2/2).$$

Consider a "bad" event $\mathcal{E}$ such that for at least one set of columns of size at most k, there are few good rows. More specifically, $X(C) \le \epsilon t/k$ for at least one set of columns C, where $|C| = k$. The probability R of event $\mathcal{E}$ happening fulfills the following inequality:

$$R \le k\binom{n}{k} exp(-Pt\delta^2/2).$$

Therefore $R < 1$ if

$$exp(-Pt\delta^2/2) < 1/\left[k\binom{n}{k}\right]$$

$$-Pt\delta^2/2 < -\ln\left(k\binom{n}{k}\right)$$

$$Pt\left(\frac{kP - \epsilon}{kP}\right)^2/2 > \ln\left(k\binom{n}{k}\right)$$

Let $c = kP$. Using $\binom{n}{k} \le \left(\frac{ne}{k}\right)^k$, provided $c \ne \epsilon$, we obtain the following:

$$t(c - \epsilon)^2/(2ck) > \ln k + \ln\left(\frac{ne}{k}\right)^k$$

$$t > \left[2ck \ln k + 2ck^2 \ln\left(\frac{ne}{k}\right)\right]/(c - \epsilon)^2$$

Therefore, as long as $0 \le \delta = \frac{c-\epsilon}{c} \le 1$ (so that we can use the Chernoff bound) and $\epsilon \ne c$, the probability of generating a random matrix M such that event $\mathcal{E}$ occurs is less than 1. Thus, there exists a matrix M such that, for every set of k columns $j_1, \ldots, j_k$, there are at least $\epsilon t/k$ rows such that $M_{i,j_1} = 1$ and $M_{i,j_2} = \cdots = M_{i,j_k} = 0$. Trivially, such matrix M guarantees the above property for any set of at most k columns. Hence, M represents a (n, k, ϵ)-universally-strong selector, provided that $\epsilon < c = kP$. Next, we calculate which values of ϵ fulfill that inequality.

Consider a sequence $a_i = (1 + 1/i)^i$. a_i is known as a lower bound on the Euler's number e (i.e., $\forall i\ a_i < e$). Note that $c = kP = (1 - 1/k)^{k-1} = 1/a_{k-1} > 1/e$ for all $k \ge 2$. This implies that any $\epsilon \le 1/e$ fulfills the requirement of $\delta > 0$ and results in the existence of a (n, k, ϵ)-universally-strong selector. □

1. Let $d = \lceil \log_k n \rceil$ and $q = c \cdot k \cdot d$ for some constant $c > 0$ such that $q^{d+1} \geq n$.
2. Consider all polynomials P_i of degree d over field $[q]$. Notice that there are q^{d+1} of such polynomials.
3. Create a matrix M' of size $q \times q^{d+1}$. Each column will represent values $P_i(x)$ of each polynomial P_i for arguments $x = 0, 1, \ldots, q-1$ (corresponding to rows of M'). Next, matrix M'' is created from M' as follows: each value $y = P_i(x)$ is represented and padded in q consecutive rows of 0s and 1s, where 1 is on y-th position, while on all other positions there are 0s. Notice that each column of M'' has q^2 rows (q rows for each argument), thus M'' has size $q^2 \times q^{d+1}$.
4. Remove $q^{d+1} - n$ arbitrary columns from matrix M'', creating matrix M with exactly n columns remaining.
5. Each row of matrix M will correspond to one set T_i of a universally strong selector $T_i \ {}_{i=1}^{q^2}$ over the set $1, \ldots, n$ of elements.

Fig. 2. The POLY-UNIVERSALLY-STRONG algorithm, given parameters n and k.

3.2 Obtaining Universally Strong Selectors of Polynomial Size in Polynomial Time

In the proof of Theorem 3.1, we have introduced a family of universally strong selectors of polynomial size. However, obtaining them by derandomizing would be very inefficient (all the approaches we know are, at least, exponential in n). Here, we present an algorithm, which we call POLY-UNIVERSALLY-STRONG, that computes universally strong selectors of polynomial size in polynomial time (they only have slightly lower values of ϵ comparing to the existential result in Theorem 3.1).

The algorithm, whose code is shown in Fig. 2, has to be executed by each node in the network taking the same polynomials, so that all nodes will obtain exactly the same matrix that defines the transmission schedule.

The next theorem shows that, indeed, it constructs a (n, k, ϵ)-universally-strong selector of polynomial size with $\epsilon = 1/(4 \log_k n)$.

Theorem 3.2. POLY-UNIVERSALLY-STRONG *constructs (by using* $c = 2$*) a* (n, k, ϵ)-*universally-strong selector of size* $4 \cdot k^2 \cdot \lceil \log_k n \rceil^2$ *with* $\epsilon = 1/(4 \log_k n)$.

Proof. First, note that two polynomials P_i and P_j of degree d with $i \neq j$, can have equal values for at most d different arguments. This is because they have equal values for arguments x for which $P_i(x) - P_j(x) = 0$. However, $P_i - P_j$ is a polynomial of degree at most d, so it can have at most d zeroes. So, $P_i(x) = P_j(x)$ for at most d different arguments x.

Take any polynomial P_i and any k polynomials P_j still represented in M (so excluding columns/polynomials removed from consideration in step 4). There are at most $k \cdot d$ different arguments where one of the k polynomials can be equal to P_i. So, for $q - k \cdot d$ different arguments, the values of the polynomial P_i are unique. Therefore, if we look at rows with 1 in column i of matrix M (there are q of those rows, one for each argument), at least $q - k \cdot d$ of them have 0s in chosen k columns. Since there are q^2 rows, so a fraction $(q - k \cdot d)/q^2$ of rows have the desired property (i.e., there is value 1 in column i and value 0 in the chosen k columns):

$$\frac{q - k \cdot d}{q^2} = \frac{(c-1) \cdot k \cdot d}{(c \cdot k \cdot d)^2} = \frac{c-1}{c^2 \cdot k \cdot d} \triangleq f(c) \ .$$

Let us find the value of c that maximizes the function f. To do it, we compute its differential

$$f'(c) = (\frac{c-1}{c^2 \cdot k \cdot d})' = \frac{1 \cdot (c^2 \cdot k \cdot d) - (c-1) \cdot k \cdot d \cdot 2c}{c^4 \cdot k^2 \cdot d^2} =$$
$$= \frac{-c^2 \cdot k \cdot d + 2c \cdot k \cdot d}{c^4 \cdot k^2 \cdot d^2} = \frac{-c+2}{c^3 \cdot k \cdot d}.$$

Thus, $f'(c) = 0$ for $c = 0$ or $c = 2$. The value $c = 2$ maximizes f, giving $f(c) \leq f(2) = 1/(4k \cdot d) = 1/(4k \cdot \log_k n)$.

Therefore, we can construct a (n, k, ϵ)-universally-strong selector with $\epsilon = f(2) \cdot k = 1/(4d) = 1/(4\log_k n)$ of length $4k^2 \cdot \lceil \log_k n \rceil^2$ (which means that an $f(2) = 1/(4k \cdot \log_k n)$ fraction of the selector's sets have the desired property). $\square$

4   A Local-Knowledge Routing Algorithm

In this section, we introduce a local-knowledge packet-oblivious routing algorithm that makes use of the family of universally strong selectors introduced in Sect. 3 as *transmission schedules* (i.e., the time instants when packets stored at each one node must be transmitted to a receiving node). As it has been mentioned previously, local-knowledge routing algorithms work without using any topological information, except for maybe some network's features that do not require full knowledge of its topology. In our particular case, that will consists of some upper bounds on the number of links and on the network's degree.

The code of the proposed algorithm, which we call USS-PLUS-ALG, is shown in Fig. 3. Given a graph G with a number of links bounded by m, and an in-degree of its conflict graph H (which we denote as Δ_{in}^H) bounded by $\Delta \geq 1$, it uses a $(m, \Delta + 1, \epsilon)$-universally-strong selector as a schedule: assuming the selector is represented by matrix M with t rows, each link $z \in E_G$ will transmit at time i iff $M_{i \bmod t, z} = 1$. Notice that here each link is assumed to have an independent queue, and therefore they will act as a sort of "nodes" (in terms of selectors, such as it has been stated in the previous section). This means that each individual link will have its own schedule.

4.1   The USS-PLUS-LIS Algorithm

Next, we show that the USS-PLUS-LIS Algorithm (i.e., the USS-PLUS-ALG algorithm where ALG is the Longest-In-System scheduling policy), guarantees stability, provided a given packets' injection admissibility condition is fulfilled. But first, we define what is a (ρ', T)-frequent schedule.

1. Choose m and Δ such that $\mid E(G) \mid \leq m$ and $\Delta_{in}^H \leq \Delta$.
2. Obtain a $(m, \Delta+1, \epsilon)$ universally strong selector (for some value of ϵ) of some length t and use it as the transmission schedule.
3. When there are several packets awaiting in a single queue, choose the packet to be transmitted according to ALG, breaking ties in any arbitrary fashion.

Fig. 3. The USS-PLUS-ALG algorithm for a network G.

Definition 4.1. A (ρ, T)-frequent schedule *for graph G is an algorithm that decides which links of graph G transmit at every round in such a way that each link is guaranteed to successfully (without radio network collisions) transmit at least $\rho \cdot T$ times in any window of length T (provided at least $\rho \cdot T$ packets await for transmission at the link at the start of the window).*

At this point, we note that the transmission schedules provided by our universally strong selectors can be seen as (ρ, T)-*frequent schedules*.

We now proceed with the main result in this section.

Theorem 4.1. *Given a network G, the* USS-PLUS-LIS *algorithm is stable against any (ρ, b)-adversary, for $\rho < \frac{\epsilon}{\Delta+1}$.*

Proof. Let us take any arbitrary link $z \in E_G$ and consider the set of all other links that conflict with link z, of which there are at most Δ. This means that there exist at least $\epsilon \cdot t/(\Delta + 1)$ rows i in M such that $M_{i \mod t,z} = 1$ and $M_{i \mod t,c_1} = \cdots = M_{i \mod t,c_j} = 0$. Therefore, at time i, link z will transmit a message, and no link that conflicts with the link z will transmit. This guarantees that each link will successfully transmit, at least, $\epsilon \cdot t/(\Delta + 1)$ messages during any schedule of length t (i.e., we obtained a $(\epsilon/(\Delta + 1), t)$-frequent schedule $\mathcal{S}$). Then, we can apply the result Lemma 3 in the Appendix A in [15] to deduce that such an algorithm is stable against any (ρ, b)-adversary, where $\rho < \frac{\epsilon}{\Delta+1}$. □

By using the selectors provided by the POLY-UNIVERSALLY-STRONG algorithm in USS-PLUS-LIS, we have the following result:

Corollary 4.1. *Given a network G, the* USS-PLUS-LIS *algorithm using a universally strong selector computed by the* POLY-UNIVERSALLY-STRONG *algorithm is a stable algorithm against any (ρ, b)-adversary, for $\rho < \frac{1}{4(\Delta+1)\log_{\Delta+1} m}$.*

If instead of the selectors provided by the POLY-UNIVERSALLY-STRONG algorithm, we use a selector from Theorem 3.1, we have that:

Corollary 4.2. *Given a network G, there exists a universally strong selector that, used in the* USS-PLUS-LIS *algorithm, provides a stable algorithm against any (ρ, b)-adversary, for $\rho < \frac{1}{e \cdot (\Delta+1)}$.*

As it can be readily seen, the USS-PLUS-LIS algorithm for a network G requires some knowledge of the value of the in-degree of its conflict graph H

(i.e., of Δ_{in}^H). In order to obtain H it is necessary to gather the whole topology of G. However, as the next lemma shows, Δ_{in}^H can be bounded by the in-degree of the network G (denoted Δ_G).

Lemma 4.1. $\Delta_{in}^H \leq \Delta_G^2 + \Delta_G - 1$, provided $\Delta_G > 0$.

Proof. If $\Delta_{in}^H = 0$, then the lemma is trivially true. Otherwise, consider a vertex e in H of maximum in-degree $deg(e) = \Delta_{in}^H$. Since $\Delta_{in}^H \neq 0$, there is at least one edge $(e', e) \in H$ such that, in G, e cannot successfully transmit at the same time instant when e' transmits. Let us denote $e = (u, v)$ and $e' = (u', v')$, and let us consider the different scenarios where e and e' may conflict.

Now, we make a case analysis regarding the possible conflicts in G (note that its in-degree is equal to its out-degree, since G is symmetric):

1. $u' = u$ and $v' \neq v$ (a node $u = u'$ cannot transmit messages to 2 different receivers): there are at most $\Delta_G - 1$ such links e', given fixed link e.
2. $u' = v$ (if u' transmits, it cannot listen at the same time): there are at most Δ_G such links e', given fixed link e.
3. $u' \neq u$ is a neighbor of v (i.e., v can hear both from u and u'): there are at most $\Delta_G - 1$ neighbors of v different than node u, and each of them has, at most, Δ_G different links. This gives $\Delta_G^2 - \Delta_G$ such links e', given fixed link e.

Therefore, in overall there are at most $(\Delta_G - 1) + \Delta_G + (\Delta_G^2 - \Delta_G) = \Delta_G^2 + \Delta_G - 1$ such links. □

The previous lemma shows that USS-PLUS-LIS can be seen as a local-knowledge algorithm, in the sense that it only requires some knowledge about two basic system parameters: the number of links and the network's in-degree. In Sect. 5 we will also look at a solution that requires some global-knowledge of G.

4.2 Optimality of the USS-PLUS-LIS Algorithm

In the next theorem, we show an impossibility result regarding routing algorithms, either based on selectors or not, that only make use of upper bounds on the number of links and on the network's degree.[2]

Theorem 4.2. *No routing algorithm that only makes use of upper bounds on the number of links and on the network's degree guarantees stability for all networks of degree at most Δ, provided the injection rate $\rho = \omega(1/\Delta^2)$.*

Proof. Assume, to the contrary, that there exists a routing algorithm ALG such that, given any network of which it is aware of both its number of links and its degree, it guarantees that there are no more than Q_{max} packets in the system at all times against all adversaries with injection rate $\rho = \omega(1/\Delta^2)$. Note that Q_{max} could be a function on ρ, n, but a constant with respect to time.

---

[2] To be strict, it is also necessary that each node v decides whether or not to an outgoing edge $e = (u, v)$ should transmit at time t based on t and on the ID of node u. •

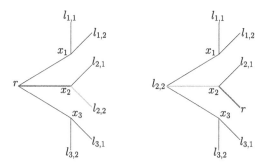

Fig. 4. Example of a tree T (on the left) and tree $T_{2,2}$ (on the right) for $\Delta = 3$. Nodes r and $l_{2,2}$ swapped places, which means that edges (x_2, r) and $(x_2, l_{2,2})$ (marked in blue and red, respectively) swapped their places as well. (Color figure online)

Consider a complete Δ-regular tree T of depth 2, rooted at r. Let us denote the nodes at distance 1 from r as x_i, for $i = 1, \ldots, \Delta$ and leaves adjacent to x_i as l_i^j for $j = 1, \ldots, \Delta - 1$. Let us generate a family $\mathcal{F}$ of trees $T_{i,j}$ as follows: swap the root r of T with leaf $l_{i,j}$ of T (see Fig. 4). Note that edges (x_i, r) and $(x_i, l_{i,j})$ swapped places, edges (x_k, r) for $k \neq i$ were removed and in their place edges $(x_k, l_{i,j})$ appeared. Other edges, i.e., $(x_k, l_{k,a})$ for $a = 1, \ldots, \Delta - 1$ and $(x_i, l_{i,b})$ for $b \neq j$, remain in the same place in both T and $T_{i,j}$.

Note that edges $(x_k, l_{k,a})$ (for $k = 1, \ldots, \Delta$ and $a = 1, \ldots, \Delta - 1$) exist in every tree in $\mathcal{F} \cup \{T\}$. Let us denote the set of these edges as E.

Consider an adversary $\mathcal{A}$ that, starting from round 0, injects 1 packet into every edge outgoing from x_i (for $i = 1, \ldots, \Delta$) every $1/\rho$ rounds. Such adversary is a $(\rho, 1)$-adversary in each tree in $\mathcal{F} \cup \{T\}$.

Note that each packet injected into an edge incoming into the root of a tree $T' \in \mathcal{F} \cup \{T\}$ cannot be simultaneously transmitted with any other packet injected by $\mathcal{A}$. In particular, it cannot be simultaneously transmitted with any other packet on edges in E.

Consider a time prefix of length τ rounds. Consider any edge $e \in E$. Edge e is incident to the root in some tree $T' \in \mathcal{F} \cup \{T\}$. ALG must successfully transmit from e in T' in at least $\rho\tau - Q_{max}$ rounds during the considered prefix, since ALG is stable. This means that all other edges in E must not transmit in those rounds. Since there are $\Delta(\Delta - 1)$ possible choices of edge $e \in E$, each choice requiring all other edges in E not to transmit in $\rho\tau - Q_{max}$ rounds, we get that each edge in E must not transmit in $\Delta(\Delta - 1) \cdot (\rho\tau - Q_{max})$ rounds and must transmit in $\rho\tau - Q_{max}$ rounds, for a total of $\Delta^2 \cdot (\rho\tau - Q_{max})$ rounds in the prefix of length τ. For $\rho = \omega(1/\Delta^2)$, we can choose τ such that $\Delta^2 \cdot (\rho\tau - Q_{max}) > \tau$, which gives us a contradiction. □

If we apply Theorem 4.2 to USS-PLUS-LIS, then our goal is to find how close to $\rho = O(1/\Delta_G^2)$ is its maximum injection rate for which it guarantees stability.

If we consider Theorem 4.1 with $\Delta = \Delta_{in}^H$, we have that USS-PLUS-LIS can be stable for $\rho = O(1/\Delta_{in}^H)$. Furthermore, by Lemma 4.1 we know that Δ_{in}^H can be as large as $\Theta(\Delta_G^2)$. Then, we have that USS-PLUS-LIS guarantees stability for $\rho = O(1/\Delta_G^2)$ for all networks G, which matches the result in Theorem 4.2. This proves that the USS-PLUS-LIS algorithm is asymptotically optimal regarding the injection rate for which stability is guaranteed.

5   A Global-Knowledge Routing Algorithm

In this section, we introduce a global-knowledge packet-oblivious routing algorithm, which we call COLORING-PLUS-ALG, that is based on using graph coloring as *transmission schedules*. Such an algorithm does not take into account any historical information. However, it has to be seeded by some information about the network topology (i.e., it is a global-knowledge routing algorithm).

Next, we show that the COLORING-PLUS-LIS algorithm (i.e., the COLORING-PLUS-ALG algorithm where ALG is the Longest-In-System scheduling policy), guarantees stability, provided a given packets' injection admissibility condition is fulfilled.

Now, we proceed withe the main results in this section. But before we introduce the COLORING-PLUS-ALG routing algorithm, we state the following fact regarding the relationship between vertex coloring in a conflict graph, and its use as a transmission schedule.

Fact 1 ([10]). *Vertex coloring of the conflict graph $H(G)$ using x colors is equivalent to a schedule of length x that successfully transmits a packet via each directed link of network G.*

Note that every set of vertices of same color can be extended to a maximal independent set. The resulting family of independent sets is still a feasible schedule that guarantees no conflicts and is no worse than just coloring. In fact, it may allow some links to transmit more than once during the schedule, without increasing the length of the schedule.

Following, we show that coloring of a collision graph can be used to obtain a transmission schedule, where each link is guaranteed to regularly transmit.

Lemma 5.1. *A k-coloring of collision graph H provides a $(1/k, k)$-frequent schedule.*

Proof. Let us split the vertices V_H of the graph H into sets V_H^i for $i = 0, 1, \ldots, k-1$, where every vertex in V_H^i is assigned the i-th color in the vertex coloring of graph H. Each link in the graph G is represented by one vertex in V_H, and therefore each link is assigned a unique color. According to the definition of the conflict graph H, if there is no edge $(u, v) \in E_H$, then links $u \in E_G$ and $v \in E_G$ can deliver their packets simultaneously, without a collision. Therefore, if at a given round t only links of $(t \mod i)$-th color transmit, then no collision occurs. Since each link has a color $i \in \{0, 1, \ldots, k-1\}$ assigned to it, then each link will successfully transmit a packet once each k consecutive rounds (as far as there is one packet waiting in its queue). $\square$

---

1. Use optimal coloring of graph H as the transmission schedule, and repeat it indefinitely.
2. When there are several packets awaiting in a single queue, choose the packet to be transmitted according to ALG, breaking ties in any arbitrary fashion.

Fig. 5. The COLORING-PLUS-ALG algorithm for graph G.

Since $\chi(H)$-coloring is an optimal coloring of graph H, we have the following result.

Corollary 5.1. *An optimal coloring of collision graph H provides a $(1/\chi(H), \chi(H))$-frequent schedule.*

Once we have made it clear that coloring of a collision graph can be used to obtain a transmission schedule, the code of the COLORING-PLUS-ALG algorithm is shown in Fig. 5.

5.1 The COLORING-PLUS-LIS Algorithm

Now, we show that COLORING-PLUS-LIS (i.e., the COLORING-PLUS-ALG algorithm where ALG is the Longest-In-System scheduling policy), guarantees stability, provided a given packets' injection admissibility condition is fulfilled.

Theorem 5.1. *The COLORING-PLUS-LIS algorithm is stable provided $\rho < 1/\chi(H)$, where $\chi(H)$ is the chromatic number of the conflict graph H of the network G.*

Proof. We start the proof with referring to Corollary 5.1, which shows that coloring of a collision graph can be used to obtain a $(1/\chi(H), \chi(H))$-frequent schedule $\mathcal{C}$.

Let us take any $\rho = 1/\chi(H) - \epsilon$, for some $\epsilon > 0$. We can use Lemma 3 in the Appendix A in [15] with $\mathcal{S} = \mathcal{C}$ (so, $\rho' = 1/\chi(H)$) to show that COLORING-PLUS-LIS is stable against any (ρ, b)-adversary in the radio network model.   □

Observe that, contrary to the USS-PLUS-LIS protocol, the COLORING-PLUS-LIS algorithm requires global-knowledge of the structure of the graph: first, to construct H, and then to obtain its optimal coloring.

5.2 Optimality of the COLORING-PLUS-LIS Algorithm

Now, we show that the COLORING-PLUS-LIS algorithm is optimal regarding the injection rate, in the sense that no algorithm can guarantee stability for a higher injection rate that provide by it.

Theorem 5.2. *No algorithm can be stable for all networks against a (ρ, b)-adversary for $\rho > 1/\chi(H)$.*

Proof. Let us consider a network graph G on n nodes that is a clique. For such network, the collision graph H is also a clique, since each link is in conflict with each other link. Collision graph H has $n^2 - n$ vertices and requires $n^2 - n$ colors to be colored, i.e., $\chi(H) = n^2 - n$.

Consider a $(1/\chi(H) + \varepsilon, 2)$-adversary for some $\varepsilon > 0$ that after every $\chi(H)$ rounds injects one packet into each link (starting in round 0) and simultaneously after each $1/\varepsilon$ rounds injects another packet into each link (starting in round 0). Therefore, in any prefix of $T = k \cdot \chi(H)$ rounds for $k \in \mathbb{N}$, the adversary injects $(k + 1) + \lfloor T/\varepsilon \rfloor + 1$ packets into each link, i.e., $I = (k + \lfloor T/\varepsilon \rfloor + 2) \cdot (n^2 - n)$ packets into the system.

On the other hand, since G is a clique, any algorithm can successfully transmit at most 1 packet per round in the entire network. Therefore, in $T = k \cdot \chi(H) = k \cdot (n^2 - n)$ rounds at most $k \cdot (n^2 - n)$ packets can be transmitted. So, at the end of a prefix of length T, there are at least $I - k \cdot (n^2 - n) = (\lfloor T/\varepsilon \rfloor + 2) \cdot (n^2 - n)$ packets remaining in the system. For T approaching infinity, the number of packets remaining in the queues grows to infinity. This means that the queues are not bounded and the algorithm is not stable. □

5.3 Global-Knowledge Vs Local-Knowledge Routing Protocols

Regarding the COLORING-PLUS-LIS protocol, by the Brooks' theorem [9], we have that $\chi(H) \leq \Delta^H + 1$. Let $indeg^H(e)$ (and $outdeg^H(e)$) denote the indegree (outdegree) of node e in graph H. Recall that each edge in the network graph was replaced by two oppositely directed links. This means that, if a link e blocks $outdeg^H(e)$ other links, then the opposite link e' is blocked by $indeg^H(e') = outdeg^H(e)$ links. Therefore, $\Delta^H = \Theta(\Delta_{in}^H)$. Then, Theorem 5.1 guarantees stability for $\rho = O(1/\Delta_{in}^H)$.

On the other hand, from the result in Corollary 4.2, we have that USS-PLUS-LIS can only guarantee stability for $\rho = O(1/(e \cdot \Delta_{in}^H))$. This implies that, by using the COLORING-PLUS-LIS protocol, it is possible to guarantee stability for a wider range of injection rates than by using the USS-PLUS-LIS protocol: namely, the injection rate for which stability is guaranteed is e times higher.

6 Extension of the Results to Other Scheduling Policies

In this section, we show that the results obtained in Sect. 5 and 4 for routing combined with LIS (Longest In System) can be extended to other scheduling policies; namely, NFS (Nearest-From-Source), SIS (Shortest-In-System) and FTG (Farthest-To-Go). Indeed, for such a scheduling policies, Theorems 6.1 and 6.2 respectively parallelize the analogous results in Theorems 4.1 and 5.1 obtained for LIS.

Theorem 6.1. *Given a network G, the* USS-PLUS-ALG *algorithm, where* ALG $\in \{NFS, SIS, FTG\}$, *is stable against any (ρ, b)-adversary, for $\rho < \frac{\epsilon}{\Delta + 1}$.*

Proof. The proof is similar to that in Theorem 4.1. The only difference is that, instead of Lemma 3, we can apply the results in Lemma 6 for NFS, SIS and FTG (see Appendix D in [15]) to deduce that such an algorithm is stable against any (ρ, b)-adversary, where $\rho < \frac{\epsilon}{\Delta+1}$.   □

Theorem 6.2. *The* COLORING-PLUS-ALG *algorithm, where* ALG $\in$ {*NFS,SIS, FTG*}, *is stable provided* $\rho < 1/\chi(H)$, *where* $\chi(H)$ *is the chromatic number of the conflict graph* H *of the network* G.

Proof. We will reduce the packet scheduling in radio network problem to the problem of packet scheduling in the wired failure model [1], in which these policies are known to be stable.

We start the proof with referring to Corollary 5.1, which shows that coloring of a collision graph can be used to obtain a $(1/\chi(H), \chi(H))$-frequent schedule $\mathcal{C}$.

Let us take any $\rho = 1/\chi(H) - \epsilon$, for some $\epsilon > 0$. Now, we can use Lemma 6 with $\mathcal{S} = \mathcal{C}$ (so, $\rho' = 1/\chi(H)$) and ALG $\in$ {NFS,SIS,FTG} (with $\rho'' = 1 - \epsilon$) to show that we can build an algorithm that is stable against any (ρ, b)-adversary in the radio network model (see Appendix D in [15]). Note that COLORING-PLUS-ALG is a special case of the algorithm built in the proof of Lemma 6 with $\mathcal{S} = \mathcal{C}$. Therefore COLORING-PLUS-ALG with ALG $\in$ {NFS,SIS,FTG} is stable against any (ρ, b)-adversary in the radio network model.   □

7   Conclusions and Future Work

In this work, we studied the fundamental problem of stability in multi-hop wireless networks.

We introduced a new family of combinatorial structures, which we call *universally strong selectors*, that are used to provide a set of transmission schedules. Making use of these structures, combined with some known queuing policies, we propose a packet-oblivious routing algorithm which is working without using any global topological information, and guarantees stability for certain injection rates. We show that this protocol is asymptotically optimal regarding the injection rate for which stability is guaranteed.

Furthermore, we also introduced a packet-oblivious routing algorithm that guarantees stability for higher traffic. We also show that this protocol is optimal regarding the injection rate for which stability is guaranteed. However, it needs to use some global information of the system topology.

A natural direction would be to study other classes of protocols; for instance, when packets are injected without pre-defined routes. Universally strong selectors are interesting on its own right – finding more applications for them is a promising open direction. Finally, exploring the reductions between various settings of adversarial routing could lead to new discoveries, as demonstrated in the last part of this work.

References

1. Àlvarez, C., Blesa, M.J., Díaz, J., Serna, M.J., Fernández, A.: Adversarial models for priority-based networks. Networks **45**(1), 23–35 (2005)
2. Àlvarez, C., Blesa, M.J., Serna, M.J.: A characterization of universal stability in the adversarial queuing model. SIAM J. Comput. **34**(1), 41–66 (2004)
3. Anantharamu, L., Chlebus, B.S., Kowalski, D.R., Rokicki, M.A.: Packet latency of deterministic broadcasting in adversarial multiple access channels. J. Comput. Syst. Sci. **99**, 27–52 (2019)
4. Andrews, M., Awerbuch, B., Fernández, A., Leighton, F.T., Liu, Z., Kleinberg, J.M.: Universal-stability results and performance bounds for greedy contention-resolution protocols. J. ACM **48**(1), 39–69 (2001)
5. Andrews, M., Zhang, L.: Scheduling over a time-varying user-dependent channel with applications to high-speed wireless data. J. ACM **52**(5), 809–834 (2005)
6. Andrews, M., Zhang, L.: Routing and scheduling in multihop wireless networks with time-varying channels. ACM Trans. Algorithms **3**(3), 33 (2007)
7. Bhattacharjee, R., Goel, A., Lotker, Z.: Instability of FIFO at arbitrarily low rates in the adversarial queueing model. SIAM J. Comput. **34**(2), 318–332 (2004)
8. Borodin, A., Kleinberg, J.M., Raghavan, P., Sudan, M., Williamson, D.P.: Adversarial queuing theory. J. ACM **48**(1), 13–38 (2001)
9. Brooks, R.L.: On colouring the nodes of a network. Math. Proc. Cambridge Philos. Soc. **37**(2), 194–197 (1941)
10. Chlebus, B.S., Cholvi, V., Garncarek, P., Jurdziński, T., Kowalski, D.R.: Routing in wireless networks with interferences. IEEE Commun. Lett. **21**(9), 2105–2108 (2017)
11. Chlebus, B.S., Cholvi, V., Kowalski, D.R.: Universal routing in multi hop radio network. In: Proceedings of the 10th ACM International Workshop on Foundations of Mobile Computing (FOMC), pp. 19–28. ACM (2014)
12. Chlebus, B.S., Kowalski, D.R., Pelc, A., Rokicki, M.A.: Efficient distributed communication in ad-hoc radio networks. In: Aceto, L., Henzinger, M., Sgall, J. (eds.) ICALP 2011. LNCS, vol. 6756, pp. 613–624. Springer, Heidelberg (2011). https://doi.org/10.1007/978-3-642-22012-8_49
13. Chlebus, B.S., Kowalski, D.R., Rokicki, M.A.: Adversarial queuing on the multiple access channel. ACM Trans. Algorithms **8**(1), 5:1–5:31 (2012)
14. Cholvi, V., Echagüe, J.: Stability of FIFO networks under adversarial models: state of the art. Comput. Netw. **51**(15), 4460–4474 (2007)
15. Cholvi, V., Garncarek, P., Jurdzinski, T., Kowalski, D.R.: Packet-oblivious stable routing in multi-hop wireless networks. ArXiv, abs/1909.12379 (2019)
16. Cholvi, V., Kowalski, D.R.: Bounds on stability and latency in wireless communication. IEEE Commun. Lett. **14**(9), 842–844 (2010)
17. Clementi, A.E.F., Monti, A., Silvestri, R.: Distributed broadcast in radio networks of unknown topology. Theor. Comput. Sci. **302**(1), 337–364 (2003)
18. Echagüe, J., Cholvi, V., Fernández, A.: Universal stability results for low rate adversaries in packet switched networks. IEEE Commun. Lett. **7**(12), 578–580 (2003)
19. Gamarnik, D.: Stability of adaptive and nonadaptive packet routing policies in adversarial queueing networks. SIAM J. Comput. **32**(2), 371–385 (2003)
20. Goel, A.: Stability of networks and protocols in the adversarial queueing model for packet routing. Networks **37**(4), 219–224 (2001)

21. Koukopoulos, D., Mavronicolas, M., Nikoletseas, S.E., Spirakis, P.G.: The impact of network structure on the stability of greedy protocols. Theory Comput. Syst. **38**(4), 425–460 (2005)
22. Lotker, Z., Patt-Shamir, B., Rosén, A.: New stability results for adversarial queuing. SIAM J. Comput. **33**(2), 286–303 (2004)

Stateless Information Dissemination Algorithms

Volker Turau[(✉)]

Institute of Telematics Hamburg University of Technology, 21073 Hamburg, Germany
turau@tuhh.de

Abstract. Stateless protocols are advantageous in high volume applications, increasing performance by removing the load caused by retention of session information and by providing crash tolerance. In this paper we present an optimal stateless information dissemination algorithm for synchronous distributed systems. The termination time is considerable lower than that of a recently proposed stateless dissemination protocol. Apart from a special case the new algorithm achieves the minimum possible termination time. The problem of selecting k dissemination nodes with minimal termination time is NP-hard. We prove that unless $NP = P$ there is no approximation algorithm for this problem with approximation ratio $3/2 - \epsilon$. We also prove for asynchronous systems that deterministic stateless information dissemination is only possible if a large enough part of the message can be updated by each node.

1 Introduction

A stateless protocol is a communications protocol in which no session information is retained by the participating nodes, i.e., communication does not depend on one or more preceding events in a sequence of interactions. Thus, stateless protocols do not utilize local storage. This is a big advantage in high volume applications, increasing performance by removing the load caused by retention of session information. In addition they provide fault tolerance after node crashes. In this paper we focus on stateless information dissemination algorithms for distributed systems. These algorithms distribute information initially stored at some node to all nodes of the network in finite time. The most basic algorithm for this purpose is the deterministic flooding algorithm for asynchronous systems. The originator of the information sends a message containing the information to all neighbors and whenever a node receives the message for the first time, it sends it to all its neighbors in the communication graph. This algorithm is a stateful algorithm, it requires each node to keep a record of which messages have already arrived at the node. This requires storage proportional to the number of disseminated messages per node, which is a problem for resource-constrained devices. Furthermore, since the termination of the flooding algorithm cannot be detected by the nodes, the storage requirements grow over time. These disadvantages motivate the search for stateless information dissemination algorithms.

A. W. Richa and C. Scheideler (Eds.): SIROCCO 2020, LNCS 12156, pp. 183–199, 2020.
https://doi.org/10.1007/978-3-030-54921-3_11

A way to circumvent the restrictions caused by excluding the usage of local storage is to add session information to messages. This way session information is permanently in transit and thus available at receiving nodes. An extreme example is to include in the message the identifiers of all nodes that have already received the message. This way the repeated circulation of messages in closed loops can be intercepted. Obviously, this leads to messages of size $O(n \log n)$. Thus, an operative definition of a stateless algorithm must also refer to the size of messages. A stateless algorithm that is allowed to update $O(f(n))$ bits of a message before forwarding is called $f(n)$-stateless. If no bits can be updated it is called a *truly stateless* algorithm. A formal definition is given in Sect. 2.

A recently introduced variant of the classic flooding algorithm for synchronous systems that does not need to remember whether a messages has been received is called *amnesiac flooding* [13,14]. In this algorithm every time a node receives a message, it forwards it to those neighbors from which it didn't receive the message in the current round. In contrast to classic flooding, a node may forward a message several times. Amnesiac flooding is truly stateless and avoids the mentioned storage issues. Hussak et al. have analyzed the termination time of amnesiac flooding with a single initiator v_0 [14]. They showed that it terminates on any finite graph; for bipartite graphs after $\epsilon_G(v_0)$ rounds, i.e., one round less than in the classic algorithm, where $\epsilon_G(v_0)$ denotes the eccentricity of v_0 in G. For non-bipartite graphs amnesiac flooding requires at least $\epsilon_G(v_0) + 1$ and at most $\epsilon_G(v_0) + Diam(G) + 1$ rounds, where $Diam(G)$ denotes the diameter of G. The authors also provide examples showing that the bounds are sharp. Thus, the relinquishment of the usage of storage leads to a prolonged termination time. Amnesiac flooding was also analyzed for sets of initiators and upper resp. lower bounds for the termination time are proved in [18].

The results of [14] and [18] show that the termination time of amnesiac flooding in synchronous systems can be significantly larger then that of the classic flooding algorithm. Hence, the question arises, whether there exists a truly stateless flooding algorithm that has the same termination time as the classic flooding algorithm. In asynchronous systems the requirement to store information about received messages cannot be offhand dropped. The reason is that since messages can be arbitrarily delayed they can arrive multiple times at a node. If a node is oblivious of arriving messages then a message can circulate forever and the algorithm would not terminate. Thus, deterministic truly stateless algorithms are impossible in asynchronous systems.

In this paper we propose stateless information algorithms for synchronous and asynchronous systems. For the synchronous case we propose a new flooding algorithm $\mathcal{A}_{SF}$ that has the same termination time as the classic flooding algorithm. Apart from a special case it is optimal and significantly faster than amnesiac flooding. In particular, for a single initiator v_0 the algorithm terminates after $\epsilon_G(v_0) + 1$ rounds for any graph G. This is an improvement by $Diam(G)$ compared to the result of [14]. We also analyze the case for a set S with k initiators and prove that the algorithm terminates after $d_G(S, V) + 1$ rounds. Here, $d_G(S, V)$ denotes the maximal distance of a node in V to those in S.

This implies that there exists a set S of initiators for which $\mathcal{A}_{SF}$ terminates in $r_k(G) + 1$ rounds, here $r_k(G)$ denotes the k-radius of G.

We also define the *stateless-flooding problem* (SF problem) for synchronous systems. An instance of this problem is given by a connected graph $G = (V, E)$ and a positive integer $k \leq |V|$. The stateless-flooding problem is to find a subset S of V of size k, such that algorithm $\mathcal{A}_{SF}$ when started concurrently by all nodes of S terminates in a minimal number of rounds. We prove that unless $P = NP$ there is no approximation algorithm for the stateless-flooding problem which has an approximation ratio less than $3/2$.

In the last section we focus on asynchronous systems. We show that no deterministic stateless information dissemination algorithm exists that can only update a constant number of bits of the message. We show the existence of an algorithm that is allowed to update $O(\log n)$ bits.

2 Notation

In this paper $G(V, E)$ is always a finite, connected, undirected, unweighted graph with $n = |V| > 2$ and $m = |E|$. Denote by $deg(v)$ the *degree* of a node $v \in V$. A node with degree 0 is called *isolated*. For $u, v \in V$ denote by $d_G(v, u)$ the *distance* in G between v and u, i.e., the number of edges of a shortest path between v and u. For $U \subseteq V$ and $v \in V$ let $d_G(v, U) = \min\{d_G(v, u) \mid u \in U\}$ and $d_G(U, V) = \max\{d_G(v, U) \mid v \in V\}$. For $v \in V$ denote by $\epsilon_G(v)$ the *eccentricity* of v in G, i.e., the greatest distance between v and any other node in G, i.e., $\epsilon_G(v) = \max\{d_G(w, v) \mid w \in V\}$. The *radius* $Rad(G)$ (resp. *diameter* $Diam(G)$) of G is the minimum (resp. maximum) eccentricity of any node of G. A node v is called *central* if $\epsilon_G(v) = Rad(G)$. Let $n \geq k \geq 1$ be an integer. We call $r_k(G) = \min\{d_G(U, V) \mid |U| = k\}$ the *k-radius* of G, i.e., $r_1(G) = Rad(G)$. Each subset $U \subseteq V$ with $|U| = k$ and $r_k(G) = d_G(U, V)$ is called a *k-center* of G.

An edge $(u, w) \in E$ is called a *cross edge* with respect to a node v_0 if $d_G(v_0, u) = d_G(v_0, w)$. Any edge of G that is not a cross edge with respect to v_0 is called a *forward edge* for v_0. Furthermore, for $U \subseteq V$ the graph induced by U is denoted by $G[U]$.

A synchronous distributed algorithm is executed in rounds of fixed lengths and all messages sent by all nodes in a particular round are received in the next round. In particular, in each round all nodes first receive messages sent in the previous round and then send messages. An asynchronous distributed algorithm decides immediately upon receiving a message which messages to send and to which neighbors. Throughout the paper we assume that no messages are lost or corrupted. We use the following definition of statelessness in this paper.

Definition 1. *A synchronous information dissemination algorithm is called truly stateless if a node decides only on the basis of the messages received in the current round which messages to send in this round. Furthermore, nodes are not allowed to change the content of a received message before forwarding.*

The definition implies that synchronous truly stateless algorithms do not rely on information that is stored beyond the duration of a single round. In addition, nodes are only allowed to read the content of incoming messages, but cannot alter the content before forwarding, e.g., they are not allowed to add information related to the forwarding process or routing in general to outgoing messages.

Definition 2. *Let f be a function from $\mathbb{N}$ to $\mathbb{N}$. An asynchronous information dissemination algorithm is called $f(n)$-stateless if a node decides only on the basis of each received message which messages to send as a reaction. Nodes are allowed to update up to $O(f(n))$ bits of a message before forwarding it.*

The definition for the asynchronous case implies that the usage of local storage for prospective decisions about method forwarding is completely excluded.

3   State of the Art

Different facets of stateless programming received a lot of attention in recent years: MapReduce framework, monads in functional programming, and reentrant code [5]. Stateless protocols are very popular in client-server applications because of their high degree of scalability. They simplify the design of the server and require less resources because servers do not need to keep track of session information or a status about each communicating partner for multiple requests. In stateless protocols each message travels on it's own without reference to any previous message. Despite their significance for practical application, stateless protocols have only received limited attention in theoretical research. According to Awerbuch et al. statelessness implies various desirable properties of a distributed algorithm, such as: asynchronous updates and self-stabilization [2].

Motivated by the Border Gateway Protocol (BGP) Dolev et al. define a model of stateless computation in which processors do not have an internal state, but rather interact by repeatedly mapping incoming messages to outgoing messages and output values [4]. The authors consider a distributed network, in which every node receives an external input x_i and nodes have to compute a global function $f(x_1, \ldots, x_n)$, by repeatedly exchanging messages. Dolev et al. provide general upper bounds on the round complexity for any function f, showing that a linear number of rounds is sufficient to compute any function. While our work basically uses the same model, we focus on the problem of information dissemination.

Broadcast in computer networks has been the subject of extensive research. The survey paper [11] covers early work. In the standard flooding algorithm, each node that receives the message for the first time forwards to it all other neighbors. If v_0 is the originating node then after $\epsilon_G(v_0)$ rounds each node has received the message. This algorithm uses $2m$ messages and terminates in $\epsilon_G(v_0) + 1$ rounds. These bounds hold in the synchronous and in the asynchronous case [17]. The number of messages can be reduced if a received message is not forwarded to the sender of the message. This has no influence on the run-time of the algorithm and the number of messages sent is still $O(m)$. The number of messages can be reduced to $O(n)$ if flooding is performed via the edges of a spanning tree only.

The flooding algorithm can be generalized to the case of a set S of originating nodes, also called *initiators*.

The standard flooding algorithm is a stateful algorithm. Each node needs to maintain for each message a marker that the message has been forwarded. This requires storage per node proportional to the number of disseminated messages. Standard flooding is therefore of limited suitability for resource-constrained devices as those used in the Internet of Things. A stateless version of flooding was proposed by Hussak and Trehan [13]. Their algorithm – called *amnesiac flooding* – works as follows: Whenever a node receives a message it forwards the message to those neighbors from which it did not receive the message in the current round. Amnesiac flooding has a lower memory requirement since markers are only kept for one round. Note, that a node may forward a message more than once. The authors prove that in synchronous networks amnesiac flooding when started by a node v_0 terminates after at most $\epsilon_G(v_0) + Diam(G) + 1$ rounds. They also show that a node forwards each message at most twice. Amnesiac flooding for sets of initiators was analyzed in [18].

A problem related to selecting initiators of information dissemination leading to a minimal termination time is the k-center problem. The task of this problem is to find a k-center of a graph [15]. The problem and many variants of it including some approximations are known to be NP-hard [3,12]. It is shown in [18] that choosing the nodes of a k-center for amnesiac flooding does not lead to a minimal termination time. This observation is not true for the stateless flooding algorithm presented in this work.

A problem related to broadcast is rumor spreading that describes the dissemination of information in networks through pairwise interactions. A simple model for rumor spreading is to assume that in each round, each vertex that knows the rumor, forwards it to a randomly chosen neighbor. Thus, rumor spreading is truly stateless. For many topologies, this strategy is a very efficient way to spread a rumor. With high probability the rumor is received by all vertices in time $\Theta(\log n)$, if the graph is a complete graph or a hypercube [7,8]. New results about rumor spreading can be found in [16].

4   Synchronous Stateless Information Dissemination

A lower bound for the round complexity of any information dissemination algorithm with k initiators is $r_k(G) + 1$. Consider the case $S = \{v_0\}$ and G a cycle with an odd number of nodes. Let v_1, v_2 be nodes with the maximal distance in G to v_0. The message reaches these two nodes at the earliest in round $\epsilon_G(v_0)$. Since nodes v_1 and v_2 do not know whether their neighbors already have received the message they have to forward it in round $r_k(G) + 1$. The lower bound is achieved by the classic flooding algorithm which is not stateless. This raises the question whether truly stateless information dissemination can also be realized in $r_k(G) + 1$ rounds. This question will be answered affirmatively for synchronous systems in the following.

The goal of the new flooding algorithm is to distribute information – initially stored at the nodes of a set S – to all nodes of the network. In the first round each node of S sends the message to all its neighbors. A node from S that does not receive one of these messages sent in the first round sends in round two again the message to all its neighbors. In each of the following rounds – including round two – each node that receives a message forwards this message to all of its neighbors from which it did not receive this message in this round. The algorithm terminates, when no more messages are sent.

Algorithm 1 shows a formal definition of algorithm $\mathcal{A}_{SF}$. A node in S has to retain the information that it has initiated a broadcast for one round. We regard this as belonging to the initiation of the broadcast process and therefore this information is available to algorithm $\mathcal{A}_{SF}$. Hence, algorithm $\mathcal{A}_{SF}$ is truly stateless. The code shows the handling of a single message m. If several different messages are disseminated concurrently, each of them requires its own handling, i.e., its own set M.

Algorithm 1: Algorithm $\mathcal{A}_{SF}$ distributes a message in the graph G

---

input: A graph $G = (V, E)$, a subset S of V, and a message m.

Round 1: Each node $v \in S$ sends message m to each neighbor in G;
Round 2: Each node $v \in S$ that does not receive a message in round 1 sends message m to each neighbor in G;
Round $i > 1$: *Each node v executes*

> $M := N(v)$;
> **foreach** *receive*(w, m) **do**
> > $M := M \setminus \{w\}$
>
> **if** $M \neq N(v)$ **then**
> > **forall** $u \in M$ **do** send(u, m);

To illustrate the flow of messages of algorithm $\mathcal{A}_{SF}$ we consider an example with a cycle C_n with seven nodes as depicted in Fig. 1. The top row shows the flow of messages for the case $|S| = 1$. $\mathcal{A}_{SF}$ terminates after four rounds in this case. In the next two rows S consists of three nodes. The difference is that in the second row each node in S has a neighbor in S while this is not the case in the last row. Therefore, algorithm $\mathcal{A}_{SF}$ behaves in the second row exactly as amnesiac flooding and terminates after three rounds. In the third example the top node does not receive a message in the first round hence it sends the message to all neighbors in round two again. In this case the algorithm terminates after two rounds. These examples demonstrate that the termination time of $\mathcal{A}_{SF}$ highly depends on S. This is captured by the following definition.

Definition 3. *Denote by $SF_G(S)$ the number of rounds algorithm $\mathcal{A}_{SF}$ requires to terminate for graph G when started by all nodes in S. For $1 \leq k \leq n$ define*

$$SF_k(G) = \min\{SF_G(S) \mid S \subseteq V \text{ with } |S| = k\}.$$

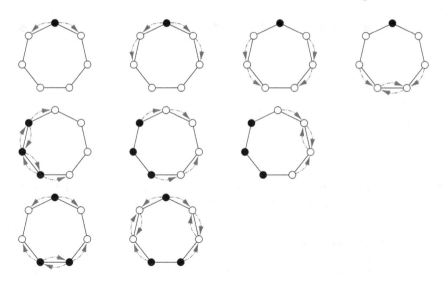

Fig. 1. Three executions of algorithm $\mathcal{A}_{\mathsf{SF}}$ for different choices of S (one per row). Nodes in S are depicted in black and the flow of messages is indicated by arrows.

For a complete graph K_n with $n > 2$ we have $SF_i(K_n) = 2$ for $1 \leq i < n$, and $SF_n(K_n) = 1$. If P_n is a path then $SF_k(P_n) = \lfloor \frac{n+k-1}{2k} \rfloor + 1$. For a cycle graph C_n, and $1 \leq k \leq n$ we also have $SF_k(C_n) = \lfloor \frac{n+k-1}{2k} \rfloor + 1$.

Lemma 1. *Let G be a connected graph. Then $SF_k(G) = 1$ if and only if $k = n$.*

Proof. Obviously, $SF_n(G) = 1$ for each graph G. Let $SF_k(G) = 1$. Then there exists $S \subseteq V$ with $|S| = k$ and $SF_G(S) = 1$. Since $\mathcal{A}_{\mathsf{SF}}$ terminates after the first round $N(v) \subseteq S$ for all $v \in S$. Since G is connected this yields $S = V$ and thus, $k = n$. □

4.1   Reduction to Amnesiac Flooding

In this section we prove that an execution of algorithm $\mathcal{A}_{\mathsf{SF}}$ on a graph G with initiators S is equivalent to the execution of amnesiac flooding on a graph $\hat{G}$ with

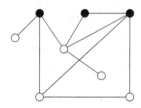

Fig. 2. A connected graph G and a set S of initiators (depicted in black).

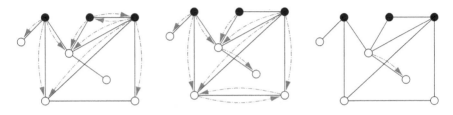

Fig. 3. Messages sent by algorithm $\mathcal{A}_{\mathsf{SF}}$ for the graph of Fig. 2.

initiators $\hat{S}$. Throughout this section we will use the graph of Fig. 2 as an example. Algorithm $\mathcal{A}_{\mathsf{SF}}$ terminates after three rounds for this graph (see Figs. 3).

4.2   Amnesiac Flooding

Before explaining the construction of $\hat{G}$ and $\hat{S}$ we recap algorithm $\mathcal{A}_{\mathsf{AF}}$ for amnesiac flooding from [14]. Algorithm 2 shows a formal definition of algorithm $\mathcal{A}_{\mathsf{AF}}$.

Algorithm 2: Algorithm $\mathcal{A}_{\mathsf{AF}}$ distributes a message in the graph G

input: A graph $G = (V, E)$, a subset S of V, and a message m.

Round 1: Each node $v \in S$ sends message m to each neighbor in G;
Round $i > 1$: *Each node v executes*

> $M := N(v)$;
> **foreach** *receive*(w, m) **do**
> > $M := M \setminus \{w\}$
>
> **if** $M \neq N(v)$ **then**
> > **forall** $u \in M$ **do** send(u, m);

The two algorithms $\mathcal{A}_{\mathsf{AF}}$ and $\mathcal{A}_{\mathsf{SF}}$ only behave differently during the second round. In this round nodes from S with no neighbor in S send in $\mathcal{A}_{\mathsf{SF}}$ a message to each neighbor, whereas in $\mathcal{A}_{\mathsf{AF}}$ they do not send a message at all in the second round. Algorithm $\mathcal{A}_{\mathsf{AF}}$ terminates after four rounds for the graph G and the set S shown in Fig. 2, i.e., one more round than algorithm $\mathcal{A}_{\mathsf{SF}}$. The behavior in the first round is identical to that of $\mathcal{A}_{\mathsf{SF}}$. Figure 4 shows the flow of messages during the last three rounds.

Even so $r_k(G) + 1$ is a general lower bound for information dissemination, there are graphs, for which $\mathcal{A}_{\mathsf{AF}}$ terminates in $r_k(G)$ rounds. According to Theorem 4 and 5 of [18] amnesiac flooding with k initiators terminates in $r_k(G)$ rounds if and only if G is a bipartite graph G with $V = V_1 \cup V_2$ such that G has a k-center that is either contained in V_1 or V_2.

4.3   The Reduction

In this section we describe the reduction process in detail. In the light of Lemma 1 we assume that $|S| < n$. The reduction consists of a sequence of extensions of

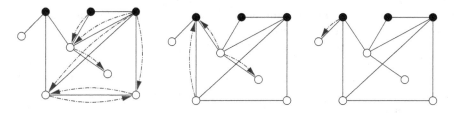

Fig. 4. Messages sent in rounds 2, 3, and 4 by algorithm $\mathcal{A}_{\mathsf{AF}}$ for the graph of Fig. 2.

the original graph and the set of initiators. In the final stage there will be just one initiator and the graph will be bipartite.

Let $G = (V, E)$ and $S \subseteq V$. Define $S_I = \{v \in S \mid N(v) \cap S = \emptyset\}$, i.e., S_I is the set of isolated nodes in $G[S]$. Let $V_S = \{w_v \mid v \in S_I\}$ be a set of new vertices and $E_S = \{(v, w_v) \mid v \in S_I\}$ a new set of edges.

Definition 4. *Denote by $\hat{G}(S)$ be the undirected graph with node set $\hat{V} = V \cup V_S$ and edge set $\hat{E} = E \cup E_S$.*

Furthermore, let $\hat{S} = S \cup V_S$. For the graph of Fig. 2 S_I consists of a single node only, the leftmost black node. Figure 5 shows the graph $\hat{G}(S)$.

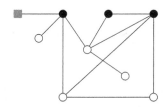

Fig. 5. The graph $\hat{G}(S)$ for the graph G of Fig. 2. Note that in $G[S]$ only the top left node is isolated. Thus, only one node is added (depicted by a gray box).

Lemma 2. *Let $\mathcal{E}_{SF}$ be an execution of algorithm $\mathcal{A}_{SF}$ on a graph G with initiators S and $\mathcal{E}_{AF}$ an execution of algorithm $\mathcal{A}_{AF}$ on graph $\hat{G}(S)$ with initiators $\hat{S}$. If a node $v \in V$ sends in round i a message to a node w in $\mathcal{E}_{SF}$ then node v also sends in round i a message to w in $\mathcal{E}_{AF}$. Moreover, the length of $\mathcal{E}_{SF}$ is equal to the length of $\mathcal{E}_{AF}$.*

Proof. All nodes of S send in both algorithms during the first round a message to all their neighbors. In algorithm $\mathcal{A}_{AF}$ in addition the nodes in V_S also send a message to each neighbor. Thus, each node in S_I receives in the second round of $\mathcal{A}_{AF}$ a message from a node in V_S and no other node. Thus, each node in S_I sends in the second round of $\mathcal{A}_{AF}$ a message to each neighbor in V. This implies that these nodes behave in the same way as for algorithm $\mathcal{A}_{SF}$. This is also true for the nodes in $S \setminus S_I$. The remaining nodes in $V \setminus S$ also send the same messages in round two for both algorithms.

Note that from round three on the algorithms are identical with one minor exception. When a node v in S_I receives a message it will in addition send in algorithm $\mathcal{A}_{AF}$ a message to node w_v. Since v is the only neighbor of w_v this triggers no further message. These arguments also yield that the executions of $\mathcal{A}_{SF}$ on G and $\mathcal{A}_{AF}$ on $\hat{G}(S)$ terminate after the same round.   □

The last lemma implies that algorithm $\mathcal{A}_{SF}$ is correct, i.e., all nodes receive the message in finite time, since algorithm $\mathcal{A}_{AF}$ is correct.

In the next step of the reduction, the number of initiators is reduced to one. This simplifies the analysis of the behavior of $\mathcal{A}_{AF}$ on graph $\hat{G}(S)$ with initiators $\hat{S}$. Next, an auxiliary graph $\mathcal{G}(S)$ is introduced in two steps. Let v^* be a new node and define $V^\circ = \hat{V} \cup \{v^*\}$ and $E^\circ = \hat{E} \cup \{(v^*, v) \mid v \in \hat{S}\}$.

Definition 5. *Denote by $G^\circ(S)$ the undirected graph with node set V° and edge set E° and by $\mathcal{F}(v^*)$ the subgraph of $G^\circ(S)$ with node set V° and all edges of $G^\circ(S)$ that are not cross edges with respect to v^*.*

Figure 6 shows on the left the graph $G^\circ(S)$ and on the right the subgraph $\mathcal{F}(v^*)$ for the graph G depicted in Fig. 2. Obviously $\mathcal{F}(v^*)$ is always bipartite.

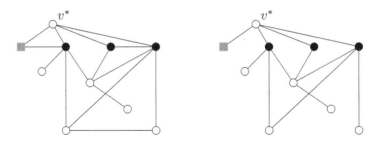

Fig. 6. The new node v^* is connected to the four nodes of $\hat{S}$. The subgraph $\mathcal{F}(v^*)$ depicted on the right contains the forward edges of $G^\circ(S)$ only.

Using $G^\circ(S)$ we next define the auxiliary graph $\mathcal{G}(S)$. This graph roughly consists of two copies of $\mathcal{F}(v^*)$ with edges linking both copies. These edges are induced by the cross edges of $G^\circ(S)$.

Definition 6. *Let $^\circ V$ be a copy of $\hat{V}$. Denote by $\mathcal{G}(S)$ the graph with node set $V^\circ \cup {}^\circ V$ and the following edges:*

1. *Edges of $\mathcal{F}(v^*)$ connecting nodes of V°.*
2. *Edges between nodes of $^\circ V$ if there exits an edge in $\mathcal{F}(v^*)$ between the corresponding nodes in $\hat{V}$.*
3. *Edges $(u^\circ, {}^\circ w)$ and $(w^\circ, {}^\circ u)$ with $u^\circ \in V^\circ$ and $^\circ w \in {}^\circ V$ if (u, w) is a cross edge of $G^\circ(S)$. Here u (resp. w) corresponds to the copies of u° and $^\circ u$ (resp. w° and $^\circ w$).*

Figure 7 demonstrates this construction for the graph G depicted in Fig. 6. $\mathcal{G}(S)$ consists of $2(|V| + |S_I|) + 1$ nodes and $2|E| + 3|S_I| + |S|$ edges. For each $v \in V^\circ$ we have $deg_{G^\circ(S)}(v) = deg_{\mathcal{G}(S)}(v)$.

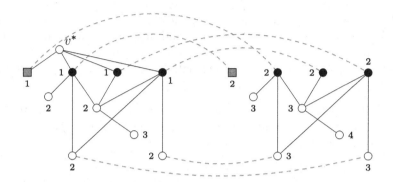

Fig. 7. The graph $\mathcal{G}(S)$ for the graph $G^\circ(S)$ of Fig. 6. Note that $G^\circ(S)$ has three cross edges for v^*. Therefore, there are six edges in $\mathcal{G}(S)$ connecting nodes in V° with nodes in $^\circ V$ (depicted as dashed lines). The numbers next to the nodes indicate the number of the round, in which the node receives a message in execution $\mathcal{E}_{AF}^{v^*}$.

Denote by L_i the nodes of $\mathcal{F}(v^*)$ with distance i in $\mathcal{F}(v^*)$ from v^*. Furthermore denote by R_i the nodes of the copy of $\mathcal{F}(v^*)$ in $\mathcal{G}(S)$ that have distance i to the copy of the node set $\hat{S}$ in $^\circ V$. Clearly, the sets L_i and R_i are independent sets. Let $V_L = L_0 \cup R_0 \cup L_2 \cup R_2 \cup \ldots$ and $V_R = L_1 \cup R_1 \cup L_3 \cup R_3 \cup \ldots$.

Lemma 3. *The graph $\mathcal{G}(S)$ is bipartite with $^\circ V \cup V^\circ = V_L \cup V_R$ and $d_G(S, V) = \epsilon_{\mathcal{G}(S)}(v^*) - 1$.*

Proof. Note that the graph $\mathcal{F}(v^*)$ and its copy in $\mathcal{G}(S)$ are bipartite. Also $V_L \cap V_R = \emptyset$. For $i > 0$ the edges edges between nodes in V° and nodes in $^\circ V$ connect nodes in L_i with nodes in R_{i-1}. Thus, V_L and V_R are independent sets and hence $\mathcal{G}(S)$ is bipartite. For the last statement note that $d_G(v, S) = d_{\mathcal{F}(v^*)}(v^*, S)$ for $v \in V$. □

Lemma 4. *Let $\mathcal{E}_{AF}$ be an execution of algorithm $\mathcal{A}_{AF}$ on graph $\hat{G}(S)$ with initiators $\hat{S}$ and $\mathcal{E}_{AF}^{v^*}$ an execution of algorithm $\mathcal{A}_{AF}$ on graph $\mathcal{G}(S)$ with initiator v^*. The length of $\mathcal{E}_{AF}$ is one round less than the length of $\mathcal{E}_{AF}^{v^*}$.*

Proof. We show by induction on the round number i that messages sent in round $i > 1$ in $\mathcal{E}_{AF}^{v^*}$ correspond with messages sent in round $i - 1$ in $\mathcal{E}_{AF}$. Let $i = 2$. In the first round of $\mathcal{E}_{AF}^{v^*}$ node v^* sends a message to each node of $\hat{S}$. Nodes of $\hat{S}$ will therefore send in the second round of $\mathcal{E}_{AF}^{v^*}$ a message to all their neighbors except v^*. The construction of $G^\circ(S)$ and $\mathcal{G}(S)$ implies that all copies of the nodes $\hat{S}$ in $^\circ V$ receive a message. The same is true for all nodes in V° that are reachable via forward edges from nodes in S (see Fig. 7 for an example). These messages correspond precisely to the messages sent in the first round of $\mathcal{E}_{AF}$ (via forward and cross edges). This proves the base case.

For the induction step let $i > 2$. $\mathcal{G}(S)$ is by Lemma 3 bipartite and messages are never sent from nodes in $°V$ to nodes in $V°$ only in the opposite direction. Thus, in round i the nodes in $\mathcal{G}(S)$ that receive a message are exactly those in $L_i \cup R_{i-2}$. By induction hypothesis the nodes in $\hat{G}(S)$ that receive a message in round $i - 1$ are those nodes that correspond in V to the nodes in $L_i \cup R_{i-2}$. Thus, the messages sent in round $i + 1$ in $\mathcal{E}_{AF}^{v^*}$ correspond with messages sent in round i in $\mathcal{E}_{AF}$. This yields that the length of $\mathcal{E}_{AF}$ is one round less than the length of $\mathcal{E}_{AF}^{v^*}$.   □

Theorem 1. *Let $G = (V, E)$ be a connected graph and $\emptyset \neq S \subseteq V$. Algorithm $\mathcal{A}_{SF}$ is truly stateless, distributes a message stored at the nodes of S to all nodes, and terminates after $d_G(S, V) + 1$ rounds.*

Proof. First consider the execution of algorithm $\mathcal{A}_{AF}$ for graph $\hat{G}(S)$ with initiators $\hat{S}$. By Lemma 3 $\hat{G}(S)$ is bipartite. Obviously, $\mathcal{A}_{AF}$ terminates for bipartite graphs. In particular it terminates after $t = \epsilon_{\mathcal{G}(S)}(v^*)$ rounds. Now Lemma 2 and Lemma 4 yield that Algorithm $\mathcal{A}_{SF}$ terminates in $t - 1$ rounds on graph G with initiators S. Note that $d_G(S, V) = \epsilon_{\mathcal{G}(S)}(v^*) - 1$ by Lemma 3.   □

Corollary 1. *Let G be a connected graph with $n > 2$ and $1 \leq k < n$. Then $SF_k(G) = r_k(G) + 1$. In particular $SF_1(G) = Rad(G) + 1$.*

This result demonstrates that $\mathcal{A}_{SF}$ terminates significantly faster than amnesiac flooding in case G is non-bipartite. $\mathcal{A}_{AF}$ with single initiator v_0 terminates in this case after at most $Diam(G) + \epsilon_G(v_0) + 1$ rounds by Theorem 12 of [14]. Theorem 1 implies that $\mathcal{A}_{SF}$ terminates already in $\epsilon_G(v_0) + 1$ rounds. The bound for $\mathcal{A}_{AF}$ is sharp as a cycle C_n of odd length demonstrates. $\mathcal{A}_{AF}$ terminates after n rounds whereas $\mathcal{A}_{SF}$ terminates already after $(n+1)/2$ rounds. Similar examples can be found in case $|S| > 1$, e.g., in [18].

Corollary 2. *$\mathcal{A}_{SF}$ is a truly stateless information dissemination algorithm. Its time complexity is optimal unless G is bipartite with $V = V_1 \cup V_2$ such that V_1 or V_2 contains a k-center $\mathcal{A}_{SF}$. In this case $\mathcal{A}_{AF}$ requires one round less.*

As stated in Subsect. 4.1 amnesiac flooding with k initiators terminates in $r_k(G)$ rounds if and only if G is a bipartite with $V = V_1 \cup V_2$ and a k-center of G is either contained in V_1 or in V_2. In this case, algorithm $\mathcal{A}_{SF}$ still terminates only after $r_k(G) + 1$ rounds. The reason is that the extended behavior in round 2 in Algorithm 1 is not needed in this constellation. Thus, $\mathcal{A}_{SF}$ is only optimal if this constellation is not given (see Fig. 8 for an example). In particular, $\mathcal{A}_{SF}$ is not optimal if $k = 1$ and G is bipartite, in this case $\mathcal{A}_{AF}$ is optimal.

4.4   Approximation Algorithm

Theorem 1 shows the strong link between the k-center problem and this problem. This relationship immediately gives rise to an approximation algorithm for the

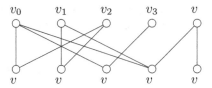

Fig. 8. A bipartite graph for which neither V_1 nor V_2 contains a 2-center. The set $\{v_0, v_8\}$ is a 2-center with radius 2. Note that $SF_2(G) = 3$, e.g., $SF_G(\{v_0, v_1\}) = 3$. After removing node v_9, $\{v_0, v_1\}$ is a 2-center of the remaining graph contained in V_1. Hence, $\mathcal{A}_{\mathsf{AF}}$ (resp. $\mathcal{A}_{\mathsf{SF}}$) terminates after 2 (resp. 3) rounds for the remaining graph.

SF problem: If S is a k-center with minimal radius then $SF_G(S) = SF_k(G)$. There exist several approximation algorithms for the k-center problem. The algorithm of Gonzalez has a run-time of $O(nk)$ [10]. The algorithm of Feder and Greene requires only $O(n \log k)$ time [6]. Both algorithms achieve an approximation ratio of 2 and this is also the best possible ratio assuming $P \neq NP$ [12]. For the SF problem we have the following result.

Theorem 2. *There exits a polynomial time approximation algorithm for the stateless-flooding problem with approximation ratio $2 - 1/SF_k(G)$.*

Proof. Let $\mathcal{A}$ be any 2-approximation algorithm for the k-center problem and S a k-center computed by $\mathcal{A}$. Then $d(S, V) \leq 2r_k(G)$. Hence

$$\frac{SF_G(S)}{SF_k(G)} \leq \frac{2r_k(G) + 1}{r_k(G) + 1} = 2 - \frac{1}{SF_k(G)} \leq 2.$$

There is little hope to find an approximation algorithm with approximation ratio $3/2 - \epsilon$ with $\epsilon > 0$.

Theorem 3. *Assuming $P \neq NP$, there is no polynomial time algorithm achieving a factor of $3/2 - \epsilon$, $\epsilon > 0$, for the stateless-flooding problem.*

Proof. We will show that such an algorithm can solve the dominating set problem in polynomial time. Given a graph $G = (V, E)$ and an integer $k < n$, the dominating set problem is to decide whether there exist a set D of k nodes with the property that each of the remaining nodes has a neighbor in D. According to [9] this problem is NP-complete. Note that a graph G has a dominating set of size k if and only if $r_k(G) = 1$. Hence, by Corollary 1 G has a dominating set of size k if and only if $SF_k(G) = 2$.

Suppose there exists a polynomial time algorithm $\mathcal{A}$ that computes a subset S of V with $|S| = k$ such that $SF_G(S) \leq (3/2 - \epsilon)SF_k(G)$. We will prove that $\mathcal{A}$ solves the dominating set problem. First consider the case $SF_G(S) > 2$. Assume there exists a dominating set D of G of size k then $SF_k(G) = 2$. Hence, $SF_G(S) \leq (3/2 - \epsilon)SF_k(G) < 3$. Thus, $SF_G(S) \leq 2$ since $SF_G(S)$ is an integer. Contradiction. Therefore, $SF_G(S) > 2$ yields that there exists no dominating set S of G of size k. On the other hand $SF_G(S) \leq 2$ implies $r_k(G) = 1$ by

Corollary 1. Hence, there exists a dominating set S of G of size k. This implies that $\mathcal{A}$ solves the dominating set problem. Contradiction to $P \neq NP$.   □

5   Asynchronous Stateless Information Dissemination

The definition of a synchronous stateless algorithm inherently relies on the concept of a round. It is used to define the unit of time, information can be kept by a node. Several papers proposed a definition of the round concept in asynchronous systems. Hussak et al. introduce what they call as the *round-asynchronous model* where the computation still proceeds in global synchronous rounds but an adversary can decide the delay of message delivery on any link [14]. Messages will be eventually delivered but the adversary decides which round to deliver the message in. The authors show with a simple example that $\mathcal{A}_{\mathsf{AF}}$ may not terminate in this model. This argument also holds for algorithm $\mathcal{A}_{\mathsf{SF}}$. Dolev et al. introduce *r-fair* asynchronous systems [5]. Compared to the round-asynchronous model it is assumed that messages are delivered at least r rounds after they have been sent. Thus, 1-fair systems are precisely the synchronous systems. It is easy to see that $\mathcal{A}_{\mathsf{AF}}$ will also not terminate in this model with $r > 1$. Both proposals do not lead to a practical definition of stateless algorithm for asynchronous systems.

Coordination in asynchronous systems is not achieved via the concept of a round but purely via events such as message arrival. An extreme definition of statelessness would be that nodes have to react upon each incomming message and cannot retain information between two message receptions. Such systems interact by repeatedly mapping incoming messages to outgoing messages and output values. The computational power of this type of stateless algorithm strongly depends on the maximal allowed size of a message. With messages of size $\Omega(n \log n)$ all state information can be encoded in a single message. Thus, in principle it should be possible to solve all problems that are also solvable by stateful algorithms, possibly at the cost of a high number of messages.

More interesting problems arise if only a limited part of each message can be used to encode session information, e.g., of size $O(1)$ or $O(\log n)$.

Theorem 4. *There is no deterministic 1-stateless information dissemination algorithm for asynchronous systems.*

Proof. Assume there exists a 1-stateless information dissemination algorithm $\mathcal{A}$ that can update up to d bits in each message. Let G be a graph that has a node v_0 with $\epsilon_G(v_0) > 2^d$. Consider an execution of $\mathcal{A}$ with initiator v_0. Then there exists a message flow $\mathcal{S}: v_0 \xrightarrow{m_0} v_1 \xrightarrow{m_1} v_2 \xrightarrow{m_2} \ldots$ with nodes $v_0, v_1, v_2, \ldots$ and $v_i \in N(v_{i+1})$ for $i \geq 0$ such that v_i sends a message to v_{i+1} as a reaction of receiving a message from v_{i-1} for $i > 0$ and the length of $\mathcal{S}$ is greater than 2^d. Thus, there are two nodes v_s and v_t in this flow with $s < t$ which receive identical messages. Hence, as a reaction they also send identical messages. Thus, $\mathcal{S}$ is infinite. This yields that $\mathcal{A}$ does not terminate. Contradiction.   □

Clearly, in asynchronous anonymous networks no deterministic stateless information dissemination algorithm exists. Hence, we assume unique identifiers. According to Definition 2 an algorithm is stateless if a node decides only on the basis of each received message which messages to send as a reaction. This definition can be relaxed by allowing a node to include its own identifier in its decision process. In this case we can prove the following theorem.

Theorem 5. *There exists a $\log n$-stateless information dissemination algorithm for asynchronous systems terminating in n^{c+1} rounds provided each node has a unique identifier in the range $0, \ldots, n^c$ with $c \geq 1$.*

Proof. We sketch a simple $\log n$-stateless information dissemination algorithm for a single originator. Each message consists of two values each of size $O(\log n)$. The originator v_0 sends the pair $(v_0.id, v_0.id)$ to all neighbors. A node v receiving a message (a, b) reacts as follows. If its own id is strictly larger than a it sends $(v.id, v.id)$ to all neighbors except the one from which the message came. If its own id is strictly less than a and $b \neq 0$ its sends $(a, b-1)$ to all neighbors except the one from which the message came. In all other cases no message is sent.

Obviously the information reaches all nodes. Assume false. Among all nodes which are not informed by the algorithm choose v such that $d(v, S)$ is minimal. Let $w \in N(v)$ such that $d(w, S) < d(v, S)$. Then w is informed with a message $(a, 0)$ and $a > w.id$ otherwise v would be informed. Consider a shortest path from the node u with $u.id = a$ to w. Since u sent the messages (a, a) the second component was a-times decreased by nodes with an id less than a. Thus, there must be $a + 1$ nodes with an id less than a. Contradiction.

To prove that the algorithm terminates consider a sequence S of nodes $v_0, v_1, \ldots$, where v_i forwards a message to v_{i+1}. Then either the first component in each message increases or the second decreases. The first case can happen at most $n-1$ times. After each such event at a node v_i the second case can happen at most $v_i.id$ times. Thus, there exists a constant $C \leq n^{c+1}$ such that S has length at most C. Hence, the algorithm terminates in at most C rounds.   □

6   Conclusion and Future Work

We presented an optimal truly stateless information dissemination algorithm with k initiators for synchronous systems. The algorithm terminates in $r_k(G)+1$ rounds. We proved that unless $P = NP$ there is no approximation algorithm for the SF problem with an approximation ratio less than $3/2$. It remains an open problem to design a $3/2$-approximation or disprove its existence.

There are also open questions related to stateless asynchronous information dissemination. The algorithm in the proof of Theorem 5 terminates in $O(n^{c+1})$ rounds, but the number of messages grows exponentially with n. It remains to design a more efficient stateless information dissemination algorithm in this case. Another open problem is whether Theorem 4 still holds for the relaxed definition of statelessness. Lastly, it is unknown whether there exits a deterministic $f(n)$-stateless asynchronous information dissemination algorithm with $f \in o(\log n)$.

A well-known procedure to execute synchronous algorithms in asynchronous systems are *synchronizers*. They simulate an execution of a failure-free synchronous system in a failure-free asynchronous system [1]. Thus, a truly stateless synchronizer would transform $\mathcal{A}_{\mathsf{AF}}$ into a truly stateless asynchronous information dissemination algorithm, this contradicts Theorem 4. Thus, there is no truly stateless synchronizer. Does there exist a $\log n$-stateless synchronizer?

References

1. Attiya, H., Welch, J.L.: Distributed Computing - Fundamentals, Simulations, and Advanced Topics. Series on Parallel and Distributed Computing, 2nd edn. Wiley, Hoboken (2004)
2. Awerbuch, B., Khandekar, R.: Stateless distributed gradient descent for positive linear programs. In: Proceedings of the 40th Symposium on Theory of Computing, pp. 691–700. ACM (2008)
3. Calik, H., Labbé, M., Yaman, H.: p-center problems. In: Laporte, G., Nickel, S., da Gama, F.S. (eds.) Location Science, pp. 79–92. Springer, Cham (2015). https://doi.org/10.1007/978-3-319-13111-5_4
4. Dolev, D., Erdmann, M., Lutz, N., Schapira, M., Zair, A.: Brief announcement: stateless computation. In: Schiller, E., Schwarzmann, A. (eds.) Proceedings of the Symposium on Principles of Distributed Computing, (PODC), pp. 419–421. ACM (2017)
5. Dolev, S., Kahil, R.M., Yagel, R.: Stateless stabilization bootstrap (extended abstract). In: Felber, P., Garg, V. (eds.) SSS 2014. LNCS, vol. 8756, pp. 180–194. Springer, Cham (2014). https://doi.org/10.1007/978-3-319-11764-5_13
6. Feder, T., Greene, D.: Optimal algorithms for approximate clustering. In: Proceedings of the 20th Annual ACM Symposium on Theory of Computing, STOC 1988, pp. 434–444. ACM, New York (1988)
7. Feige, U., Peleg, D., Raghavan, P., Upfal, E.: Randomized broadcast in networks. In: Asano, T., Ibaraki, T., Imai, H., Nishizeki, T. (eds.) SIGAL 1990. LNCS, vol. 450, pp. 128–137. Springer, Heidelberg (1990). https://doi.org/10.1007/3-540-52921-7_62
8. Frieze, A.M., Grimmett, G.R.: The shortest-path problem for graphs with random arc-lengths. Discrete Appl. Math. **10**(1), 57–77 (1985)
9. Garey, M.R., Johnson, D.S.: Computers and Intractability: A Guide to the Theory of NP-Completeness, 1st edn. W. H. Freeman, San Francisco (1979)
10. Gonzalez, T.F.: Clustering to minimize the maximum intercluster distance. Theoret. Comput. Sci. **38**, 293–306 (1985)
11. Hedetniemi, S.M., Hedetniemi, S.T., Liestman, A.L.: A survey of gossiping and broadcasting in communication networks. Networks **18**(4), 319–349 (1988)
12. Hsu, W.L., Nemhauser, G.L.: Easy and hard bottleneck location problems. Discrete Appl. Math. **1**(3), 209–215 (1979)
13. Hussak, W., Trehan, A.: Brief announcement: on termination of a flooding process. In: Proceedings ACM Symposium on Principles of Distributed Computing, PODC, New York, pp. 153–155 (2019)
14. Hussak, W., Trehan, A.: On the termination of flooding. In: Paul, C., Bläser, M. (eds.) 37th International Symposium on Theoretical Aspects of Computer Science (STACS). LIPIcs, vol. 154, pp. 17:1–17:13. Dagstuhl, Germany (2020)

15. Kariv, O., Hakimi, S.L.: An algorithmic approach to network location problems. I: the p-centers. SIAM J. Appl. Math. **37**(3), 513–538 (1979)
16. Mocquard, Y., Sericola, B., Anceaume, E.: Probabilistic analysis of rumor-spreading time. INFORMS J. Comput. **32**(1), 172–181 (2020)
17. Peleg, D.: Distributed Computing: A Locality-Sensitive Approach. SIAM Society for Industrial and Applied Mathematics, Philadelphia (2000)
18. Turau, V.: Analysis of amnesiac flooding. CoRR abs/2002.10752 (2020). https://arxiv.org/abs/2002.10752

Multi-agent Systems

Cops and Robbers on Dynamic Graphs: Offline and Online Case

Stefan Balev[1], Juan Luis Laredo Jiménez[1], Ioannis Lamprou[2(✉)],
Yoann Pigné[1], and Eric Sanlaville[1]

[1] Normandie Univ., UNIHAVRE, UNIROUEN, INSA Rouen, LITIS,
Le Havre, France
{stefan.balev,juanlu.jimenez,yoann.pigne,eric.sanlaville}@univ-lehavre.fr
[2] Department of Informatics and Telecommunications,
National and Kapodistrian University of Athens, Zografou, Greece
ilamprou@di.uoa.gr

Abstract. We examine the classic game of Cops and Robbers played on models of dynamic graphs, that is, graphs evolving over discrete time steps. At each time step, a graph instance is generated as a subgraph of the *underlying graph* of the model. The cops and the robber take their turns on the current graph instance. The cops win if they can capture the robber at some point in time. Otherwise, the robber wins.

In the *offline* case, the players are fully aware of the evolution sequence, up to some finite time horizon T. We provide a $\mathcal{O}(n^{2k+1}T)$ algorithm to decide whether a given evolution sequence for an underlying graph with n vertices is k-cop-win via a reduction to a reachability game.

In the *online* case, there is no knowledge of the evolution sequence, and the game might go on forever. Also, each generated instance is required to be connected. We provide a nearly tight characterization for sparse underlying graphs, i.e., with at most linear number of edges. We prove $\lambda + 1$ cops suffice to capture the robber in any underlying graph with $n - 1 + \lambda$ edges. Further, we define a family of underlying graphs with $n - 1 + \lambda$ edges where $\lambda - 1$ cops are necessary (and sufficient) for capture.

Keywords: Cops and robbers · Dynamic graphs · Offline · Online

1 Introduction

Cops and robbers is a classic pursuit-evasion combinatorial game played on a graph. There are two opposing players aiming to win the game. A cop player controlling k cop tokens and a robber player controlling one robber token. Initially, the k cops are placed at vertices of the graph. Subsequently, the robber is also placed at a graph vertex. The two players proceed (possibly ad infinitum) by taking turns alternately commencing with the cops. During a cops' turn, each

---

This work was partially funded by the Normandy region via the ASTREOS project.

A. W. Richa and C. Scheideler (Eds.): SIROCCO 2020, LNCS 12156, pp. 203–219, 2020.
https://doi.org/10.1007/978-3-030-54921-3_12

cop may move to a vertex adjacent to its current one; note that cops are presumed to move simultaneously. Similarly, during a robber's turn, the robber may move to a vertex adjacent to its current placement. The cops win, if at least one of them manages to eventually lie at the same vertex as the robber, i.e., captures the robber. Otherwise, the robber wins if it can indefinitely avoid capture.

Thus far, Cops and Robbers literature has focused on (several variations of) the game taking place on static graphs. Very little is known with respect to Cops and Robbers games taking place on *dynamic graphs*. In this paper, we consider the above described Cops and Robbers game taking place in models of dynamic graphs, otherwise referred to as *temporal graphs/networks* [18,27]. In our case, a dynamic graph is represented by a (possibly infinite) sequence of subgraphs of the same static graph, which is the *underlying graph* of the model. In other words, the underlying graph evolves over a series of discrete time steps under a set of evolution rules. In this setting, we introduce the problem of Cops and Robbers taking place in some standard models of dynamic graphs.

Related Work. The preliminary question in mind is to compute the minimum number of cops needed to capture the robber on some (static) graph family.

Definition 1. *The cop number of a graph G, denoted $c(G)$, is the minimum number of cops needed to ensure that the robber is eventually captured, regardless of the robber's strategy.*

Problems related to the cop number have been studied heavily over the last four decades. Originally, Quillot [31], and independently Nowakowski and Winkler [29], characterized graphs with cop number equal to 1, otherwise referred to as *cop-win* graphs. The set of (di)graphs with cop number equal to $k > 1$ was characterized in [13,17]. Building on these notions, a general framework for characterizing discrete-time pursuit-evasion games was developed in [9].

There is a lot of literature regarding the cop number of specific graph classes. Aigner and Fromme [1] proved $c(G) \leq 3$ for any planar graph G. Frankl [16] proved a lower bound for graphs of large girth. Other works include [2,15,25].

Moving onto general graphs, Meyniel conjectured $\sqrt{n}$ cops are always sufficient to capture the robber in any graph. The current state of the art is $\mathcal{O}(n/2^{(1-o(1))\sqrt{\log n}})$ proved independently in [23,32]. Yet, the conjecture remains unresolved. On the contrary, the conjecture was proved positive for random graphs [30]; relevant works include [6,10,24]. Finally, there exists a book capturing all the activity on Cops & Robber until recently; see [8].

The computational complexity of the corresponding decision problem is also worth a note. *Given a graph G and an integer k, does $c(G) \leq k$ hold?* Recently, Kinnersley [21] answered the question by proving EXPTIME-completeness. With respect to algorithmic results, for a fixed constant k, there is a polynomial time algorithm to determine whether $c(G) \leq k$ [3]. Other algorithmic results include [7] (capture from a distance), and [9] (generalized Cops and Robbers).

Cops and robbers applications are found in many fields, for example in motion planning [33], routing [22], network security [10], and distributed computing [5].

Dynamic graphs, sometimes called temporal networks [27] or time-varying graphs [12], have received a lot of attention as they capture realistic scenarios where the underlying graph changes over time periodically or intermittently. Several models for such constructs have been considered, for example, [11, 28].

With respect to Cops and Robbers games played on dynamic graphs, we were unaware of any work until the recent paper by Erlebach and Spooner [14]. They examine the game on edge-periodic graphs, where each edge e is present at time steps indicated by a bit-pattern of length l_e used periodically and ad infinitum as evolution rule. Let LCM(L) denote the least common multiple for input lengths l_e. The paper presents a $\mathcal{O}(\text{LCM(L)} \cdot n^3)$ algorithm to determine whether the graph is 1-cop-win as well as some other results on cycle graphs.

Our Results. We consider two dynamic graph models and present preliminary results for a (classic-style) Cops and Robbers game taking place in them. At each discrete time step of evolution, the current graph instance is fixed, then the cops take their turn, and finally the robber takes its turn. Note that movement may be restricted due to the possibly limited topology of each instance.

In the offline case, the cop and the robber know the whole evolution sequence (up to some finite time horizon T) a priori. For an underlying graph with n vertices, we prove that deciding whether it is cop-win can be done in time complexity $O(n^3 T)$; see Theorem 3. To do so, we employ a reduction to another game, a reachability game, played now on the configuration graph (Lemma 1). Our results extend to deciding k-cop-win graphs (Corollary 1), and an exponential time algorithm to determining the exact value of the cop-number (Corollary 2).

In the online case, no knowledge is given to the players regarding graph dynamics. The only restriction imposed is that, at each time step, the realized graph instance needs to be connected. We consider sparse graphs and show that the cop number is at most $\lambda + 1$ for underlying graphs with $n - 1 + \lambda$ edges; see Theorem 4. Moreover, we demonstrate a (nearly tight) graph family where $\lambda - 1$ cops are necessary (and sufficient) to ensure cop victory; see Theorem 5.

Outline. In Sect. 2, we present introductory notions and notation on the dynamic graph models used and on the game of Cops and Robbers played on said models. In each of Sects. 3 and 4, we formalize our definitions for the respective model considered: In Sect. 3, we consider the offline case, whereas in Sect. 4, we consider the online case. In Sect. 5, we cite concluding remarks.

2   Preliminaries

Dynamic Graphs. Let $G = (V, E)$ stand for a (static) graph to which we refer to as the *underlying graph* of the model. We assume G is *simple*, i.e., not containing loops or multi-edges, and *connected*, i.e., there exists a path between any two vertices in G. No further assumptions are made on the topology of G. An edge from vertex $v \in V$ to vertex $u \in V$ is denoted as $(v, u) \in E$, or equivalently $(u, v) \in E$. We refer to the edges of the underlying graph as the *possible edges* of our model. We denote the number of vertices of G by $n = |V|$ and the number of its edges by $m = |E|$. For any vertex $v \in V$, we denote its *open neighborhood* by $N(v) = \{u : (v, u) \in E\}$ and its *closed neighborhood* by $N[v] = N(v) \cup \{v\}$. The (static) *degree* of $v \in V$ in G is given by $d(v) = |N(v)|$.

The dynamic graph evolves over a sequence of discrete time steps $t \in \mathbb{N}$. We consider two cases with respect to time evolution. First, $t = 1, 2, 3, \ldots, T$, that is, t takes consecutive values starting from time 1 up to a *time horizon* $T \in \mathbb{N}$ given as part of the input. In this case, we define a dynamic graph $\mathcal{G}$ with a time horizon T as $\mathcal{G} = (G_1, G_2, \ldots, G_T)$ (Sect. 3). Second, $t = 1, 2, 3, \ldots$, that is, we consider the sequence of time steps t evolving ad infinitum (Sect. 4).

For any t, let $G_t = (V_t, E_t)$ be the graph instance realized at time step t, where $V_t = V$ and $E_t \subseteq E$: all vertices of the underlying graph G are present at each time step, whereas a possible edge $e \in E$ may either be *present/alive*, i.e., $e \in E_t$, or *absent/dead*, i.e., $e \notin E_t$ at time t. For any vertex $v \in V$, we denote by $N_t(v) = \{u : (v, u) \in E_t\}$ its *available neighborhood* at time t. Also, similarly to before, let $N_t[v] = N_t(v) \cup \{v\}$ refer to the closed neighborhood at time t.

Cops and Robber on Dynamic Graphs. We play a game of Cops and Robbers on a dynamic graph evolving under the general model defined above. There are two players: $\mathcal{C}$ controlling $k \geq 1$ ($k \in \mathbb{N}$) cop tokens and $\mathcal{R}$ controlling one robber token. Initially, $\mathcal{C}$ places its k tokens on the vertices of the underlying graph. Notice that we allow *multiple cops* to lie *on the same vertex*. Afterward, $\mathcal{R}$ chooses an initial placement for the robber. Round 0 is over. From now on, for every $t \geq 1$, first, the current graph instance G_t is fixed and, second, a round of the game takes place. A *round* consists of two *turns*, one for $\mathcal{C}$ and one for $\mathcal{R}$, in this order of play. $\mathcal{C}$ may move any of its cops lying on a vertex v to any vertex in $N_t[v]$. Note that all cops controlled by $\mathcal{C}$ move simultaneously during $\mathcal{C}$'s turn. After $\mathcal{C}$'s turn is over, $\mathcal{R}$ may move the robber lying on a vertex u to any vertex in $N_t[u]$. $\mathcal{C}$ wins the game if, after any player's turn, the robber lies on the same vertex as a cop. $\mathcal{R}$ wins if it can perpetually prevent this from happening.

A *cop-strategy*, respectively a *robber-strategy*, is a set of movement decisions for the cops, respectively the robber. Having knowledge of all past turns, the current positions, and the current graph instance G_t, the cops/robber decide on a move for round t according to the rules of the game. A dynamic graph is called *k-cop-win*, if there exists a cop-strategy such that k cops win the game against any robber-strategy. For $k = 1$, we say that such a dynamic graph is *cop-win*.

3   Offline Case

In the offline case, we are given a dynamic graph $\mathcal{G}$ with a time horizon T, namely $\mathcal{G} = (G_1, G_2, \ldots, G_T)$, where both $\mathcal{C}$ and $\mathcal{R}$ have complete knowledge of the evolution sequence. That is, both players are aware of $G_t = (V, E_t)$, for all $t = 1, 2, \ldots, T$, *a priori*. Let $c_{off}(\mathcal{G})$ stand for the *temporal cop number (offline case)*, the worst-case minimum number of cops required to capture a robber when the whole sequence $\mathcal{G} = (G_1, G_2, \ldots, G_T)$ is given as input to both players. If the robber is not captured within the T rounds, for any cop strategy, then the dynamic graph is robber-win. Overall, this is a model of a Cops and Robbers game on a time-horizon bound dynamic graph. From now on, we refer to it as the *offline model*. The results presented in this section can be viewed by the reader as an extension/completion of the work in [14] on the model of edge-periodic graphs.

Configuration Graph. The first task we tackle in the offline model is to characterize the set of given inputs $(\mathcal{G}, T)$, which are cop-win, i.e., one cop can always capture the robber within the T rounds of play. To do so, we first construct a directed *configuration graph* capturing the cop and robber motion on $\mathcal{G}$. Then, we can play another game, i.e., a *reachability game* [20] to be defined later, on the configuration graph which corresponds to the original cop and robber game played on $\mathcal{G}$ and derive our result this way. We define the directed configuration graph as $P = (S, A)$, where S refers to configuration *states* (vertices) and A to arcs from one state to another state which is a feasible potential next state.

The vertex set S consists of all four-tuples of the form (c, r, p, t), where $t \in \{1, 2, \ldots, T\}$ indicates the time step, i.e., round of play t, $p \in \{\mathcal{C}, \mathcal{R}\}$ indicates whether it is the cop's/robber's turn to play, $c \in V$ is the vertex/position of the cop just before p's turn takes place in round t, and $r \in V$ is the vertex/position of the robber just before p's turn takes place in round t.

The arc set A contains the arcs below, for all $x, y \in V$ and $t \in \{1, 2, \ldots, T\}$, such that both the dynamics of the graph and the game moves are represented:

(1) if $z \in N_t[x]$ and $t \in \{1, 2, \ldots, T\}$, then $((x, y, \mathcal{C}, t), (z, y, \mathcal{R}, t)) \in A$, and
(2) if $z \in N_t[y]$ and $t \in \{1, 2, \ldots, T - 1\}$, then $((x, y, \mathcal{R}, t), (x, z, \mathcal{C}, t + 1)) \in A$

Case (1) arcs represent the cop's turn at round t, where the cop moves within its closed neighborhood available at time t, the robber retains its position, and, after the cop moves, it is the robber's turn at round t. Respectively, case (2) arcs represent the robber's turn at round t, where the robber moves within its closed neighborhood available at time t, the cop retains its position, and, after the robber moves, it is the cop's turn, but at the next round, namely round $t + 1$.

Let us now consider the size of P. By the definition of the states $s \in S$, it holds for the number of vertices $|S| \in \mathcal{O}(n^2 T)$. Considering the set of arcs A, in case (1), for each time step $t = 1, 2, \ldots, T$, we get at most $\binom{n}{2}$ choices for x, z cop-move pairs, one per $(x, z) \in E_t \subseteq E$, and at most n choices for the robber position $y \in V$. A similar observation holds for the arcs considered in case (2). Put together, we obtain for the number of arcs $|A| \in \mathcal{O}(n^3 T)$.

Before we proceed utilizing the configuration graph, let us add some auxiliary, yet necessary, states and arcs to capture the round of initial cop and robber placement, i.e., round 0. This way, we ensure the full correspondence of the reachability game played on P to the cop and robber game played on $\mathcal{G}$. Note that all state and arc additions discussed hereunder do not affect the order of magnitude of the size of P. Let S contain also the states $(\emptyset, \emptyset, \mathcal{C}, 0)$, and $(x, \emptyset, \mathcal{R}, 0)$, for all $x \in V$. State $(\emptyset, \emptyset, \mathcal{C}, 0)$ captures the situation at round 0 before the cop's turn: neither the cop nor the robber have been placed yet on V.

States $(x, \emptyset, \mathcal{R}, 0)$ capture the situation at round 0 before the robber's turn: the cop has been placed and it is the robber's turn to be placed. Overall, we have added an extra $n + 1$ states in S. We now proceed adding the necessary arcs in A to make the transitions from one turn to the next. For each $x \in V$, we add an arc $((\emptyset, \emptyset, \mathcal{C}, 0), (x, \emptyset, \mathcal{R}, 0)) \in A$, that is, n extra arcs in total. For each $x, y \in V$, we add an arc $((x, \emptyset, \mathcal{R}, 0), (x, y, \mathcal{C}, 1)) \in A$, that is, n^2 extra arcs in total.

Reachability. We now employ the configuration graph P constructed above by playing another two-player game on it referred to in literature as a *reachability game* [4,20,26]. The goal is to define a reachability game, which corresponds exactly to the Cops and Robbers game (offline case), and so be able to utilize known results in this area to prove our cop-win characterization (Theorem 3). The connection of a (classic) Cops and Robbers game to a reachability game was first identified in [19]. Other results, in [10,17], on cop-win characterizations employ similar tools without explicitly stating the reduction to reachability.

A reachability game is played by two players $\mathcal{C}$, and $\mathcal{R}$, where we maintain the notation such that it corresponds to players in the game of Cops and Robbers. The two players play alternately on a directed graph $D = (V_D, A_D)$, where V_D is partitioned into two player-respective subsets, that is, $V_D = V_{\mathcal{C}} \cup V_{\mathcal{R}}$ and $V_{\mathcal{C}} \cap V_{\mathcal{R}} = \emptyset$. Moreover, $V_{\mathcal{C}}$, respectively $V_{\mathcal{R}}$, is a pairwise disjoint set of vertices, that is, for any $x, y \in V_{\mathcal{C}}$, respectively $x, y \in V_{\mathcal{R}}$, it holds $(x, y) \notin A_D$ and $(y, x) \notin A_D$. A single token is initially placed on a vertex $v \in V_D$. If $v \in V_{\mathcal{C}}$, then $\mathcal{C}$ plays and moves the token to a vertex $u \in V_{\mathcal{R}}$ for which it holds $(v, u) \in A_D$. Then, it is $\mathcal{R}$'s turn: R chooses to move the token to a vertex $w \in V_{\mathcal{C}}$ for which it holds $(u, w) \in A_D$. Note that either player has to move the token across an available arc in A_D. The game proceeds in this fashion for an indefinite number of rounds. Player $\mathcal{C}$ wins, if the token eventually arrives to a designated *target vertex set* $Tar \subseteq V_D$. Otherwise, if for any $\mathcal{C}$-strategy a vertex in Tar can never be reached, then $\mathcal{R}$ wins. In a nutshell, the reachability game played on $D = (V_D, A_D)$ is defined by the tuple $(V_{\mathcal{C}}, V_{\mathcal{R}}, Tar)$. Theorem 1 demonstrates that the game can be decided for *any input* $(V_{\mathcal{C}}, V_{\mathcal{R}}, Tar)$ and D. Moreover, by Theorem 2, it can be decided in time linear to the size of the directed graph D.

Theorem 1 ([4,26]). *Consider a reachability game* $(V_{\mathcal{C}}, V_{\mathcal{R}}, Tar)$ *played on a directed graph* $D = (V_D, A_D)$. V_D *can be partitioned into two sets* $W_{\mathcal{C}}$ *and* $W_{\mathcal{R}}$ *such that, if the token is initially placed on* $w \in W_p$, *then there exists a winning strategy for player* $p \in \{\mathcal{C}, \mathcal{R}\}$.

Theorem 2 ([4,20]). *There exists an algorithm deciding a reachability game* $(V_{\mathcal{C}}, V_{\mathcal{R}}, Tar)$ *played on a directed graph* $D = (V_D, A_D)$ *in time* $\mathcal{O}(|V_D| + |A_D|)$.

Let us now consider a reachability game taking place in our constructed configuration graph P. In this respect, let $D = P$, $V_D = S$, and $A_D = A$. For any $(c, r, p, t) \in S$, let $(c, r, p, t) \in V_p$ where $p \in \{\mathcal{C}, \mathcal{R}\}$. Finally, let $Tar = \{(x, x, p, t) \mid x \in V, p \in \{\mathcal{C}, \mathcal{R}\}, t \in \{1, \ldots, T\}\}$. We can now use the just defined sets $V_{\mathcal{C}}, V_{\mathcal{R}}, Tar$ to prove Lemma 1 and, as a consequence, our main result for this section, which is given in Theorem 3.

Lemma 1. $\mathcal{G} = (G_1, G_2, \ldots, G_T)$ *is cop-win, if and only if, for a reachability game* $(V_\mathcal{C}, V_\mathcal{R}, Tar)$ *played on* $P = (S, A)$, *where* $V_p = \{(c, r, p, t) \in S \mid c, r \in V, t \in [T]\}$, *for* $p \in \{\mathcal{C}, \mathcal{R}\}$, *and* $Tar = \{(x, x, p, t) \in S \mid x \in V, p \in \{\mathcal{C}, \mathcal{R}\}, t \in [T]\}$, *it holds* $(\emptyset, \emptyset, \mathcal{C}, 0) \in W_\mathcal{C}$.

Theorem 3. *Given a dynamic graph* $\mathcal{G} = (G_1, G_2, \ldots, G_T)$ *in the offline model, we can decide if* $c_{off}(\mathcal{G}) = 1$, *that is, if* $\mathcal{G}$ *is cop-win, in time* $\mathcal{O}(n^3 T)$.

Proof. By Lemma 1, $c_{off}(\mathcal{G}) = 1$ holds, if and only if, for a reachability game $(V_\mathcal{C}, V_\mathcal{R}, Tar)$ played on $P = (S, A)$, where $V_\mathcal{C}, V_\mathcal{R}, Tar$ are defined according to the statement of Lemma 1, it holds $(\emptyset, \emptyset, \mathcal{C}, 0) \in W_\mathcal{C}$. By Theorem 2, we decide whether $(\emptyset, \emptyset, \mathcal{C}, 0) \in W_\mathcal{C}$ in time $\mathcal{O}(|S| + |A|) = \mathcal{O}(n^2 T + n^3 T) = \mathcal{O}(n^3 T)$.   □

An important remark is that, in Theorem 1 [4,20], the winning strategy derived for player $p \in \{\mathcal{C}, \mathcal{R}\}$ is *memoryless*; see Proposition 2.18 in [26]. In other words, it only depends on the current position of the token, and not on any past moves. By the reduction presented in Lemma 1, the winning strategy for the cop/robber is also memoryless, i.e., it only depends on the current positions of the cop and the robber and the time step of evolution.

Let us conclude this part by explaining how the above framework can be generalized, and therefore used to determine whether a dynamic graph $\mathcal{G} = (G_1, G_2, \ldots, G_T)$ is k-cop-win, where $k > 1$. Since k cops are placed on V throughout the game, it suffices to expand our definition of states by substituting the cop position by a k-tuple of cop positions. That is, the state set S of the configuration graph P now contains tuples of the form $((c_1, c_2, \ldots, c_k), r, p, t)$, where c_i, for $i \in \{1, 2, \ldots, k\}$, denotes the location of the i-th cop in V. For the arc set, with respect to $\mathcal{C}$, we add $(((c_1, c_2, \ldots, c_k), r, \mathcal{C}, t), ((c'_1, c'_2, \ldots, c'_k), r, \mathcal{R}, t)) \in A$, if for each $i \in \{1, 2, \ldots, k\}$ it holds $c'_i \in N_t[c_i]$. With respect to $\mathcal{R}$, we get $(((c_1, c_2, \ldots, c_k), r, \mathcal{R}, t), ((c_1, c_2, \ldots, c_k), r', \mathcal{C}, t + 1)) \in A$, if it holds $r' \in N_t[r]$. Again, we include auxiliary states and arcs to cater for the initial cops placement. Overall, we now get $|S| \in \mathcal{O}(n^{k+1} T)$, and $|A| \in \mathcal{O}(n^{2k+1} T)$, since for the dominant-in-magnitude number of $\mathcal{C}$-turn arcs there exist at most n^{2k} cop transitions from $(c_1, c_2, \ldots, c_k)$ to $(c'_1, c'_2, \ldots, c'_k)$. By reapplying the whole framework with $Tar = \{((c_1, c_2, \ldots, c_k), r, p, t) \mid c_i = r \text{ for some } 1 \le i \le k\}$ we conclude.

Corollary 1. *Given a dynamic graph* $\mathcal{G} = (G_1, G_2, \ldots, G_T)$ *and an integer* $k \ge 1$ *in the offline model, we can decide if* $c_{off}(\mathcal{G}) \le k$, *i.e., if* $\mathcal{G}$ *is* k*-cop-win, in time* $\mathcal{O}(n^{2k+1} T)$.

We may now run a search utilizing the result in Corollary 1 and derive an exponential time algorithm to determine the exact value of $c_{off}(\mathcal{G})$.

Corollary 2. *For some dynamic graph* $\mathcal{G}$, *with an associated time horizon* T, *the problem of determining the exact value of* $c_{off}(\mathcal{G})$ *is in EXPTIME.*

4   Online Case

In the online case, we are given an underlying graph $G = (V, E)$ and an indefinite number of discrete time steps of evolution $t = 1, 2, 3, \ldots$, that is, time evolution may take place ad infinitum. At each time step t, an instance $G_t = (V_t, E_t)$ is realized, where $V_t = V$, $E_t \subseteq E$. The *only assumption* we make on the topology of generated instances, is that we require each G_t to be *connected*. Note that this a widely used assumption in several dynamic graph models that appear in literature [11,27]. Removing this assumption could lead to trivial cases where, for instance, the k cops or the robber lie indefinitely on isolated vertices.

Initially, the cops and then the robber place themselves on V before the appearance of G_1. In the general case, neither the cops nor the robber have any knowledge about the evolution sequence. The cops and the robber, taking turns in this order, make their respective moves in G_t, then G_{t+1} is generated, and so forth. Similarly to the offline case, a token at vertex v moves to a vertex in $N_t[v]$ (all the cops move simultaneously). Let $c_t(G)$ stand for the *temporal cop number*, i.e., the worst-case minimum number of cops required to capture a robber for an underlying graph evolving like described above. In our analysis, we consider worst-case scenarios for the temporal cop number; a different type of analysis is left for future work. In other words, for our bounds to follow, one may assume the robber controls the dynamics of G to its advantage. Hence, at round t, the robber defines instance G_t according to the aforementioned restrictions.

Preliminary Bounds. As a warm up, let us consider two special cases for the topology of the underlying graph: a tree, and a complete graph.

Proposition 1. *For any tree T, it holds $c_t(T) = 1$.*

Proposition 2. *For any complete graph K_n, $n \geq 2$, it holds $c_t(K_n) = n - 1$.*

The above propositions cast some intuition on the relationship between the (static/classical) cop number $c(G)$ and our introduced temporal cop number $c_t(G)$. For the static case, it is easy to see that if G is either a tree or a clique then $c(G) = 1$. However, in the temporal case $c_t(T) = 1$ for a tree T, and $c_t(K_n) = n - 1$ for any clique on n vertices. Intuitively, the denser the underlying graph is, the more leeway there is for the robber due to worst-case dynamics. Overall, for any graph G, $c_t(G) \leq n - 1$, since initially placing the $n - 1$ cops on distinct vertices guarantees an edge between a cop-vertex and the robber-vertex in G_1 due to connectedness of the model. Thus, for the ratio of the two cop numbers, we get $1 \leq c_t(G)/c(G) \leq n - 1$.

We now provide preliminary bounds on $c_t(\cdot)$ by considering a subset of sparse graphs, that is, underlying graphs with at most linear number of edges.

Theorem 4. *For any graph $G = (V, E)$, $m = n - 1 + \lambda$, it holds $c_t(G) \leq \lambda + 1$.*

Proof. To describe the cop-winning strategy, let us define a partition of the vertices into V_C and V_R such that $V = V_C \cup V_R$ and $V_C \cap V_R = \emptyset$. Intuitively,

V_C stands for the *cop-secured* vertices, i.e., vertices the robber will never be able to visit, whereas V_R stands for the vertices (possibly) still within the eventual reach of the robber. More precisely, the cop strategy below builds a sequence of partitions (V_C, V_R) where V_C is a set of vertices the robber will never be able to visit, V_R contains the other vertices and the cardinality of V_R strictly decreases at each time step. This strategy may not be the fastest as V_R may contain unreachable vertices, but this is not required for the proof.

Consider the situation before some round t. Let T be some (arbitrary) spanning tree of G. We refer to the edges of T as the *black edges* and to any path consisting only of black edges as a *black path*. We refer to all other edges, which are exactly λ, as the *blue edges*. Suppose there is one cop at one extremity of each blue edge. Note that several cops may lie on the same vertex. We refer to these cops as the *blue cops*. One last cop, the *black cop*, is placed on some other (blue-cop free) vertex, say $x \in V$. The robber is on a cop-free vertex, say $r \in V$.

Consider the spanning tree T: there exists a unique (black) path from x to r. Let (x, x') be the first edge of this path. If this edge is removed from the black tree T, T is split into two black subtrees containing x and x' respectively, namely T_x and $T_{x'}$. Then, let $V_C = V(T_x)$ and $V_R = V \backslash V_C = V(T_{x'})$. Notice that it holds $r \in V_R$ with this partition.

By construction, the cut associated to (V_C, V_R) contains exactly one black edge, (x, x'), plus (possibly) some blue edges. Since G_t is connected for all time steps t, at least one edge associated to the cut is present in E_t. If the black edge (x, x') is present, then the black cop moves from x to x' during the cops turn. Otherwise, if only a blue edge, say (v, v'), where $v \in V_C$, is present, then the associated cop moves from v to v' (or remains at v' if it were already there). Now, we *swap* the role of the two edges. That is, (v, v') becomes a black edge, and its associated cop becomes the black cop, whereas (x, x') becomes a blue edge, and its associated cop becomes a blue cop. By construction, the set of black edges defines a new black spanning tree T': the unique black path from v to v' is replaced by the new black edge (v, v'). (Notice that, in the previous swap-less case, we trivially had $T' = T$). Afterwards, the robber may move at its turn; we still refer to its position by r. Even after the robber moves, it holds $r \in V_R \backslash \{v'\}$: there is no edge the robber could use to reach V_C since all cut-edges are protected, and v' is occupied by the black cop.

Before the next round of the game, let us now reapply the method used to obtain the partition on G_{t+1}. Let $x = v'$ stand for the black cop's current position, and set $T = T'$. Consider again the unique black path from x to r, and denote by (x, x') its first edge. By construction of T, there is a unique black path from x to all vertices of V_C. Hence, if T is split as before into two subtrees after removing edge (x, x'), the resulting subtree T_x contains x and also all vertices of former V_C (and possibly more vertices). Then, let us set $V_C = V(T_x)$ and $V_R = V \backslash V_C = V(T_{x'})$. As there is still one cop on one extremity of each blue edge, the vertices of V_C are unreachable by the robber.

Let us now consider the very first step. We start from an arbitrary spanning tree, denoted by T, whose edges are the black ones, the other being the blue ones. For the initial positions, let us place one cop at one extremity of each blue edge. One last cop, the *black cop*, is placed on some other (blue-cop free) vertex, say $x \in V$. Then, the robber chooses a cop-free vertex, say $r \in V$, for its initial place. Edge (x, x'), and sets V_C and V_R are similarly defined, hence the vertices of V_C are unreachable by the robber. The cardinality of V_R is at most $n - 1$.

If we inductively apply the above method for the cops, it follows that after each round, the number of vertices of V_R, which contains the vertices reachable by the robber, is strictly decreased. It will eventually reach the value of zero and the robber will be captured in at most n rounds.   □

The above result provides a better upper bound than the easy to see $c_t(G) \leq n - 1$, for sparse graphs when $\lambda \leq n - 3$. Moreover, we demonstrate it is nearly tight for certain graph families, see Theorem 5 in the next part of this section.

A Nearly Tight Graph Family for Sparse Graphs. We hereby consider the graph family $\mathbb{G} = \{G_\lambda \mid \lambda \mod 2 = 1 \land \lambda \geq 5\}$ for sufficiently large *odd* values of $\lambda \in \mathbb{N}$. We define the vertex set as $V(G_\lambda) = \{v_1, v_2, \ldots, v_{2\lambda-2}, v'_1, v'_2, \ldots, v'_{2\lambda-2}\}$. For $i = 1, 2, \ldots, 2\lambda - 2$, let $(v_i, v_{(i+1) \mod (2\lambda-2)}) \in E(G_\lambda)$ and $(v_i, v'_i) \in E(G_\lambda)$. Also, for $i = 1, 3, 5, \ldots, 2\lambda - 1$, let $(v'_i, v'_{i+1}) \in E(G_\lambda)$. Overall, it holds $n = |V(G_\lambda)| = 4\lambda - 4$ and $m = |E(G_\lambda)| = 2(2\lambda - 2) + (2\lambda - 2)/2 = 5(2\lambda - 2)/2 = 5\lambda - 5 = n - 1 + \lambda$. Intuitively, G_λ is a cycle on $2\lambda - 2$ vertices where another $\lambda - 1$ disjoint C_4-cycles are attached. An example depiction is given in Fig. 1. Notice that G_λ becomes a tree by the removal of λ edges, for example, the $\lambda - 1$ edges (v'_i, v'_{i+1}), for $i = 1, 3, 5, \ldots, 2\lambda - 1$ and another edge (v_j, v_{j+1}) for some $j \in \{1, 2, \ldots, 2\lambda - 1\}$.

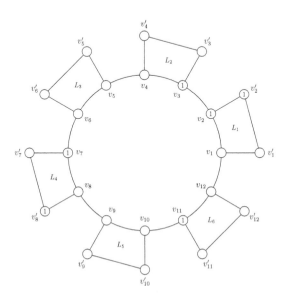

Fig. 1. An example initial placement of $\lambda - 2 = 5$ cops on G_7.

Theorem 5. *For any $G_\lambda \in \mathbb{G}$, it holds $c_t(G_\lambda) = \lambda - 1$.*

This theorem is a direct consequence of the two lemmata that follow, which demonstrate the corresponding (worst-case) upper and lower bound strategies.

Lemma 2. *For any $G_\lambda \in \mathbb{G}$, it holds $c_t(G_\lambda) \leq \lambda - 1$.*

Proof. We present a strategy for $\lambda - 1$ cops to win against the robber under any dynamics and/or robber strategy. Initially, the $\lambda - 1$ cops are placed as follows: place one cop at v_i for each $i = 2, 6, 10, \ldots, 2\lambda - 4$ and for each $i = 3, 7, 11, \ldots, 2\lambda - 3$. To verify, since there are two sequences of cops using a distance 4 step, overall the number of cops is $2(2\lambda - 2)/4 = \lambda - 1$. For an example placement on G_7, see Fig. 2. Then, the robber places itself at some cop-free vertex. By symmetry of G_λ and cop placement, without loss of generality, we assume the robber places itself on some vertex in $R := \{v_4, v_5, v_3', v_4', v_5', v_6'\}$. In the cop strategy we will now propose, the robber will never be able to escape this set of vertices. Therefore, we restrict the proof to the subgraph induced by $\{v_2, v_3, v_4, v_5, v_6, v_7, v_3', v_4', v_5', v_6'\}$, see Fig. 3a, and will demonstrate how the four cops in this subgraph can always capture a robber with an initial placement within R. For all robber turns below, we assume the robber always remains within R; by our strategy, it is impossible for the robber to move outside R since it would mean "jumping" over a cop.

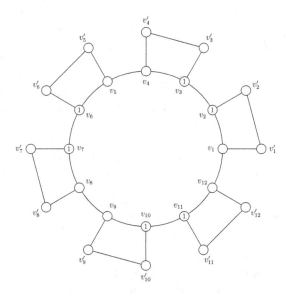

Fig. 2. The initial positions for cops in graph $G_7 \in \mathbb{G}$. In all figures, an integer within a vertex stands for the number of cops currently placed on the vertex.

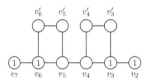

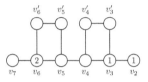

(a) Initial positions for cops in subgraph.

(b) After the first move: a cop moved from v_7 to v_6.

Fig. 3. The first move of the cop strategy

The cops' strategy is the following. Since the instance needs to be connected at each time step, at least one edge in $\{(v_2, v_3), (v_7, v_6)\}$ is available. By symmetry, without loss of generality, let us assume the corresponding cop moves to v_6, see Fig. 3b. It suffices to prove that the cops have a winning strategy starting from this configuration. The rest of the proof with all the necessary case analysis is left to the reader. □

To help us with the matching lower bound to follow, we hereby provide some useful definitions and claims on cop movement restrictions on G_λ incurred by worst-case dynamics. From now on, all vertex indices are assumed to be modulo $2\lambda - 2$. For $i = 1, 2, \ldots, \lambda - 1$, let *loop* L_i refer to the 4-cycle with $V(L_i) = \{v_{2i-1}, v_{2i}, v'_{2i-1}, v'_{2i}\}$ and let its edge-set be defined as $E(L_i) = \{(v_{2i-1}, v_{2i}), (v'_{2i-1}, v'_{2i}), (v_{2i-1}, v'_{2i-1}), (v_{2i}, v'_{2i})\}$. We say that L_i is *cop-occupied* if at least one cop lies at some vertex in $V(L_i)$, otherwise, L_i is *cop-free*. We say that a cop *crosses* L_i if, starting from vertex v_{2i-1} (*cross-start* vertex), it can eventually arrive to vertex v_{2i} (*cross-end* vertex), or vice versa. We refer to a (counterclockwise-movement) crossing from v_{2i-1} to v_{2i} as a *cc-crossing* and to a (clockwise-movement) crossing from v_{2i} to v_{2i-1} as a *c-crossing*. A cop trivially cc-crosses L_i if it already lies on v_{2i} or v_{2i+1}, i.e., the counterclockwise neighbor of v_{2i}. Respectively, a cop trivially c-crosses L_i if it already lies on v_{2i-1} or v_{2i-2}, i.e., the clockwise neighbor of v_{2i-1}. The intuition behind Proposition 3 is that, while a number of cops crosses a loop, at least one of them must *stay behind*, that is, will not be able to ever cross the loop due to worst-case dynamics.

Proposition 3. *Assume we focus on a given loop $L_i \subset V(G_\lambda)$ and at most one edge in $E(L_i)$ is not present at each time step of evolution. In the worst case, at most ρ cops can cross L_i, if $\rho + 1$ cops are present at the cross-start vertex.*

Proof. Without loss of generality, consider loop $L_1 = \{v_1, v_2, v'_1, v'_2\}$ and suppose $\rho + 1$ cops lie on v_1 and wish to cross to v_2. The dynamics of the graph evolve as follows: for each t, if at the end of round t there is at least one cop on v_1, then $(v_1, v_2) \notin E_{t+1}$. Otherwise, $(v'_1, v'_2) \notin E_{t+1}$. In other words, as long as there is a cop on v_1, the edge to v_2 is blocked. The cops could take advantage of this situation such that at most ρ of them reach v_2 via the available path v_1, v'_1, v'_2, v_2. If at any time v_1 is cop-free, then the above path is blocked and (v_1, v_2) is available, however no cop is there to traverse it and cross the loop.

The last remaining cop cannot cross since it would mean that, at some point in time, either v_1 is cop-occupied and (v_1, v_2) is available or v_1 is cop-free and (v_1', v_2') is available, a contradiction to the specified dynamics.   □

Assume strictly fewer than $\lambda - 1$ cops initially place themselves at the vertices of G_λ. Since there are $\lambda - 1$ loops, there exists at least one cop-free loop L_i. In general, after the cops are initially positioned, G_λ can be partitioned into alternating sequences of cop-occupied and cop-free loops L_i. Let $O_1, \ldots, O_p$, respectively $F_1, \ldots, F_p$, stand for the sequences of cop-occupied, respectively cop-free loops, where F_1 is set arbitrarily, and we assume F_i is between O_i (clockwise) and O_{i+1} (counterclockwise). Moreover, for $i = 1, 2, \ldots, p$, let $|F_i| = f_i$ and $|O_i| = o_i$. The cardinality p of the two sequence sets is the same, since two non-maximal adjacent cop-occupied subsequences, i.e., with no cop-free loop between them, form one bigger cop-occupied sequence; a similar observation holds for cop-free sequences. By the reasoning above, it holds $p \geq 1$. An example initial placement on G_7 is given in Fig. 1: The sequences of cop-occupied and cop-free loops formed are $O_1 = \{L_6, L_1, L_2\}$, $F_1 = \{L_3\}$, $O_2 = \{L_4\}$, and $F_2 = \{L_5\}$.

The following proposition provides us with a necessary condition in order for the cops to win against a robber placed on some cop-free sequence of loops. For a sequence $F = \{L_1, L_2, \ldots, L_f\}$, let $V(F) = \cup_{i=1}^{f} V(L_i)$.

Proposition 4. *Let $F = \{L_1, L_2, \ldots, L_f\}$ be a cop-free sequence of loops with the robber lying on some vertex within $V(F)$ not adjacent to a cop. Let L_{cc}, respectively L_c, stand for the cop-occupied loop adjacent to the counter clockwise of F, respectively to the clockwise of F. At least $f+1$ cops must be able to c-cross L_{cc}, and another $f+1$ cops to cc-cross L_c, in order for the cops to win.*

Proof. In contradiction, and without loss of generality, assume at most f cops can c-cross L_{cc} and $f + 1$ cops can cc-cross L_c. Assume L_f is counterclockwise adjacent to L_{cc}, L_{i+1} is counterclockwise adjacent to L_i for $i = 1, 2, \ldots, f - 1$, and L_c is counterclockwise adjacent to L_1; see Fig. 4. From now on, consider that the dynamics of the graph force the single edge connecting L_c to L_1 to be unavailable in all graph instances. Therefore, no cop from L_c can ever reach L_1 in F, while each graph instance remains connected.

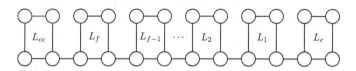

Fig. 4. Loop numbering for a cop-free sequence $F = \{L_1, L_2, \ldots, L_f\}$. Left stands for counterclockwise direction in G_λ, respectively right for clockwise direction.

Having crossed L_{cc}, the f cops may all move to L_f. By Proposition 3, at most $f - 1$ cops can c-cross L_f. For some $i \geq 1$, assume $f - i$ cops have c-crossed

L_{f+1-i}. Then, by Proposition 3, at most $f - i - 1 = f - (i+1)$ cops can c-cross $L_{f+1-(i+1)}$. Overall, by induction, at most $f - i$ cops are able to c-cross L_{f+1-i}. For $i = f - 1$, at most $f - (f - 1) = 1$ cop is able to c-cross L_2.

To win, the robber has a feasible strategy to cross L_1, that is, to be placed at the vertex in L_1 connected by the (always unavailable) edge to L_c as discussed above. No cop can arrive to L_1 from L_c due to the missing edge, and since at most 1 cop can c-cross L_2, no cop can ever c-cross L_1.   □

Now, we are ready to show, in Proposition 5, how the robber can identify a cop-free sequence to employ the winning strategy demonstrated in Proposition 4.

For integers i, w_i, w_i', where $1 \leq i \leq \lambda - 1$, $0 \leq w_i \leq p - 1$, $0 \leq w_i' \leq p - 1$, let $F_c(i, w_i) = \{F_i, F_{i-1}, \ldots, F_{i-w_i}\}$ stand for the set including F_i and the w_i cop-free sequences nearer to F_i in clockwise fashion, and $F_{cc}(i, w_i') = \{F_i, F_{i+1}, \ldots, F_{i+w_i'}\}$ stand for the set including F_i and the w_i' cop-free sequences nearer to F_i in counterclockwise fashion. In a similar manner, for the cop-occupied sequences, let $O_c(i, w_i) = \{O_i, O_{i-1}, \ldots O_{i-w_i}\}$ and $O_{cc}(i, w_i') = \{O_i, O_{i+1}, \ldots, O_{i+w_i'}\}$. For a cop-occupied sequence O_i, we say that $o_i = |O_i|$ cops (choosing one per loop in O_i) are its *occupant* cops. If strictly more than o_i cops lie at vertices of O_i, then this surplus of cops are referred to as *extra* cops.

Proposition 5. *If there exists a cop-free sequence F_i in G_λ such that at least one of the following holds:*

(a) strictly fewer than $\sum_{F_j \in F_c(i, w_i)} f_j$ extra cops lie within vertices in $O_c(i, w_i)$, for all integers w_i, where $0 \leq w_i \leq p - 1$,

(b) strictly fewer than $\sum_{F_j \in F_{cc}(i, w_i')} f_j$ extra cops lie within vertices in $O_{cc}(i + 1, w_i')$, for all integers w_i', where $0 \leq w_i' \leq p - 1$,

then the robber wins.

Proposition 6. *Assume strictly fewer than $\lambda - 1$ cops are initially placed on G_λ. Then, there exists a cop-free sequence F_i in G_λ such that at least one of conditions (a) and (b) in Proposition 5 holds.*

Proof. Since strictly fewer than $\lambda - 1$ cops are initially placed on G_λ, and G_λ has exactly $\lambda - 1$ loops L_i, then by pigeonhole principle there exists at least one loop with no cop on its vertices, and so at least one cop-free sequence in G_λ.

By contradiction, suppose that for every cop-free sequence F_i in G_λ (i) there exists w_i such that at least $\sum_{F_j \in F_c(i, w_i)} f_j$ extra cops lie within $O_c(i, w_i)$, and (ii) there exists w_i' such that at least $\sum_{F_j \in F_{cc}(i, w_i')} f_j$ extra cops lie within $O_{cc}(i + 1, w_i')$. Consider some cop-free sequence, say F_{i_1}, without loss of generality. By (ii), there exists some (*minimum-value*) w_{i_1} such that at least $\sum_{F_j \in F_{cc}(i_1, w_{i_1})} f_j$ extra cops lie within $O_{cc}(i_1 + 1, w_{i_1})$. Let $F_{i_2} = F_{i_1+1+w_1}$ be the first cop-free sequence to the counterclockwise of $O_{cc}(i_1 + 1, w_{i_1})$. Then, by (ii), there exists some (minimum-value) w_{i_2} such that at least $\sum_{F_j \in F_{cc}(i_2, w_{i_2})} f_j$ extra cops lie within $O_{cc}(i_2 + 1, w_{i_2})$. We proceed with such statements, inductively, until we reach F_{i_l}, for which there exists (minimum-value) w_{i_l} such that at least $\sum_{F_j \in F_{cc}(i_l, w_{i_l})} f_j$ extra cops lie within $O_{cc}(i_l + 1, w_{i_l})$ and $i_l + 1 + w_{i_l} \geq i_1 \pmod{p}$. That is, we have performed a full round on G_λ. There are three cases to consider with respect to the value $i_l + 1 + w_{i_l}$.

- If $i_l + 1 + w_{i_l} = i_1$, then, for $i_z = i_1, i_2, \ldots, i_l$, sets $O_{cc}(i_z + 1, w_{i_z})$ form a partition of the cop-occupied space in G_λ. By assumption, for each such i_z, at least $\sum_{F_j \in F_{cc}(i_z, w_{i_z})} f_j$ extra cops lie within $O_{cc}(i_z + 1, w_{i_z})$. Summing it all, $\sum_{z=1}^{l} \sum_{F_j \in F_{cc}(i_z, w_{i_z})} f_j = \sum_{j=1}^{p} f_j$ extra cops lie within the cop-occupied sequences, since $\cup_{z=1}^{l} F_{cc}(i_z, w_{i_z})$ contains all cop-free loops in G_λ.
- If $i_l + 1 + w_{i_l} = i_y$, for some $y > 1$, then the last interval fully covers some already defined intervals starting at $i_1, i_2, \ldots, i_{y-1}$. In this case, for $i_z = i_y, i_{y+1}, \ldots i_l$, sets $O_{cc}(i_z + 1, w_{i_z})$ form a partition of the cop-occupied space in G_λ. By assumption, for each such i_z, at least $\sum_{F_j \in F_{cc}(i_z, w_{i_z})} f_j$ extra cops lie within $O_{cc}(i_z + 1, w_{i_z})$. Summing it all together as in the previous case, at least $\sum_{z=y}^{l} \sum_{F_j \in F_{cc}(i_z, w_{i_z})} f_j = \sum_{j=1}^{p} f_j$ extra cops lie within the cop-occupied sequences.
- If $i_y < i_l + 1 + w_{i_l} < i_{y+1}$, for some $y > 1$, then the last interval fully contains intervals starting at $i_1, i_2, \ldots, i_{y-1}$ and partially overlaps with interval i_y. Let $i_l + 1 + w_{i_l} = i_y + x$ for some $1 \leq x < i_{y+1} + 1 - i_y$. There are strictly fewer than $f_{i_y} + f_{i_y+1} + \ldots + f_{i_y+x-1}$ extra cops within vertices in $O_{i_y+1}, \ldots, O_{i_y+x}$, otherwise the choice of w_{i_y} would not be minimum. As an implication, there are at least $f_{i_y+x} + \ldots + f_{i_y+w_{i_y}}$ extra cops within $O_{i_y+x+1}, \ldots, O_{i_y+w_{i_y}}$. Also, by assumption, at least $f_{i_l} + f_{i_l+1} + \ldots + f_1 + \ldots + f_{i_y+x-1}$ extra cops lie within vertices in $O_{i_l+1}, O_{i_l+2}, \ldots, O_1, \ldots, O_{i_y+x}$. Overall, at least $f_{i_l} + f_{i_l+1} + \ldots + f_1 + \ldots + f_{i_y+x-1} + f_{i_y+x} + \ldots + f_{i_y+w_{i_y}}$ extra cops lie within $O_{i_l+1}, O_{i_l+2}, \ldots, O_1, \ldots, O_{i_y+x}, O_{i_y+x+1}, \ldots, O_{i_y+w_{i_y}}$. For the rest of the graph, for $z = y + 1, \ldots, l - 1$, at least $\sum_{F_j \in F_{cc}(i_z, w_{i_z})} f_j$ extra cops lie within $O_{cc}(i_z + 1, w_{i_z})$. Summing it all together, at least $\sum_{j=1}^{p} f_j$ extra cops lie within the cop-occupied sequences, since each f_j is considered once in the above calculations.

In all three cases, considering occupant guards and extra guards together, it follows there are at least $\sum_{i=1}^{p}(o_i + f_i) = \lambda - 1$ cops in G_λ, since the number of loops in all the sequences is exactly the number of loops in G_λ. $\square$

Lemma 3. *For any $G_\lambda \in \mathbb{G}$, it holds $c_t(G_\lambda) \geq \lambda - 1$.*

Proof. Follows by the combination of Propositions 5 and 6. $\square$

5   Conclusions

In this paper, we consider the (barely studied) topic of playing Cops and Robbers games in models of dynamic graphs. We show how the cop number can be computed in the offline case, where all the graph dynamics are known a priori, via a reduction to a reachability game. In the online case with a connectedness restriction, we show a nearly tight bound on the cop number of sparse graphs.

In future, one could improve the bounds in this paper and also consider dense graphs for online models. There exists an abundance of dynamic graph models in literature: it would be interesting to consider them and compare among them.

References

1. Aigner, M., Fromme, M.: A game of cops and robbers. Discrete Appl. Math. **8**, 1–12 (1984)
2. Andrae, T.: Note on a pursuit game played on graphs. Discrete Appl. Math. **9**, 111–115 (1984)
3. Berarducci, A., Intrigila, B.: On the cop number of a graph. Adv. Appl. Math. **14**(4), 389–403 (1993)
4. Berwanger, D.: Graph games with perfect information, Preprint (2013)
5. Blin, L., Fraigniaud, P., Nisse, N., Vial, S.: Distributed chasing of network intruders. Theor. Comput. Sci. **399**, 12–37 (2008)
6. Bollobás, B., Kun, G., Leader, I.: Cops and robbers in a random graph. J. Comb. Theory Ser. B **103**(2), 226–236 (2013)
7. Bonato, A., Chiniforooshan, E., Pralat, P.: Cops and Robbers from a distance. Theor. Comput. Sci. **411**, 3834–3844 (2010)
8. Bonato, A., Nowakowski, R.J.: The Game of Cops and Robbers on Graphs. American Mathematical Society, Providence (2011)
9. Bonato, A., MacGillivray, G.: Characterizations and algorithms for generalized cops and robbers games arXiv:1704.05655 (2017)
10. Bonato, A., Pralat, P., Wang, C.: Pursuit-evasion in models of complex networks. Internet Math. **4**, 419–436 (2007)
11. Casteigts, A.: A journey through dynamic networks (with excursions). Université de Bordeaux, Habilitation à diriger des recherches (2018)
12. Casteigts, A., Flocchini, P., Quattrociocchi, W., Santoro, N.: Time-varying graphs and dynamic networks. In: Frey, H., Li, X., Ruehrup, S. (eds.) ADHOC-NOW 2011. LNCS, vol. 6811, pp. 346–359. Springer, Heidelberg (2011). https://doi.org/10.1007/978-3-642-22450-8_27
13. Clarke, N.E., MacGillivray, G.: Characterizations of k-copwin graphs. Discrete Math. **312**, 1421–1425 (2012)
14. Erlebach, T., Spooner, J.T.: A game of cops and robbers on graphs with periodic edge-connectivity. In: Chatzigeorgiou, A., et al. (eds.) SOFSEM 2020. LNCS, vol. 12011, pp. 64–75. Springer, Cham (2020). https://doi.org/10.1007/978-3-030-38919-2_6
15. Fitzpatrick, S.L., Nowakowski, R.J.: Copnumber of graphs with strong isometric dimension two. Ars Comb. **59**, 65–73 (2001)
16. Frankl, P.: Cops and robbers in graphs with large girth and Cayley graphs. Discrete Appl. Math. **17**(3), 301–305 (1987)
17. Hahn, G., MacGillivray, G.: A note on k-cop, l-robber games on graphs. Discrete Math. **306**, 2492–2497 (2006)
18. Holme, P., Saramäki, J.: Temporal networks. Phys. Rep. **519**, 97–125 (2012)
19. Kehagias, A., Konstantinidis, G.: Cops and robbers, game theory and Zermelo's early results, preprint, arXiv:1407.1647 (2014)

20. Khaliq, I., Imran, G.: Reachability games revisited. In: SOFTENG, pp. 129–132 (2016)
21. Kinnersley, W.: Cops and robbers is EXPTIME-complete. J. Comb. Theory Ser. B **111**, 201–220 (2015)
22. Kosowski, A., Li, B., Nisse, N., Suchan, K.: k-chordal graphs: from cops and robber to compact routing via treewidth. In: Czumaj, A., Mehlhorn, K., Pitts, A., Wattenhofer, R. (eds.) ICALP 2012. LNCS, vol. 7392, pp. 610–622. Springer, Heidelberg (2012). https://doi.org/10.1007/978-3-642-31585-5_54
23. Lu, L., Peng, X.: On Meyniel's conjecture of the cop number. J. Graph Theory **71**, 192–205 (2012)
24. Luczak, T., Pralat, P.: Chasing robbers on random graphs: zigzag theorem. Random Struct. Algorithms **37**, 516–524 (2010)
25. Maamoun, M., Meyniel, H.: On a game of policemen and robber. Discrete Appl. Math. **17**(3), 307–309 (1987)
26. Mazala, R.: Infinite games. In: Grädel, E., Thomas, W., Wilke, T. (eds.) Automata Logics, and Infinite Games. LNCS, vol. 2500, pp. 23–38. Springer, Heidelberg (2002). https://doi.org/10.1007/3-540-36387-4_2
27. Michail, O.: An introduction to temporal graphs: an algorithmic perspective. Internet Math. **12**, 239–280 (2016)
28. Michail, O., Spirakis, P.G.: Elements of the theory of dynamic networks. Commun. ACM **61**(2), 72–72 (2018)
29. Nowakowski, R., Winkler, P.: Vertex-to-vertex pursuit in a graph. Discrete Math. **43**, 235–239 (1983)
30. Pralat, P., Wormald, N.: Meyniel's conjecture holds for random graphs. Random Struct. Algorithms **48**(2), 396–421 (2016)
31. Quillot, A.: Étude de quelques problémes sur les graphes et hypergraphes et applications à la théorie des jeux à information compléte. Thése, UPMC, Paris (1980)
32. Scott, A., Sudakov, B.: A bound for the cops and robbers problem. SIAM J. Discrete Math. **25**, 1438–1442 (2011)
33. Sugihara, K., Suzuki, I.: On a pursuit-evasion problem related to motion coordination of mobile robots. In: 21st Annual Hawaii International Conference on System Sciences, vol. 4, pp. 218–226 (1988)

Black Virus Decontamination
of Synchronous Ring Networks
by Initially Scattered Mobile Agents

Nikos Giachoudis[1], Maria Kokkou[2], and Euripides Markou[1]([✉])

[1] University of Thessaly, Lamia, Greece
{ngiachou,emarkou}@dib.uth.gr
[2] Chalmers University of Technology, Gothenburg, Sweden
kokkou@student.chalmers.se

Abstract. We study the Black Virus Decontamination problem in ring topologies for initially scattered mobile agents. In this problem a number of mobile agents operate in a network where one of its nodes u is hostile (contaminated) in the following way: when u is visited by an agent, it is decontaminated, the agent vanishes without leaving any trace, and all adjacent nodes of u which are unoccupied by agents are now contaminated. The goal is to find the minimum number of agents that can decontaminate a given network with a black virus at an unknown location and design a fast distributed algorithm for a certain (preferably weak) model of mobile agents. The problem has been introduced by J., Cai et al in 2014 and combines details from two widely studied problems: the Black Hole Search problem and the Intruder Capture problem.

We study here the problem for initially scattered mobile agents in synchronous ring topologies. We prove that ten initially scattered agents with a common chirality (i.e., agreement in a global sense of orientation) are necessary and sufficient to solve the problem. If the agents do not have a common chirality then twelve scattered agents with distinct identities are necessary and sufficient, while for anonymous agents the problem is unsolvable. To the best of our knowledge these are the first results concerning the problem for initially scattered agents.

1 Introduction

We often need to solve problems in networks where hostile entities are present which may harm the system. Hence, models of such networks have appeared in the literature. Especially in the distributed computing literature one such hostile static entity which is called *black hole* has been extensively studied. A black hole is a node infected by a process that destroys any incoming agent without leaving any trace, and the goal is for a group of agents to locate the black hole within finite time. Another type of a malicious mobile entity is the *intruder*. In the *intruder capture* problem (also known as *graph decontamination* and *connected graph search*) a harmful agent, called the intruder, can move in the network and

© Springer Nature Switzerland AG 2020
A. W. Richa and C. Scheideler (Eds.): SIROCCO 2020, LNCS 12156, pp. 220–236, 2020.
https://doi.org/10.1007/978-3-030-54921-3_13

infect the visited nodes. The objective is for a team of mobile agents, that cannot be harmed by the intruder, to decontaminate the network. Hence a black hole is harmful to the mobile agents but not to other nodes of the network, while an intruder damages nodes but not agents.

We study here a model of another hostile entity called *black virus* which has been introduced by J., Cai et al in 2014. The black virus is a malicious entity similar to a *black hole* which is initially located at a node u of the network and has the following behaviour: when a mobile agent enters node u, the agent is removed from the network without leaving any trace. The black virus spreads to all unoccupied by agents neighbouring nodes of u, effectively expanding its contamination over the network, but node u is now clean (decontaminated). The goal of the *Black Virus Decontamination* (BVD) problem is to find the minimum number of agents that can decontaminate a given network with a black virus initially located at an unknown place and design a fast distributed algorithm for a certain (preferably weak) model of mobile agents. The black virus combines the threat of a harmful node with the need for decontamination and the ability to infect additional nodes. In order to prevent the black virus from spreading, the agents must occupy the node(s) where the black virus is going to spread to. Hence the only way to clear a contaminated node and at the same time prevent it from spreading the contamination, is to have the node visited by an agent while at the same time all adjacent nodes are occupied by agents.

1.1 Related Work

The Black Virus Decontamination (BVD) problem combines the Black Hole Search problem and the Intruder Capture problem. The Black Hole Search problem has been extensively studied for various topologies and communication models (e.g., in [8,9]) and a recent survey of the results on this problem can be found in [12]. The Intruder Capture problem has been also extensively studied (e.g., in [2,3,13]) and a recent survey can be found in [14]. The BVD problem was introduced in [5] where the problem was studied in specific topologies, namely *q-grids, q-tori* and *hypercubes*. The problem was considered for a team of initially *co-located* mobile agents that are injected somewhere in the network and a different solution protocol was given for each topology. The main strategy used in the paper was the following: one agent (EA) visits a previously unexplored node v, while another agent (LA) waits on a safe node u, incident to v. All the already explored neighbours of v are guarded by agents. If v is not infected then agent EA returns to u and reports that v is safe. If agent EA is destroyed, then agent LA learns the location of the black virus and the remaining agents clean the network. In [6] the BVD problem was studied in arbitrary networks for co-located agents, where the agents have a map of the network. The strategy which was used here is similar to the one that was used in [5] with the addition that this time the agents also compute an optimal *exploration sequence*. Furthermore, in [6] two types of black virus clones are considered; *fertile clones*, that maintain the same capabilities as the original black virus and *sterile clones* that cannot spread when visited by an agent. In [7] a protocol providing a distributed optimal solution for arbitrary graphs and co-located agents is presented. The main difference between [6] and [7] is that in [7] the agents have a '2-hop visibility' capability instead of a map. Moreover, only the case of *sterile clones* is considered in [7]. In [4] the problem was considered for co-located agents in arbitrary networks that contain multiple black viruses. The agents in [4] are provided with a '2-hop visibility' capability and both the cases of *sterile* and *fertile* clones are investigated. The BVD problem has been also studied for co-located agents in chordal rings in [1]. Finally the problem has been studied in [11] for co-located agents and selected (by the algorithm) agents' configurations for which parallel strategies for the decontamination of the network were given. To the best of our knowledge, the BVD problem has only been studied before for initially co-located agents with distinct identities that cannot leave messages (or tokens) at nodes of the network [1,4–7] and for agents with distinct identities that can be initially placed in specific positions (i.e., selected by the algorithm) and either cannot leave messages at nodes or they can write messages at whiteboards at nodes [11]. The techniques presented in those papers cannot be directly used if the agents are initially *scattered* in the network (i.e., at most one agent at a node). If the agents are not initially co-located, then they cannot communicate in the face to face communication model (until they meet at a node) and divide duties so that to occupy all adjacent nodes of a contaminated node u before an agent visits u. Hence, in that case the agents first need to meet which may not be easy given the presence of the black virus and a limited knowledge about the network. For example, four initially co-located agents with distinct identities can easily decontaminate a synchronous ring with one black virus when the agents agree on the orientation of the ring as follows. The agents select a direction and move using the technique presented in [5]: one agent moves to an adjacent node v and if this node is safe then it returns to meet the other agents that wait; then all agents move to v. They repeat this procedure until one agent vanishes at a node u while all other agents wait at node w, adjacent to u. The black virus clones itself to a node z adjacent to u (node u is cleaned and w was occupied). Then one of the remaining three agents moves to node u where it stays while the other two agents explore the ring on the other direction using the same technique until eventually one of the two agents visits node z while both neighbours of z are occupied and the ring is completely decontaminated. If however the agents are initially scattered in the ring (i.e., at most one agent at a node), then this technique cannot be applied before the agents meet. In this paper we are interested to study the problem on the ring when the agents are initially scattered and investigate the impact of such a scenario in the minimum number of agents needed to decontaminate the network. In a model similar to the one described above but for initially scattered agents we show that ten agents are necessary and sufficient to solve the problem.

1.2   Model

We study the black virus decontamination problem in synchronous ring topologies. More specifically, a ring topology is an undirected graph $G = (V, E)$ where each vertex $u \in V$ has exactly two neighbours. The nodes of the ring are anonymous. The ring initially contains a number of mobile agents and one black virus. All edges incident to a node have distinct port labels, visible to an agent at the node. These port labels are globally consistent, which means that an agent exiting any node always via the same port label can eventually traverse the whole ring. However it might happen that not all agents agree on which is the *clockwise* direction between the two directions of the ring. In such a case (using the standard terminology) we say that the agents do not have a common *chirality*, or there is no global sense of orientation. If all agents agree on the clockwise direction, we say the that the agents have a common chirality, or there is a global sense of orientation.

The ring is synchronous meaning that an agent needs one unit of time to traverse an edge. The time an agent needs in order to compute, communicate, release or pick-up a token (see below) is negligible.

The mobile agents are computational entities that operate in the network and are able to move from one node to another via the connecting edge of the two nodes. The agents may have distinct identities taken from the set of natural numbers and each agent knows its identity number. However, those identities (even if they are distinct) are selected by an adversary and are not assigned by an algorithm. Most of our negative results hold even for agents with distinct identities. Our algorithm for agents with common chirality, although it is initially presented (for more convenience) assuming that the agents have distinct identities, as we later describe, it can be slightly modified to work for anonymous agents. The agents are identical (apart from their identities when they are distinct) and they are equipped with movable identical tokens, which they can leave at (or pick-up from) nodes. In all our algorithms each agent has one token. An agent can only communicate with other co-located agents (face to face communication). More specifically, when two (or more) agents are at the same node they can read each other's state and identity. They cannot exchange messages. Their memory, although it is enough for reading other agents' identities when needed, is not related to the number of nodes of the ring and therefore for example the agents cannot count more than a constant number of nodes. The agents can only meet at nodes and not inside edges, i.e., two agents traversing the same edge do not notice each other. All the agents are initially scattered in the network (i.e., there is at most one agent at a node) and they execute the same deterministic protocol starting at the same time. The agents do not initially have any knowledge about the size of the ring or their configuration (unless clearly stated).

The *black virus* is a malicious entity which is initially placed at an unknown location (i.e., node) in the ring. The interaction between the agents and the black virus is the following. When an agent decides at time t to move from a node u to an adjacent node w where the black virus resides three events occur. At time

$t + 1$ the agent vanishes without a trace, node w is cleaned, and the virus *clones* itself to any neighbouring nodes of w which are unoccupied by agents at time $t + 1$. If an agent tries to move to a node v and at the same time a clone of the black virus tries to move to v, then the agent moves to v while the clone does not. If a clone of the black virus moves to a node (unoccupied by agents) where a clone already exists (or moves there at the same time), then the two clones merge to one. In other words any node can contain at most one clone of the black virus. If a token is located at a node v where the black virus spreads to, and no agent is located at v or moves to v at the same time, then the token is destroyed (if an agent is already at v, or moves to v at the same time then the token survives).

The goal of the Black Virus Decontamination problem is to design an algorithm which completely cleans (i.e., eliminates any black virus clones from) the network within finite time and at least one agent survives.

1.3   Main Contributions

We first examine the problem in a synchronous ring for scattered agents with a common chirality. We prove that a ring cannot be decontaminated by less than ten scattered agents even if the agents have distinct identities, an unlimited number of tokens, unlimited memory and a common chirality. We then present an algorithm, for ten scattered agents with a common chirality, one token each and constant memory that decontaminates any synchronous ring within $O(n)$ time units, where n is the number of nodes of the ring. The algorithm can be slightly modified to work for anonymous agents. A preliminary version of those results has appeared in the Bachelor's Thesis of M. Kokkou [10].

Next we study the problem for scattered agents without a common chirality. We prove that a ring cannot be decontaminated by any number of scattered agents if the agents are anonymous and do not have a common chirality. We also prove that it cannot be decontaminated by less than twelve scattered agents with distinct identities and one token each if the agents do not have a common chirality. On the positive side, we present an algorithm, for twelve scattered agents with distinct identities (which are natural numbers) and one token each, that decontaminates any ring within $O(n^2 + L)$ time units, where n is the number of nodes of the ring and L is the maximum identity number, even when the agents do not have a common chirality.

Due to space limitations some proofs, figures, and formal algorithms will appear in the full version of the paper.

2   Agents with Common Chirality

2.1   Impossibility Results

The only way to decontaminate a ring with a black virus located at a node u is having an agent visiting node u, while the two adjacent nodes of u are occupied by agents. Hence the problem is unsolvable by less than three agents.

Consider an interval of $x \geq 2$ consecutive nodes (i.e., an interval of length $x - 1 \geq 1$ edges) in a ring, so that its two endpoints u, v are occupied by clones of the virus and no node of the interval is occupied by an agent. We call such an interval a *contaminated interval* (notice that any other node of the interval apart from its endpoints could be either clean or not). Since at least one agent has to vanish in order to decrease by one the length of the interval, and three agents are needed to decontaminate the last infected node the following lemma holds.

Lemma 1. *Consider an interval of $x \geq 2$ consecutive nodes in a synchronous and labelled ring so that the two endpoints of the interval are occupied by clones of the virus and no node of the interval is occupied by an agent. Then $x+1$ agents with common chirality are not enough to decontaminate the interval, even if they have distinct ids, an unlimited number of tokens, unlimited memory and know the initial configuration on the ring (i.e., they have an exact map of the ring and know the initial locations of all agents) and the exact location of the contaminated interval.*

An immediate consequence of Lemma 1 gives us a lower bound of 5 agents when the location of the black virus is not initially known, even for a strong model of agents, for all initial configurations of agents such that each node is initially occupied by at most one agent and the initial distance between any two agents is greater than 3 edges. In such a scenario an adversary arranges the location of the black virus so that the first agent that moves (or one of the first agents that move) vanishes creating a contaminated interval of 3 nodes. Hence by Lemma 1, an additional 4 agents are not sufficient for decontaminating the 3-nodes interval.

Moving to weaker agent models we can increase the lower bound from 5 to 9 when the agents do not know their initial configuration.

Lemma 2. *Nine initially scattered agents with common chirality are not enough to decontaminate a synchronous and labelled ring consisting of $n \geq 36$ nodes, even if they have distinct ids, an unlimited number of tokens, unlimited memory and know the size of the ring.*

Proof. Suppose for the sake of contradiction that there is an algorithm $\mathcal{A}$ that solves the problem in any ring consisting of $n \geq 36$ nodes and for any initial configuration. Consider an initial configuration of nine scattered agents. Initially the agents have exactly the same input apart from their identities. Hence, any first different decisions taken by the agents have to be solely based on their different identities. Since the identities of the agents are selected by an adversary (from the set of natural numbers), even if they are distinct, the agents cannot immediately use them so that to take different decisions with respect to the direction of moving or releasing at least one of their tokens. For example algorithm $\mathcal{A}$ might instruct an agent with identity l_i to wait for $f(l_i)$ time units before it moves, to release $f'(l_i) \in \mathbb{N}$ of its tokens and to select $f''(l_i) \in \{\texttt{clockwise}, \texttt{counter} - \texttt{clockwise}\}$ direction for moving. However an adversary may always select identities l_i so that $\forall i, j$, where $i \neq j$: $f''(l_i) = f''(l_j)$ and either $(\forall i: f'(l_i) = 0)$ or $(\forall i: f'(l_i) > 0)$. In other words, algorithm $\mathcal{A}$ cannot instruct: i) one agent to release some of its tokens before its first move and another one to release no tokens, and ii) one agent to take its first move clockwise and another agent to take its first move counter-clockwise[1].

Therefore for any two agents following algorithm $\mathcal{A}$ and deciding their first move: i) either only one or both move at the same time, ii) if they move they do it towards the same direction, and iii) they either do not release any tokens or both of them release some tokens (although the numbers of tokens they release might be different).

Consider the first two agents A, B which are instructed to move by Algorithm $\mathcal{A}$ (if more than two move at the same time take any two of them). Agents A, B should either simultaneously move, or one of them first moves and then, within a finite time, a second one moves. Notice that an algorithm that does not move a second agent within a finite time, cannot be a correct algorithm, since the adversary can initially place the agents so that the first one immediately vanishes and then nobody moves. As we noted before, the agents should take their first step towards the same direction and both of them should either not release any tokens or release at least one token.

First consider the case where the agents A, B do not move simultaneously. Let A be the first agent that moves from node u to an adjacent node v towards a counter-clockwise direction without loss of generality. The adversary initially places the black virus at v and agent A vanishes along with its tokens (even if some of them were placed at u before the agent moves). Now the contaminated interval extends to the two neighbouring nodes u, w of v (since the adversary can arrange so that no other agents are initially located at nodes u or w). Agent B moves in a counter-clockwise direction within a finite time. The adversary can select its initial position to be at the adjacent node of u different than v, and therefore B vanishes after its move at time t along with its tokens (even if some of them were placed at its initial position). The contaminated interval at time t has increased to 4 nodes. A third agent should be instructed to move towards a counter-clockwise direction either simultaneously with agent B or within a finite time after agent B's move. If the remaining agents have been initially placed so that the distance between any two of them is more than 3 edges, and two of them C, D occupy nodes at a distance of at most two edges at time t from the endpoints of the current contaminated interval, then the first agent that moves a distance of at least 2 edges towards one direction (there must be such an agent, otherwise the agents will never meet other agents or tokens different than their tokens, or approach the contaminated interval) is selected to be one of the agents C or D and therefore vanishes. The contaminated interval increases to 5 nodes, while the remaining agents are 6. Hence due to Lemma 1, the 6 remaining agents cannot solve the problem.

---

[1] In fact for any function $g(l_i) : \mathbb{N} \to S$ that an algorithm $\mathcal{A}$ uses, where S is a finite set, obviously the adversary can always select all identities l_i so that $\forall i, j$, where $i \neq j$: $g(l_i) = g(l_j)$.

Now consider the remaining case where there are at least two agents A, B that move simultaneously at time t towards the same direction (say counter-clockwise without loss of generality) and might release some of their tokens. The adversary can arrange the initial positions of the agents so that agent A vanishes along with its tokens, while the tokens that agent B had left (if any), are vanished and B is now located next to a contaminated node. A possible resulting configuration is shown in Fig. 1, where Δ denotes the nodes where the agents were initially located[2]. Hence at time t some of the agents (at least two of them) moved in the counter-clockwise direction and some of them did not move. Now consider the first agent H that moves for a second time at a time $t' > t$ (or one of them if more than one move for a second time at t') either clockwise or counter-clockwise.

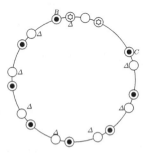

Fig. 1. Nine agents are initially scattered on a ring containing a black virus. This is a resulting configuration after one agent has met the black virus. The nodes denoted with Δ clockwise next to each agent are the initial homebases of the respective agents. The initial homebase of agent B is now contaminated and therefore any token that had been left there by B has disappeared.

If H does its second move clockwise then the adversary can select $H = B$ and the agent vanishes along with its tokens. Suppose the initial configuration was selected so that among the remaining 7 agents, the two closest to the two endpoints of the contaminated interval (which now has a length 4) are at a distance $y \geq 1$, while any two of the remaining agents are at a distance $x \geq 2y$. The algorithm should eventually move at least one of the remaining agents for at least y nodes on the same direction in order to meet with another agent, meet another agent's token or approach the contaminated area. The agent D that first moves such a distance can always be selected by the adversary to be one of the agents closest to the contaminated interval in such a way that this agent vanishes (possibly leaving many tokens), and the contaminated interval now consists of at least 5 nodes, since the closest agent to D towards the safe

---

[2] We remind the reader that we study the case where agents A and B first move simultaneously (possibly together with other agents) towards the same direction. A specific scenario where *all* agents moved simultaneously is shown in Fig. 1.

area is at a distance greater or equal to $x \geq 2y \geq 2$. In view of Lemma 1, the remaining 6 agents cannot decontaminate the interval consisting of 5 nodes and therefore the algorithm cannot decontaminate the network.

If H does its second move counter-clockwise then the adversary can select $H = C$ (that was initially located at a distance 2 clockwise from agent A) which vanishes and the contaminated interval consists of 4 nodes. Now each of the remaining agents can safely move clockwise until one node before its respective initial homebase (Δ) (without reaching it) and counter-clockwise until the closer (another agent's) initial homebase. Notice that all those intervals do not have a node in common. Therefore at least one agent should eventually move to a node outside its safe area (otherwise no agent can meet any other agent or the black virus). The first agent that moves clockwise (respectively counter-clockwise) to a node outside its safe area is selected to be agent B (respectively the next agent clockwise of C) and vanishes. Due to Lemma 1 the remaining 6 agents cannot clean a contaminated interval consisting of 5 nodes.

Since in every case all agents have been initially placed so that the distance between any two of them is no more than 4 edges, the lemma holds for any ring consisting of $n \geq 36$ nodes. □

2.2   An Algorithm with Ten Agents

We present here an algorithm that cleans any ring using ten scattered agents with common chirality. In order to make the presentation easier, we first describe the algorithm for agents with distinct identities and then we discuss how this algorithm can be modified to work with anonymous scattered agents.

Before describing the algorithm, we define the Cautious-Move procedure, which we use in all our algorithms. This procedure is a combination of an attacking and guarding action which is used by a group of at least two co-located agents where one of them has the label 'leader' and the other(s) the label 'companion'. One of the agents (leader) of the group located at a node u at time t, moves to an adjacent node v while the remaining agents of the group (companion) wait at u. Hence at time $t + 1$ the leader agent is at v while the remaining agents of the group are at u. In the next time unit all remaining agents of the group move to v while the leader (if it is still alive) waits at v.

Procedure Cautious-Move(dir)

---

1 **if** *leader* **then**
2 | Move 1 step *dir*
3 | Wait(1)
4 **else if** *companion* **then**
5 | Wait(1)
6 | Move 1 step *dir*
7 **end**

It is easy to see that if node v was safe when the leader moved there then all agents of the group gather at v within two time units. If there was a black virus at v when the leader moved there, then the leader has vanished but the virus did not spread to node u, and node v is now safe. Hence if the companion agents do not meet the leader at node v, they can conclude that the virus was at v. Furthermore, if v was an endpoint of a contaminated interval consisting of at least two nodes, now its length has been decreased by one.

The main idea of the algorithm which cleans any ring using ten scattered agents with common chirality is the following. Each agent releases its token and moves clockwise until it finds a token. Now each agent moves back until it meets a token (i.e., to its homebase), collects its token, and moves clockwise again until it finds a token. This time it waits at the node with the token until another agent comes. Eventually at least two agents meet and form a group. The group scans the ring, using Procedure Cautious-Move, until it meets an infected node. Then the remaining agent of the group waits forever at a node u (adjacent to one endpoint of the contaminated interval or to the infected node if there is only one node infected). The rest of the agents eventually meet each-other either at u or elsewhere and form groups of at least two agents. Then they repeatedly scan the ring using Procedure Cautious-Move, in order to find the other endpoint of the contaminated interval (i.e., not the one adjacent to u) or to approach the infected node (if there is only one infected node) from its other neighbour (different than u). Hence the length of the contaminated interval constantly decreases.

The above simple algorithm exploits the fact that after the described moves of the agents, exactly one of the agents which met the black virus and vanished, left its token on the ring. This token will be used by the remaining agents as a meeting point (if they do not manage to meet earlier). This idea works for many initial configurations. It manages to place an agent at a node u, adjacent to one endpoint of the contaminated interval, and form at least one group of at least two agents while at most three agents have been vanished and the contaminated interval consists of at most five nodes. The remaining seven agents are enough to clean the contaminated interval of five nodes. However, there are some initial configurations with agents located at adjacent nodes for which this algorithm might lead to the loss of four agents and to the creation of a contaminated interval consisting of six nodes. An example of such a scenario is the following. Consider the sequence of nodes $<v, x, u, y, w, z>$. Suppose there are agents initially located at nodes v, u, w, z and the black virus is located at node y. If the agents would just release their tokens and execute the main algorithm, then the agents which initially start at nodes u, w and z would soon vanish with this order together with their tokens. The agent which initially starts at node v will also eventually vanish (this one will leave its token). Since the remaining six agents cannot decontaminate the interval of six infected nodes, this simple algorithm is not enough. Hence we implement a procedure by which two surviving initially adjacent agents manage to meet before they apply the above algorithm. The algorithm takes also care of situations where an agent which belongs to a group meets another agent. The complete algorithm and its correctness analysis will appear in the full version of the paper.

Lemma 3. *The black virus decontamination problem can be solved in any synchronous ring consisting of n nodes within $O(n)$ time units, using ten or more scattered agents with common chirality, constant memory and one token each.*

The algorithm above has been described for agents with distinct identities. In the algorithm, the agents use their distinct identities in order to assign different roles to themselves only when they are co-located. It is easy to modify the algorithm so that it works for anonymous agents. By exploiting the fact that each node is initially occupied by at most one agent, we can derive a mechanism that could be used to help co-located agents to assign themselves different roles. For example, as soon as (i.e., the first time that) two agents occupy the same node u and they are at the same state, they may differentiate themselves according to the direction by which they entered node u and the actions they were doing one time unit before. Notice that if the agents entered u from the same direction and they have the same state and they were not co-located before, then exactly one of them was moving to u one time unit before, while the other one was already at u.

3   Agents Without Common Chirality

3.1   Impossibility Results

Suppose that the agents do not agree on the clockwise orientation of the ring. Naturally, all impossibility results for agents with common chirality still hold. We first show that ten initially scattered agents without common chirality cannot solve the problem even if they have distinct identities and an unlimited number of tokens. We then show that eleven agents with distinct identities and one token each are also not enough. We also show that if the agents are anonymous then they cannot solve the problem no matter how many they are and how many tokens they have.

Lemma 4. *Ten scattered agents without common chirality are not enough to decontaminate any synchronous ring consisting of $n \geq 40$ nodes, even if they have distinct identities, an unlimited number of tokens and unlimited memory.*

Proof. Consider ten scattered agents in a ring consisting of $n \geq 40$ nodes. If only one agent moves first then the adversary can select an initial configuration so that this agent vanishes (together with all its tokens) and the nine remaining agents have to clean a contaminated interval consisting of three nodes. Now the adversary can select the initial locations of the next two agents that move to be the ones adjacent to the two endpoints of the contaminated interval. The adversary can also select the directions of their movements and the initial locations of the remaining agents so that those two agents vanish along with their tokens and the contaminated interval consists of 5 nodes. Now among the remaining agents consider the first agent C that tries to move to a node located at least two edges away from its initial location (clearly there must be such an agent, otherwise the remaining agents will not approach the contaminated area). The adversary can select the initial configuration so that agent C vanishes and the contaminated interval expands to 6 nodes. In view of Lemma 1, the remaining six agents cannot solve the problem.

If at least two agents move simultaneously the adversary can select the initial configuration so that two of them vanish (together with their tokens). Now consider the next two agents that try to move to some nodes located at least two edges away from their initial locations (clearly there must be such agents, otherwise the remaining agents will not approach the contaminated area). The adversary can select the initial configuration so that those agents vanish and the contaminated interval expands to 5 nodes. In view of Lemma 1, the remaining six agents cannot solve the problem. Notice that in all those initial configurations that are selected by the adversary, the distance between any two agents is at most 4 edges. □

By selecting and analyzing a few initial configurations, we can show the following result.

Lemma 5. *For any algorithm using eleven scattered agents with distinct identities, one token each and no common chirality, there is a positive number n such that the algorithm can not solve the problem in any synchronous ring of size at least n.*

Proof. Suppose for the sake of contradiction that there is an algorithm $\mathcal{A}$ that solves the problem in every ring consisting of at least n nodes, for some n.

If at most two agents first move simultaneously the adversary selects an initial configuration where the agents that move first simultaneously, meet the black virus and vanish (together with their tokens) after their first step, while the contaminated area expands to 3 nodes. Moreover the adversary can arrange so that the remaining at most 10 agents have an initial configuration for which in view of Lemma 4 the problem is unsolvable.

Suppose now that at least three agents first start to move simultaneously. Similarly as explained in the proof of Lemma 2, the adversary can select the agents' (distinct) identities so that before their first move either all agents release their tokens or no agent releases its token.

First consider the case where no agent releases its token before its first move. The adversary initially places one of the agents (say A) that moves first so that this agent vanishes after its first move. Now among the remaining agents consider the first one (say B) that makes two traversals, either traversing two edges towards the same direction or traversing the same edge back and forth (clearly there must be such an agent, otherwise no other agent will approach the contaminated area). In any case and no matter whether agent B released its token before its second move, the adversary can initially place B so that it vanishes along with its token. Now there are 9 remaining agents and the contaminated area consists of 4 nodes. Among the remaining agents consider the first agent (say C) that tries to traverse at least 3 edges towards the same direction (it is again clear that if there is no such an agent, then no other agent will approach the contaminated area for a sufficiently big size of ring n). Then the adversary can initially place agent C so that it vanishes during those traversals (possibly leaving a token somewhere close to the contaminated area). Similarly, there must be another agent D which also tries to traverse at least 3 edges towards the same direction. The adversary initially places agent D so that it approaches the other endpoint of the contaminated area (i.e., not the one where agent C had vanished). Hence, the contaminated area has been expanded to 6 nodes, while the remaining agents are only 7 and in view of Lemma 1, the problem cannot be solved.

Finally let us analyze the remaining case where at least 3 agents start moving simultaneously and all agents release their tokens before their first move. Algorithm $\mathcal{A}$ should always instruct at least one of the remaining agents to move and eventually cover a distance of at least $\lceil \frac{x}{2} \rceil$ edges away from its initial location, where x is the minimum distance between any two initial locations of the remaining agents, otherwise no agent will meet another agent, or another agent's token or the contaminated area for a sufficiently big size of ring. Suppose that when an agent meets another agent's token at a node v by traversing an edge (u, v), it never traverses the edge (v, w) (where $u \neq w$). Suppose also that an agent never comes back to its initial location. If both above conditions hold, then the adversary can arrange a configuration so that all the remaining agents start moving towards the same direction, they never meet other agents and each agent is forever trapped to traverse the path between its initial location (where never comes back after its first move) and the first node with a token (which the agent could move it inside this path) that it meets in the direction it moves. Hence the agents can never approach the contaminated area and solve the problem. Therefore either an agent has to come back to its initial location, or it has to move further when it meets a token (or both).

First suppose that algorithm $\mathcal{A}$ instructs an agent to move further than a token it meets and consider the following initial configuration. One of the agents (say A) that first moves has been initially placed by the adversary at a node u adjacent to the black virus at node v so that it vanishes after its first move along with its token and the interval consisting of the nodes u, v, w is contaminated. Consider the first agent (say C) that tries to traverse at least two distinct edges (clearly there must be such an agent, otherwise no other agent will approach the contaminated area). Agent C has been initially placed by the adversary so that it starts moving at the same time as A and vanishes during those traversals as follows. If agent C visits the two nodes at distance one from its initial location the adversary initially places agent C at a node $y \neq v$ adjacent to node w, arranging that it first moves to a node $z \neq w$. If agent C visits a node at distance two from its initial location the adversary initially places agent C at node z, arranging that it first moves to node y. Hence in any case agent C vanishes while its token has been either destroyed or left at node z which is now adjacent to one of the endpoints of the contaminated area which consists of nodes u, v, w, y. The third agent that moves simultaneously with agent A is selected by the adversary similarly as agent C and initially placed at a node x adjacent to u (instead of w)

or at a node p at distance two from u (instead of w). This third agent moves and vanishes in a similar way as agent C, possibly leaving its token at node p while the contaminated interval extends now to nodes x, u, v, w, y. Since at least one of the remaining agents should move beyond another agent's token that it meets, one more agent can be initially placed and selected by the adversary to vanish at nodes x or y. Due to Lemma 1, the remaining 7 agents cannot clean the contaminated interval consisting of 6 nodes.

For the remaining case suppose that algorithm $\mathcal{A}$ never instructs an agent to move further than a token it meets and therefore, as we argued before, it should instruct an agent to come back to its initial location.

In the initial configuration which is selected by the adversary, agent A is placed exactly as in the previous case. The adversary initially places an agent B at node w. Agent B moves simultaneously with agent A and towards the same direction. After the move the token released by agent B has been destroyed. An agent will try to go back to its initial location either after it meets a token or before it meets a token (i.e., after moving a number of times). If an agent returns to its initial location only after it meets a token the adversary can initially place the third agent D that moves simultaneously with agent A at node y and selecting its first move towards the same direction as agents A, B. Agent D has released its token at node y and it will not return to y before it meets another token. Hence agent B vanishes when it returns at node w and the token at node y is destroyed since the contaminated interval now extends to nodes u, v, w, y. When agent D tries to return to node y is also vanished and the contaminated interval consists of 5 nodes. In the second subcase where an agent can try to go back to its initial location before it meets a token (i.e., returns back after some time d), the adversary initially places agent D at a location q which is at a distance at least $d + 2$ from node w and selects its movement towards the opposite direction of the first move of agent A. Agent B will vanish (as before) at w extending the contaminated interval to nodes u, v, w, y and eventually agent D will vanish at node y and the contaminated interval will extend to 5 nodes as in the previous subcase. One more agent can be initially placed and selected by the adversary to vanish at node u. Due to Lemma 1, the remaining 7 agents cannot clean the contaminated interval consisting of 6 nodes.   □

If the agents are anonymous then we can find initial configurations where the agents either initially or after the first move are located at symmetric positions and they cannot break their symmetry while they move, and they eventually vanish.

Lemma 6. *For any number of anonymous scattered agents without common chirality there is a synchronous ring which cannot be cleaned even if the agents have an unlimited number of tokens and unlimited memory.*

Proof. Assume a ring with k anonymous agents without common chirality and consider the starting configuration of Fig. 2 (if k is even) or Fig. 3 (if k is odd). If the initial number of agents is even, the number of nodes of the ring is fixed (by an adversary) as $n = k + 1$ nodes and each node is initially occupied by one agent, except the node that contains the black virus (see Fig. 2). If the initial number of agents is odd, the number of nodes of the ring is fixed (by an adversary) as $n = k + 2$ nodes and each node is initially occupied by one agent, except the node that contains the black virus and one of the nodes incident to the black virus (see Fig. 3). In either case, each agent has the same number of tokens and neighbouring agents are forced by the adversary to move in different directions: agent A_1 moves clockwise and agent A_{i+1} moves opposite than A_i, $\forall i \geq 1$.

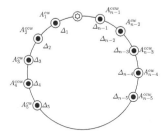

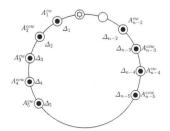

Fig. 2. An initial configuration consisting of $n - 1$ anonymous agents, where $n - 1$ is an even number.

Fig. 3. An initial configuration consisting of $n - 2$ anonymous agents, where $n - 2$ is an odd number.

If the initial number of agents is even, two agents, A_1 and A_{n-1}, along with their tokens are destroyed after the first step. If the initial number of agents is odd, one agent, A_1, along with its tokens is destroyed after the first step. In both cases the resulting configuration is symmetric where each of the remaining agents reaches a node containing the same number of tokens. The agents cannot stop moving after finding a token, otherwise, the ring will not be cleaned. If the agents continue moving in the same direction, two additional agents and two piles of tokens are lost immediately and once again all the remaining agents reach the same number of tokens. If the agents change direction and move until they find a token, all agents return to their respective homebases and no agent is destroyed. However, the agents cannot keep moving between any two nodes with tokens indefinitely, otherwise the ring will not be decontaminated. Therefore, the agents eventually have to move to a node they have not visited before. Consequently, in any case two additional agents and two piles of tokens are destroyed, leaving $k - 4$ agents, $k - 4$ nodes with (the same number of) tokens and five nodes that need to be cleaned.

The resulting configuration is similar to the previous one with the difference that there are two more nodes that need to be cleaned and two agents less. The agents in the new configuration do not have any further knowledge of the network and cannot move in a more effective/different way than before. Each time the agents move to a new node, the two agents next to the two endpoints of the contaminated interval are destroyed along with their tokens. Each time an agent is lost, a pile of tokens is lost as well. Thus, any two agents have always the same input and therefore they cannot break this symmetry and they all eventually vanish.   □

3.2   An Algorithm with Twelve Agents

We present here an algorithm for twelve scattered agents with distinct identities and one token each for the decontamination of any synchronous ring even when the agents do not have a common chirality.

The main idea of the algorithm is the following. Each agent releases its token h, selects a direction and moves until it finds a token h'. Then repeatedly bounces back and forth between the two tokens h and h', each time moving its own token h towards the other token h' until the distance between the two tokens is at most one edge. Eventually, at least two agents meet and start to move trying to find an endpoint of the contaminated interval using Procedure `Cautious-Move`. Eventually all surviving agents meet using the above technique and using Procedure `Cautious-Move` manage to decontaminate the ring. The algorithm needs to take care a few details:

- Before the agents execute the above algorithm they move their tokens one step to guarantee that, apart from at most two agents that might vanish together with their tokens, for any other agent of the remaining at least ten agents, even if it vanishes while trying to bounce, its token is preserved.
- When two agents move their tokens closer and closer to each other and do not meet until their tokens occupy adjacent nodes, they use their distinct labels (natural numbers) to break ties in their movements and meet.
- When there are more than one agents at the same node they use their distinct identities to assign different roles to themselves.
- An agent or group of agents that meets a guard (i.e., an agent adjacent to one of the endpoints of the contaminated interval) for the first time, changes direction.

The complete algorithm along with its correctness analysis will appear in the full version of the paper.

Lemma 7. *Twelve scattered agents with distinct identities (taken from the set of natural numbers), constant memory and one token each can decontaminate any $n-$node, synchronous, ring within $O(n^2 + L)$ time units, where L is the maximum identity label, even when the agents do not have a common chirality.*

4   Open Problems

An interesting question is whether the presented algorithms can be extended to handle asynchronous networks. It is also interesting to investigate what is the minimum number of scattered agents and the weakest model under which the problem can be solved in other graph topologies.

References

1. Alotaibi, M.: Black virus disinfection in chordal rings. Master's thesis, Université d'Ottawa/University of Ottawa (2014)
2. Barrière, L., Flocchini, P., Fraigniaud, P., Santoro, N.: Capture of an intruder by mobile agents. In: Proceedings of the Fourteenth Annual ACM Symposium on Parallel Algorithms and Architectures, pp. 200–209. ACM (2002)
3. Blin, L., Fraigniaud, P., Nisse, N., Vial, S.: Distributed chasing of network intruders. Theor. Comput. Sci. **399**(1–2), 12–37 (2008)
4. Cai, J., Flocchini, P., Santoro, N.: Decontamination of an arbitrary network from multiple black viruses. In: 32nd International Conference on Computers and Their Applications, (CATA), pp. 231–237 (2017)
5. Cai, J., Flocchini, P., Santoro, N.: Decontaminating a network from a black virus. Int. J. Netw. Comput. **4**(1), 151–173 (2014)
6. Cai, J., Flocchini, P., Santoro, N.: Black virus decontamination in arbitrary networks. In: Rocha, A., Correia, A.M., Costanzo, S., Reis, L.P. (eds.) New Contributions in Information Systems and Technologies. AISC, vol. 353, pp. 991–1000. Springer, Cham (2015). https://doi.org/10.1007/978-3-319-16486-1_98
7. Cai, J., Flocchini, P., Santoro, N.: Distributed black virus decontamination and rooted acyclic orientations. In: 2015 IEEE International Conference on Computer and Information Technology; Ubiquitous Computing and Communications; Dependable, Autonomic and Secure Computing; Pervasive Intelligence and Computing (CIT/IUCC/DASC/PICOM), pp. 1681–1688. IEEE (2015)
8. Cooper, C., Klasing, R., Radzik, T.: Locating and repairing faults in a network with mobile agents. Theor. Comput. Sci. **411**(14–15), 1638–1647 (2010)
9. Dobrev, S., Flocchini, P., Prencipe, G., Santoro, N.: Searching for a black hole in arbitrary networks: optimal mobile agents protocols. Distrib. Comput. **19**(1), 1–99999 (2006)
10. Kokkou, M.: Distributed computing - fault tolerant distributed algorithms. Bachelor's thesis, University of Thessaly (2019)
11. Lin, Y.: Decontamination from black viruses using parallel strategies. Master's thesis, Université d'Ottawa/University of Ottawa (2018)
12. Markou, E., Shi, W.: Dangerous graphs. In: Flocchini, P., Prencipe, G., Santoro, N. (eds.) Distributed Computing by Mobile Entities. LNCS, vol. 11340, pp. 455–515. Springer, Cham (2019). https://doi.org/10.1007/978-3-030-11072-7_18
13. Nisse, N.: Connected graph searching in chordal graphs. Discrete Appl. Math. **157**(12), 2603–2610 (2009)
14. Nisse, N.: Network decontamination. In: Flocchini, P., Prencipe, G., Santoro, N. (eds.) Distributed Computing by Mobile Entities. LNCS, vol. 11340, pp. 516–548. Springer, Cham (2019). https://doi.org/10.1007/978-3-030-11072-7_19

The Power of Global Knowledge
on Self-stabilizing Population Protocols

Yuichi Sudo[1]([✉]), Masahiro Shibata[2], Junya Nakamura[3], Yonghwan Kim[4],
and Toshimitsu Masuzawa[1]

[1] Osaka University, Osaka, Japan
y-sudou@ist.osaka-u.ac.jp
[2] Kyushu Institute of Technology, Fukuoka, Japan
[3] Toyohashi University of Technology, Aichi, Japan
[4] Nagoya Institute of Technology, Aichi, Japan

Abstract. In the population protocol model, many problems cannot
be solved in a self-stabilizing way. However, global knowledge, such as
the number of nodes in a network, sometimes allow us to design a self-
stabilizing protocol for such problems. In this paper, we investigate the
effect of global knowledge on the possibility of self-stabilizing population
protocols in arbitrary graphs. Specifically, we clarify the solvability of
the leader election problem, the ranking problem, the degree recognition problem, and the neighbor recognition problem by self-stabilizing
population protocols with knowledge of the number of nodes and/or the
number of edges in a network.

1 Introduction

We consider the *population protocol* (PP) model [2] in this paper. A network
called *population* consists of a large number of finite-state automata, called
agents. Agents make *interactions* (*i.e.,* pairwise communication) with each other
by which they update their states. The interactions are opportunistic, that is,
they are unpredictable for the agents. Agents are strongly anonymous: they do
not have identifiers and they cannot distinguish their neighbors with the same
states. One example represented by this model is a flock of birds where each bird
is equipped with a sensing device with a small transmission range. Two devices
can communicate (*i.e.,* interact) with each other only when the corresponding
birds come sufficiently close to each other. Therefore, an agent cannot predict
when it has its next interaction.

In the field of population protocols, many efforts have been devoted to
devising protocols for a complete graph, that is, a population where every
pair of agents interacts infinitely often. On the other hand, several works
[2,4,5,8–10,15,16,19,20] study the population represented by a general graph

This work was supported by JSPS KAKENHI Grant Numbers 17K19977, 18K18000,
18K18029, 18K18031, 19H04085, and 20H04140 and JST SICORP Grant Number
JPMJSC1606.

A. W. Richa and C. Scheideler (Eds.): SIROCCO 2020, LNCS 12156, pp. 237–254, 2020.
https://doi.org/10.1007/978-3-030-54921-3_14

$G = (V, E)$ where V is the set of agents and E specifies the set of interactable pairs. Each pair of agents $(u, v) \in E$ has interactions infinitely often, while each pair of agents $(u', v') \notin E$ never has an interaction.

Self-stabilization [11] is a fault-tolerant property that, even when any transient fault (*e.g.*, memory crash) hits a network, it can autonomously recover from the fault. Formally, self-stabilization is defined as follows: (i) starting from an arbitrary configuration, a network eventually reaches a *safe configuration* (*convergence*), and (ii) once a network reaches a safe configuration, it keeps its specification forever (*closure*). Self-stabilization is of great importance in the PP model because self-stabilization tolerates any finite number of transient faults, and this is a necessary property in a network consisting of a huge number of cheap and unreliable nodes.

Consequently, many studies have been devoted to self-stabilizing population protocols [4,5,7–9,12,14,16,17,19–21]. Angluin *et al.* [4] gave self-stabilizing protocols for a variety of problems: the leader election in the rings whose size are not multiples of a given integer k (in particular, the rings of odd size), the token circulation in rings with a pre-selected leader, the 2-hop coloring in degree-bounded graphs, the consistent global orientation in undirected rings, and the spanning-tree construction in regular graphs. The protocols for the first four problems use only a constant space of agent memory, while the protocol for the last problem requires $O(\log D)$ bits of agent memory, where D is (a known upper bound[1] on) the diameter of the graph. Chen and Chen [9] gave a constant-space and self-stabilizing protocol for the leader election in rings with arbitrary size.

On the negative side, Angluin *et al.* [4] proved that the self-stabilizing leader election (SS-LE) is impossible for arbitrary graphs. In particular, it immediately follows from their theorem that no protocol solves SS-LE in complete graphs with three different sizes, *i.e.*, in all of K_i, K_j, and K_k for any distinct integers $i, j, k \geq 2$, where K_l is a complete graph with size l. Cai *et al.* [7] proved that no protocol solves SS-LE both in K_i and in K_{i+1} for any integer $i \geq 2$. In almost the same way, we can easily observe that no protocol solves SS-LE both in K_i and K_j for any distinct integers $i, j \geq 2$. (See a more detailed explanation in the second page of [21].) In other words, SS-LE is impossible unless the exact number of agents in the population is known to the agents. Because Cai *et al.* [7] also gave a protocol that solves SS-LE in K_l for a given integer l, the knowledge of the exact number of agents is necessary and sufficient to solve SS-LE in a complete graph.

In addition to [4,7,9], many works have been devoted to SS-LE. This is because the leader election is one of the most fundamental and important problems in the PP model: several important protocols [2–4] require a pre-selected

---

[1] In [4], D is defined as the diameter of the graph, not a known upper bound on it. However, since we must take into account an arbitrary initial configuration, we require an upper bound on the diameter; Otherwise, the agents need the memory of unbounded size. Fortunately, the knowledge of the upper bound is not a strong assumption in this case: any upper bound which is polynomial in the true diameter is acceptable since the space complexity is $O(\log D)$ bits.

unique leader, especially, it is shown by Angluin *et al.* [3] that if we have a unique leader, all semi-linear predicates can be solved very quickly. However, we have strong impossibility as mentioned above: SS-LE can not be solved unless the knowledge of the exact number of agents is given to the agents. In the literature, there are three approaches to overcome this impossibility. One approach [6,7] is to assume that every agent knows the exact number of agents. Cai *et al.* [7] took this approach for the first time. Their protocol uses $O(\log n)$ bits (n states) of memory space per agent and converges within $O(n^3)$ steps in expectation in the complete graph of n agents under *the uniformly random scheduler*, which selects a pair of agents to interact uniformly at random from all pairs at each step. Burman *et al.* [6] gave three *faster* SS-LE protocols than the protocol of Cai *et al.* [7], also for the complete graph of n agents. These self-stabilizing protocols in [6,7] solve not only the leader election problem but also *the ranking problem*, which requires ranking the n agents by assigning them the different integers from $0, 1, \ldots, n - 1$. There are other two approaches to over come the impossibility of SS-LE in the PP model. One is *loose-stabilization* [14,16,17,19–21], which may deviate from the legitimate behavior even after a safe configuration is reached, but guarantee that the expected time before the deviation occurs is sufficiently long. The other one is to design SS-LE protocols with *oracles* [5,8,12].

1.1 Our Contribution

As mentioned above, if we have knowledge of the exact number of agents, we can solve the self-stabilizing leader election in complete graphs, which we can never solve otherwise. In this paper, we investigate in detail how powerful global knowledge, such as the exact number of agents in the population, is to design self-stabilizing population protocols for *arbitrary graphs*. Specifically, we consider two kinds of global knowledge, the number of agents and the number of edges (*i.e.*, interactable pairs) in the population, and clarify the relationships between the knowledge and the solvability of the following four problems:

- leader election (**LE**): Elect exactly one leader,
- ranking (**RK**): Assign the agents in the population $G = (V_G, E_G)$ distinct integers (or *ranks*) from 0 to $|V_G| - 1$,
- degree recognition (**DR**): Let each agent recognize its degree in the graph,
- neighbor recognition (**NR**): Let each agent recognize the set of its neighbors in the graph. Since the population is anonymous, this problem also requires having 2-hop coloring, that is, all agents must be assigned integers (or *colors*) such that all neighbors of any agent have different colors.

In addition to the above specifications, we require that no agent change its outputs (e.g., its *rank* in **RK**) after the population converges, that is, it reaches a safe configuration.

We denote $A_1 \preceq A_2$ if problem A_1 is reducible to A_2. We have **LE** $\preceq$ **RK** and **DR** $\preceq$ **NR**. The first relationship holds because if the agents are labeled $0, 1, \ldots, |V_G| - 1$, **LE** is immediately solved by selecting the agent with label 0 as the unique leader. The second relationship is trivial.

To describe our contributions, we formally define the global knowledge that we consider. Define $\mathcal{G}_{n,m}$ as the set of all the simple, undirected, and connected graphs with n nodes and m edges. Let ν and μ be any sets of positive integers such that $\nu \subseteq \mathbb{N}_{\geq 2} = \{n \in \mathbb{N} \mid n \geq 2\}$ and $\mu \subseteq \mathbb{N}_{\geq 1} = \{m \in \mathbb{N} \mid m \geq 1\}$. Then, we define $\mathcal{G}_{\nu,\mu} = \bigcup_{n \in \nu, m \in \mu} \mathcal{G}_{n,m}$. For simplicity, we define $\mathcal{G}_{\nu,*} = \mathcal{G}_{\nu,\mathbb{N}_{\geq 1}}$ and $\mathcal{G}_{*,\mu} = \mathcal{G}_{\mathbb{N}_{\geq 2},\mu}$ for any $\nu \subseteq \mathbb{N}_{\geq 2}$ and $\mu \subseteq \mathbb{N}_{\geq 1}$. We consider that ν and μ are global knowledge on the population: ν is the set of the possible numbers of agents and μ is the set of the possible numbers of interactable pairs. In other words, when we are given ν and μ, our protocol has to solve a problem only in the populations represented by the graphs in $\mathcal{G}_{\nu,\mu}$. We say that protocol P solves problem A *in arbitrary graphs* given knowledge ν and μ if P solves A in all graphs in $\mathcal{G}_{\nu,\mu}$.

In this paper, we investigate the solvability of **LE**, **RK**, **DR**, and **NR** for arbitrary graphs with the knowledge ν and μ. Specifically, we prove the following propositions assuming that the agents are given knowledge ν and μ:

1. When the agents know nothing about the number of interactable pairs, *i.e.*, $\mu = \mathbb{N}_{\geq 1}$, there exists a self-stabilizing protocol that solves **LE** and **RK** in arbitrary graphs if and only if the agents know the exact number of agents *i.e.*, $\mathcal{G}_{\nu,\mu} = \mathcal{G}_{n,*}$ for some $n \in \mathbb{N}_{\geq 2}$.
2. There exists a self-stabilizing protocol that solves **NR** ($\succeq$**DR**) in arbitrary graphs if the agents know the exact number of agents and the exact number of interactable pairs *i.e.*, $\mathcal{G}_{\nu,\mu} = \mathcal{G}_{n,m}$ holds for some $n \in \mathbb{N}_{\geq 2}$ and $m \in \mathbb{N}_{\geq 1}$.
3. The knowledge of the exact number of agents is not enough to design a self-stabilizing protocol that solves **DR** ($\preceq$**NR**) in arbitrary graphs if the agents do not know the number of interactable pairs *exactly*. Specifically, no self-stabilizing protocol solves **DR** in all graphs in $\mathcal{G}_{\nu,\mu}$ if $\mathcal{G}_{n,m_1} \cup \mathcal{G}_{n,m_2} \subseteq \mathcal{G}_{\nu,\mu}$ holds for some $n \in \mathbb{N}_{\geq 2}$ and some distinct $m_1, m_2 \in \mathbb{N}_{\geq 1}$ such that $\mathcal{G}_{n,m_1} \neq \emptyset$ and $\mathcal{G}_{n,m_2} \neq \emptyset$.

In standard distributed computing models, generally, each node always has its local knowledge, *e.g.*, its degree and the set of its neighbors. In the PP model, the agents does not have the local knowledge *a priori*, and many impossibility results (*e.g.*, the impossibility of SS-LE in complete graphs [4,7]) come from the lack of the local knowledge. Interestingly, the third proposition yields that, for self-stabilizing population protocols, obtaining some local knowledge (degree recognition of each agent) is at least as difficult as obtaining the corresponding global knowledge (the number of interactable pairs). It is also worthwhile to mention that the PP model is empowered greatly if **LE** and **NR** are solved. After the agents recognize their neighbors correctly, the population can simulate one of the most standard distributed computing models, the message passing model, if each agent maintains a variable corresponding to a *message buffer* for each neighbor. Moreover, we have the unique leader in the population, by which we can easily break the symmetry of a graph and solve many important problems even in a self-stabilizing way. For example, we can construct a spanning tree rooted by the leader. This fact and the above propositions show how powerful this kind of global knowledge is when we design self-stabilizing population protocols.

2   Preliminaries

A *population* is represented by a simple and connected graph $G = (V_G, E_G)$, where V_G is the set of the *agents* and $E_G \subseteq V_G \times V_G$ is the set of the *interactable pairs* of agents. If $(u, v) \in E_G$, two agents u and v can interact in the population G, where u serves as the *initiator* and v serves as the *responder* of the interaction. In this paper, we consider only *undirected* populations, that is, we assume that, for any population G, $(u, v) \in E_G$ yields $(v, u) \in E_G$ for any $u, v \in V_G$. We define the set of the neighbors of agent v as $N_G(v) = \{u \in V_G \mid (v, u) \in E_G\}$.

A *protocol* $P(Q, Y, T, \pi_{out})$ consists of a finite set Q of states, a finite set Y of output symbols, a transition function $T : Q \times Q \to Q \times Q$, and an output function $\pi_{out} : Q \to Y$. When two agents interact, T determines their next states according to their current states. The *output* of an agent is determined by π_{out}: the output of an agent in state q is $\pi_{out}(q)$. As mentioned in Sect. 1, we assume that the agents can use knowledge ν and μ. Therefore, the four parameters of protocol P, i.e., Q, Y, T, and π_{out}, may depend on ν and μ. We sometimes write $P(\nu, \mu)$ explicitly to denote protocol P with knowledge ν and μ.

A *configuration* on population G is a mapping $C : V_G \to Q$ that specifies the states of all the agents in G. We denote the set of all configurations of protocol P on population G by $\mathcal{C}_{all}(P, G)$. We say that a configuration C changes to C' by an interaction $e = (u, v)$, denoted by $C \overset{P,e}{\to} C'$, if $(C'(u), C'(v)) = T(C(u), C(v))$ and $C'(w) = C(w)$ for all $w \in V \backslash \{u, v\}$. We also denote $C \overset{P,G}{\to} C'$ if $C \overset{P,e}{\to} C'$ holds for some $e \in E_G$. We also say that a configuration C' is reachable from C by P on population G if there is a sequence of configurations $C_0, C_1, \ldots, C_k$ such that $C_i \overset{P,G}{\to} C_{i+1}$ for $i = 0, 1, \ldots, k-1$. We say that a set $\mathcal{S}$ of configurations is *closed* if no configuration out of $\mathcal{S}$ is reachable from a configuration in $\mathcal{S}$.

An *execution* of protocol P on population G is an infinite sequence of configurations $\Xi = C_0, C_1, \ldots$ such that $C_i \overset{P,G}{\to} C_{i+1}$ for $i = 0, 1, \ldots$. We call C_0 the initial configuration of the execution Ξ. We have to assume some kind of fairness of an execution. Otherwise, for example, we cannot exclude an execution such that only one pair of agents have interactions in a row and no other pair has an interaction forever. Unlike most distributed computing models in the literature, *the global fairness* is usually assumed in the PP model. We say that an execution $\Xi = C_0, C_1, \ldots$ of P on population G satisfies the global fairness (or Ξ is globally fair) if for any configuration C that appears infinitely often in Ξ, every configuration C' such that $C \overset{P,G}{\to} C'$ also appears infinitely often in Ξ.

A problem is specified by a predicate on the outputs of the agents. We call this predicate the *specification* of the problem. We say that a configuration C satisfies the specification of a problem if the outputs of the agents satisfy it in C. We consider the following four problems in this paper.

Definition 1 (LE). *The specification of the leader election problem (LE) requires that exactly one agent outputs L and all the other agents output F.*

Definition 2 (RK). *The specification of the ranking problem (**RK**) requires that in the population $G = (V_G, E_G)$, the set of the outputs of the agents in the population equals to $\{0, 1, \ldots, |V_G| - 1\}$.*

Definition 3 (DR). *The specification of the degree recognition problem (**DR**) requires that in the population $G = (V_G, E_G)$, every agent $v \in V_G$ outputs $|N_G(v)|$.*

Definition 4 (NR). *The specification of the neighbor recognition problem (**NR**) requires that in the population $G = (V_G, E_G)$, every agent $v \in V_G$ outputs a two-tuple $(c_v, S_v) \in \mathbb{Z} \times 2^{\mathbb{Z}}$ such that, for all $v \in V_G$, we have $S_v = \{c_u \mid u \in N_G(v)\}$ and $|S_v| = |N_G(v)|$.*

Note that the second condition in the definition of **NR**, *i.e.*, $|S_v| = |N_G(v)|$, requires that the population is 2-hop colored, that is, every two distinct neighbors u and w of agent v must have different integers c_u and c_w.

Now, we define self-stabilizing protocols in Definitions 5 and 6, where we use the definitions given in Sect. 1.1 for knowledge ν and μ and the set $\mathcal{G}_{\nu,\mu}$ of graphs. Note that Definition 5 is not enough if we consider *dynamic problems* such as the token circulation, where the specifications must be defined as predicates not on configurations but on executions. However, we consider only static problems in this paper, thus this definition is enough for our purpose.

Definition 5 (Safe configuration). *Given a protocol P and a population G, we say that a configuration $C \in \mathcal{C}_{\mathrm{all}}(P(\nu, \mu), G)$ is safe for problem A if (i) C satisfies the specification of problem A, and (ii) no agent changes its output in any execution of P on G starting from C.*

Definition 6 (Self-stabilizing protocol). *For any ν and μ, we say that a protocol P is a self-stabilizing protocol that solves problem A in arbitrary graphs given knowledge ν and μ if every globally-fair execution of $P(\nu, \mu)$ on any population G, which starts from any configuration $C_0 \in \mathcal{C}_{\mathrm{all}}(P(\nu, \mu), G)$, reaches a safe configuration for A.*

Finally, we define *the uniformly random scheduler*, which has been considered in most of the works [1–3,13,16–21] in the PP model. Under this scheduler, exactly one ordered pair $(u, v) \in E_G$ is chosen to interact uniformly at random from all interactable pairs. We need this scheduler to evaluate *time complexities* of protocols because global fairness only guarantees that an execution makes progress *eventually*. Formally, the uniformly random scheduler is defined as a sequence of interactions $\mathbf{\Gamma} = \Gamma_0, \Gamma_1, \ldots$, where each Γ_t is a random variable such that $\Pr(\Gamma_t = (u, v)) = 1/|E_G|$ for any $t \geq 0$ and any $(u, v) \in E_G$. Given a population G, a protocol $P(\nu, \mu)$, and an initial configuration $C_0 \in \mathcal{C}_{\mathrm{all}}(P(\nu, \mu), G)$, the execution under the uniformly random scheduler is defined as $\Xi_{P(\nu,\mu)}(G, C_0, \mathbf{\Gamma}) = C_0, C_1, \ldots$ such that $C_t \stackrel{P(\nu,\mu), \Gamma_t}{\rightarrow} C_{t+1}$ for all $t \geq 0$. When we assume this scheduler, we can evaluate time complexities of a population protocol, for example, the expected number of steps required to reach a safe configuration. We have the following observation.

Observation 1. *A protocol $P(\nu, \mu)$ is self-stabilizing for a problem A if and only if $\Xi_{P(\nu,\mu)}(G, C_0, \Gamma)$ reaches a safe configuration for A with probability 1 for any configuration $C_0 \in \mathcal{C}_{\text{all}}(P(\nu, \mu), G)$.*

Proof. Remember that we do not allow a protocol to have an infinite number of states. According to [2], we say that a set $\mathcal{C}$ of configurations is *final* if $\mathcal{C}$ is closed, and all configurations in $\mathcal{C}$ are reachable from each other. We also say that a configuration C is *final* if it belongs to a final set. It is trivial that protocol P is self-stabilizing if and only if all final configurations are safe. Thus, it suffices to show that all final configurations of $P(\nu, \mu)$ are safe for A if and only if execution $\Xi = \Xi_{P(\nu,\mu)}(G, C_0, \Gamma)$ reaches a safe configuration for A with probability 1 for any $C_0 \in \mathcal{C}_{\text{all}}(P(\nu, \mu), G)$. The sufficient condition is trivial because Ξ reaches a final configuration with probability 1 regardless of C_0. We prove the necessary condition below. Suppose that there is a final configuration C that is not safe. By definition, C belongs to a final set $\mathcal{C}$. Since C is reachable from all configurations in $\mathcal{C}$, no configuration in $\mathcal{C}$ is safe. Since $\mathcal{C}$ is closed, Ξ will never reaches a safe configuration if $C_0 = C$.

Due to the lack of space, we omit the proofs for a few propositions regarding the random walk in the PP model, which are used in the proofs of the lemmas in Sect. 3 and 4. See the preprint [22] for the complete proofs.

3   Leader Election and Ranking

The goal of this section is to give a necessary and sufficient condition to solve **RK** and **LE** on knowledge ν, provided that μ gives no information, *i.e.*, $\mu = \mathbb{N}_{\geq 1}$. For a necessary condition, we have the following lemma.

Lemma 1 ([4,7,21]). *Given knowledge ν and μ, there exists no self-stabilizing protocol that solves **LE** in arbitrary graphs if $\mathcal{G}_{n_1,*} \cup \mathcal{G}_{n_2,*} \subseteq \mathcal{G}_{\nu,\mu}$ for some two distinct $n_1, n_2 \in \mathbb{N}_{\geq 2}$.*

Proof. The lemma immediately follows from the fact that there exists no self-stabilizing protocol that solves **LE** in complete graphs of two different sizes, *i.e.*, both in K_{n_1} and K_{n_2} for any two integers $n_1 > n_2 \geq 2$. As mentioned in Sect. 1, Sudo *et al.* [21] gave how to prove this fact based on the proofs of [4,7].   □

To give a sufficient condition, we give a self-stabilizing protocol P_{rank}, which solves the ranking problem (**RK**) in arbitrary graphs given the knowledge of the exact number of agents in a population. Specifically, this protocol assumes that the given knowledge ν satisfies $|\nu| = 1$ while it does not care about the number of interactable pairs, that is, $P_{\text{rank}}(\nu, \mu)$ works even if μ does not give any knowledge (*i.e.*, $\mu = \mathbb{N}_{\geq 1}$). Let n be the integer such that $\nu = \{n\}$.

If we focus only on complete graphs, the following simple algorithm [7] is enough to solve self-stabilizing ranking with the exact knowledge n of agents:

Algorithm 1. $P_{\mathrm{rank}}(\nu, \mu)$

Assumption: $|\nu| = 1$. (Let $\nu = \{n\}$.)

Variables:

  $\mathrm{id}_A, \mathrm{id}_T \in \{0, 1, \ldots, n-1\}$
  $\mathrm{color}_A \in \{W, R, B\}$, $\mathrm{color}_T \in \{R, B\}$, $\mathrm{timer}_T \in \{0, 1, \ldots, U_T\}$

Output function π_{out}: id_A

Interaction between initiator a_0 and responder a_1:

1:   $(a_0.\mathrm{id}_T, a_0.\mathrm{color}_T, a_0.\mathrm{timer}_T) \leftrightarrow (a_1.\mathrm{id}_T, a_1.\mathrm{color}_T, a_1.\mathrm{timer}_T)$
            // Execute the random walk of two tokens
2:   **if** $a_0.\mathrm{id}_T = a_1.\mathrm{id}_T$ **then** $a_1.\mathrm{id}_T \leftarrow a_1.\mathrm{id}_T + 1 \pmod{n}$ **endif**
3:   **for all** $i \in \{0, 1\}$ **do** $a_i.\mathrm{timer}_T \leftarrow \max(0, a_i.\mathrm{timer}_T - 1)$ **endfor**

4:   **for all** $i \in \{0, 1\}$ such that $a_i.\mathrm{id}_A = a_i.\mathrm{id}_T$ **do**
5:    **if** $a_i.\mathrm{color}_A = W$ **then** $a_i.\mathrm{color}_A \leftarrow a_i.\mathrm{color}_T$ **endif**
6:    **if** $a_i.\mathrm{color}_A \neq a_i.\mathrm{color}_T$ **then**
7:     $a_i.\mathrm{id}_A \leftarrow a_i.\mathrm{id}_A + 1 \pmod{n}$
8:     $a_i.\mathrm{color}_A \leftarrow W$
9:    **else if** $a_i.\mathrm{timer}_T = 0$ **then**
10:     $a_i.\mathrm{timer}_T \leftarrow U_T$
11:     **if** $a_i.\mathrm{color}_A = R$ **then** $a_i.\mathrm{color}_A \leftarrow a_i.\mathrm{color}_T \leftarrow B$ **endif**
12:     **if** $a_i.\mathrm{color}_A = B$ **then** $a_i.\mathrm{color}_A \leftarrow a_i.\mathrm{color}_T \leftarrow R$ **endif**
13:    **end if**
14:   **end for**

- Each agent v has only one variable $v.\mathrm{id} \in \{0, 1, \ldots, n-1\}$, and
- Every time two agents with the same id meet, one of them (the initiator) increases its id by one modulo n.

Since this algorithm assumes complete graphs, every pair of agents in the population eventually has interactions. Therefore, as long as two agents have the same identifiers, they eventually meet and the collision of their identifiers is resolved. However, this algorithm does not work in arbitrary graphs, even if the exact number of agents is given. This is because some pair of agents may not be interactable in an arbitrary graph, then they cannot resolve the conflicts of their identifiers by meeting each other.

 Protocol P_{rank} detects the conflicts between any (possibly non-interactable) two agents by traversing n tokens in a population where each agent always has exactly one token. This protocol is inspired by a self-stabilizing leader election protocol with *oracles* given by Beauquier *et al.* [5], where the agents traverse exactly one token in a population.

 The pseudocode of P_{rank} is shown in Algorithm 1. Our goal is to assign the agents the distinct labels $0, 1, \ldots, n-1$. Each agent v stores its label in a variable $v.\mathrm{id}_A \in \{0, 1, \ldots, n-1\}$ and outputs it as it is. To detect and resolve the conflicts of the labels in arbitrary graphs, each agent maintains four other variables $\mathrm{id}_T \in \{0, 1, \ldots, n-1\}$, $\mathrm{color}_A \in \{W, R, B\}$, $\mathrm{color}_T \in \{R, B\}$, and $\mathrm{timer}_T \in \{0, 1, \ldots, U_T\}$, where U_T is a sufficiently large $\Omega(mn)$ value and m

is the number of interactable pairs in the population. We will explain later how to assign U_T such a value. We say that v has a token labeled x if $v.\mathrm{id}_T = x$. Each agent v has one color, white (W), red (R), or blue (B), while v's token has one color, red (R) or blue (B), maintained by variables $v.\mathrm{id}_A$ and $v.\mathrm{id}_T$, respectively.

The tokens always make *the random walk*: two agents swap their tokens whenever two agents interact (Line 1). If the two tokens have the same label, one of them increments its label modulo n (Line 2). Since all tokens meet each other infinitely often by the random walk, they eventually have mutually distinct labels (id_T), after which they never change their labels. Thereafter, the conflicts of labels among the agents are resolved by using the tokens. Let x be any integer in $\{0, 1, \ldots, n-1\}$ and denote the token labeled x by T_x. Ideally, an agent labeled x always has the same color as that of T_x. Consider the case that an agent labeled x, say v, meets T_x, and v and T_x have different colors, blue and red. Then, v suspects that there is another agent labeled x, and v increases its label by one modulo n (Line 7). The agent v, now labeled $x+1 \pmod{n}$, changes its color to white (Line 8). When v meets $T_{x+1 \pmod{n}}$ the next time, it copies the color of the token to its color to synchronize a color with $T_{x+1 \pmod{n}}$. Token T_x changes its color periodically. Specifically, T_x decreases its $\mathtt{timer}_T$ whenever it moves unless $\mathtt{timer}_T$ already reaches zero (Line 3). If token T_x meets an agent labeled x, they have the same color, and the timer of the token is zero, then they change their color from blue to red or from red to blue (Lines 11–12). If there are two or more agents labeled x, this multiplicity is eventually detected because T_x makes a random walk forever: T_x eventually meets an agent labeled x with a different color. By repeating this procedure, the population eventually reaches a configuration where all the agents have distinct labels and the agent labeled x has the same color as that of T_x for all $x = 0, 1, \ldots, n-1$. No agent changes its label thereafter.

Note that this protocol works even if we do not use variable $\mathtt{timer}_T$ and color W. We introduce them to make this protocol faster under the uniformly random scheduler. In the rest of this section, we prove the following theorem.

Theorem 1. *Given knowledge ν and μ, $P_{\mathrm{rank}}(\nu, \mu)$ is a self-stabilizing protocol that solves* **RK** *in arbitrary graphs if $\nu = \{n\}$ for some integer n, regardless of μ. Starting from any configuration C_0 on any population $G = (V_G, E_G) \in \mathcal{G}_{n,*}$, the execution of $P_{\mathrm{rank}}(\nu, \mu)$ under the uniformly random scheduler (i.e., $\Xi_{P_{\mathrm{rank}}(\nu,\mu)}(G, C_0, \Gamma)$) reaches a safe configuration within $O(mn^3 d \log n + n^2 U_T)$ steps in expectation, where $m = |E_G|/2$ and d is the diameter of G. Each agent uses $O(\log n)$ bits of memory space to execute $P_{\mathrm{rank}}(\nu, \mu)$.*

Recall that we require parameter U_T to be a sufficiently large $\Omega(mn)$ value. If an upper bound M of m such that $M = \Theta(m)$ is obtained from knowledge μ, we can substitute a sufficiently large $\Theta(mn)$ value for U_T. Then, $P_{\mathrm{rank}}(\nu, \mu)$ converges in $O(mn^3 d \log n)$ steps in expectation. Even if such M is not obtained from μ, e.g., $\mu = \mathbb{N}_{\geq 1}$, we can substitute a sufficiently large $\Theta(n^3)$ value for U_T. Then, $P_{\mathrm{rank}}(\nu, \mu)$ converges in $O(mn^3 d \log n + n^5)$ steps in expectation.

In the rest of this section, we fix a population $G = (V_G, E_G) \in \mathcal{G}_{n,*}$, let $m = |E_G|/2$, and let d be the diameter of G. To prove Theorem 1, we define three sets $\mathcal{S}_{\text{token}}$, $\mathcal{S}_{\text{sync}}$, and $\mathcal{S}_{\text{rank}}$ of configurations in $\mathcal{C}_{\text{all}}(P_{\text{rank}}(\nu, \mu), G)$ as follows.

- $\mathcal{S}_{\text{token}}$: the set of all the configurations in $\mathcal{C}_{\text{all}}(P_{\text{rank}}(\nu, \mu), G)$ where all tokens have distinct labels, i.e., $\forall u, v \in V_G : u.\text{id}_T \neq v.\text{id}_T$. In a configuration in $\mathcal{S}_{\text{token}}$, there exists exactly one token labeled x in the population for each $x \in \{0, 1, \ldots, n - 1\}$. We use notation T_x both to denote the unique token labeled by x and to denote the agent on which this token *currently* stays.
- $\mathcal{S}_{\text{sync}}$: the set of all the configurations in $\mathcal{S}_{\text{token}}$ where proposition $Q_{\text{token}}(x) \overset{\text{def}}{\equiv} V_G(x) \neq \emptyset \Rightarrow (\exists u \in V_G(x) : u.\text{color}_A = T_x.\text{color}_T \lor u.\text{color}_A = W)$ holds for any $x \in \{0, 1, \ldots, n - 1\}$, where $V_G(x) \overset{\text{def}}{=} \{v \in V \mid v.\text{id}_A = x\}$.
- $\mathcal{S}_{\text{rank}}$: the set of all the configurations in $\mathcal{S}_{\text{sync}}$ where all the agents in V_G have distinct labels, that is, $\forall u, v \in V_G : u.\text{id}_A \neq v.\text{id}_A$.

Lemma 2. *The set $\mathcal{S}_{\text{token}}$ is closed for $P_{\text{rank}}(\nu, \mu)$.*

Proof. A token changes its label only if it meets another token with the same label. Hence, no token changes its label in an execution starting from a configuration in $\mathcal{S}_{\text{token}}$. □

Lemma 3. *Let $x \in \{0, 1, \ldots, n-1\}$. In an execution of $P_{\text{rank}}(\nu, \mu)$ starting from a configuration in $\mathcal{S}_{\text{token}}$, once $Q_{\text{token}}(x)$ holds, it always holds thereafter.*

Proof. This lemma holds because (i) an agent must be white just after it changes its label from $x - 1 \pmod{n}$ to x, (ii) a white agent labeled x changes its color only when token T_x visits it at an interaction, at which this white agent gets the same color as that of T_x, (iii) an agent labeled x with the same color as that of T_x changes its color only when token T_x visits it at an interaction, at which this agent and T_x get the same new color. □

Lemma 4. *The set $\mathcal{S}_{\text{sync}}$ is closed for $P_{\text{rank}}(\nu, \mu)$.*

Proof. The lemma immediately follows from Lemma 3. □

Lemma 5. *Let $x \in \{0, 1, \ldots, n - 1\}$. In an execution of $P_{\text{rank}}(\nu, \mu)$ starting from a configuration in $\mathcal{S}_{\text{sync}}$, once at least one agent is labeled x, the number of agents labeled x never becomes zero thereafter.*

Proof. This lemma holds in the same way as the proof of Lemma 3. □

Lemma 6. *The set $\mathcal{S}_{\text{rank}}$ is closed for $P_{\text{rank}}(\nu, \mu)$.*

Proof. The lemma immediately follows from Lemmas 4 and 5. □

The following lemma is useful to analyze the expected number of steps required to reach a configuration in $\mathcal{S}_{\text{rank}}$ in an execution of $P_{\text{rank}}(\nu, \mu)$.

Lemma 7. *Consider the following game with n players $p_0, p_1, \ldots, p_{n-1}$. Each player always has one state in $\{0, 1, \ldots, n-1\}$. At each step, an arbitrary pair of players is selected and they check the states of each other. If they have the same state, one of them increases its state by one modulo n. Otherwise, they do not change their states. Starting this game from any configuration (i.e., any combination of the states of all players), there is at least one state $z \in \{0, 1, \ldots, n-1\}$ such that no player changes its state from $z - 1$ (mod n) to z. The set of such states is uniquely determined by a configuration from which the game starts.*

Proof. Fix an initial configuration $\psi_0 = (k_0, k_1, \ldots, k_{n-1})$, where k_i represents the number of agents in state i in the configuration. In this proof, we make every addition and subtraction in modulo n and omit the notation "(mod n)". It is trivial that for any $x \in \{0, 1, \ldots, n-1\}$, no player changes its state from $x - 1$ to x if and only if x satisfies $\sum_{j=1}^{i} k_{x-j} \leq i$ for all $i \in \{1, 2, \ldots, n-1\}$. Therefore, the set of states z such that no player changes its state from $z - 1$ to z is uniquely determined by the initial configuration ψ_0.

By the uniqueness of the above set, it suffices to show that for any execution Ξ of this game starting from ψ_0, there is a state $z \in \{0, 1, \ldots, n-1\}$ such that no player changes its state from $z - 1$ to z in Ξ. We say that a state $x \in \{0, 1, \ldots, n-1\}$ is *filled* if at least one player is in state x. By definition of this game, once x is filled, x is always filled thereafter. If there is a state z that is never filled in Ξ, no player changes its state from $z - 1$ to z. Suppose the other case and let z be the state that is filled for the last time in execution Ξ. By definition, when z gets filled, all the n states are filled, which yields that all the n players have mutually distinct states at this time. Therefore, no player never changes its state from $z - 1$ to z in execution Ξ. □

Lemma 8. *Starting from any configuration $C_0 \in \mathcal{C}_{\mathrm{all}}(P_{\mathrm{rank}}(\nu, \mu), G)$, an execution of $P_{\mathrm{rank}}(\nu, \mu)$ under the uniformly random scheduler (i.e., $\Xi_{P(\nu,\mu)}(G, C_0, \Gamma)$) reaches a configuration in $\mathcal{S}_{\mathrm{token}}$ within $O(mn^3 d \log n)$ steps in expectation.*

Proof. By Lemma 7, there exists an integer $z \in \{0, 1, \ldots, n-1\}$ such that no *token* changes its label from $z - 1$ (mod n) to z. Then, the number of tokens labeled z becomes exactly one before or when all the tokens meet each other. Since Sudo et al. [20] proved that n tokens making random walks in arbitrary graphs meet each other within $O(mn^2 d \log n)$ steps in expectation, the number of tokens labeled z becomes exactly one within $O(mn^2 d \log n)$ steps in expectation. Thereafter, no token changes its label from z to $z+1$ (mod n). Hence, the number of tokens labeled $z + 1$ (mod n) becomes one in the next $O(mn^2 d \log n)$ steps in the same way. Repeating this procedure, all the tokens have distinct labels within $O(mn^3 d \log n)$ steps in expectation. □

Lemma 9. *Starting from any configuration $C_0 \in \mathcal{S}_{\mathrm{token}}$, an execution of $P_{\mathrm{rank}}(\nu, \mu)$ under the uniformly random scheduler (i.e., $\Xi_{P(\nu,\mu)}(G, C_0, \Gamma)$) reaches a configuration in $\mathcal{S}_{\mathrm{sync}}$ within $O(mn^3)$ steps in expectation.*

Proof. By Lemmas 2 and 3, it suffices to show that for each $x \in \{0, 1, \ldots, n-1\}$, $Q_{\mathrm{token}}(x)$ becomes true within $O(mn^2)$ steps in expectation in an execution of

$P_{\text{rank}}(\nu, \mu)$ starting from C_0. We have $Q_{\text{token}}(x) = \textit{false}$ if and only if there exists at least one agent labeled x and all of them have colors different from that of T_x (*i.e.*, the token labeled x). Even if $Q_{\text{token}}(x) = \textit{false}$ in C_0, $Q_{\text{token}}(x)$ becomes true before or when T_x meets all of them. With slight modification of the analysis in [20], we can show that T_x visits (*i.e.*, meets) all agents within $O(mn^2)$ steps in expectation (See the preprint [22]), from which the lemma follows. □

Lemma 10. *Assume that U_T is sufficiently large $\Omega(mn)$ value. Starting from any configuration $C_0 \in \mathcal{S}_{\text{sync}}$, an execution of $P_{\text{rank}}(\nu, \mu)$ under the uniformly random scheduler (i.e., $\Xi_{P(\nu,\mu)}(G, C_0, \Gamma)$) reaches a configuration in $\mathcal{S}_{\text{rank}}$ within $O(mn^3 + n^2 U_T)$ steps in expectation.*

Proof. By Lemmas 5 and 7, there exists an integer $z \in \{0, 1, \ldots, n-1\}$ such that no *agent* changes its label from $z - 1 \pmod{n}$ to z. Therefore, at least one agent is labeled z in C_0. All of them get non-white color, *i.e.*, blue or red, or get a new label $z + 1 \pmod{n}$ before or when T_z meets all agents, which requires only $O(mn^2)$ steps in expectation. (See the preprint [22].) Without loss of generality, we assume that token T_z is red at this time. By Lemma 4, there is at least one *red* agent labeled z. After that, the $\texttt{timer}_T$ of T_z becomes zero within $O(nU_T)$ steps in expectation. (See the preprint [22].) In the next $O(mn^2)$ steps in expectation, T_z meets a red agent labeled z, at which T_z and this agent changes their colors to *blue*, and T_z resets its $\texttt{timer}_T$ to U_T. It is well known that a token making the random walk visits all nodes of any undirected graph within $O(mn)$ moves in expectation. Since a token decreases its $\texttt{timer}_T$ only by one every time it moves, T_z meets all agents and makes each agent labeled z blue or pushes it to the next label (*i.e.*, $z + 1 \pmod{n}$) before its $\texttt{timer}_T$ reaches zero again from $U_T = \Omega(mn)$, with probability $1 - p$ for any small constant p, by Markov's inequality. This requires only $O(mn^2)$ steps in expectation. (See the preprint [22].) Similarly, (i) the $\texttt{timer}_T$ of T_z becomes zero again in the next $O(nU_T)$ steps, (ii) T_x meets a blue agent labeled z, say v, in the next $O(mn^2)$ steps, at which T_x and v become red, and (iii) T_x meets all agents and pushes all agents labeled z except for v to the next label in the next $O(mn^2)$ steps in expectation and with probability $1 - p$ for any small constant p. Therefore, the number of agents labeled z becomes one within $O(mn^2 + nU_T)$ steps in expectation. After that, no agent changes its label from z to $z + 1 \pmod{n}$. Hence, the number of agents labeled $z + 1 \pmod{n}$ becomes one in the next $O(mn^2 + nU_T)$ steps in expectation by the same reason. Repeating this procedure, all agents get mutually distinct labels (*i.e.*, $\texttt{id}_A$) within $O(mn^3 + n^2 U_T)$ steps in expectation. □

Proof (of Theorem 1). By Lemmas 8, 9, and 10, $\Xi_{P_{\text{rank}}(\nu,\mu)}(G, C_0, \Gamma)$ reaches a configuration in $\mathcal{S}_{\text{rank}}$ within $O(mn^3 d \log n + n^2 U_T)$ steps in expectation. By Lemma 6, every configuration in $\mathcal{S}_{\text{rank}}$ is a safe configuration for the ranking problem. □

Theorem 2. *Let ν be any subset of $\mathbb{N}_{\geq 2}$ and let $\mu = \mathbb{N}_{\geq 1}$. Given knowledge ν and μ ($= \mathbb{N}_{\geq 1}$), there exists a self-stabilizing protocol that solves* **LE** *and* **RK** *in* *arbitrary graphs if and only if the agents know the exact number of agents i.e.,* $\mathcal{G}_{\nu,\mu} = \mathcal{G}_{n,*}$ *for some* $n \in \mathbb{N}_{\geq 2}$.

Proof. The theorem immediately follows from Lemma 1, Theorem 1, and the fact that $\mathbf{LE} \preceq \mathbf{RK}$. □

4   Degree Recognition and Neighbor Recognition

Our goal is to prove the negative and positive propositions for **DR** and **NR** introduced in Sect. 1. First, we prove the negative proposition.

Lemma 11. *Let ν and μ be any sets such that $\nu \subseteq \mathbb{N}_{\geq 2}$ and $\mu \subseteq \mathbb{N}_{\geq 1}$. There exists no self-stabilizing protocol that solves **DR** in all graphs in $\mathcal{G}_{\nu,\mu}$ if $\mathcal{G}_{n,m_1} \cup \mathcal{G}_{n,m_2} \subseteq \mathcal{G}_{\nu,\mu}$ holds for some $n \in \mathbb{N}_{\geq 2}$ and some distinct $m_1, m_2 \in \mathbb{N}_{\geq 1}$ such that $\mathcal{G}_{n,m_1} \neq \emptyset$ and $\mathcal{G}_{n,m_2} \neq \emptyset$.*

Proof. Assume $m_1 < m_2$ without loss of generality. By definition, there must exist two graphs $G' = (V_{G'}, E_{G'}) \in \mathcal{G}_{n,m_1}$ and $G'' = (V_{G''}, E_{G''}) \in \mathcal{G}_{n,m_2}$ such that $V_{G'} = V_{G''}$ and $E_{G'} \subset E_{G''}$. Then, there exists at least one agent $v \in V_{G''}$ such that its degree differs in G' and G''. Let δ' and δ'' be the degrees of v in G' and G'', respectively. Assume for contradiction that there is a self-stabilizing protocol $P(\nu, \mu)$ that solves **DR** both in G' and G''. By definition, there must be at least one safe configuration S of protocol $P(\nu, \mu)$ on G'' for **DR**. In every execution of $P(\nu, \mu)$ starting from S on G'', agent v must always output δ'' as its degree. The configuration S can also be a configuration on G' because $V_{G'} = V_{G''}$. Since $P(\nu, \mu)$ is self-stabilizing in G', there must be a finite sequence of interactions $\gamma_0, \gamma_1, \ldots, \gamma_t$ of G' that put configuration S to a configuration where v outputs δ' as its degree. Since $E_{G'} \subset E_{G''}$, $\gamma_0, \gamma_1, \ldots, \gamma_t$ is also a sequence of interactions in G''. This implies that this sequence changes the output of v from δ'' to δ' starting from a *safe* configuration, a contradiction. □

To prove the positive proposition, we give a self-stabilizing protocol P_{neigh}, which solves the neighbor recognition problem (**NR**) in arbitrary graphs given the knowledge of the exact number of agents and the exact number of interactable pairs, that is, given knowledge ν and μ such that $|\nu| = |\mu| = 1$. In the rest of this section, let n and m be the integers such that $\nu = \{n\}$ and $\mu = \{m\}$.

The pseudocode of P_{neigh} is shown in Algorithm 2. Our goal is to let the agents recognize the set of their neighbors. Each agent v stores its label in a variable $v.\text{id}_A \in \{0, 1, \ldots, n-1\}$ and the set of the labels assigned to its neighbors in a variable $\text{neighbors} \in 2^{\{0,1,\ldots,n-1\}}$. Each agent v outputs $(v.\text{id}_A, v.\text{neighbors})$.

We use P_{rank} as a sub-algorithm to assign the agents the distinct labels $0, 1, \ldots, n-1$ and to let the n tokens make the random walk. Specifically, we first execute P_{rank} whenever two agents have an interaction (Line 1), substituting a sufficiently large $\Theta(mn)$ value for U_T. We do not update the variables used in P_{rank} in the other lines (Lines 2–17). Therefore, by Theorem 1, an execution of P_{neigh} starting from any configuration reaches a configuration in $\mathcal{S}_{\text{rank}}$ within

---

Algorithm 2. $P_{\text{neigh}}(\nu, \mu)$

Assumption: $|\nu| = 1$ and $|\mu| = 1$. (Let $\nu = \{n\}$ and $\mu = \{m\}$.)

Variables:

$\text{id}_A, \text{id}_T \in \{0, 1, \ldots, n-1\}$ // Updated only by P_{rank}
$\text{degree}_T \in \{0, 1, \ldots, n\}$, $\text{sum} \in \{0, 1, \ldots, 2m+1\}$
$\text{reset}_E \in \{0, 1, \ldots, U_E\}$, $\text{timer}_P \in \{0, 1, \ldots, U_P\}$
$\text{neighbors}, \text{counted} \in 2^{\{0,1,\ldots,n-1\}}$

Output function π_{out}: $(\text{id}_A, \text{neighbors})$

Interaction between initiator a_0 and responder a_1:

1: Execute P_{rank} with substituting sufficiently large $\Theta(mn)$ value for U_T.
2: $a_0.\text{degree}_T \leftrightarrow a_1.\text{degree}_T$
               // Execute the random walk of two tokens with P_{rank}

3: $a_0.\text{reset}_E \leftarrow a_1.\text{reset}_E \leftarrow \max(0, a_0.\text{reset}_E - 1, a_1.\text{reset}_E - 1)$
4: **if** $a_0.\text{reset}_E > 0$ **then** $a_0.\text{neighbors} \leftarrow a_1.\text{neighbors} \leftarrow \emptyset$ **endif**

5: **for all** $i \in \{0, 1\}$ **do**
6:     $a_i.\text{timer}_P \leftarrow \max(0, a_i.\text{timer}_P - 1)$
7:     **if** $a_i.\text{timer}_P = 0$ **then**
8:        $(a_i.\text{sum}, a_i.\text{counted}, a_i.\text{timer}_P) \leftarrow (0, \emptyset, U_P)$
9:     **end if**
10:    $a_i.\text{neighbors} \leftarrow a_i.\text{neighbors} \cup \{a_{1-i}.\text{id}_A\}$
11:    **if** $a_i.\text{id}_A = a_i.\text{id}_T$ **then** $a_i.\text{degree}_T \leftarrow |a_i.\text{neighbors}|$ **endif**

12:    **if** $a_i.\text{id}_T \notin a_i.\text{counted}$ **then**
13:       $a_i.\text{sum} \leftarrow \min(2m+1, a_i.\text{sum} + a_i.\text{degree}_T)$
14:       $a_i.\text{counted} \leftarrow a_i.\text{counted} \cup \{a_i.\text{id}_T\}$
15:    **end if**
16:    **if** $a_i.\text{sum} = 2m+1$ **then** $a_i.\text{reset}_E \leftarrow U_E$ **endif**
17: **end for**

$O(mn^2 d \log n)$ steps in expectation. Hence, we need to consider only an execution after reaching a configuration in $\mathcal{S}_{\text{rank}}$. Then, we can assume that the population always has exactly one agent labeled x and exactly one token labeled x for each $x = \{0, 1, \ldots, n-1\}$. We denote them by A_x and T_x, respectively.

The agents compute their **neighbors** in a simple way: every time two agents u and v have an interaction, u adds $v.\text{id}_A$ to $u.\text{neighbors}$ and v adds $u.\text{id}_A$ to $v.\text{neighbors}$ (Line 10). However, this simple way to compute **neighbors** is not enough to design a self-stabilizing protocol because we consider an arbitrary initial configuration. Specifically, in an initial configuration, $v.\text{neighbors}$ may include $u.\text{id}_A$ for some $u \notin N_G(v)$. We call such $u.\text{id}_A$ a *fake label*. To compute $v.\text{neighbors}$ correctly, in addition to the above simple mechanism, it suffices to detect the existence of a fake label and reset the **neighbors** of all agents to the empty set if a fake label is detected.

Using the knowledge $\mu = \{m\}$, we achieve the detection of fake labels with the following strategy. Each token T_x carries $|A_x.\text{neighbors}|$ in a variable

$\text{degree}_T \in \{0, 1, \ldots, n\}$ (Line 2). Whenever T_x meet A_x, the value of $T_x.\text{degree}_T$ is updated by the current value of $|A_x.\text{neighbors}|$ (Line 11). Each agent always tries to estimate $\sum_{v \in V_G} |v.\text{neighbors}|$ using variables $\text{sum} \in \{0, 1, \ldots, 2m + 1\}$, $\text{counted} \in 2^{\{0,1,\ldots,n-1\}}$, and $\text{timer}_P \in \{0, 1, \ldots, U_P\}$, where U_P is a sufficiently large $\Theta(mnd \log n)$ value. It uses timer_P as a count-down timer to reset sum and counted periodically. Specifically, an agent v decreases $v.\text{timer}_P$ by one every time it has an interaction and resets $v.\text{sum}$, $v.\text{counted}$, and $v.\text{timer}_P$ to 0, $\emptyset$, and U_P, respectively, when $v.\text{timer}_P$ reaches zero (Lines 6–9). Whenever agent v meets T_x such that $x \notin v.\text{counted}$, v executes $v.\text{sum} \leftarrow \min(2m + 1, v.\text{sum} + T_x.\text{degree}_T)$ and adds x to $v.\text{counted}$. (Lines 12–15) We expect $v.\text{sum} = \sum_{v \in V_G} |v.\text{neighbors}|$ when v meets all of $T_0, T_1, \ldots, T_{n-1}$. If $v.\text{sum}$ reaches $2m + 1$, agent v concludes that at least one agent has a fake label, i.e., $u.\text{neighbors} \not\subseteq \{w.\text{id}_A \mid w \in N_G(u)\}$ for some u.

When the existence of a fake label is detected, we reset the neighborss of all agents using a variable $\text{reset}_E \in \{0, 1, \ldots, U_E\}$, where U_E is a sufficiently large $\Theta(n^2)$ value. Specifically, when $v.\text{sum} = 2m + 1$ holds, v emits the error signal by setting variable $v.\text{reset}_E$ to U_E (Line 16). Thereafter, the error signal is propagated to the whole population via *the larger value propagation*: when two agents u and v meet, they substitute $\max(0, u.\text{reset}_E - 1, v.\text{reset}_E - 1)$ for their reset_Es. (Line3). Whenever an agent v receives the error signal, i.e., $v.\text{reset}_E > 0$ holds, it resets its neighbors to the empty set (Line 4).

Thus, even if some agent has fake labels at the beginning of an execution, the population eventually reaches a configuration where no agent has fake labels after the occurrence of the following events: the existence of a fake label is detected, the error signal propagates to the whole population, and all agents reset their neighborss to the empty set. Thereafter, for any $x \in \{0, 1, \ldots, n-1\}$, T_x eventually meets A_x, after which $T_x.\text{degree}_T \leq |N_G(A_x)|$ always hold. Hence, by the periodical reset of sum and counted, the population eventually reach a configuration from which no agent emits the error signal. Thereafter, the population will soon reach a configuration that satisfies $v.\text{neighbors} = \{u.\text{id}_A \mid u \in N_G(v)\}$ for all $v \in V_G$ by the above simple computation of neighbors (Line 10). Once it reaches such a configuration, no agent changes its neighbors.

Theorem 3. *Given knowledge ν and μ, $P_{\text{neigh}}(\nu, \mu)$ is a self-stabilizing protocol that solves* **NR** *in arbitrary graphs if $\nu = \{n\}$ and $\mu = \{m\}$ for some integers n and m. Starting from any configuration C_0 on any population $G = (V_G, E_G) \in \mathcal{G}_{n,m}$, the execution of $P_{\text{neigh}}(\nu, \mu)$ under the uniformly random scheduler (i.e., $\Xi_{P_{\text{neigh}}(\nu,\mu)}(G, C_0, \Gamma)$) reaches a safe configuration within $O(mn^3 d \log n)$ steps in expectation, where $m = |E_G|/2$ and d is the diameter of G. Each agent uses $O(n)$ bits of memory space to execute $P_{\text{neigh}}(\nu, \mu)$.*

Proof. Define $L_{\text{neigh}}(v) = \{u.\text{id}_A \mid u \in N_G(v)\}$ and define $\mathcal{S}_{\text{noFake}}$ as the set of all configurations in $\mathcal{S}_{\text{rank}}$ where no agent has a fake label in its neighbors, that is, $v.\text{neighbors} \subseteq L_{\text{neigh}}(v)$ holds for all $v \in V_G$.

First, we show that execution $\Xi = \Xi_{P_{\text{neigh}}(\nu,\mu)}(G, C_0, \Gamma)$ reaches a configuration in $\mathcal{S}_{\text{noFake}}$ within $O(mn^3 d \log n)$ steps in expectation. By Theorem 1,

Ξ reaches a configuration C' in $\mathcal{S}_{\text{rank}}$ within $O(mn^3d\log n)$ steps in expectation because $U_T = \Theta(mn)$. We assume $C' \notin \mathcal{S}_{\text{noFake}}$ because otherwise we need not discuss anything. Interactions happen between all interactable pairs within $O(m\log n)$ steps in expectation. Therefore, after reaching C', Ξ reaches within $O(m\log n)$ steps in expectation a configuration C'' where $L_{\text{neigh}}(v) \subseteq v.\text{neighbors}$ for all $v \in V_G$ or a configuration where $u.\text{reset}_E > 0$ for some $u \in V_G$. In the former case, $\sum_{v \in V} |v.\text{neighbors}| > 2m$ holds in C'' since at least one agent has one or more fake labels in its neighbors. Thereafter, some agent v decreases its timer_P to zero and resets it to U_P in the next $O(mU_P) = O(m^2nd\log n) \subseteq O(mn^3d\log n)$ steps in expectation. After that, v meets all tokens within $O(mnd\log n)$ steps in expectation. (See the preprint [22].) As a result, $v.\text{sum}$ reaches $2m+1$ and v emits the error signal. To conclude, after Ξ reaches C', some agent emits the error signal, *i.e.*, it substitutes U_E for its reset_E. Since we set U_E to a sufficiently large $\Theta(n^2)$ value, the error signal is propagated to the whole population within $O(mn)$ steps with probability $1 - O(1/n)$. (See Lemma 5 in [19].) Every time an agent receives the error signal, it resets its neighbors to the empty set. Therefore, Ξ reaches a configuration in $\mathcal{S}_{\text{noFake}}$ within $O(mn^3d\log n)$ steps in expectation.

After entering $\mathcal{S}_{\text{noFake}}$, Ξ reaches within $O(mnd\log n)$ steps in expectation a configuration where $\sum_{x=0,1,\ldots,n-1} T_x.\text{degree}_T \le 2m$ holds; because every T_x meets A_x within $O(mnd)$ steps in expectation for every $x \in \{0, 1, \ldots, n-1\}$. Similarly, all agents reset their sum and counted in the next $O(mU_P) \subseteq O(mn^3d\log n)$ step in expectation. Thereafter, no agent sees $\text{sum} = 2m+1$, hence no agent emits the error signal, after which the error signal disappears from the population in the next $O(U_E \cdot m\log m) = O(mn^2\log n)$ steps in expectation. Therefore, interactions happen between all interactable pairs in the next $O(m\log n)$ steps in expectation, by which $v.\text{neighbors} = L_{\text{neigh}}(v)$ holds for all $v \in V_G$. After that, no agent v changes $v.\text{neighbors}$, which yields that Ξ has reached a safe configuration.

Each agent uses only $O(n)$ bits: both variables neighbors and counted require n bits and all other variables used in P_{neigh} require $O(\log n)$ bits.   □

5   Conclusion

In this paper, we clarified the solvability of the leader election problem, the ranking problem, the degree recognition problem, and the neighbor recognition problem by self-stabilizing population protocols with knowledge of the number of nodes and/or the number of edges in a network. The protocols we gave in this paper require *exact* knowledge on the number of agents and/or the number of interactable pairs. It is interesting and still open whether *ambiguous* knowledge such as "the number of interactable pairs is at most M" and "the number of agents is not a prime number" is useful to design self-stabilizing population protocols.

References

1. Alistarh, D., Gelashvili, R.: Polylogarithmic-time leader election in population protocols. In: Halldórsson, M.M., Iwama, K., Kobayashi, N., Speckmann, B. (eds.) ICALP 2015. LNCS, vol. 9135, pp. 479–491. Springer, Heidelberg (2015). https://doi.org/10.1007/978-3-662-47666-6_38
2. Angluin, D., Aspnes, J., Diamadi, Z., Fischer, M.J., Peralta, R.: Computation in networks of passively mobile finite-state sensors. Distrib. Comput. 18(4), 235–253 (2006)
3. Angluin, D., Aspnes, J., Eisenstat, D.: Fast computation by population protocols with a leader. Distrib. Comput. 21(3), 183–199 (2008)
4. Angluin, D., Aspnes, J., Fischer, M.J., Jiang, H.: Self-stabilizing population protocols. ACM Trans. Auton. Adapt. Syst. 3(4), 13 (2008)
5. Beauquier, J., Blanchard, P., Burman, J.: Self-stabilizing leader election in population protocols over arbitrary communication graphs. In: Baldoni, R., Nisse, N., van Steen, M. (eds.) OPODIS 2013. LNCS, vol. 8304, pp. 38–52. Springer, Cham (2013). https://doi.org/10.1007/978-3-319-03850-6_4
6. Burman, J., Doty, D., Nowak, T., Severson, E.E., Xu, C.: Efficient self-stabilizing leader election in population protocols. arXiv preprint arXiv:1907.06068 (2019)
7. Cai, S., Izumi, T., Wada, K.: How to prove impossibility under global fairness: on space complexity of self-stabilizing leader election on a population protocol model. Theory Comput. Syst. 50(3), 433–445 (2012)
8. Canepa, D., Potop-Butucaru, M.G.: Stabilizing leader election in population protocols (2007). http://hal.inria.fr/inria-00166632
9. Chen, H.-P., Chen, H.-L.: Self-stabilizing leader election. In: Proceedings of the 2019 ACM Symposium on Principles of Distributed Computing, pp. 53–59 (2019)
10. Cordasco, G., Gargano, L.: Space-optimal proportion consensus with population protocols. In: Spirakis, P., Tsigas, P. (eds.) SSS 2017. LNCS, vol. 10616, pp. 384–398. Springer, Cham (2017). https://doi.org/10.1007/978-3-319-69084-1_28
11. Dijkstra, E.: Self-stabilizing systems in spite of distributed control. Commun. ACM 17(11), 643–644 (1974)
12. Fischer, M., Jiang, H.: Self-stabilizing leader election in networks of finite-state anonymous agents. In: Shvartsman, M.M.A.A. (ed.) OPODIS 2006. LNCS, vol. 4305, pp. 395–409. Springer, Heidelberg (2006). https://doi.org/10.1007/11945529_28
13. Gąsieniec, L., Stachowiak, G., Uznanski, P.: Almost logarithmic-time space optimal leader election in population protocols. In: The 31st ACM on Symposium on Parallelism in Algorithms and Architectures, pp. 93–102. ACM (2019)
14. Izumi, T.: On space and time complexity of loosely-stabilizing leader election. In: Scheideler, C. (ed.) SIROCCO 2014. LNCS, vol. 9439, pp. 299–312. Springer, Cham (2015). https://doi.org/10.1007/978-3-319-25258-2_21
15. Mertzios, G.B., Nikoletseas, S.E., Raptopoulos, C.L., Spirakis, P.G.: Determining majority in networks with local interactions and very small local memory. In: Esparza, J., Fraigniaud, P., Husfeldt, T., Koutsoupias, E. (eds.) ICALP 2014. LNCS, vol. 8572, pp. 871–882. Springer, Heidelberg (2014). https://doi.org/10.1007/978-3-662-43948-7_72
16. Sudo, Y., Masuzawa, T., Datta, A.K., Larmore, L.L.: The same speed timer in population protocols. In: The 36th IEEE International Conference on Distributed Computing Systems, pp. 252–261 (2016)

17. Sudo, Y., Nakamura, J., Yamauchi, Y., Ooshita, F., Kakugawa, H., Masuzawa, T.: Loosely-stabilizing leader election in a population protocol model. Theor. Comput. Sci. **444**, 100–112 (2012)

18. Sudo, Y., Ooshita, F., Izumi, T., Kakugawa, H., Masuzawa, T.: Logarithmic expected-time leader election in population protocol model. In: Ghaffari, M., Nesterenko, M., Tixeuil, S., Tucci, S., Yamauchi, Y. (eds.) SSS 2019. LNCS, vol. 11914, pp. 323–337. Springer, Cham (2019). https://doi.org/10.1007/978-3-030-34992-9_26

19. Sudo, Y., Ooshita, F., Kakugawa, H., Masuzawa, T.: Loosely-stabilizing leader election on arbitrary graphs in population protocols. In: Aguilera, M.K., Querzoni, L., Shapiro, M. (eds.) OPODIS 2014. LNCS, vol. 8878, pp. 339–354. Springer, Cham (2014). https://doi.org/10.1007/978-3-319-14472-6_23

20. Sudo, Y., Ooshita, F., Kakugawa, H., Masuzawa, T.: Loosely stabilizing leader election on arbitrary graphs in population protocols without identifiers or random numbers. IEICE Trans. Inf. Syst. **103**(3), 489–499 (2020)

21. Sudo, Y., Ooshita, F., Kakugawa, H., Masuzawa, T., Datta, A.K., Larmore, L.L.: Loosely-stabilizing leader election with polylogarithmic convergence time. Theor. Comput. Sci. **806**, 617–631 (2020)

22. Sudo, Y., Shibata, M., Nakamura, J., Kim, Y., Masuzawa, T.: The power of global knowledge on self-stabilizing population protocols. arXiv preprint arXiv:2003.07491 (2020)

Phase Transition of a Non-linear Opinion Dynamics with Noisy Interactions
(Extended Abstract)

Francesco d'Amore[1](✉), Andrea Clementi[2], and Emanuele Natale[1]

[1] Université Côte d'Azur, Inria, CNRS, I3S, Sophia Antipolis, France
`{francesco.d-amore,emanuele.natale}@inria.fr`
[2] University of Rome Tor Vergata, Rome, Italy
`clementi@mat.uniroma2.it`

Abstract. In several real *Multi-Agent Systems* (MAS), it has been observed that only weaker forms of *metastable consensus* are achieved, in which a large majority of agents agree on some opinion while other opinions continue to be supported by a (small) minority of agents. In this work, we take a step towards the investigation of metastable consensus for complex (non-linear) *opinion dynamics* by considering the famous UNDECIDED-STATE dynamics in the binary setting, which is known to reach consensus exponentially faster than the VOTER dynamics. We propose a simple form of uniform noise in which each message can change to another one with probability p and we prove that the persistence of a *metastable consensus* undergoes a *phase transition* for $p = \frac{1}{6}$. In detail, below this threshold, we prove the system reaches with high probability a metastable regime where a large majority of agents keeps supporting the same opinion for polynomial time. Moreover, this opinion turns out to be the initial majority opinion, whenever the initial bias is slightly larger than its standard deviation. On the contrary, above the threshold, we show that the information about the initial majority opinion is "lost" within logarithmic time even when the initial bias is maximum. Interestingly, using a simple coupling argument, we show the equivalence between our noisy model above and the model where a subset of agents behave in a *stubborn* way.

1 Introduction

We consider a fully-decentralized *Multi-Agent Systems* (for short, MAS) formed by a set of n agents (i.e. nodes) which mutually interact by exchanging messages over an underlying communication graph. In this setting, *opinion dynamics* are mathematical models to investigate the way a fully-decentralized MAS is able to reach some form of *Consensus*. Their study is a hot topic touching several research areas such as MAS [11,18], Distributed Computing [4,16,25], Social Networks [1,33], and System Biology [7,8]. Typical examples of opinion dynamics are the Voter Model, the averaging rules, and the majority rules. Some of such

© Springer Nature Switzerland AG 2020
A. W. Richa and C. Scheideler (Eds.): SIROCCO 2020, LNCS 12156, pp. 255–272, 2020.
https://doi.org/10.1007/978-3-030-54921-3_15

dynamics share a surprising efficiency and resiliency that seem to exploit common *computational principles* [4,16,25].

Within such framework, the tasks of *(valid) Consensus* and *Majority Consensus* have attracted a lot of attention within different application domains in social networks [33], in biological systems [23], passively-mobile sensor networks [2] and chemical reaction networks [12]. In the Consensus task, the system is required to converge to a stable configuration where all agents supports the same opinion and this opinion must be *valid*, i.e., it must be supported by at least one agent in the initial configuration. While, in the Majority Consensus task, starting from an initial configuration where there is some positive bias towards one *majority opinion*, the system is required to converge to the configuration where all agents support the initial majority opinion. Here, the *bias* of a configuration is defined as the difference between the number of agents supporting the majority opinion (for short, we name this number as *majority*) and the number of agents supporting the second-largest opinion.

Different opinion dynamics have been studied in a variety of settings [13,22], and then used as subroutine to solve more complex computational tasks [6,15,35].

In the aforementioned applicative scenarios, it has been nevertheless observed that only weaker forms of *metastable* consensus are achieved, in which the large majority of agents rapidly achieves a consensus (while other opinions continue to be supported by a small set of agents), and this setting is preserved for a relatively-long regime. Models that have been considered to study such phenomenon include MAS where: i) agents follow a linear dynamics, such as the VOTER model or the AVERAGING dynamics and ii) a small set of *stubborn agents* are present in the system [31,32,37], or the local interactions are affected by *communication noise* [29].

We emphasize that the VOTER model has a slow (i.e. polynomial in the number n of agents) convergence time even in a fully-connected network (i.e. in the complete graph) and it does not guarantee a high probability to reach consensus on the initial majority opinion, even starting from a large initial bias (i.e. $\Theta(n)$, where n is the number of the agents of the system) [26]. On the other hand, averaging dynamics requires agents to perform numerical operations and, very importantly, to have a large local memory (to guarantee a good-enough approximation of real numbers). For the reasons above, linear opinion dynamics cannot explain fast and reliable metastable consensus phenomena observed in some MAS [6,12,23].

The above discussion naturally leads us to investigate the behaviour of other, non-linear dynamics in the presence of stubborn agents and/or communication noise. Over a MAS having the n-node complete graph as the underlying graph, we introduce a simple model of *communication noise* in the stochastic process yielded by a popular dynamics, known as the UNDECIDED-STATE dynamics. In some previous papers [34], this protocol has been called the *Third-State Dynamics*. We here prefer the term "undecided" since it well captures the role of this additional state.

According to this simple dynamics, the state of every agent can be either an opinion (chosen from a finite set Σ) or the *undecided state*. At every discrete-time step (i.e., round), every agent "pulls" the state of a random neighbor and updates its state according to the following rule: if a non-undecided agent pulls a different opinion from its current one, then it will get undecided, while in all other cases it keeps its opinion; moreover, if the node is undecided then it will get the state of the pulled neighbor.

This non-linear dynamics is known to compute Consensus (and Majority Consensus) on the complete network within a logarithmic number of rounds [2,9] and, very importantly, it is optimal in terms of local memory since it requires just one extra state/opinion [30].

While communication noise is a common feature of real-world systems and its effects have been thoroughly investigated in physics and information theory [14], its study has been mostly focused on settings in which communication happens over *stable links* where the use of error-correcting codes is feasible since message of large size are allowed; it has been otherwise noted that when interactions among the agents are random and opportunistic and consists of very-short messages, classical information-theoretic arguments do not carry on and new phenomena calls for a theoretical understanding [7].

Our Contribution. In this work, we show that, under a simple model of uniform noise, the UNDECIDED-STATE dynamics exhibits an interesting *phase transition*.

We consider the binary case (i.e., $|\Sigma| = 2$) together with an *oblivious* and *symmetric* action of noise over messages: any sent message is changed upon being received to any other value, independently and uniformly at random with probability p (where p is any fixed positive constant smaller than $1/2$).

On one hand, if $p < 1/6$, starting from an arbitrary configuration of the complete network of n agents, we prove that the system *with high probability*[1] (*w.h.p.*, for short) reaches, within $\mathcal{O}(\log n)$ rounds, a metastable almost consensus regime where the bias towards one fixed valid opinion keeps large, i.e. $\Theta(n)$, for at least a *poly(n)* number of rounds (see Theorem 3). In particular, despite the presence of random communication noise, our result implies that the UNDECIDED-STATE dynamics is able to rapidly break the initial symmetry of any balanced configuration and reach a metastable regime of almost consensus (e.g., the perfectly-balanced configuration with $n/2$ agents having one opinion and the other $n/2$ agents having the other opinion).

Importantly enough, our probabilistic analysis also shows that, for any $p < 1/6$, the system is able to "compute" the task of almost Majority Consensus. Indeed, in Theorem 1, starting from an arbitrary configuration with bias $\Omega(\sqrt{n \log n})$,[2] we prove that the system w.h.p. reaches, within $\mathcal{O}(\log n)$ rounds,

---

[1] An event E holds *with high probability* if a constant $\gamma > 0$ exists such that $\mathbf{P}(E) \geq 1 - (1/n)^{\gamma}$.

[2] We remark that, when every agent chooses its initial binary opinion uniformly at random, the standard deviation of the bias is $\Theta(\sqrt{n})$.

a metastable regime where the bias towards the initial majority opinion keeps large, i.e. $\Theta(n)$, for at least a $poly(n)$ number of rounds (see Theorem 1). For instance, our analysis for $p = 1/10$ implies that the process rapidly reaches a metastable regime where the bias keeps size larger than $n/3$.

On the other hand, if $p > 1/6$, even when the initial bias is maximum (i.e., when the system starts from any full-consensus configuration), after a logarithmic number of rounds, the information about the initial majority opinion is "lost": in Theorem 2, we indeed show that the system w.h.p. enters into a regime where the bias keeps bounded by $\mathcal{O}(\sqrt{n \log n})$. We also performed some computer simulations that confirm our theoretical results, showing that the majority opinion switches continuously during this regime (see Sect. 5 for further details).

Interestingly, in Subsect. 2.1 we show that our noise model is equivalent to a noiseless setting in which *stubborn* agents are present in the system [37] (that is, agents that never change their state): we thus obtain an analogous phase transition in this setting. The obtained phase transition thus separates qualitatively the behavior of the UNDECIDED-STATE dynamics from that of the VOTER model which is, to the best of our knowledge, the only opinion dynamics (with a finite opinion set) which has been rigorously analyzed in the presence of communication noise or stubborn agents [31,37]: this hints at a more general phenomenon for dynamics with fast convergence to some metastable consensus.

We believe this work contributes to the research endeavour of exploring the interplay between communication noise and stochastic interaction pattern in MAS. As we will discuss in the Related Work, despite the fact that these two characteristics are quite common in real-world MAS, their combined effect is still far from being understood and poses novel mathematical challenges. Within such framework, we have identified and rigorously analyzed a phase transition behaviour of the famous UNDECIDED-STATE process in the presence of communication noise (or, of stubborn agents) on the complete graph.

Related Work. The UNDECIDED-STATE dynamics has been originally studied as an efficient majority-consensus protocol by [2] and independently by [5] for the binary case (i.e. with two initial input values). They proved that w.h.p., within a logarithmic number of rounds, all agents support the initial majority opinion. Some works have then extended the analysis of the UNDECIDED-STATE dynamics to non-complete topologies. In the Poisson-clock model (formally equivalent to the Population Protocol model), [20] derive an upper bound on the expected convergence time of the dynamics that holds for arbitrary connected graphs, which is based on the location of eigenvalues of some contact rate matrices. They also instantiate their bound for particular network topologies. Successively, [30] provided an analysis when the initial states of agents are assigned independently at random, and they also derive "bad" initial configurations on certain graph topologies such that the initial minority opinion eventually becomes the majority one. As for the use of UNDECIDED-STATE as a generic consensus protocol, [10] recently proved that, in the synchronous uniform PULL model in which all agents update their state in parallel by observing the state of a random other node the convergence time of the UNDECIDED-STATE dynamics is w.h.p. logarithmic.

Notably, communication noise in random-interacting MAS appears to be a neglected area of investigation [11,29]. Such shortage of studies contrasts with the vast literature on communication noise over *stable* networks.[3] More recently, in [23], the authors consider a settings in which agents interact uniformly at random by exchanged binary messages which are subject to noise. In detail, the authors provide simple and efficient protocols to solve the classical distributed-computing problems of Broadcast (a.k.a Rumor Spreading) and Majority Consensus, in the Uniform-PUSH model with binary messages, in which each message can be changed upon being received with probability $1/2 - \epsilon$. Their results have been generalized to the Majority Consensus Problem for the multi-valued case in [24]. When the noise is constant, [23] proves that in their noisy version of the Uniform-PUSH model, the Broadcast Problem can be solved in *logarithmic* time. Rather surprisingly, [7] and [9] prove that solving the Broadcast Problem in the Uniform-PULL model takes linear time, while the time to perform Majority Consensus remains logarithmic in both models.

The fact that real-world systems such as social networks fail to converge to consensus has been extensively studied in various disciplines; formal models developed to investigate the phenomenon include the multiple-state Axelrod model [3] and the bounded-compromise model by Weisbuch et al. [36]; the failure to reach consensus in these models is due to the absence of interaction among agent opinions which are *"too far apart"*. A different perspective is offered by models which investigate the effect of stubborn agents (also known as *zealotry* in the literature), in which some *stubborn/zealot* agents never update their opinion.

Finally, several works have been devoted to study such effect under linear models of opinion dynamics [31,32,37].

2   Preliminaries

We study the discrete-time, parallel version of the UNDECIDED-STATE dynamics on the complete graph in the binary setting [10]. In detail, there is an additional state/opinion, i.e. the *undecided state*, besides the two possible opinions (say, opinion *Alpha* and opinion *Beta*) an agent can support, and, in the absence of noise, the updating rule works as follows: at every round $t \geq 0, t \in \mathbb{N}$, each agent u chooses a neighbor v (or, possibly, itself) independently and uniformly at random and, at the next round, it gets a new opinion according to the rule given in Table 1.[4] The definition of noise we consider is the following.

Definition 1 (Definition of noise). *Let p be a real number in the interval $(0, 1/2)$. When an agent u chooses a neighbor v and looks at (pulls) its opinion,*

---

[3] For stable networks, we here mean a network where communication between agents can be modeled as a classical *channel* the agents can use to exchange messages at will [14].

[4] Notice that this dynamics requires no labeling of the agents, i.e., the network can be anonymous.

Table 1. The update rule of the USD.

$u \setminus v$	Undecided	Alpha	Beta
Undecided	Undecided	Alpha	Beta
Alpha	Alpha	Alpha	Undecided
Beta	Beta	Undecided	Beta

it sees v's opinion with probability $1 - 2p$, and, with probability p, it sees one of the two other opinions.

For instance, if v supports opinion $Alpha$, then u sees $Alpha$ with probability $1 - 2p$, it sees $Beta$ with probability p, and it sees the undecided state with probability p. In this work, the terms *agent* and *node* are interchangeable.

Notation, Characterization, and Expected Values. Let us name C the set of all possible configurations; notice that, since the graph is complete and its nodes are anonymous, a configuration $\mathbf{x} \in C$ is uniquely determined by giving the number of $Alpha$ nodes, $a(\mathbf{x})$ and the number of $Beta$ nodes, $b(\mathbf{x})$. Accordingly to this notation, we call $q(\mathbf{x})$ the number of undecided nodes in configuration $\mathbf{x}$, and $s(\mathbf{x}) = a(\mathbf{x}) - b(\mathbf{x})$ the *bias* of the configuration $\mathbf{x}$. When the configuration is clear from the context, we will omit $\mathbf{x}$ and write just a, b, q, and s instead of $a(\mathbf{x}), b(\mathbf{x}), q(\mathbf{x})$, and $s(\mathbf{x})$. The UNDECIDED-STATE dynamics defines a finite-state non reversible Markov chain $\{\mathbf{X}_t\}_{t \geq 0}$ with state space C and no absorbing states.

The stochastic process yielded by the UNDECIDED-STATE dynamics, starting from a given configuration, will be denoted as UNDECIDED-STATE process. Once a configuration $\mathbf{x}$ at a round $t \geq 0$ is fixed, i.e. $\mathbf{X}_t = \mathbf{x}_t$, we use the capital letters A, B, Q, and S to refer to random variables $a(\mathbf{X}_{t+1}), b(\mathbf{X}_{t+1}), q(\mathbf{X}_{t+1})$, and $s(\mathbf{X}_{t+1})$. Notice that we consider the bias as $a(\mathbf{x}) - b(\mathbf{x})$ instead of $|a(\mathbf{x}) - b(\mathbf{x})|$ since the expectation of $|A - B|$ is much more difficult to evaluate than that of $A - B$.

The expected values of the above key random variables can be written as follows:

$$\mathbb{E}\left[A \mid \mathbf{x}\right] = \frac{a}{n}(a + 2q)(1 - 2p) + [a(a + b) + (a + q)(b + q)]\frac{p}{n}, \tag{1}$$

$$\mathbb{E}\left[B \mid \mathbf{x}\right] = \frac{b}{n}(b + 2q)(1 - 2p) + [b(a + b) + (a + q)(b + q)]\frac{p}{n}, \tag{2}$$

$$\mathbb{E}\left[S \mid \mathbf{x}\right] = s\left(1 - p + (1 - 3p)\frac{q}{n}\right), \tag{3}$$

$$\mathbb{E}\left[Q \mid \mathbf{x}\right] = pn + \frac{1 - 3p}{2n}\left[2q^2 + (n - q)^2 - s^2\right]. \tag{4}$$

2.1 Oblivious Noise and Stubborn Agents

We can now consider the following more general message-*oblivious* model of noise. We say that the communication is affected by *oblivious noise* if the value of any sent message changes according to the following scheme:

(i) with probability $1 - p_{noise}$ independent from the value of the sent message, the message remains unchanged;

(ii) otherwise, the noise acts on the message and it changes its value according to a fixed distribution $\mathbf{p} = p_1, \ldots, p_m$ over the possible message values $1, \ldots, m$.

In other words, according to the previous definition of noise (Definition 1), the probability that the noise changes any message to message i is $p_{noise} \cdot p_i$. It is immediate to verify that the definition of noise adopted in Theorems 1 and 2 corresponds to the aforementioned model of oblivious noise in the special case $m = 3$, $p_{noise} = p$, and $p_{Alpha} = p_{Beta} = p_{undecided} = \frac{1}{3}$.

Recalling that an agent is said to be *stubborn* if it never updates its state [37], we now observe that the above noise model is in fact equivalent to consider the behavior of the same dynamics *in a noiseless setting with stubborn agents*.

Lemma 1. *Consider the* UNDECIDED-STATE *dynamics on the complete graph in the binary setting. The following two processes are equivalent:*

(a) the UNDECIDED-STATE *process with n agents in the presence of oblivious noise with parameters p_{noise} and $\mathbf{p} = (p_{Alpha}, p_{Beta}, p_{undecided})$;*

(b) the UNDECIDED-STATE *process with n agents and $n_{stub} = \frac{p_{noise}}{1-p_{noise}} n$ additional stubborn agents present in the system, of which: $n_{stub} \cdot p_{Alpha}$ are stubborn agents supporting opinion* Alpha, *$n_{stub} \cdot p_{Beta}$ are stubborn agents supporting opinion* Beta, *and $n_{stub} \cdot p_{undecided}$ are stubborn agents supporting the undecided state.*

Proof of Lemma 1. The equivalence between the two processes is showed through a coupling. Consider the complete graph of n nodes, K_n, over which the former process runs. Consider also the complete graph $K_{n+n_{stub}}$, which contains a subgraph isomorphic to K_n we denote as $\tilde{K}_n$. Let $H = K_{n+n_{stub}} \setminus \tilde{K}_n$. The nodes of H are such that $n_{stub} \cdot p_{Alpha}$ are stubborn agents supporting opinion *Alpha*, $n_{stub} \cdot p_{Beta}$ are stubborn agents supporting opinion *Beta*, and $n_{stub} \cdot p_{undecided}$ are stubborn agents supporting the undecided state. Observe that $p_{Alpha} + p_{Beta} + p_{undecided} = 1$, so this partition of $K_{n+n_{stub}}$ is well defined.

The UNDECIDED-STATE dynamics behaves in exactly the same way over $K_{n+n_{stub}}$, with the exception that the stubborn agents never change their opinion and that there is no noise perturbing communications between agents. Let C and $\tilde{C}$ be the set of all possible configurations of, respectively, K_n and $K_{n+n_{stub}}$. Let $\phi : K_n \to \tilde{K}_n$ be any bijective function. The coupling is a bijection $f : C \to \tilde{C}$ such that, for any node $v \in K_n$ in the configuration $\mathbf{x} \in C$, the corresponding node $\phi(v) \in \tilde{K}_n$ in the configuration $f(c) \in \tilde{C}$ supports v's opinion. Consider the two resulting Markov processes $\{\mathbf{X}_t\}_{t \geq 0}$ over K_n and $\{\mathbf{X}'_t\}_{t \geq 0}$ over $K_{n+n_{stub}}$, denoting the opinion configuration at time t in K_n and in $K_{n+n_{stub}}$, respectively. It is easy to see that the two transition matrices are exactly the same, namely the probability to go from configuration $c \in C$ to configuration $c' \in C$ for $\mathbf{X}_t$ is the same as that to go from configuration $f(c) \in \tilde{C}$ to configuration $f(c') \in \tilde{C}$ for $\mathbf{X}'_t$.

Indeed, in the former model (a), the probability an agent pulls opinion $j \in$ *Alpha, Beta, undecided* at any given round is

$$(1 - p_{noise})\frac{c_j}{n} + p_{noise} \cdot p_j,$$

where c_j is the size of the community of agents supporting opinion j; in the model defined in (b), the probability a non-stubborn agent pulls opinion j at any given round is

$$\frac{c_j + n_{stub} \cdot p_j}{n + n_{stub}} = \frac{c_j + \frac{p_{noise}}{1 - p_{noise}} n \cdot p_j}{n + \frac{p_{noise}}{1 - p_{noise}} n} = (1 - p_{noise}) \cdot \frac{c_j}{n} + p_{noise} \cdot p_j.$$

□

Basically, this equivalence implies that any result we state for the process defined in (a) has an analogous statement for the process defined in (b).

Probabilistic Tools. Our analysis makes use of the following probabilistic results. The first one is the additive form of the well-known Chernoff bound (for an overview on the Chernoff bounds see [21] or [19]).

Lemma 2 (Additive forms of Chernoff bounds). *Let* $X_1, X_2, \ldots, X_n$ *be independent* $\{0,1\}$ *random variables. Let* $X = \sum_{i=1}^{n} X_i$ *and* $\mu = \mathbb{E}[X]$. *Then:*

(i) for any $0 < \lambda < n$ *and* $\mu \leq \mu_+ \leq n$, *it holds that*

$$P(X \geq \mu_+ + \lambda) \leq e^{-\frac{2}{n}\lambda^2}, \tag{5}$$

(ii) for any $0 < \lambda < \mu_-$ *and* $0 \leq \mu_- \leq \mu$, *it holds that*

$$P(X \leq \mu_- - \lambda) \leq e^{-\frac{2}{n}\lambda^2}. \tag{6}$$

The next standard result states that the intersection of some polynomial number of events, each of them holding w.h.p., is still an event which holds w.h.p.

Lemma 3 *Consider any family of events* $\{\xi_i\}_{i \in I}$ *with* $|I| \leq n^\lambda$, *for some* $\lambda > 0$. *Suppose that each event* ξ_i *holds with probability at least* $1 - n^\eta$, *with* $\eta > \lambda$. *Then, the intersection* $\cap_{i \in I} \xi_i$ *holds w.h.p.*

Proof of Lemma 3. By the union bound, $\Pr(\cap_{i \in I} \xi_i) = 1 - \Pr(\cup_{i \in I} \bar{\xi}_i) \geq 1 - \sum_{i \in I} n^{-\eta} = 1 - n^{\lambda - \eta} \geq 1 - n^{-\delta}$, where $\bar{\xi}_i$ denotes the negation of ξ_i and $\delta = \frac{\eta - \lambda}{2}$. □

Finally, we will use the well-known Berry-Eseen theorem. The Berry-Eseen theorem is well treated in [28], and it gives an estimation on "how far" is the distribution of the normalized sum of i.i.d. random variables from the standard normal distribution.

Lemma 4 (Berry-Eseen). *Let $X_1, \ldots, X_n$ be n i.i.d. (either discrete or continuous) random variables with zero mean, variance $\sigma^2 > 0$, and finite third moment. Let Z the standard normal random variable, with zero mean and variance equal to 1. Let $F_n(x)$ be the cumulative function of $\frac{S_n}{\sigma\sqrt{n}}$, where $S_n = \sum_{i=1}^n X_i$, and $\Phi(x)$ that of Z. Then, there exists a positive constant $C > 0$ such that, for each $n \geq 1$, $\sup_{x\in\mathbb{R}} |F_n(x) - \Phi(x)| \leq C/\sqrt{n}$.*

Remark. Due to lack of space, most of the technical proofs will be omitted, while they are available in the full version of the paper [17].

3   Process Analysis for Biased Initial Configurations

In this section, we analyze the UNDECIDED-STATE process when the system starts from biased configurations. The following two theorems show the phase transition exhibited by this process. We remind that our notion of noise is that of Definition 1.

Theorem 1 (Almost Majority Consensus). *Let $\mathbf{x}$ be any initial configuration having bias $s(\mathbf{x}) \geq \gamma\sqrt{n\log n}$ for some constant $\gamma > 0$, and let $\epsilon \in (0, 1/6)$ be some absolute constant. If $p = 1/6 - \epsilon$ is the noise probability, then the UNDECIDED-STATE process reaches a configuration $\mathbf{y}$ having bias $s(\mathbf{y}) \in \Delta = \left[\frac{2\sqrt{\epsilon}}{1+6\epsilon}n, \left(1 - 2\left(\frac{1-6\epsilon}{12}\right)^3\right)n\right]$ within $\mathcal{O}(\log n)$ rounds, w.h.p. Moreover, starting from $\mathbf{y}$, the UNDECIDED-STATE process enters a (metastable) phase of length $\Omega\left(n^\lambda\right)$ rounds (for some constant $\lambda > 0$)[5] where the bias remains in the range Δ, w.h.p.*

Observe that if the theorem is true, then it also holds analogously for the symmetrical case in which $s(\mathbf{x}) \leq -\gamma\sqrt{n\log n}$.

Theorem 2 (Victory of Noise). *Let $p = 1/6 + \epsilon$ be the noise probability for some absolute constant $\epsilon \in (0, 1/3]$. Assume the system starts from any configuration $\mathbf{x}$ with $|s(\mathbf{x})| \geq \gamma\sqrt{n\log n}$, for some constant $\gamma > 0$. Then, the UNDECIDED-STATE process reaches a configuration $\mathbf{y}$ having bias $|s(\mathbf{y})| = \mathcal{O}(\sqrt{n\log n})$ in $\mathcal{O}(\log n)$ rounds, w.h.p. Furthermore, starting from such a configuration, the UNDECIDED-STATE process enters a (metastable) phase of length $\Omega\left(n^{\lambda'}\right)$ rounds (for some constant $\lambda' > 0$) where the absolute value of the bias keeps bounded by $\mathcal{O}(\sqrt{n\log n})$, w.h.p.*

The next subsection is devoted to the proof of Theorem 1, while we refer to the full version for the proof of Theorem 2. We here just remark that the adopted arguments in the two proofs are similar.

Let us now consider the equivalent model with stubborn agents according to Lemma 1, in which $p_{noise} = 3p$ and $p_{Alpha} = p_{Beta} = p_{undecided} = \frac{1}{3}$. We

---

[5] The constant λ depends only on the values of ϵ and γ. The same holds for the constant λ' in Theorem 2.

thus have $n_{stub} = \frac{3p}{1-3p}n$ additional stubborn nodes, of which $n_{stub} \cdot \frac{1}{3} = \frac{p}{1-3p}n$ support opinion *Alpha*, $n_{stub} \cdot \frac{1}{3} = \frac{p}{1-3p}n$ opinion *Beta*, and $n_{stub} \cdot \frac{1}{3} = \frac{p}{1-3p}n$ are undecided. On this new graph of $n + n_{stub}$ nodes, let the UNDECIDED-STATE dynamics run and call the resulting process the STUB process. The next result is an immediate corollary of the two previous theorems.

Corollary 1. *Let $\frac{1}{2} > p > 0$ be a constant, and let the STUB process start from any configuration having bias $s \geq \gamma\sqrt{n\log n}$ for some constant $\gamma > 0$. If $p < \frac{1}{6}$, then, in $\mathcal{O}(\log n)$ rounds, the STUB process enters a metastable phase of almost consensus of length $\Omega\left(n^{\lambda}\right)$ for some constant $\lambda > 0$, in which the bias is $\Theta(n)$, w.h.p. If $p \in (\frac{1}{6}, \frac{1}{2}]$, then, in $\mathcal{O}(\log n)$ rounds, the STUB process enters a metastable phase of length $\Omega\left(n^{\lambda'}\right)$ for some constant $\lambda' > 0$ where the absolute value of the bias keeps bounded by $\mathcal{O}(\sqrt{n\log n})$, w.h.p.*

Proof of Theorem 1. Informally, while the analysis is technically involved, it can be appreciated from it that the phase transition phenomenon at hand relies ultimately on the exponential drift of the UNDECIDED-STATE towards the majority opinion in the absence of noise: as long as the noise is kept within a certain threshold, the dynamics manages to quickly amplify and sustain the bias towards the majority opinion; as soon as the noise level reaches the threshold, the expected increase of the majority bias abruptly decreases below the standard deviation of the process and the ability of the dynamics to preserves a *signal* towards the initial majority rapidly vanishes.

We now proceed with the formal analysis. Wlog, in the sequel, for a given starting configuration $\mathbf{x}$, we will assume $a(\mathbf{x}) \geq b(\mathbf{x})$. Indeed, as it will be clear from the results, if $s(\mathbf{x}) \geq \gamma\sqrt{n\log n}$, then the plurality opinion does not change for $\Omega(n^{\lambda})$ rounds, w.h.p., and the argument for the case $b(\mathbf{x}) > a(\mathbf{x})$ is symmetric. First notice that, for any fixed $\epsilon \in (0, 1/6)$ and $p = 1/6 - \epsilon$, Eqs. (3) and (4) become

$$\mathbb{E}\left[S \mid \mathbf{x}\right] = s\left(\frac{5}{6} + \epsilon + \frac{1}{2}(1 + 6\epsilon)\frac{q}{n}\right), \tag{7}$$

$$\mathbb{E}\left[Q \mid \mathbf{x}\right] = \frac{3}{4}\left(\frac{1+6\epsilon}{n}\right)q^2 - \frac{1+6\epsilon}{2}q + \frac{5+6\epsilon}{12}n - \frac{1+6\epsilon}{n}\left(\frac{s}{2}\right)^2. \tag{8}$$

The key-point to prove the first claim of the theorem is to show that, if the bias of the configuration is less than βn (for some suitable constant β), and the number of undecided nodes is some constant factor of n, then the bias at the next round increases by a constant factor, w.h.p. At the same time, as long as the bias is below βn, the number of undecided nodes in the next round is sufficiently large, w.h.p.

Lemma 5. *Let $\mathbf{x}$ be a configuration such that $q \geq \frac{1-4\epsilon}{3(1+6\epsilon)}n$ and $s \geq \gamma\sqrt{n\log n}$ for some constant $\gamma > 0$. Then, in the next round, $S \geq s\left(1 + \frac{\epsilon}{6}\right)$, w.h.p.*

Proof of Lemma 5. We first notice that Eq. (7) implies $\mathbb{E}\left[S \mid \mathbf{x}\right] \geq s\left(1 + \epsilon/3\right)$. Then, consider the events

$$E_1 = \left\{A \leq \mathbb{E}\left[A \mid \mathbf{x}\right] - \frac{\epsilon}{12}\gamma\sqrt{n\log n}\right\} \text{ and } E_2 = \left\{B \geq \mathbb{E}\left[B \mid \mathbf{x}\right] + \frac{\epsilon}{12}\gamma\sqrt{n\log n}\right\}$$

For the additive form of Chernoff bound (Lemma 2), it holds that

$$\mathbb{P}\left(E_1 \mid \mathbf{x}\right) \leq e^{-\frac{2n\log n}{144n}} = n^{-\frac{1}{77}} \text{ and } \mathbb{P}\left(E_2 \mid \mathbf{x}\right) \leq e^{-\frac{2n\log n}{144n}} = n^{-\frac{1}{77}}.$$

It follows that

$$
\begin{aligned}
\mathbb{P}&\left(S \geq s\left(1 + \tfrac{\epsilon}{6}\right) \mid \mathbf{x}\right) = \mathbb{P}\left(S \geq s\left(1 + \tfrac{\epsilon}{3}\right) - \tfrac{\epsilon}{6}s \mid \mathbf{x}\right) \\
&\geq \mathbb{P}\left(S \geq \mathbb{E}\left[S \mid \mathbf{x}\right] - \tfrac{\epsilon}{6}\gamma\sqrt{n\log n} \mid \mathbf{x}\right) \\
&= \mathbb{P}\left(A - B \geq \mathbb{E}\left[A - B \mid \mathbf{x}\right] - 2\tfrac{\epsilon}{12}\gamma\sqrt{n\log n} \mid \mathbf{x}\right) \geq \mathbb{P}\left(E_1^C \cap E_2^C \mid \mathbf{x}\right) \\
&= \mathbb{P}\left(E_1^C \mid \mathbf{x}\right) + \mathbb{P}\left(E_2^C \mid \mathbf{x}\right) - \mathbb{P}\left(E_1^C \cup E_2^C \mid \mathbf{x}\right) \geq 1 - 2n^{-\frac{1}{77}},
\end{aligned}
$$

where in the last inequality we bounded the probability of the union with 1. $\square$

We now fix $\beta = \frac{2\sqrt{3\epsilon}}{1+6\epsilon}$ and show the following bound.

Lemma 6. *Let* $\mathbf{x}$ *be a configuration such that* $s \leq \beta n$. *Then, in the next round,* $Q \geq \frac{1-4\epsilon}{3(1+6\epsilon)}n$, *w.h.p.*

Proof of Lemma 6. Since Eq. (8) has its minimum in $\bar{q} = \frac{n}{3}$,

$$
\begin{aligned}
\mathbb{E}\left[Q \mid \mathbf{x}\right] &\geq (1 + 6\epsilon)\frac{n}{12} - (1 + 6\epsilon)\frac{n}{6} + (5 + 6\epsilon)\frac{n}{12} - (1 + 6\epsilon)\left(\frac{\beta}{2}\right)^2 n \\
&= \frac{n}{12}\left(1 + 6\epsilon - 2 - 12\epsilon + 5 + 6\epsilon - \frac{36\epsilon}{1 + 6\epsilon}\right) = \frac{1 - 3\epsilon}{3(1 + 6\epsilon)}n.
\end{aligned}
$$

Hence, we can apply the additive form of Chernoff bound (Lemma 2), and get $Q \geq \frac{1-4\epsilon}{3(1+6\epsilon)}n$, w.h.p. (actually, with probability $1 - \exp(\Theta(n))$). Formally,

$$
\begin{aligned}
\mathbb{P}&\left(Q \leq \frac{1-4\epsilon}{3(1+6\epsilon)}n \mid \mathbf{x}\right) = \mathbb{P}\left(Q \leq \frac{1-3\epsilon}{3(1+6\epsilon)}n - \frac{\epsilon}{3(1+6\epsilon)}n \mid \mathbf{x}\right) \\
&\leq \mathbb{P}\left(Q \leq \mathbb{E}\left[Q \mid \mathbf{x}\right] - \frac{\epsilon}{3(1+6\epsilon)}n \mid \mathbf{x}\right) \leq e^{-\frac{2}{n}\frac{\epsilon^2}{9(1+6\epsilon)^2}n^2} = e^{-\frac{2\epsilon^2}{9(1+6\epsilon)^2}n}. \quad \square
\end{aligned}
$$

The two lemmas above ensure that the system eventually reaches a configuration $\mathbf{y}$ with bias $s(\mathbf{y}) > \beta n$ within $\mathcal{O}(\log n)$ rounds, w.h.p. (see the proof of Theorem 1). We now consider configurations in which $s > \beta n$ and derive a useful bound on the possible decrease of s.

Lemma 7. *Let* $\mathbf{x}$ *be any configuration such that* $s \geq \gamma\sqrt{n\log n}$ *for some constant* $\gamma > 0$. *Then, in the next round, it holds that* $S \geq s\left(\frac{5}{6} + \frac{\epsilon}{2}\right)$ *w.h.p.*

Proof of Lemma 7. Observe that Eq. (7) implies $\mathbb{E}\left[S \mid \mathbf{x}\right] \geq s\left(5/6 + \epsilon\right)$. By the additive form of Chernoff bound and the union bound (as we did in the proof of Lemma 5), we get $S \geq s\left(\frac{5}{6} + \frac{\epsilon}{2}\right)$, w.h.p. $\qquad\square$

Lemma 7 is used to show the metastable phase of almost consensus, which lasts for a polynomial number of rounds and in which the bias keeps lower bounded by $\frac{2\sqrt{\epsilon}}{1+6\epsilon}n$ (see the proof of Theorem 1). The next two lemmas provide an upper bound on the bias during this phase.

Lemma 8. *Let* $\mathbf{x}$ *be any configuration. Then, in the next round,* $Q \geq \frac{n}{12}(1-6\epsilon)$, *w.h.p.*

Proof of Lemma 8. From Eq. (8)

$$
\begin{aligned}
\mathbb{E}\left[Q \mid \mathbf{x}\right] &\geq \frac{3}{4}\left(\frac{1+6\epsilon}{n}\right)q^2 - \frac{1+6\epsilon}{2}q + \frac{5+6\epsilon}{12}n - \frac{1+6\epsilon}{n}\left(\frac{n-q}{2}\right)^2 \\
&\geq \frac{1}{2}\left(\frac{1+6\epsilon}{n}\right)q^2 + \frac{1-6\epsilon}{6}n \geq \frac{1-6\epsilon}{6}n,
\end{aligned}
$$

where we used $s \leq n - q$. For the additive form of Chernoff bound (Lemma 2), we get $Q \geq \frac{1-6\epsilon}{12}n$, w.h.p. $\qquad\square$

Lemma 9. *Let* $\mathbf{x}$ *be a configuration with* $q \geq \frac{n}{12}(1 - 6\epsilon)$. *Then, in the next round,* $B \geq \left(\frac{1-6\epsilon}{12}\right)^3 n$, *w.h.p.*

Proof of Lemma 9. From the last term of Eq. (2), we have

$$
\mathbb{E}\left[B \mid \mathbf{x}\right] \geq \frac{1}{6}\left(\frac{1-6\epsilon}{n}\right)(q^2) \geq \frac{n}{6 \cdot 12^2}(1-6\epsilon)^3.
$$

The additive form of Chernoff bound (Lemma 2) implies that $B \geq \left(\frac{1-6\epsilon}{12}\right)^3 n$, w.h.p. $\qquad\square$

Proof of Theorem 1. Let $\mathbf{x}$ be the initial configuration. We now prove that the bias keeps upper bounded by the value $\left(1 - 2[(1 - 6\epsilon)/12]^3\right)n$. Indeed, Lemma 8 ensures that the number of undecided nodes keeps at least $\frac{n}{12}(1 - 6\epsilon)$, w.h.p. Thus, applying Lemmas 3 and 9, we get that $b(\mathbf{X}_t) \geq [(1-6\epsilon)/12]^3 n$, w.h.p., for a polynomial number of rounds .

As for the lower bound of the bias, we distinguish two initial cases.

Case $s(\mathbf{x}) \geq \beta n$. From Lemma 7, we know that as long as the bias is of magnitude $\Omega(\sqrt{n \log n})$, then it cannot decrease too fast w.h.p., namely $s(\mathbf{X}_{t+1}) \geq s(\mathbf{X}_t)(5/6 + \epsilon/2)$, w.h.p. Notice that

$$
\left(\frac{5}{6} + \frac{\epsilon}{2}\right)^2 \cdot \beta n \geq \frac{2\sqrt{\epsilon}}{1 + 6\epsilon}n,
$$

which means that, if at some round t the bias goes below the value βn, then it remains at least $\frac{2\sqrt{\epsilon}}{1+6\epsilon}n$ and it will not decrease below that value for at least another round, w.h.p. Then, by Lemma 6 we know that at round $t+1$ the number of undecided nodes is at least $\frac{1-4\epsilon}{3(1+6\epsilon)}n$, w.h.p., which means that the bias starts increasing again each round due to Lemma 5, w.h.p., as long as it is still below βn. Indeed, the number of undecided nodes keeps greater than $\frac{1-4\epsilon}{3(1+6\epsilon)}n$ as long as the bias is below βn, w.h.p. (Lemma 6). This phase, in which the bias keeps greater than $\frac{2\sqrt{\epsilon}}{1+6\epsilon}n$, lasts for a polynomial number of rounds, w.h.p. (see Lemma 3);

Case $\gamma\sqrt{n\log n} \le s(\mathbf{x}) < \beta n$. Thanks to Lemma 7, in the next round, the bias is greater than $\gamma'\sqrt{n\log n}$, w.h.p., while the number of undecided nodes gets greater than $\frac{1-4\epsilon}{3(1+6\epsilon)}n$, w.h.p. (Lemma 6). Then, Lemmas 5 and 6 guarantee that, within the next $\mathcal{O}(\log n)$ rounds, the bias reaches the value βn, w.h.p. (Lemma 3), and so the process turns to be in the first Case.

We finally remark that our analysis above shows that the polynomial length of the metastable phase, i.e. n^λ, has the exponent λ that (only) depends on the (constant) parameters γ and ϵ of the considered process. □

4   Symmetry Breaking from Balanced Configurations

In this section, we consider the UNDECIDED-STATE process starting from arbitrary initial configurations: in particular, from configurations having no bias. Interestingly enough, we show a transition phase similar to that proved in the previous section. Informally, the next theorem states that when $p < 1/6$, the UNDECIDED-STATE process is able to break the symmetry of any perfectly-balanced initial configuration and to compute almost consensus within $O(\log n)$ rounds, w.h.p.

Theorem 3. *Let $\mathbf{x}$ be any initial configuration, and let $\epsilon \in (0, 1/6)$ be some absolute constant. If $p = 1/6 - \epsilon$ is the noise probability, then the UNDECIDED-STATE process reaches a configuration $\mathbf{y}$ having bias s toward some opinion $j \in \{\text{Alpha}, \text{Beta}\}$ such that $|s(\mathbf{y})| \in \Delta = \left[\frac{2\sqrt{\epsilon}}{1+6\epsilon}n, \left(1 - 2\left(\frac{1-6\epsilon}{12}\right)^3\right)n\right]$ within $\mathcal{O}(\log n)$ rounds, w.h.p. Moreover, once reached configuration $\mathbf{y}$, the UNDECIDED-STATE process enters a (metastable) phase of length $\Omega(n^\lambda)$ rounds (for some constant $\lambda > 0$) where the majority opinion is j and the bias keeps within the range Δ, w.h.p.*

Outline of Proof of Theorem. 3 If the initial configuration $\mathbf{x}$ has bias $s = \Omega(\sqrt{n\log n})$ then the claim of the theorem is equivalent to that of Theorem 1, so we are done. Hence, we next assume the initial bias s be $o(\sqrt{n\log n})$: for this case, our proof proceeds along the following main steps.

Step I. Whenever the bias s is small, i.e. $o(n)$, we prove that, within the next $\mathcal{O}(\log n)$ rounds, the number of undecided nodes turns out to keep always in a suitable linear range: roughly speaking, we get that this number lies in $(n/3, n/2]$, w.h.p.

Step II. Whenever s is very small, i.e. $s = o(\sqrt{n})$, there is no effective drift towards any opinion. However, we can prove that, thanks to Step I, the random variable S, representing the bias in the next round, has *high variance*, i.e. $\Theta(n)$. The latter holds since S can be written as a suitable sum whose addends include some random variables having binomial distribution of expectation 0: so, we can apply the Berry-Essen Theorem (Lemma 4 in the preliminaries) to get a lower bound on the variance of S. Then, thanks to this large variance, standard arguments for the standard deviation imply that, in this parameter range, there is a positive constant probability that S will get some value of magnitude $\Omega(\sqrt{n})$. Not surprisingly, in this phase, we find out that the variance of S is not decreased by the communication noise. We can thus claim that the process, at every round, has positive constant probability to reach a configuration having bias $s = \omega(\sqrt{n})$ and $q \in (n/3, n/2]$. Then, after $O(\log n)$ rounds, this event will happen w.h.p.

Step III. Once the process reaches a configuration with $s = \omega(\sqrt{n})$ and $q \in (n/3, n/2]$, we then prove that the expected bias increases by a constant factor (which depends on ϵ). Observe that we cannot use here the same round-by-round concentration argument that works for bias over $\sqrt{n \log n}$ (this is in fact the minimal magnitude required to apply the Chernoff's bounds [21]). We instead exploit a useful general tool [10], which bounds the stopping time of some class of Markov chains having rather mild conditions on the drift towards their absorbing states. This tool in fact allows us to consider the two phases described, respectively, in Step II and Step III as a unique *symmetry-breaking* phase of the process. Our final technical contribution here is to show that the conditions required to apply this tool hold whenever the communication noise parameter is such that $p \in (0, 1/6)$. This allows us to prove that, within $O(\log n)$ rounds, the process reaches a configuration with bias $s = \Omega(\sqrt{n} \log n)$, w.h.p. □

Large Communication Noise (The Case $p > 1/6 + \epsilon$). When $p > 1/6 + \epsilon$, Theorem 2 a fortiori holds when the initial bias is small, i.e. $s = o(\sqrt{n \log n})$: thus, we get that, in this case, the system enters into a long regime of non consensus, starting from any initial configuration. Then, by combining the results for biased configurations in Sect. 3 with those in this section, we can observe the phase transition of the UNDECIDED-STATE process starting from any possible initial configuration.

Theorem 4. *Let* $\mathbf{x}$ *be any initial configuration, and let* $\epsilon \in (0, 1/3]$ *be some absolute constant. If* $p = 1/6 + \epsilon$ *is the noise probability, then the* UNDECIDED-STATE *process reaches a configuration* $\mathbf{y}$ *having bias* $|s(\mathbf{y})| = \mathcal{O}(\sqrt{n \log n})$ *within* $\mathcal{O}(\log n)$ *rounds, w.h.p. Furthermore, starting from such a configuration, the* UNDECIDED-STATE *process enters a (metastable) phase of length* $\Omega\left(n^{\lambda'}\right)$ *rounds (for some constant* $\lambda' > 0$*) where the absolute value of the bias keeps bounded by* $\mathcal{O}(\sqrt{n \log n})$*, w.h.p.pg*

Stubborn Agents. We conclude this section by observing that the equivalence result shown in Lemma 1 holds independently of the choices of the noise parameter $p \in (0, 1/2]$, and of the initial bias: the phase transition of the UNDECIDED-STATE process in the presence of stubborn agents thus holds even in the case of unbiased configurations.

Corollary 2. *Let $\frac{1}{2} > p > 0$ be a constant, and let the STUB process start from any initial configuration. If $p < \frac{1}{6}$, then, in $\mathcal{O}(\log n)$ rounds, the STUB process enters a metastable phase of almost consensus towards some opinion $j \in \{\text{Alpha}, \text{Beta}\}$ of length $\Omega\left(n^{\lambda}\right)$ for some constant $\lambda > 0$, in which the absolute value of the bias is $\Theta(n)$, w.h.p. If $p \in (\frac{1}{6}, \frac{1}{2}]$, then, in $\mathcal{O}(\log n)$ rounds, the STUB process enters a metastable phase of length $\Omega\left(n^{\lambda'}\right)$ for some constant $\lambda' > 0$ where the absolute value of the bias keeps bounded by $\mathcal{O}(\sqrt{n \log n})$, w.h.p.*

5   Simulations

We made computer simulations with values of the input size n ranging from 2^{10} to 2^{17}, and for noise probabilities of $p = 1/12$, $p = 1/8$, $p = 1/7$, and $p = 1/5$. Besides confirming the phase transition predicted by our theoretical analysis, the outcomes show this behaviour emerges even for reasonable sizes (i.e. n) of the system. Indeed, we made the UNDECIDED-STATE dynamics run for 400 rounds for the above values of p. In the first three settings of p, we started from complete balanced configurations (i.e. when both opinions are supported by, respectively, $\frac{n}{2}$ agents) we found a fast convergence to the meta-stable regime of almost consensus, which then did not break for all the rest of the simulation. Furthermore, we have noticed that the symmetry is always broken when the bias is "roughly" $10\sqrt{n \log n}$. As for the case $p = 1/5$, we started from a configuration of complete consensus and we observed that, within a short time, the system looses any information on the majority opinion (say, the bias becomes less than $10\sqrt{n \log n}$)

Table 2. The left-hand table shows the average time to reach a meta-stable almost-consensus phase, while the right-hand table shows the average time the bias goes below $10\sqrt{n \log n}$, and the number of switches.

Size n	Average times			Size n	$p = 1/5$	
	$p = 1/12$	$p = 1/8$	$p = 1/7$		Average time	Number of switches
2^{10}	24	Failed	Failed	2^{10}	1	39
2^{11}	24	39	Failed	2^{11}	4	42
2^{12}	28	41	Failed	2^{12}	7	42
2^{13}	27	53	Failed	2^{13}	10	37
2^{14}	32	52	77	2^{14}	14	38
2^{15}	32	54	88	2^{15}	18	38
2^{16}	36	57	96	2^{16}	22	44
2^{17}	39	68	103	2^{17}	27	39

and it keeps this meta-stable phase with many switches of the majority opinion. In the left-hand part of Table 2, we can see the average time (computed over 100 trials and approximated to the closest integer) in which the system enters the predicted meta-stable phase of almost consensus for any value of $p = 1/12$, $p = 1/8$, and $p = 1/7$, for different input sizes. We also see that, when p gets close to $1/6$, the emergent behaviour is observed only for large values of n and some of the experiments fail. In the right-hand part of Table 2, we see the average times in which the bias of the system goes below $10\sqrt{n \log n}$ for different input sizes, and the corresponding number of switches of majority opinion during the remaining time.

6   Conclusions

While our mathematical analysis for the UNDECIDED-STATE dynamics does not directly apply to other opinion dynamics, it suggests that a general phase-transition phenomenon may hold for a large class of dynamics characterized by an *exponential drift* towards consensus configurations. Our work thus naturally poses the general question of whether it is possible to provide a characterization of opinion dynamics with stochastic interactions, in terms of their critical behavior with respect to uniform communication noise.

As for the specific mathematical questions that follow from our results, our assumption of a complete topology as underlying graph is, for several real MAS, a rather strong condition. However, two remarks on this issue follow. On one hand, we observe that, according to the adopted communication model, at every round, every agent can pull information from just one other agent: the *dynamic* communication pattern is thus random and sparse. This setting may model opportunistic MAS where mobile agents use to meet randomly, at a relatively-high rate. On the other hand, we believe that a similar transition phase does hold even for sparse topologies having good expansion/conductance [27]: this is an interesting question left open by this work.

References

1. Acemoglu, D., Como, G., Fagnani, F., Ozdaglar, A.E.: Opinion fluctuations and disagreement in social networks. Math. Oper. Res. **38**(1), 1–27 (2013)
2. Angluin, D., Aspnes, J., Eisenstat, D.: A simple population protocol for fast robust approximate majority. Distrib. Comput. **21**(2), 87–102 (2008)
3. Axelrod, R.: The dissemination of culture: a model with local convergence and global polarization. J. Conflict Resolut. **41**(2), 203–226 (1997)
4. Becchetti, L., Clementi, A.E.F., Natale, E.: Consensus dynamics: an overview. SIGACT News **51**(1), 58–104 (2020). https://doi.org/10.1145/3388392.3388403
5. Benezit, F., Thiran, P., Vetterli, M.: Interval consensus: from quantized gossip to voting. In: ICASSP 2009, pp. 3661–3664 (2009)
6. Boczkowski, L., Korman, A., Natale, E.: Minimizing message size in stochastic communication patterns: fast self-stabilizing protocols with 3 bits. Distrib. Comput. **32**, 173–191 (2018)

7. Boczkowski, L., Natale, E., Feinerman, O., Korman, A.: Limits on reliable information flows through stochastic populations. PLoS Comput. Biol. **14**(6), e1006195 (2018)
8. Cardelli, L., Csikász-Nagy, A.: The cell cycle switch computes approximate majority. Sci. Rep. **2**, 656 (2012)
9. Clementi, A., Gualà, L., Natale, E., Pasquale, F., Scornavacca, G., Trevisan, L.: Consensus needs broadcast in noiseless models but can be exponentially easier in the presence of noise. Report, CNRS (2018)
10. Clementi, A.E.F., Ghaffari, M., Gualà, L., Natale, E., Pasquale, F., Scornavacca, G.: A tight analysis of the parallel undecided-state dynamics with two colors. In: MFCS 2018, pp. 28:1–28:15 (2018)
11. Coates, A., Han, L., Kleerekoper, A.: A unified framework for opinion dynamics. In: AAMAS 2018, pp. 1079–1086 (2018)
12. Condon, A., Hajiaghayi, M., Kirkpatrick, D., Maňuch, J.: Approximate majority analyses using tri-molecular chemical reaction networks. Nat. Comput. **19**(1), 249–270 (2019). https://doi.org/10.1007/s11047-019-09756-4
13. Cooper, C., Radzik, T., Rivera, N., Shiraga, T.: Fast plurality consensus in regular expanders. In: DISC 2017. LIPIcs, vol. 91, pp. 13:1–13:16 (2017)
14. Cover, T.M., Thomas, J.A.: Elements of Information Theory, 2nd edn. Wiley, Hoboken (2006)
15. Cruciani, E., Natale, E., Nusser, A., Scornavacca, G.: Phase transition of the 2-choices dynamics on core-periphery networks. In: AAMAS 2018, pp. 777–785 (2018)
16. Cruciani, E., Natale, E., Scornavacca, G.: Distributed community detection via metastability of the 2-choices dynamics. In: AAAI 2019, Honolulu, Hawaii, United States, January 2019
17. d'Amore, F., Clementi, A., Natale, E.: Phase transition of a non-linear opinion dynamics with noisy interactions. Technical report (2020). https://hal.archives-ouvertes.fr/hal-02487650
18. Deffuant, G., Neau, D., Amblard, F., Weisbuch, G.: Mixing beliefs among interacting agents. Adv. Complex Syst. **03**(01n04), 87–98 (2000)
19. Doerr, B.: Probabilistic tools for the analysis of randomized optimization heuristics. CoRR abs/1801.06733 (2018)
20. Draief, M., Vojnovic, M.: Convergence speed of binary interval consensus. SIAM J. Control Optim. **50**(3), 1087–1109 (2012)
21. Dubhashi, D.P., Panconesi, A.: Concentration of Measure for the Analysis of Randomized Algorithms. Cambridge University Press, Cambridge (2009)
22. Natale, E.: On the computational power of simple dynamics. Ph.D. thesis, Sapienza University of Rome (2017)
23. Feinerman, O., Haeupler, B., Korman, A.: Breathe before speaking: efficient information dissemination despite noisy, limited and anonymous communication. Distrib. Comput. **30**(5), 339–355 (2015). https://doi.org/10.1007/s00446-015-0249-4
24. Fraigniaud, P., Natale, E.: Noisy rumor spreading and plurality consensus. Distrib. Comput. **32**, 257–276 (2018). https://doi.org/10.1007/s00446-018-0335-5
25. Ghaffari, M., Lengler, J.: Nearly-tight analysis for 2-choice and 3-majority consensus dynamics. In: PODC 2018, pp. 305–313 (2018)
26. Hassin, Y., Peleg, D.: Distributed probabilistic polling and applications to proportionate agreement. Inf. Comput. **171**(2), 248–268 (2001)
27. Hoory, S., Linial, N., Wigderson, W.: Expander graphs and their applications. Bull. Amer. Math. Soc. (N.S) **43**, 439–561 (2006)
28. Korolev, V., Shevtsova, I.: On the upper bound for the absolute constant in the Berry–Esseen inequality. Theory Probab. Appl. **54**, 638–658 (2010)

29. Lin, W., Zhixin, L., Lei, G.: Robust consensus of multi-agent systems with noise. In: 2007 Chinese Control Conference, pp. 737–741, July 2007
30. Mertzios, G.B., Nikoletseas, S.E., Raptopoulos, C.L., Spirakis, P.G.: Determining majority in networks with local interactions and very small local memory. Distrib. Comput. **30**(1), 1–16 (2016). https://doi.org/10.1007/s00446-016-0277-8
31. Mobilia, M., Petersen, A., Redner, S.: On the role of zealotry in the voter model. J. Stat. Mech: Theory Exp. **2007**(08), P08029 (2007)
32. Mobilia, M.: Does a single zealot affect an infinite group of voters? Phys. Rev. Lett. **91**(2), 028701 (2003)
33. Mossel, E., Tamuz, O.: Opinion exchange dynamics. Probab. Surv. **14**, 155–204 (2017)
34. Perron, E., Vasudevan, D., Vojnović, M.: Using three states for binary consensus on complete graphs. In: IEEE INFOCOM 2009, pp. 2527–2535, April 2009
35. Shimizu, N., Shiraga, T.: Phase transitions of best-of-two and best-of-three on stochastic block models. In: DISC 2019, July 2019
36. Weisbuch, G., Deffuant, G., Amblard, F., Nadal, J.P.: Meet, discuss, and segregate! Complexity **7**(3), 55–63 (2002)
37. Yildiz, E., Ozdaglar, A., Acemoglu, D., Saberi, A., Scaglione, A.: Binary opinion dynamics with stubborn agents. ACM Trans. Econ. Comput. **1**(4), 19:1–19:30 (2013)

Communication Complexity

Distributed Testing of Distance-k Colorings

Pierre Fraigniaud[1], Magnús M. Halldórsson[2], and Alexandre Nolin[2(✉)]

[1] IRIF, CNRS and Université de Paris, Paris, France
pierre.fraigniaud@irif.fr
[2] ICE-TCS, Department of Computer Science, Reykjavik University,
Reykjavik, Iceland
{mmh,alexandren}@ru.is

Abstract. We study the distributed decision problem related to checking distance-k coloring, defined as color assignments to the nodes such that every pair of vertices at distance at most k must receive distinct colors. While checking the validity of a distance-k coloring only requires $\lceil k/2 \rceil$ rounds in the LOCAL model, and a single round in the CONGEST model when $k \leq 2$, the task is extremely costly for higher k's in CONGEST—there is a lower bound of $\Omega(\Delta^{k/2})$ rounds in graphs with maximum degree Δ. We therefore explore the ability of checking distance-k coloring via distributed property testing. We consider several farness criteria for measuring the distance to a valid coloring, and we derive upper and lower bounds for each of them. In particular, we show that for one natural farness measure, significantly better algorithms are possible for testing distance-3 coloring than for testing distance-k coloring for $k \geq 4$.

Keywords: Distributed property testing · Graph coloring · Distributed decision

1 Introduction

We study problems related to checking whether a given distance-k coloring is proper, in the distributed CONGEST model. A valid (or proper) distance-k coloring of a graph $G = (V, E)$, for $k \geq 1$, is a coloring of each node v with integer c_v so that any two nodes u, v of distance at most k are colored differently, i.e., $c_u \neq c_v$. This is equivalent to the usual vertex coloring of the graph $G^k = (V, E^k)$, where two nodes are adjacent if they are within distance k in G.

Classical distance-1 colorings have been extensively studied in distributed computing as a tool of breaking symmetry. Let us denote by n the number of nodes, by m the number of edges, and by Δ the maximum degree of G. For the core problem of finding a $(\Delta + 1)$-coloring, there is a simple folklore

Pierre Fraigniaud is partially supported by ANR Projects DESCARTES, QuDATA, and FREDDA; Magnús M. Halldórsson and Alexandre Nolin are partially supported by Icelandic Research Foundation grant 174484-051.

A. W. Richa and C. Scheideler (Eds.): SIROCCO 2020, LNCS 12156, pp. 275–290, 2020.
https://doi.org/10.1007/978-3-030-54921-3_16

$O(\log n)$-round randomized algorithm, and recent polylog(n)-round deterministic algorithm by Bamberger, Kuhn and Maus [4] that works in CONGEST (leveraging the recent breakthrough of Rozhoň and Ghaffari [35]). The corresponding distance-2 coloring questions has recently been addressed in [25], where an $O(\log \Delta \log n)$-round randomized algorithm is given that uses $\Delta^2 + 1$ colors, as well as a polylog(n)-round deterministic algorithm that uses $(1 + \epsilon)\Delta^2$ colors, for any $\epsilon > 0$. This opens the question about distance-k coloring problems, for $k \geq 3$, which appear considerably harder.

Why Distance-k Coloring? Distributed distance-k colorings are interesting for various reasons. They appear naturally when constant-round randomized algorithms are derandomized using the method of conditional expectation [22]. They also appear in certain models of wireless models, where senders must be sufficiently separated, to limit interference. More abstractly, we can view distance-k coloring problems as a way of studying *communication capacity constraints* on nodes, where communication must go through intermediate relays. Given that distance-2 colorings can be efficiently computed, distance-3 colorings appear to lie at the frontier of what can be solved efficiently by distributed algorithms.

Deciding Distance-k Coloring. Given the apparently challenging task of *finding* an efficient distance-3 coloring, a natural question that arises is if we can at least *check* that a given coloring is valid. We can quickly dispose of that hope, as there is an easy reduction to Set Disjointness that shows that verifying a distance-k coloring requires $\Omega(\Delta^{\lfloor (k-1)/2 \rfloor})$ rounds in CONGEST. We provide a proof of this fact, for completeness, in Appendix A. Observe that the question is trivially answered in $\lceil k/2 \rceil$ rounds of the LOCAL model.

Testing Distance-k Coloring. Distributed property testing is a relaxation of distributed decision, where we seek a CONGEST algorithm that can distinguish whether the given graph satisfies a given property (e.g., having a distance-k coloring), or is *far* from having such a property. The most common notion for this is ϵ-*farness* in the *sparse* model, when the addition or deletion of up to $\epsilon \cdot m$ arbitrary edges to/from the graph $G = (V, E)$ does not result in the property being satisfied. This notion is renamed ϵ-*edge* in this paper, so as to avoid confusion as we use alternative notions of being far to a valid coloring. Distributed algorithms testing a property (here distance-k coloring) are compared according to the *error rate* $\epsilon(r)$ they can tolerate if restricted to r rounds, or equivalently, the *round complexity* $r(\epsilon)$ to distinguish between legal instances and instances ϵ-far from being legal.

1.1   Summary of Main Results

We consider several measures of distance from a valid coloring to define various notion of ϵ-farness, deduce their relationship, and bound the efficiency of testing distance-k colorings in CONGEST under these measures. As examples of such measures, we consider ϵ-*edge*, where deleting up to ϵm edges cannot result in a valid distance-k coloring, and ϵ-*middle*, where there exist more than ϵm paths of length at most $k - 2$ between two nodes with distinct neighbors of the same color. We present the following results:

1. An algorithm for any constant $k \geq 3$, with round complexity $O(1/\epsilon)$, for all our measures but ϵ-middle. We provide a matching lower bound for any algorithm under two of our considered measures.
2. An improved algorithm for distance-3 colorings under the ϵ-middle measure. The round complexity is $O(\epsilon^{-3/2}m^{-1/2})$, for $\epsilon \geq m^{-1/3}$. We prove a matching lower bound, and as well as an $\widetilde{\Omega}(\epsilon^{-1})$ lower bound for distance-4. This shows that distance-4 is strictly harder than distance-3.
3. A communication complexity lower bound of $\widetilde{\Omega}(\epsilon^{-1}(\epsilon m)^{-1})$, for any $k \geq 3$, under the ϵ-edge measure.

The results suggest that distance-3 colorings are easier to test than for larger distances. This reinforces the role of distance-3 coloring on the frontier of what is computable efficiently in CONGEST.

1.2   Related Work

Property testing has an extensive history in the sequential setting [23]. Distributed property testing was recently introduced by Brakerski and Patt-Shamir [6], and subsequently revisited and formalized more broadly by Censor-Hillel et al. [8]. As in the centralized setting, different variants of farness can be considered, but most of the efforts on distributed property testing has been carried out in the *sparse* model, that is, the model of this paper, where farness is measured by the fraction of the *number of edges* that must be added or removed for satisfying the property under consideration. In this framework, most previous work has been dedicated to checking the absence of a specific graph pattern (e.g., a cycle C_k, or a clique K_k, for some $k \geq 3$) as a subgraph of the actual network [12,18,20]. To our knowledge, this paper is the first to consider distributed testing proper distance-k coloring.

More generally, distributed property testing falls into the wide class of *distributed decision* problems, initially motivated by *fault-tolerant* distributed computing [2,3,28]. Since these early works, there has been a large body of work on distributed decision, with a range of models—see [13] for a survey. The closest to ours are *local decision* [15,17], and *local verification* [17,19,24,31]. In both cases, the nodes perform a constant number of rounds of communication before reaching a decision. Distributed property testing is a relaxed version of randomized distributed local decision, as nodes are not bounded to detect illegal instances that are "close to be legal". In distributed verification, every node is also supplied with a *certificate* string, and the collection of certificates is supposed to form a distributed *proof* that the instance is legal. Distributed property testing performs in absence of such certificates. Recently, distributed verification has been extended to distributed *interactive proofs* [30,33], involving interactions between the nodes and a powerful centralized oracle. Such mechanisms are obviously much more powerful than distributed property testing.

Overall, distributed property testing offers a tradeoff between simplicity (no need of certificates, nor of any interactions with an external entity), and efficiency (configurations that are "slightly" illegal may not be detected). It is thus an appealing lightweight alternative to complex mechanisms for distributed systems that can tolerate to be slightly faulty. This is typically the case of wireless systems, which are able to tolerate a certain level of interference, as long as these interferences do not exceed a certain threshold.

2   Model and Definitions

The input of our algorithms is a graph G, and a proposed coloring $C = (c_v)_{v \in V}$. Given an underlying distance metrics between solutions, we say that a solution is ϵ-far from being correct (or valid, or legal) if it is of distance at least ϵ from any valid solution for G. We seek a CONGEST protocol running on G to distinguish valid solutions from ϵ-far solutions. The protocol should have 1-sided error:

If C is valid, then, with probability 1, all nodes output "yes".
If C is ϵ-far from being valid, then, with probability at least $2/3$, some node outputs "no".

We explore different types of solution distances. In particular, we can divide them into two types: distance to a graph for which the given solution is valid, and distance to a valid solution for the given graph. We call two distinct nodes with the same color at distance at most k a *bad pair*, and call a path connecting a bad pair a *bad path*.

Definition 1. *An n-node m-edge graph $G = (V, E)$ and a coloring of its vertices $(c_v)_{v \in V}$ are said to be*

- *ϵ-edge, when deleting up to ϵm arbitrary edges does not result in a valid distance-k coloring.*
- *ϵ-disjoint, when there exist more than ϵm distinct pairs of similarly colored vertices linked by edge-disjoint paths of length at most k.*
- *ϵ-middle, when there exist more than ϵm paths of length at most $k-2$ between two nodes with distinct neighbors of the same color.*
- *ϵ-node, when recoloring up to ϵn arbitrary vertices does not result in a valid distance-k coloring.*
- *ϵ-conflict, when more than ϵn vertices have the same color as one of their distance-k neighbors.*

The ϵ-*edge* measure is the classical one from property testing literature [6,8,18]. The ϵ-*disjoint* measure is a variation that requires there to be many conflict pairs, not just one vertex that conflicts with many nodes (that might not conflict between themselves).

The ϵ-*middle* measure has the appearance of being contrived, but actually captures the essence of the problem. In the first round, each node learns of the colors of *all* its neighbors. Thus, what we really need is to somehow connect the

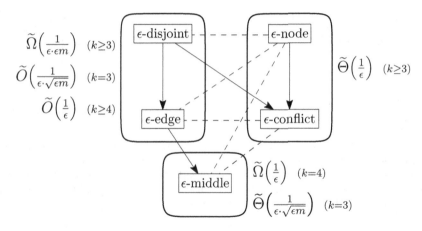

Fig. 1. The relationships between our notions of distance from a valid solution as well as our upper and lower bounds on the costs of testing for them. For two notions of farness notion$_1$ and notion$_2$, an arrow from ϵ-notion$_1$ to ϵ-notion$_2$ indicates that a solution that is ϵ-notion$_1$ is also $\Omega(\epsilon)$-notion$_2$ away from a valid solution. Dashed lines indicate incomparability.

second node on a bad k-path to the second-to-last node, and see if the sets of colors in their neighborhoods intersect.

The last two definitions correspond to natural measures of invalidity of colorings. The ϵ-*conflict* measure counts how many nodes are improperly colored (i.e., have a same-colored distance-$\leq k$ neighbor), while ϵ-*node* is more conservative, bounding the number of recolorings needed to turn the coloring into a valid one.

We say that a measure μ is more *strict* than measure μ' if $\mu(G,c) = O(\mu'(G,c))$, for all graphs G and colorings c. Thus, if (G,c) is ϵ-far in terms of measure μ', then it is $O(\epsilon)$-far in terms of measure μ (but could be much less far).

It is easy to see that ϵ-disjoint is more strict than ϵ-edge, and ϵ-node is more strict than ϵ-conflict. It also holds that ϵ-disjoint is more strict than ϵ-conflict on sparse graphs, when $|E(G)| = O(|V(G)|)$. This is illustrated in Fig. 1, where solid arrows are drawn from a stricter measures to a less stricter one.

We can also verify that other pairs of measures can be arbitrarily divergent. The examples in Fig. 2 show that for any pair of measures connected by a dotted line in Fig. 1, there is a graph where one is constant and the other is $O(1/n)$ (or $O(1/m)$), and vice versa. The same holds for the inverse direction of the solid edges.

The property of ϵ-edge and ϵ-disjoint assignments that we shall use is that there is a set of at least ϵm edges, each of which is the first edge of a bad path. For ϵ-node or ϵ-conflict assignment, it follows from the definition that there is a set of ϵn nodes that have a same-colored node within distance at most k.

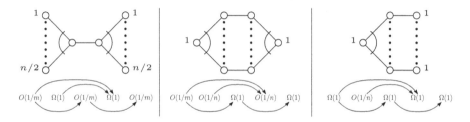

Fig. 2. Three colored graphs showing the incomparability of some of the measures of Definition 1. A number indicates a node's color, unnumbered nodes each receive a color unique to them. For each graph, the second line indicates values of ϵ for which the graph is (in order) ϵ-disjoint, ϵ-node, ϵ-edge, ϵ-conflict, and ϵ-middle. The arrows between measure values are the same as those of Fig. 1

3   Preliminaries: Set Disjointness

The *Set Disjointness* problem is a two-party communication complexity decision problem where two players each receive a subset of an universe $[N]$ and must decide whether their subsets are disjoint. This problem is known to require $\Omega(N)$ communication – as large as the players' inputs – to solve with bounded error by a randomized communication complexity protocol [5,29,34]. Doing a reduction from Set Disjointness to a task in the CONGEST model has been a fruitful source of lower bounds [1,9,10,21,27,36].

In this paper, we will use slight variations of the original Set Disjointness problem. We consider a subset of the original problem, where the players have two additional promises: that their sets are of size at most s, and that their sets' intersection is either empty or contains at least t elements, where s and t are two integer parameters.

Definition 2 (Gap Bounded Size Set Disjointness). *Let N, s, t be three integers such that $N \geq s \geq t > 0$, $\mathcal{X} = \mathcal{Y} = [N]$, and the players' set of admissible inputs $\mathcal{I}_{s,t} \subseteq \mathcal{X} \times \mathcal{Y}$ be:*

$$\mathcal{I}_{s,t} = \{(X, Y) : |X| \leq s, |Y| \leq s, |X \cap Y| \in \{0\} \cup [t, +\infty)\}$$

The Gap Bounded Size Set Disjointness *problem* $\mathbf{DISJ}_{s,t}^N : \mathcal{I}_{s,t} \to \{0,1\}$ *is defined as:*

$$\mathbf{DISJ}_{s,t}^N(X, Y) = \begin{cases} 1 & \text{if } X \cap Y = \emptyset, \\ 0 & \text{otherwise.} \end{cases}$$

The standard Set Disjointness problem corresponds to the choice of parameters $s = N, t = 1$. A commonly studied variant bounds the size of the player's sets but promises nothing about the intersection ($t = 1$). This problem is known to have randomized communication complexity $\Theta(s)$ [26]. Computing the intersection of the two sets also has randomized communication complexity only $\Theta(s)$ [7]. Leaving the players' sets unbounded ($s = N$) while keeping the promise on the intersection's size also appears in the literature, referred to as the Gap Set Disjointness problem [14,32]. In both cases, the lower bound is a simple consequence of the lower bound for the standard Set Disjointness problem.

Let us denote by $R_\epsilon(f)$ the randomized communication complexity of a problem f with error at most ϵ.

Lemma 1. *For any constant* $\epsilon \in (0, 1/2)$, $R_\epsilon\left(\mathbf{DISJ}_{s,t}^N\right) \in \Omega\left(\frac{s}{t}\right)$

Proof. Consider the $\mathbf{DISJ}_{s/t,1}^{N/t}$ Set Disjointness problem. This is known to require $\Omega(s/t)$ communication as the standard Set Disjointness problem $\mathbf{DISJ}_{s/t,1}^{s/t}$ reduces to it (it is the same problem but on a subset of its input space). Now remark that $\mathbf{DISJ}_{s/t,1}^{N/t}$ reduces to $\mathbf{DISJ}_{s,t}^N$, as the players can construct a valid input to $\mathbf{DISJ}_{s,t}^N$ from a $\mathbf{DISJ}_{s/t,1}^{N/t}$ input by making t copies of each of their set elements, which concludes the proof. $\square$

Our lower bounds on testing a distance-3 coloring use the $\mathbf{DISJ}_{s,t}^N$ problem with parameters $s \in \Theta(m)$ and $t \in \Theta(\epsilon m)$ (Theorem 3), $s \in \Theta(m)$ and $t \in \Theta(\epsilon m)$ (Theorem 4), and $s \in \Theta(\sqrt{m/\epsilon})$ and $t = 1$ (Theorem 5), while our lower bound on testing a distance-4 coloring (Theorem 6) uses parameters $s = m$ and $t = 1$. Our lower bound on verifying a distance-k coloring for an arbitrary k (Theorem 7) uses parameters $s \in \Theta(\Delta^{\lfloor (k-1)/2 \rfloor})$ and $t = 1$. Notice that the complexity of the Set Disjointness problem does not depend on N – the size of the universe – but only on s and t, the sizes of the input sets and their potential intersection.

4   Testing Distance-k Colorings

For all the measures previously introduced (Definition 1), we give upper and lower bounds on detecting being ϵ-far from a solution. Our first result is a protocol for all measures except ϵ-middle, and for any constant k (Theorem 1). This protocol is later shown to be tight for the ϵ-node and ϵ-conflict models, even for $k = 3$. In the case $k = 3$, we give a more efficient algorithm in the ϵ-middle, ϵ-edge and ϵ-disjoint models (Theorem 2). We prove that the algorithm is optimal for the ϵ-middle measure (Theorem 5) and also prove an non-matching lower bound for the ϵ-far and ϵ-disjoint measures (Theorem 3). For $k = 4$, we prove an $\widetilde{\Omega}(\epsilon^{-1})$ lower bound in the ϵ-middle model. This last lower bound is strictly higher than the complexity of the same problem when $k = 3$, demonstrating that the complexity of the problem can keep increasing as we increase k beyond 3 not just when doing verification, but also property testing.

All the lower bounds use the Set Disjointness problem (Definition 2). For a graphical summary of the results of this section, see Fig. 1.

4.1   A General Algorithm for Any k

We first give an algorithm that works for all values of k and all our farness measures except ϵ-middle. The basic building block of our algorithm is a subroutine has each node assign a random priority to its color, and has them then broadcast colors according to their assigned priorities. This idea of breaking symmetry by assigning random priorities to elements of interest has appeared previously in the literature [11,12,16].

Theorem 1. *There exists a randomized* CONGEST *algorithm running in* $O\left(\frac{1}{\epsilon}\right)$ *rounds for testing an ϵ-edge distance-k coloring. By extension this also applies to ϵ-disjoint, and a slight modification yields the same result for ϵ-node and ϵ-conflict.*

Proof. Consider the following basic algorithm BFS that runs for k rounds, which we then repeat to obtain success probability $2/3$. The edges are independently assigned a random priority (such as a random value from $[|E|^3]$, with higher values receiving precedence). Nodes use the max of the priorities of their incident edges as their own priority. In the initial round, each node transmits its color (along with its ID) to all its neighbors, along with its priority. In each subsequent round, the algorithm transmits to each neighbor the color and priority of the two highest priority colors it received in the previous round. Effectively, the color from a highest priority node gets forwarded along a breadth-first-search tree. If at the end of round k, the algorithm has received a color (from another node) that matches its own, it outputs 'invalid'; otherwise, it outputs 'valid'.

In an ϵ-edge graph, there are at least ϵm edges that are the first edge of a bad path. If any of those edges receives the highest priority in a round of the basic algorithm, a color conflict gets detected. So with probability at least ϵ, the basic algorithm detects an ϵ-edge graph.

This basic algorithm is then repeated to increase the success probability to at least $2/3$. It suffices to repeat it t times, where t satisfies $(1 - \epsilon)^t \leq 1/3$. The time complexity is then $t \cdot k$. Setting $t = \ln(3)/\epsilon$ achieves the desired result, yielding an $O(k/\epsilon)$-round algorithm.

We can simplify and adapt this algorithm for the ϵ-conflict model: Each *node* picks a random priority. There are now ϵn improperly colored nodes, and if any of them gets selected, the coloring will be found to be invalid. The rest of the argument is the same. $\square$

4.2   A Better Running Time for $k = 3$

Theorem 2. *There exists a randomized* CONGEST *algorithm running in* $O\left(\frac{1}{\epsilon \cdot \sqrt{\epsilon m}}\right)$ *rounds for testing an ϵ-middle distance-3 coloring. By extension, this also applies to the ϵ-edge and ϵ-disjoint measures.*

Proof. Let the nodes follow the following simple algorithm RANDOM: in the first round, each node informs its neighbors of its color and identifier, and in $k - 2$

subsequent rounds, for each link a node has, it picks uniformly at random one of the (color,ID) pair it received from its neighbors in the previous round and sends it on this link. Any node that receives the same color twice but with a different ID, and such that it received the pairs in two (not necessarily distinct) rounds i and j such that $i + j \leq k$, flags the coloring as invalid. This $(k - 1)$-rounds protocol is repeated T times. Let us analyze the probability of success of this protocol when $k = 3$.

Let $\sigma = 4\sqrt{m/\epsilon}$. We say that an edge uv is *good* if $\min(d(u), d(v)) \leq \sigma$, and *bad* otherwise. In an ϵ-middle graph, there is a set Π of edges, each of which is on a 3-path between same-colored nodes, with $|\Pi| \geq \epsilon m$. Observe that if a, u, v, b is a path where a and b have the same color, then this will be detected if either u forwards the ID of a to v or v forwards the ID of b to u. The probability $\bar{p}_{P_e}$ of non-detection along a path P_e with middle edge $e = (u, v) \in \Pi$ is therefore $(1 - 1/d(u))^T \cdot (1 - 1/d(v))^T \leq e^{-T/\min(d(u),d(v))}$. We say that a path $P_e \in \Pi$ is *good* if e is good. Let $\Pi' \subseteq \Pi$ be the set of good paths. Let B be the set of nodes with degree at least σ. There are at most $\sqrt{\epsilon m}/2$ nodes in B, as otherwise the total number incidences on nodes in B would exceed $2m$. Thus, there are at most $\binom{|B|}{2} \leq \epsilon m/8$ edges with both endpoints in B. Hence, there are at least $5\epsilon m/8$ good paths in Π'.

The probability that none of those good paths detect a conflict in the color assignment is:

$$\prod_{P \in \Pi'} \bar{p}_P \leq \exp\left(\frac{-T \cdot |\Pi'|}{\sigma}\right) \leq \exp\left(-\frac{5}{32}T \cdot \epsilon^{3/2} m^{1/2}\right)$$

Therefore, running the protocol for $T \in O(\epsilon^{-3/2} m^{-1/2})$ is enough to solve the problem with probability at least $2/3$.   □

In particular, this protocol runs in constant time when $\epsilon \in \Omega(m^{-1/3})$. We give a matching lower bound later in the paper (Theorem 5). This algorithm is also able to detect ϵ-edge and ϵ-disjoint graphs with the same running time because of the relationships that exist between the measures, however the lower bounds we have for these measures are weaker (Theorem 3) and do not match our upper bound.

4.3   Lower Bounds for $k \geq 3$

In this section, we prove lower bounds for the detection of ϵ-disjoint colored graphs (Theorem 3), ϵ-node colored graphs (Theorem 4) and ϵ-middle colored graphs (Theorems 5 and 6) in the CONGEST model. By the relationships that exist between the separation measures of Definition 1, the lower bound on detecting an ϵ-disjoint coloring also holds for ϵ-edge colorings. Similarly, the lower bound on the detection of ϵ-node colorings also holds for ϵ-conflict colorings.

All lower bounds use the same following classical proof architecture: we take a two-party communication complexity problem f of communication complexity $R^{cc}(f)$, and show that the players can solve an instance $f(x, y)$ of this problem by simulating a CONGEST algorithm for our testing task on a graph $G_{x,y}$ with color assignment $(c_v^{x,y})_{v \in V}$. The vertices of $G_{x,y}$ are partitioned into two sets V_A and V_B, and the edges are such that the colors and intraconnexions of V_A's vertices only depend on x, and similarly with V_B's vertices and y, while the interconnexions between a vertices of V_A and V_B are fixed and therefore independent of x and y. Let T be the number of rounds of a CONGEST algorithm for the CONGEST task, and C the number of edges between vertices of V_A and V_B. Simulating the CONGEST algorithm in the two-party communication complexity model can be done in $T \cdot C \cdot \log(n)$ bits of communication. This last quantity has to exceed $R^{cc}(f)$, which yields that any CONGEST algorithm for our testing task requires at least $T \geq \frac{R^{cc}(f)}{C \cdot \log(n)}$ rounds.

Theorem 3. *For* $k \geq 3$, *testing whether a distance-k coloring is ϵ-disjoint requires* $\widetilde{\Omega}\left(\frac{1}{\epsilon \cdot (\epsilon m)}\right)$ *rounds in the* CONGEST *model.*

Note that this lower bound matches neither our general upper bound (Theorem 1) nor our upper bound for $k = 3$ (Theorem 2), leaving open the possibility of more efficient algorithms or stronger lower bounds.

For this lower bound, we consider graphs of the form presented in Fig. 3. We conjecture that our analysis is not tight, and that detecting whether such graphs are ϵ-disjoint actually requires $\widetilde{\Omega}(\epsilon^{-3/2} m^{-1/2})$.

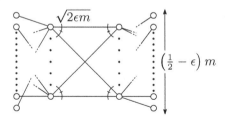

Fig. 3. The graph we use for our lower bound. It consists of 4 layers, with the outer layers having $\left(\frac{1}{2} - \epsilon\right) m$ vertices and the inner layers $\sqrt{2\epsilon m}$. There only exist edges between adjacent layers, and vertices in layers 1 and 4 have degree 1, while layers 2 and 3 form a biclique. Layers 1 and 2 are randomly connected together by $\left(\frac{1}{2} - \epsilon\right) m$ edges, as are layers 3 and 4.

Proof. Let m be an integer and $\epsilon \in \left[\frac{1}{m}, \frac{1}{2}\right)$. Set $N = m$, $s = \left(\frac{1}{2} - \epsilon\right) m$, $t = 2\epsilon m + 3$, and consider an instance of **DISJ**$_{s,t}^N$: a pair of sets (X, Y), $X, Y \subseteq [m]$. Let us consider the graph $G_{x,y} = (V, E)$ of Fig. 3. Its vertices V are partitioned into four layers $(V_i)_{i \in [4]}$. Let Alice possess the two leftmost layers ($V_A = V_1 \cup V_2$) and Bob possess the two rightmost layers ($V_B = V_3 \cup V_4$). The inner layers are of size $|V_2| = |V_3| = \sqrt{2\epsilon m}$ and form a biclique (complete bipartite graph) of $2\epsilon m$ edges, while the outer layers are of size $|V_1| = |V_4| = \left(\frac{1}{2} - \epsilon\right) m$. Let us first describe how the players color their vertices, before describing how they connect the outer layers to the inner layers.

Let all of Alice's and Bob's vertices be initially uncolored. For each element $x \in X \subseteq [m]$, Alice picks an arbitrary uncolored vertex of V_1 and colors it with x. Bob does the same with his input set Y and the layer V_4. Alice then colors her remaining uncolored vertices with distinct even numbers from $[m+1, 2m]$, while Bob colors his remaining uncolored vertices with distinct odd numbers from $[m+1, 2m]$.

Then, for each vertex $u \in V_1$, Alice connects it to a single vertex of V_2 picked uniformly at random. Bob similarly connects vertices of V_4 to vertices of V_3.

Let us now analyze the graph we constructed with respect to the Set Disjointness instance we started with. If $X \cap Y = \emptyset$, the way the players assigned colors ensures that the graph received a valid distance-3 coloring. If $|X \cap Y| \geq t$, however, there are t pairs $(u, v) \in V_1 \times V_4$ of distinct vertices that are at distance 3 and received the same color. For each pair, there is a single length-3 path connecting them, and the only way those paths can share an edge is by sharing an edge in $V_2 \times V_3$. Let us prove that with high probability, more than ϵm of those paths are edge-disjoint, and therefore the graph is ϵ-disjoint (see Definition 1).

Let S be an ϵm-sized subset of the $2\epsilon m$ edges between V_2 and V_3. The probability that none of those edges are directly connected to two vertices in layers V_1 and V_4 that received the same color is at most $\left(1 - \frac{|S|}{2\epsilon m}\right)^t = 2^{-t}$. As there are $\binom{2\epsilon m}{\epsilon m} \leq 2^{2\epsilon m}$ such subsets S, the probability that less than ϵm edges of $V_2 \times V_3$ are part of a length-3 path between similarly colored vertices of the outer layers is at most $2^{2\epsilon m - t} \leq \frac{1}{8}$ for our choice of t.

Since the graph the players constructed is well-colored when they received disjoint sets, and ϵ-disjoint with probability $\geq 7/8$ when they received intersecting sets, the players can solve the Set Disjointness problem with error at most $1/4$ by simulating a CONGEST algorithm to detect an ϵ-disjoint distance-3 coloring that makes an error at most $1/8$. Since there are $2\epsilon m$ edges between Alice's and Bob's vertices, the number of rounds T of a CONGEST algorithm detecting an ϵ-disjoint coloring with probability $\geq 7/8$ satisfies:

$$T \geq \frac{R_{1/4}^{cc}(\mathbf{DISJ}_{m/2, 2\epsilon m}^m)}{2\epsilon m \log(m)} \in \tilde{\Omega}\left(\frac{1}{\epsilon \cdot (\epsilon m)}\right)$$

$\square$

Note that since a graph that is ϵ-disjoint from a valid solution is also ϵ-edge, the lower bound also applies to testing being ϵ-edge. As corollary, we have that no constant-round algorithm can detect an ϵ-disjoint coloring when $\epsilon \in o(m^{-1/2})$.

Theorem 4. *For $k \geq 3$, testing whether a distance-k coloring is ϵ-node requires $\tilde{\Omega}\left(\frac{1}{\epsilon}\right)$ rounds in the* CONGEST *model.*

Proof sketch. We do another reduction from communication complexity, this time using the graph shown in Fig. 4, and Set Disjointness instances with sets

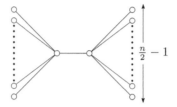

Fig. 4. The graph we use for our lower bound for the ϵ-node measure. It consists of two stars of degree $\left(\frac{n}{2} - 1\right)$, linked by their roots.

of size up to $\Theta(n)$ and intersection either empty or of size $\Omega(\epsilon n)$. The lower bound follows from the communication complexity of this type of Set Disjointness instance and the single edge between Alice's and Bob's parts of the graph.   □

Note that contrary to our previous theorem for detecting ϵ-disjoint colored graphs (Theorem 3), this lower bound is tight with respect to our first algorithm (Theorem 1).

Theorem 5. *For $k = 3$, testing whether a distance-k coloring is ϵ-middle requires* $\widetilde{\Omega}\left(\frac{1}{\epsilon \cdot \sqrt{\epsilon m}}\right)$ *rounds in the* CONGEST *model.*

Note that this lower bound matches our upper bound for $k = 3$ (Theorem 2). For this lower bound, we use graphs as depicted in Fig. 5.

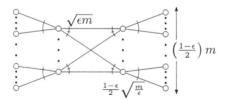

Fig. 5. The graph we use for our lower bound on testing for ϵ-middle colorings (Theorem 5), showing that the algorithm RANDOM is tight for this measure and $k = 3$. It consists of a central biclique between two layers of size $\sqrt{\epsilon m}$, and each vertex of these layers is connected to $\sqrt{m/\epsilon}$ leaves in the outer layers.

Proof. Let $m \in \mathbb{N}$ and $\epsilon \in \left[\frac{1}{m}, \frac{1}{2}\right)$. Set $N = s = \frac{1-\epsilon}{2}\sqrt{m/\epsilon}$, $t = 1$, and consider an instance of **DISJ**$_{s,t}^N$: a pair (X, Y) of subsets of $[N]$. Consider the four layer graph $G_{x,y} = (V, E)$ of Fig. 5. The vertices V_2 and V_3 of layers 2 and 3 form a biclique. Every vertex of layer 2 is connected to s degree-1 vertices in layer 1, and layers 3 and 4 are similarly connected.

Let Alice possess as V_A the vertices of layers 1 and 2 and Bob possess the rest. For any vertex $v \in V_2$, let $N_1(v) \subseteq V_1$ be the vertices of layer 1 connected to v, and similarly for any vertex $v \in V_3$, consider $N_4(v)$ the vertices of layer 4 connected to v.

For each $v \in V_2$, Alice colors the nodes of $N_1(v)$ with the elements of X as colors (without repetition, leaving uncolored nodes if necessary). Bob does the same with nodes of $N_4(v)$ for each $v \in V_3$. The coloring is then completed without creating any new distance-3 conflict, using odd large colors on Alice's side and even large colors on Bob's side.

If the players received disjoint sets, the resulting graph $G_{X,Y}$ is well-colored. If the sets' intersect, however, the coloring is ϵ-middle, because for each pair of vertices $(u, v) \in V_2 \times V_3$, there exists a pair of vertices $(u', v') \in N_1(u) \times N_4(v)$ that have the same color. Therefore, the players can solve their Set Disjointness instance by simulating a CONGEST algorithm for detection of ϵ-middle colored graphs. Since ϵm edges connect Alice's and Bob's parts of the graph, the number of rounds T of a CONGEST algorithm detecting an ϵ-middle coloring with probability $\geq 2/3$ satisfies:

$$T \geq \frac{R^{cc}_{1/3}(\mathbf{DISJ}^N_{\frac{1-\epsilon}{2}\sqrt{m/\epsilon},1})}{\epsilon m \cdot \log(m)} \in \widetilde{\Omega}\left(\frac{1}{\epsilon\sqrt{\epsilon m}}\right)$$

$\square$

Finally, we prove a lower bound on testing a distance-4 coloring in the ϵ-middle model. The lower bound we obtain is strictly higher than our upper bound on the same task with distance-3, which shows that there is a clear gap between distance-3 and distance-4 colorings.

Theorem 6. *Testing whether a distance-4 coloring is ϵ-middle requires $\widetilde{\Omega}\left(\frac{1}{\epsilon}\right)$ rounds in the* CONGEST *model.*

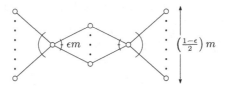

Fig. 6. The graph we use for our lower bound on testing for ϵ-middle distance-4 colorings (Theorem 6).

Proof sketch. This lower bound is again proved by a reduction from communication complexity, this time using the graph depicted in Fig. 6, and Set Disjointness instances with sets of size up to $\Theta(m)$ and no promise on the size of the intersection. $\square$

5   Conclusion

In this work, we studied the testing and verification of distance-k colorings in the
CONGEST model for $k \geq 3$ and several notions of distance from a valid solution.
We showed that the testing of distance-3 colorings admits a significantly more
efficient algorithm than distance-4 for one of our measures (ϵ-middle), and gave
indications that it might also be the case for the other edge- and path-based
measures. The node-based measures show no such gap. Our work does not give a
full picture how the complexity of the problem evolves as k increases in the edge-
and path-based models. A first open question is finding the exact complexity of
testing in the ϵ-disjoint and ϵ-edge model: we conjecture that this complexity
matches that of our algorithm for these models, rather than that of our lower
bound or something intermediate.

Another open question is what algorithm we can design in the ϵ-middle model
for arbitrary k, as the BFS algorithm does not function in it. Even tackling the
case $k = 4$ is of interest, potentially to match our lower bound. Finally, the
several measures we introduced to study this problem might be of independent
interest. Are there other problems for which the same measures would make
sense? A natural candidate here is testing edge-colorings.

A   Verifying Distance-k Colorings in Bounded-Degree Graphs

A.1   A matching lower bound for the natural algorithm

In a graph of maximum degree Δ, the nodes can learn their distance-$\lceil k/2 \rceil$
neighborhood in $O\left(\Delta^{\lceil k/2 \rceil - 1}\right)$ rounds in CONGEST. In particular, an invalid
distance-k coloring can be detected with this number of rounds in CONGEST,
since two nodes of distance at most k are both within a distance $\lceil k/2 \rceil$ of some
node. This protocol is actually close to optimal, as our next theorem shows.

Theorem 7. *For $k \geq 3$, the verification of a distance-k coloring requires*
$\widetilde{\Omega}\left(\Delta^{\lceil k/2 \rceil - 1}\right)$ *rounds in the* CONGEST *model.*

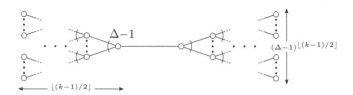

Fig. 7. The graph we use for our lower bound. It consists of 2 complete $(\Delta - 1)$-ary
trees of depth $\lceil k/2 \rceil - 1$ linked at their roots.

Proof sketch. The proof again relies on embedding a Set Disjointness instance in a graph (see Fig. 7). Here, a Set Disjointness instance with sets of size up to $\Theta(\Delta^{-\lceil k/2 \rceil - 1})$ and no promise on the intersection can be embedded, with a single edge connecting Alice's and Bob's parts of the graph.

References

1. Abboud, A., Censor-Hillel, K., Khoury, S.: Near-linear lower bounds for distributed distance computations, even in sparse networks. In: Gavoille, C., Ilcinkas, D. (eds.) DISC 2016. LNCS, vol. 9888, pp. 29–42. Springer, Heidelberg (2016). https://doi.org/10.1007/978-3-662-53426-7_3

2. Afek, Y., Kutten, S., Yung, M.: The local detection paradigm and its application to self-stabilization. Theor. Comput. Sci. **186**(1–2), 199–229 (1997)

3. Awerbuch, B., Patt-Shamir, B., Varghese, G.: Self-stabilization by local checking and correction (extended abstract). In: FOCS, pp. 268–277 (1991)

4. Bamberger, P., Kuhn, F., Maus, Y.: Efficient deterministic distributed coloring with small bandwidth. CoRR abs/1912.02814 (2019)

5. Bar-Yossef, Z., Jayram, T.S., Kumar, R., Sivakumar, D.: An information statistics approach to data stream and communication complexity. J. Comput. Syst. Sci. **68**(4), 702–732 (2004)

6. Brakerski, Z., Patt-Shamir, B.: Distributed discovery of large near-cliques. Distrib. Comput. **24**(2), 79–89 (2011). https://doi.org/10.1007/s00446-011-0132-x

7. Brody, J., Chakrabarti, A., Kondapally, R., Woodruff, D.P., Yaroslavtsev, G.: Beyond set disjointness: the communication complexity of finding the intersection. In: PODC, pp. 106–113 (2014)

8. Censor-Hillel, K., Fischer, E., Schwartzman, G., Vasudev, Y.: Fast distributed algorithms for testing graph properties. Distrib. Comput. **32**(1), 41–57 (2018). https://doi.org/10.1007/s00446-018-0324-8

9. Censor-Hillel, K., Khoury, S., Paz, A.: Quadratic and near-quadratic lower bounds for the CONGEST model. In: DISC, pp. 10:1–10:16 (2017)

10. Drucker, A., Kuhn, F., Oshman, R.: On the power of the congested clique model. In: PODC, pp. 367–376 (2014)

11. Emek, Y., Halldórsson, M.M., Mansour, Y., Patt-Shamir, B., Radhakrishnan, J., Rawitz, D.: Online set packing. SIAM J. Comput. **41**(4), 728–746 (2012)

12. Even, G., et al.: Three notes on distributed property testing. In: DISC, pp. 15:1–15:30 (2017)

13. Feuilloley, L., Fraigniaud, P.: Survey of distributed decision. Bull. EATCS **119** (2016). http://eatcs.org/beatcs/index.php/beatcs/article/view/411

14. Fischer, O., Gonen, T., Oshman, R.: Distributed property testing for subgraph-freeness revisited. CoRR abs/1705.04033 (2017)

15. Fraigniaud, P., Göös, M., Korman, A., Parter, M., Peleg, D.: Randomized distributed decision. Distrib. Comput. **27**(6), 419–434 (2014). https://doi.org/10.1007/s00446-014-0211-x

16. Fraigniaud, P., Halldórsson, M.M., Patt-Shamir, B., Rawitz, D., Rosén, A.: Shrinking maxima, decreasing costs: new online packing and covering problems. Algorithmica **74**(4), 1205–1223 (2016). https://doi.org/10.1007/s00453-015-9995-8

17. Fraigniaud, P., Korman, A., Peleg, D.: Towards a complexity theory for local distributed computing. J. ACM **60**(5), 35:1–35:26 (2013)

18. Fraigniaud, P., Olivetti, D.: Distributed detection of cycles. ACM Trans. Parallel Comput. **6**(3), 1–20 (2019)
19. Fraigniaud, P., Patt-Shamir, B., Perry, M.: Randomized proof-labeling schemes. Distrib. Comput. **32**(3), 217–234 (2018). https://doi.org/10.1007/s00446-018-0340-8
20. Fraigniaud, P., Rapaport, I., Salo, V., Todinca, I.: Distributed testing of excluded subgraphs. In: Gavoille, C., Ilcinkas, D. (eds.) DISC 2016. LNCS, vol. 9888, pp. 342–356. Springer, Heidelberg (2016). https://doi.org/10.1007/978-3-662-53426-7_25
21. Frischknecht, S., Holzer, S., Wattenhofer, R.: Networks cannot compute their diameter in sublinear time. In: SODA, pp. 1150–1162 (2012)
22. Ghaffari, M., Harris, D.G., Kuhn, F.: On derandomizing local distributed algorithms. In: FOCS, pp. 662–673 (2018)
23. Goldreich, O. (ed.): Property Testing. LNCS, vol. 6390. Springer, Heidelberg (2010). https://doi.org/10.1007/978-3-642-16367-8
24. Göös, M., Suomela, J.: Locally checkable proofs in distributed computing. Theory Comput. **12**(1), 1–33 (2016)
25. Halldórsson, M.M., Kuhn, F., Maus, Y.: Distance-2 coloring in the CONGEST model. CoRR abs/2005.06528 (2020)
26. Håstad, J., Wigderson, A.: The randomized communication complexity of set disjointness. Theory Comput. **3**(11), 211–219 (2007)
27. Holzer, S., Wattenhofer, R.: Optimal distributed all pairs shortest paths and applications. In: PODC, pp. 355–364 (2012)
28. Itkis, G., Levin, L.A.: Fast and lean self-stabilizing asynchronous protocols. In: FOCS, pp. 226–239 (1994)
29. Kalyanasundaram, B., Schnitger, G.: The probabilistic communication complexity of set intersection. SIAM J. Discrete Math. **5**(4), 545–557 (1992)
30. Kol, G., Oshman, R., Saxena, R.R.: Interactive distributed proofs. In: PODC, pp. 255–264 (2018)
31. Korman, A., Kutten, S., Peleg, D.: Proof labeling schemes. Distrib. Comput. **22**(4), 215–233 (2010)
32. Kuhn, F., Oshman, R.: The complexity of data aggregation in directed networks. In: Peleg, D. (ed.) DISC 2011. LNCS, vol. 6950, pp. 416–431. Springer, Heidelberg (2011). https://doi.org/10.1007/978-3-642-24100-0_40
33. Naor, M., Parter, M., Yogev, E.: The power of distributed verifiers in interactive proofs. In: SODA, pp. 1096–1115 (2020)
34. Razborov, A.A.: On the distributional complexity of disjointness. Theory Comput. Sci. **106**, 385–390 (1992). https://doi.org/10.1007/BF0032036
35. Rozhoň, V., Ghaffari, M.: Polylogarithmic-time deterministic network decomposition and distributed derandomization. CoRR abs/1907.10937 (2019)
36. Sarma, A.D., Holzer, S., Kor, L., Korman, A., Nanongkai, D., Pandurangan, G., Peleg, D., Wattenhofer, R.: Distributed verification and hardness of distributed approximation. SIAM J. Comput. **41**(5), 1235–1265 (2012)

Communication Complexity of Wait-Free Computability in Dynamic Networks

Carole Delporte-Gallet[1], Hugues Fauconnier[1], and Sergio Rajsbaum[2(✉)]

[1] Université de Paris, CNRS, IRIF, 75013 Paris, France
{cd,hf}@irif.fr
[2] Universidad Nacional Autónoma de México, Mexico City, Mexico
rajsbaum@im.unam.mx

Abstract. We consider a *wait-free dynamic network*. The class of solvable tasks in this model is well-known, and turns out to be the same in various similar message-passing and shared-memory models. But only *full-information* protocols have been considered, which send messages that grow with the number of rounds.

We show that for two processes, it is possible to solve any wait-free solvable task using mostly 1-bit messages, without incurring any cost in the optimal number of communication rounds. We identify an additional type of information that needs to be communicated: in some executions it is necessary to send messages of $\log c + 1$ bits, where c is the chromatic number of the distance-2 graphs of input configurations of the two processes. But on average, the size of messages of a k-round protocol is at most $1 + \frac{2 \log c}{k}$ bits. Then, we show that it is possible to solve any wait-free solvable task by exchanging only *beeps*, at a constant cost in terms of number of rounds. Finally, for 3 processes, we show that messages of constant size do not suffice to solve every task in an optimal number of rounds.

1 Introduction

The computational power of a distributed system depends on its communication, process relative speeds, and failure assumptions. A model's computational power is typically studied with respect to *tasks,* such as consensus. Each process starts with a local input value, the processes communicate with each other, and eventually each process produces an output value. The task specification is defined in terms of a relation Δ, defining which output value assignments are legal responses to each possible assignment of input values. A characterization of the tasks that are solvable in an asynchronous message passing system where at most one process may crash was initially given [3]. Another milestone was

C. Delporte-Gallet—This work was partially supported by the French ANR project DESCARTES 16-CE40-0023-03.
H. Fauconnier—This work was partially supported by the French ANR project FREDDA ANR-17-CE40-0013.
S. Rajsbaum—This work was partially supported by the UNAM-PAPIIT project IN106520.

A. W. Richa and C. Scheideler (Eds.): SIROCCO 2020, LNCS 12156, pp. 291–309, 2020.
https://doi.org/10.1007/978-3-030-54921-3_17

the Wait-Free Theorem (WFT), for characterizing the tasks, denoted $\mathcal{WF}$, solvable in an asynchronous read/write shared memory system where any number of processes may crash [15]. This paper also discovered the intimate relation with algebraic topology. Task solvability characterizations have since been described for many models, including models (such as iterated, non-iterated, snapshots, immediate snapshots, some dynamic networks) that solve the same set of tasks, $\mathcal{WF}$. Also, for other more or even less powerful models (synchronous, partially synchronous, shared-memory synchronization objects, Byzantine failures, solo-models [21], etc), for an overview see [13].

All the task solvability characterizations considered in the literature assume a *full-information* protocol, where a process keeps in its state everything it knows, and each time it communicates with other processes, it sends its entire state. While a full-information protocol is convenient to derive a task solvability characterization, the size of the messages (or the values written to the shared-memory) grows with the number of rounds. Thus, rounds become slower and slower, to implement the necessary information exchange. Furthermore, the number of rounds needed to solve a task can grow so fast, that it is undecidable if a task has a wait-free protocol, even for three processes [10,14].

The question we raise in this paper, is what is the cost in terms of communication rounds, of have constant size messages. And related questions, such as, what is the minimum number of bits per message, needed to encode a full-information protocol, without incurring a cost in communication rounds?

Indeed, the known characterizations depend on the protocol being full-information. For instance, for two processes, in an immediate snapshot model, the views of the processes after k rounds, i.e. the *protocol complex*, are represented by a path of 3^k edges (for each input configuration). And more generally, for n processes, the views are a subdivided $n - 1$-dimensional simplex. Clearly the structure of the protocol complex may change, for the case of non-full-information protocols. On the other hand, it is easy to encode a full information protocol by sending smaller messages, at the cost of extra rounds.

We initiate the study of such questions using a model that solves the same set of tasks, $\mathcal{WF}$. It is a special case of a dynamic network [4] that we call *wait-free dynamic network* (which has been studied before, using full-information protocols e.g. [1,13,20]). In the two-process case, A and B send messages to each other in synchronous rounds, and they never fail, but in each round, either of the two messages sent may be lost (but not both). The set of input configurations of a task is a graph, $\mathcal{I}$. We show that mostly single bit messages suffice, to solve any task in $\mathcal{WF}$, without an extra cost in the number of rounds, w.r.t. a full-information protocol. We also identify another type of information that needs to be exchanged, to identify the input configuration (an edge). For this, sometimes messages of $\log c$ bits must be sent, where c is the maximum of the chromatic number of the distance-2 graphs of $\mathcal{I}, \mathcal{I}_A, \mathcal{I}_B$ defined by the A-vertices and the B-vertices, respectively. On average, the size of messages of a k-round protocol is at most $1 + \frac{2\log c}{k}$ bits. A consequence of this, is that with messages of 1 bit it is possible to encode a full-information history for a given input configuration, independently of the number of rounds of the protocol.

Furthermore, we show that it is possible to solve every two-process task in $\mathcal{WF}$ without sending any bits in a *beep model*, where processes communicate only by exchanging unary signals over the dynamic network. Here there is a cost in the number of rounds, but it is still a constant (that depends on c). This result separates iterated, round-based models such as dynamic networks, from shared-memory models where the same register can be written several times repeatedly; in such models it is clearly impossible to do any useful computation with only unary signals. Yet, it is known that all these models are equivalent as far as task computability is concerned e.g. [11,13].

Finally, we show that for more than two processes, messages of constant size do not suffice to solve every task in an optimal number of rounds. For more than two processes, wait-free computability inherently requires rounds with growing information exchange. We focus on two processes because core issues appear already here, without the need of more technical algebraic topology techniques, needed for more than two processes. But we demonstrate that our techniques generalize to show the impossibility result for three processes.

Related work. Communication complexity is an important sub-area of complexity theory that studies the amount of communication needed for several distributed parties to learn something new [19]. In the basic problem there are two players, A, B, with input sets X_A, X_B, and a function $f : X_A \times X_B \longrightarrow \{0,1\}$. The goal is to evaluate f, when A holds one part of the input, $x \in X_A$, while B holds the other part $y \in X_B$. They communicate over a reliable channel by alternatively exchanging messages with each other. Our dynamic network model is similar, but with unreliable communication. Also, instead of a function, we consider a relation, in the form of a task. Furthermore, in a task not necessarily all combinations of pairs from X_A and X_B may be given as inputs to A and B, but only those specified by a graph of initial configurations $\mathcal{I}$. Finally, processes do not need to produce the same output value. In the original communication complexity setting every function is solvable, the question is what is the smallest number of bits that have to be exchanged to compute f. In our distributed setting, even central functions such as *equality*, defined by $EQ(x,y) = 1$ if and only if $x = y$ are not solvable. The class of tasks that are solvable in our two processes setting is well-known [3]. The connection between epistemic knowledge and communication complexity is studied in [17].

In the distributed computing area, the most common communication cost measure is message complexity. Based on the observation that consensus and leader election have the same message complexity in failure-free networks, bit complexity was used to distinguish between the communication costs of these tasks, using methods of communication complexity [6,7]. Bit complexity in distributed computing has been considered since then, but we are not aware of any papers considering tasks in general, nor the wait-free setting.

There are several *wait-free models* that are fundamental, and equivalent to each other, in the sense that they can solve the same set of tasks, $\mathcal{WF}$. An example of such a model (for 2 processes) is our wait-free dynamic network, but there are others, when processes may crash, either detectable (synchronous models) or undetectable (asynchronous models). Our results apply in layered models (communication-closed rounds), such as the *layered message-passing* (detectable failures) and *layered read-write* (undetectable) models described in [13]. Other important examples are when communication is by message passing [3], when it is by snapshot shared memory [15], and by read-/write shared memory. Thus, some of the basic (not related to bit complexity) results we prove here have analogues in those previously studied models.

Much work exists on dynamic networks e.g. [20], but mostly concentrating on characterizing the sequences of graphs that allow to solve various problems, and also often using full-information protocols. The case of two-process consensus is studied in [9].

It has been known since early on that binary consensus can be solved using 1-bit messages, e.g. [2]. But typically solutions that try to optimize the number of messages sent do so at the cost of larger round complexity. To reduce the size of long messages, an encoding is used to spread them over several rounds, e.g. [12].

Organization. In Sect. 2 we describe our framework. The first main result is in Sect. 3, where we show that any solvable task can be solved with 1-bit message in most executions. In Sec. 4 we describe our beep model, and the corresponding communication complexity results. In Sect. 5 we present the 3-process impossibility result, and in Sect. 6 some concluding remarks.

2   Fundamentals

The basic model is that of a dynamic network e.g. [20]. Everything in this section is already known and can be found in [13], including additional details. Essentially, the only difference is that we drop the common assumption that protocols are full-information.

We present here the two-process case, and discuss in Sect. 5 the case of three processes. Let A, B be process names, V^{in} a domain of *input values*, V^{out} a domain of *output values*, and V^{local} a domain of *local state values*. We consider graphs whose vertices are pairs of the form (id, v), where $id \in \{A, B\}$, and v belongs to either V^{in}, V^{out}, or V^{local}. We say that the vertex is *colored* by id, and *labeled* by v. The graph is *chromatic* if the vertices of each of its edges are colored with different process names. Chromatic *normal* simple paths P will be important, with one end vertex (A, a), colored A and the other (B, b), colored B. We say that the *boundary* of P is $\{(A, a), (B, b)\}$.

2.1   Tasks

Given two graphs $\mathcal{G}$ and $\mathcal{H}$, let σ, τ represent either a vertex, or a set of two vertices belonging to an edge. A *carrier map* Φ from $\mathcal{G}$ to $\mathcal{H}$ takes each vertex or edge $\sigma \in \mathcal{G}$ to a subgraph $\Phi(\sigma)$ of $\mathcal{H}$, such that Φ satisfies the following *monotonicity* property: for all $\sigma, \tau \in \mathcal{G}$, if $\sigma \subseteq \tau$, then $\Phi(\sigma) \subseteq \Phi(\tau)$. For vertex

s, $\Phi(s)$ is a (non-empty) set of vertices, and if σ is an edge, then $\Phi(\sigma)$ is a graph where each vertex is contained in an edge. Notice, that for arbitrary edges σ, τ, we have $\Phi(\sigma \cap \tau) \subseteq \Phi(\sigma) \cap \Phi(\tau)$. We say that Φ *preserves names*, if it sends a vertex to a vertex of the same name.

A *task* for A, B is a triple $(\mathcal{I}, \mathcal{O}, \Delta)$, where $\mathcal{I}$ is a chromatic *input graph* colored by $\{A, B\}$ and labeled by V^{in}; $\mathcal{O}$ is a chromatic *output graph* colored by $\{A, B\}$ and labeled by V^{out}; Δ is a name-preserving carrier map from $\mathcal{I}$ to $\mathcal{O}$.

The input graph defines all the possible ways the two processes can start the computation, the output graph defines all the possible ways they can produce an output value, and the carrier map defines which input can lead to which outputs. Each edge $\{(A, a), (B, b)\}$ in $\mathcal{I}$ defines a possible input configuration where A has input value $a \in V^{in}$ and B has input value $b \in V^{in}$. The processes communicate with one another, and each eventually decides on an output value and halts. If A decides x, and B decides y, then there is an output configuration represented by an edge $\{(A, x), (B, y)\}$ in the output graph, that should be in $\Delta(\{(A, a), (B, b)\})$. Moreover, if A *runs solo*, namely without ever hearing from B, it must decide a vertex (A, x) in $\Delta(A, a)$, and B is subject to the symmetric constraint.

2.2   Approximate Agreement Tasks

It is remarkable that wait-free task solvability is essentially a form of approximate agreement. Approximate agreement problems where processes are required to compute values that are close to each other have been thoroughly studied, since early on [8].

Given an integer $k \geq 0$, the *k-edge approximate agreement* task for A, B has an input graph $\mathcal{I}$ consisting of a single edge, $\{(A, 0), (B, 0)\}$. As we explain below, it allows processes to decide values $\epsilon = \frac{1}{3^k}$ apart, but the following outputs will be more convenient later on.

The output graph $\mathcal{O}$ consists of a chromatic normal path of 3^k edges with boundary $\{(A, 0), (B, 0)\}$, with consecutive vertices,

$$(A, 0), (B, (3^k - 1)/2), (A, 1), B, (3^k - 3)/2), (A, 2), \ldots, (B, 1), (A, (3^k - 1)/2), (B, 0).$$

The carrier map for solo executions prevents the trivial solution where processes always decide the same output values, while allowing any decisions when they both see each other:

$$\Delta((A, 0)) = \{(A, 0)\}, \Delta((B, 0)) = \{(B, 0)\} \text{ and } \Delta(\{(A, 0), (B, 0)\}) = \mathcal{O}.$$

An protocol solving k-edge approximate agreement can be used to solve a similar task, with the following *canonical* labelling: $(A, 0), (B, 1), (A, 2), \ldots (A, 3^k - 1)$, $(B, 3^k)$, as follows. Once a process p has computed its ℓ_p, it performs the following computation, to compute a canonical labeling ℓ_p^c. If $p = A$ then $\ell_A^c = 2\ell$, and if $p = B$ then $\ell_B^c = 3^k - 2\ell$. This in turn can be used to solve ϵ-*agreement*, $\epsilon = \frac{1}{3^k}$, by which processes compute values in the interval $[0, 1]$ that are ϵ apart: a process divides its output by 3^k.

Approximate agreement is the essential building block of a solution to *every* solvable task, as shown in [13, Theorem 2.5.2], using the following *k-approximate agreement* task with input $\mathcal{I}$. A solution to this task, can be used to solve any solvable task (the value of k depends on the specific task to be solved). The input graph $\mathcal{I}$ is an arbitrary chromatic input graph colored by $\{A, B\}$ and labeled by V^{in}. The output graph $\mathcal{O}$ is a k-iterated subdivision of the input graph: a collection of chromatic normal simple paths of 3^k edges, one for each $\{(A, a), (B, b)\} \in \mathcal{I}$, with boundary $\{(A, a0), (B, b0)\}$, that intersect only at their endpoints. Namely, for each input edge $\{(A, a), (B, b)\} \in \mathcal{I}$, there is a chromatic normal path P_{ab} in $\mathcal{O}$ of 3^k edges, whose vertices are: $(A, a0), (B, ab(3^k - 1)/2), (A, ab1), \ldots, (A, ab(3^k - 1)/2), (B, b0)$, and two consecutive vertices form an edge. The carrier map is $\Delta((A, a)) = \{(A, a0)\}, \Delta((B, b)) = \{(B, b0)\}$, and $\Delta(\{(A, a), (B, b)\}) = P_{ab}$.

We say that $\mathcal{O}$ is a *chromatic subdivision* of $\mathcal{I}$, iterated k times. Each input edge is *subdivided* in the output graph 3^k times, P_{ab} is a *k-subdivision* of $\{(A, a), (B, b)\}$.

2.3 Dynamic Graph Model

Here is described the *wait-free dynamic graph* model for A, B. In Sect. 4 we discuss the variant with beeps, and in Sect. 5 the version for three processes.

A *local state* is a vertex, (id, v), where $id \in \{A, B\}$, and $v \in V^{local}$. A *global state* is an edge $\{(A, a), (B, b)\}$, a pair of local states. We assume processes never fail, and hence all global states consists of the states of both processes. The set of *initial states* is given by an input graph $\mathcal{I}$. Thus, we identify an initial local state with an input value.

An *event* is a directed graph on two vertices, A, B, with at least one arc; the set of possible events is $EV = \{A \rightarrow B, B \rightarrow A, A \leftrightarrow B\}$. There is an arc from id to id' if and only if the message sent by id arrives at id'. We assume that if a process does not receive a message in an event (because there is no arc into the process), the process is aware that the message was lost; namely, executions consist of synchronous rounds.

A *schedule* is a sequence of events from EV. We assume all sequences of events from EV are possible (an oblivious adversary). Recall that a dynamic graph model is round based, and the processes never fail. An *execution* is an alternating sequence of global states and events, starting in an initial global state. Thus, we may think of a *round* as two consecutive global states of an execution, together with the intermediate event. An execution is *solo for A*, if A never gets a message from B, namely, if it contains only events from $\{A \rightarrow B\}$, and conversely for B.

A *protocol* runs for k rounds. In each round a process decides the contents of the message it sends in the next round, and its new state, as specified by its transition relation. Thus, given an input state $I \in \mathcal{I}$ and a schedule of k events, the execution is uniquely determined (the protocol is deterministic).

The *protocol graph* $\mathcal{P}$ after k rounds is a description of all *final* global states of the protocol. It consists of edges of the form $\{(A, a), (B, b)\}$, each one corresponding to the last global state of an execution. Thus, for each execution there is such an edge in the protocol graph, although different executions may correspond to the same edge.

The protocol graph has some structure, which identifies for each vertex or each edge, from which input vertex or edge it comes from. Namely, for each vertex $(id, v) \in \mathcal{I}$ we can associate a vertex of $\mathcal{P}$, that corresponds to the final state of id in the (unique) *solo* execution where process id gets no messages from the other process (recall protocols are deterministic). Thus, there is a *protocol carrier map* $\varXi$ from $\mathcal{I}$ to $\mathcal{P}$, defined on vertices in this way. And on edges, $\varXi(e)$ is equal to the union of all edges of $\mathcal{P}$ that correspond to executions starting in $e \in \mathcal{I}$. We therefore may represent a protocol as a triple $(\mathcal{I}, \mathcal{P}, \varXi)$.

A protocol *solves* a task $(\mathcal{I}, \mathcal{O}, \varDelta)$ if and only if there is a simplicial map $\delta : \mathcal{P} \to \mathcal{O}$ carried by $\varDelta$. The map δ is called the *decision map*, and *carried* by $\varDelta$ means that for each input vertex s, $\delta(\mathcal{P}(s)) \subseteq \varDelta(s)$, and for each input edge σ, $\delta(\mathcal{P}(\sigma)) \subseteq \varDelta(\sigma)$.

Remark 1. To simplify the presentation, we assume that a protocol always executes the same number of rounds, k. Similarly, each input edge is subdivided the same number of times. But all our results hold as well for non-uniform protocols and subdivisions, in which each edge e is subdivided k_e times.

2.4 Basic Characterization

We refine the well-known task solvability result to include round optimality. By solving a task in an "optimal number of rounds" we mean in the smallest number of rounds, among all protocols that solve it (recall that we consider only protocols that always execute the same number of rounds, Remark 1).

Theorem 1. *A protocol that can solve k-approximate agreement on $\mathcal{I}$ in k rounds (for any given $k \geq 0$, and input graph $\mathcal{I}$), can be used to solve any solvable task, in an optimal number of rounds.*

The idea of the proof is as follows. First, it is well-known that the protocol graph for any k-round full-information protocol with input graph $\mathcal{I}$ is a subdivision of $\mathcal{I}$, where each edge is subdivided 3^k times, e.g. [13, Fact 2.5.1]. Thus, a k-rounds full-information protocol trivially solves the k-approximate agreement task, and vice versa, a protocol that can solve k-approximate agreement on $\mathcal{I}$ in k rounds can solve any task that the k-round full-information protocol can solve, since the full-information protocol graph and the approximate agreement output graph are isomorphic. If a task is solvable, it is solvable by the full-information protocol (or equivalently, by the approximate agreement solution) see e.g. proof of [13, Theorem 2.5.2]. Clearly, a protocol that is not full-information could not solve the task in fewer rounds, k', as it can be simulated (by ignoring extra information) by a full-information protocol using the same number of rounds, k'.

In the rest of the paper we concentrate on solving k-approximate agreement, since this is what is needed to solve any other task. We aim at doing so with the fewest possible number of bits, with a number of rounds as close as possible to k (it is impossible to go below k).

3   Approximate Agreement with Few Bits

Here we describe our first main result. It is possible to solve k-approximate agreement in k rounds on $\mathcal{I}$, by sending 1-bit messages, in most executions.

3.1   Approximate Agreement on a Single Edge

We begin with the k-edge approximate agreement protocol in Fig. 1, which sends messages of a single bit, and executes k rounds. Then we will extended it to a general $\mathcal{I}$.

In the protocol of Fig. 1, each process $p \in \{A, B\}$ executes k rounds, to compute a label ℓ_p. The output vertices for the task are then (A, ℓ_A), (B, ℓ_B).

In each round of the protocol, each edge is subdivided into 3 edges. Fig. 2 illustrates the way a new labelling is computed in round k, and Fig. 3 shows the labelling of executions for the first three rounds. In Fig. 2, if A (red in the figures) with label i does not receive a message in the round, its new label becomes $3i$ (vertical line) and if it receives a message from the other process its new label is either $3i + 1$ or $3i - 1$ (diagonal lines). For this, A has to distinguish between receiving a message from its left in the path (here B with j) or, from its right (here B with label $j+1$). Hence it is sufficient for A to receive a different bit from B when the label of B is j or $j+1$. To ensure this property, two successive B (resp. A) vertices in the path have to send different bits. At round r, our construct will label the right vertex (B, m) for (A, i) such that $i + m = (3^r - 1)/2$ (right), and the left vertex (B, m') such that $i + m' = (3^r + 1)/2$. The following lemma implies that successive vertices corresponding to the same process send different bits.

Lemma 1. $r \bmod 2 = 0$ *iff* $(3^r - 1)/2 \bmod 2 = 0$ *iff* $(3^r + 1)/2 \bmod 2 = 1$.

Let $P(r)$ be the conjunction of the following properties:

$$0 \le \ell_A \le (3^r - 1)/2 \tag{3.1}$$
$$0 \le \ell_B \le (3^r - 1)/2 \tag{3.2}$$
$$(\ell_B = (3^r - 1)/2 - \ell_A) \vee (\ell_B = (3^r + 1)/2 - \ell_A) \tag{3.3}$$

Lemma 2. $P(r)$ *is true at the end of round* r *(* $0 \le r \le k$*)*.

Proof. We show by induction on r that $P(r)$ is true at the end of the round r with $0 \le r \le k$ (the round $r = 0$ is the initialization).

$P(0)$ *is true:* At the end of the initialization $\ell_A = \ell_B = 0 = (3^0 - 1)/2$, thus $P(0)$ is true.

For $0 \le r < k$, $P(r) \Rightarrow P(r + 1)$: Let x_A and x_B be the values of ℓ_A and ℓ_B resp. at the beginning of round $r + 1$. In round $r + 1$, each process p sends (Line 4) $x_p \bmod 2$. At least one of these messages is received (by definition of the wait-free dynamic graph model), Line 5.

If A received $\bot$ (no message): A executes Line 7 and then $\ell_A = 3x_A$. By $P(r)$: $0 \le x_A \le (3^r - 1)/2$, we have $0 \le \ell_A \le 3(3^r - 1)/2$. Since $3(3^r - 1)/2 = (3^{r+1} - 1)/2 - 1 \le (3^{r+1} - 1)/2$, inequality (3.1) is true.

CODE FOR PROCESS p

k-EDGE APPROXIMATE AGREEMENT

Local variables:

```
1   ℓ = 0
```

Code:

```
2   round r from 1 to k do
3       prop = ℓ mod 2
4       send(prop)
5       m = receive()
6       if m = ⊥
7           then ℓ = 3 * ℓ                                  /*receive nothing */
8           else
9               if (prop = m and r mod 2 = 1) or (prop ≠ m and r mod 2 = 0)
10                  then ℓ = 3 * ℓ + 1
11                  else ℓ = 3 * ℓ - 1
12  end round
13  ouput(p, ℓ)
```

Fig. 1. Solving k-edge approximate agreement with 1-bit messages, in k rounds. Each process $p \in \{A, B\}$ runs this code. The input vertex is $(p, 0)$, the output vertex computed is (p, ℓ).

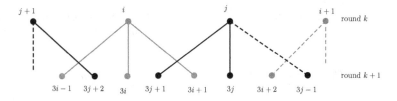

Fig. 2. Execution of one round, how the labels are computed (Color figure online)

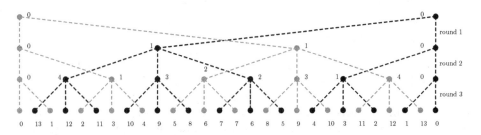

Fig. 3. Executions of the protocol in Fig. 1 for $k = 3$ (Color figure online)

If A received a message, A executes either Line 10 or Line 11, and the value of ℓ_A becomes $3x_A + d$ with $d = 1$ or $d = -1$. If $x_A \neq 0$ with $P(r)$, $1 \leq x_A \leq (3^r - 1)/2$, we have $0 \leq \ell_A \leq 3(3^r - 1)/2 + d$. Since $3(3^r - 1)/2 - 1 + d \leq (3^{r+1} - 1)/2$, inequality (3.1) is true.

If $x_A = 0$, by $P(r)$, $x_B = (3^r - 1)/2$. By Lemma 1, $(r + 1) \mod 2 = 1$ implies that $x_B \mod 2 = 0$, then $x_A \mod 2 = x_B \mod 2$, and $(r + 1) \mod 2 = 0$ implies that $x_B \bmod 2 = 1$. Then $x_A \bmod 2 \neq x_B \bmod 2$. Therefore the condition in Line 9 is true and A executes Line 10, $\ell_A = 1$, and inequality (3.1) is true.

In both cases, inequality (3.1) is true. Similarly, inequality (3.2) is also true.

By the induction hypothesis for $P(r)$, we have $x_a + x_b = (3^r - 1)/2$ or $x_a + x_b = (3^r + 1)/2$. By Lemma 1, the condition in Line 9 is true if and only if $x_a + x_b = (3^r - 1)/2$.

If $x_a + x_b = (3^r - 1)/2$: If A receives a value from B then $\ell_A = 3x_A + 1$, if B receives a value from A then $\ell_B = 3x_B + 1$. If A and B receive a value from each other, $\ell_A + \ell_B = 3(x_a + x_b) + 2 = (3^{r+1} + 1)/2$. If one of them does not receive the value from the other, $\ell_A + \ell_B = 3(x_a + x_b) + 1 = (3^{r+1} - 1)/2$, proving that Eq. (3.3) is true.

If $x_a + x_b = (3^r + 1)/2$: if A receives a value from B then $\ell_A = 3x_A - 1$. If B receives a value from A then $\ell_B = 3x_B - 1$. If A and B receive a value from each other, $\ell_A + \ell_B = 3(x_a + x_b) - 2 = (3^{r+1} - 1)/2$. If one of them does not receive the value from the other $\ell_A + \ell_B = 3(x_a + x_b) - 1 = (3^{r+1} + 1)/2$, proving Eq. (3.3). In both cases Eq. (3.3) is true.

We conclude that at the end of round $r + 1$, Eqs. (3.1), (3.2) and (3.3) are true, proving $P(r + 1)$.

By Lemma 2, the values computed ℓ_p, $p \in \{A, B\}$, satisfy $0 \leq \ell_p \leq (3^k - 1)/2$. Also, in each of the 3^k possible executions, distinct values are computed by A, B, such that $\ell_B = (3^k + d)/2 - \ell_A$ with $d = -1$ or $d = +1$. Thus, we have the following.

Theorem 2. *The protocol in Fig. 1 solves k-edge approximate agreement in k rounds.*

The protocol is optimal in the number of rounds, by Theorem 1, taking $\mathcal{I}$ to consist of a single edge.

3.2   Approximate Agreement on $\mathcal{I}$

When the input graph $\mathcal{I}$ is arbitrary, and process A starts with input a, if there are several possible inputs of B, which are neighbors of (A, a) in $\mathcal{I}$, then A needs to identify the input of B to know on which input edge to work with B to solve edge approximate agreement. Except of course in the case of a solo execution, where A does not receive any messages from B, and its output will be $(A, a0)$. For any x, if B considers possible that A has not received its input in the first x rounds, its label is $\ell = (3^x - 1)/2$, and B must continue to try to communicate its input in round $x + 1$. Otherwise its input is already known by the other process, so its label is $\ell \neq (3^x - 1)/2$ and it does not need to send information concerning its input. This is implemented in Line 4 of the protocol in Fig. 4.

Theorem 3. *The protocol in Fig. 4 solves k-approximate agreement on $\mathcal{I}$, in k rounds.*

CODE FOR PROCESS

k-APPROXIMATE ON $\mathcal{I}$ WITH $(input)$

Local variables:

1    $\ell = 0$

2    $I_{other} := \bot$

Code:

```
3    round r from 1 to k do
4        if ℓ = (3^(r-1) - 1)/2 then i = input else i = ⊥
5        prop = ℓ mod 2
6        send(i, prop)
7        m = receive()              /* if m = (i, l) then m.input denotes i and m.label denotes l */
8        if m = ⊥
9            then ℓ = 3 * ℓ                                                    /*receive nothing */
10           else
11           if m.input ≠ ⊥ and I_other = ⊥ then I_other = m.input
12           if (prop = m.label and r mod 2 = 1) or (prop ≠ m.label and r mod 2 = 0)
13               then ℓ = 3 * ℓ + 1
14               else ℓ = 3 * ℓ - 1
15   end round
16   ouput(ℓ; I_other)                              /*if ℓ ≠ 0 then I_other ≠ ⊥*/
```

Fig. 4. Solving k-Approximate on $\mathcal{I}$. Each process $p \in \{A, B\}$ runs this code, with initial value $input$. The input vertex is $(p, input)$, the output vertex computed is $(p, input \cdot I_{other} \cdot \ell)$.

The basic proof is similar to Theorem 2, with the following additional argument.

Lemma 3. *In the protocol of Fig. 4, each process p outputs a value (ℓ_p, I_{other}) s.t.*

(1) ℓ_p is between 0 and $(3^k - 1)/2$,
(2) if $\ell_p \neq 0$ then I_{other} is the colored input of the other process, and
(3) $\ell_B = (3^k - 1)/2 - \ell_A$ or $\ell_B = (3^k + 1)/2 - \ell_A$.

Proof. Lemma 2 implies this property except for part (2). First notice that I_{other} is modified at most once in Line 11. Consider process A. Let $Q(r)$ be the following property at the end of round r: if $\ell_A \neq 0$ then I_{other} is the colored input of the other process. We prove $Q(r)$, $r \geq 0$, by induction.

After the initialization (round 0) $\ell_A = 0$ so $Q(0)$ is trivially true.

Assume $Q(r)$ is true. Let x_A (resp. x_B) be the value of ℓ_A (resp. ℓ_B) at the end of round r. From $Q(r)$, if $x_A \neq 0$ then I_{other} is the colored input of the other process. As I_{other} is modified at most once, in Line 11, it remains the input of the other process.

If $x_A = 0$ and $\ell_A \neq 0$ at the end of round $r+1$, then A has received a message from B. By Lemma 2, $x_B = (3^k - 1)/2$, at Line 6, B sends its input. So $Q(r)$ is true and the lemma is established.

Remark 2. In the protocol of Fig. 4, at each round, a process sends a bit $(prop)$ and sometimes its input. It is possible to save some bits as follows: when the process sends its input color, it does not send *prop* explicitly (if there are at least 2 colors). Since a process needs from 1 to $\log(c)$ bits to send its color, we can code this color such that, in round r, if the process has to send its color α, it sends $(r + \alpha) \bmod c$.

3.3   Optimizing the Communication Complexity

In our protocol, if B considers possible that A has not received its input in the first x rounds, B continues to try to communicate its input in round $x + 1$. We first show that this is essentially necessary. Let $b_1, b_2, \ldots, b_d$ be an enumeration of the input values of B on edges incident to a vertex (A, a).

Theorem 4. *Let d be the degree of a vertex (A, a) in $\mathcal{I}$ with $d \geq 2$. Every k-round k-approximate agreement protocol on $\mathcal{I}$ has d executions (starting in the d edges incident on (A, a) with schedule $(A \rightarrow B)^k$), where messages of at least $\frac{\log_2(d+2)}{2} - 1$ bits are sent by B at each round on average over these d executions. At least $\frac{k \log_2(d+2)}{2}$ bits are sent on average over these d executions.*

Proof (sketch). Let sch_x be the schedule $(A \rightarrow B)^x (A \leftarrow B)(A \rightarrow B)^{(k-x-1)}$, for $0 \leq x \leq k - 1$ and $k \geq 1$. We prove by induction on $x \geq 0$, that B must send a different message in every round $r \leq x + 1$, for the execution starting in $\{(A, a), (B, b_j)\} \in \mathcal{I}$, with schedule sch_x, and the execution starting in $\{(A, a), (B, b_i)\} \in \mathcal{I}$, with schedule sch_x, for every $i \neq j$.

For the basis, consider $sch_0 = (A \leftarrow B)(A \rightarrow B)^{(k-1)}$, and assume for contradiction that B sends the same message in the first round, on some input, $b \in \{b_0, \ldots b_d\}$ and $b' \in \{b_0, \ldots b_d\} \setminus \{b\}$. A cannot distinguish b and b'. Thus, A cannot output a vertex (A, abx), for some $x > 0$, where b is the input of B. Its only choice is then to output $(A, a0)$. But producing $(A, a0)$ in an execution where A has received a message, implies that the distance from $(A, a0)$ to $(B, b0)$, is less than 3^k. The proof of the inductive step is similar.

We now analyze how large the message has to be. Since A has to identify the input b of B, based on a single message by B, B has to send different messages at round $x + 1$, in all the possible d input vertices (B, b_j) that are neighbours of (A, a). A naive solution would be for B to send messages of size $\log_2 d$. In this case, the total number of bits sent in each execution with schedule $(A \rightarrow B)^k$ is $k + k \log_2 d$, because A sends messages of at least 1 bit.

However, if it is possible for B to send messages of different sizes, it can send less bits. Namely, it is sufficient that in each round, it ensures that with different inputs b, b', $\{(A, a), (B, b)\} \in \mathcal{I}$, $\{(A, a), (B, b')\} \in \mathcal{I}$, it sends different messages.

In this case, the size of a message varies from 1 to $\log_2 d$ bits. If we consider the d executions with schedule $(A \rightarrow B)^k$, at each round to send d different messages, we need 2 messages of size one $(0, 1)$, 2^2 message of size two $(00, 01, 10, 11)$, etc. We assume that d is between $2^{i+1} - 3$ and $2^i - 2$ i.e $2^i - 2 \leq d \leq 2^{i+1} - 3$. Over all the executions, the process sends at least $\sum_{j=1}^{j=i-1} j2^j$ bits, i.e $2^i(i - 2) + 2$ bits. Namely, on average over these d executions at least $\frac{\log_2(d+2)}{2} - 1$ per message.

And this has to be sent at each round of $(A \to B)^k$ (A sends at least 1 bit per message), thus the average number of bits over these d executions is at least $\frac{k \log_2(d+2)}{2}$.

The previous theorem is stated in terms of the vertex degree d, but actually, a chromatic number is the more appropriate notion. Indeed, it is not necessary for the processes to send their inputs, they need only to identify them locally. Namely, A, with input a, does not need to know the input of B, it is sufficient that it distinguishes the different possible inputs of B, with respect to vertex (A, a). Consider the distance-2 graphs $\mathcal{I}_A$, $\mathcal{I}_B$ with vertices colored A, B resp. in $\mathcal{I}$. The edges of $\mathcal{I}_B$ are $\{(B, v), (B, w)\}$ if there exits a vertex (A, u), adjacent to both $(B, v), (B, w)$ in $\mathcal{I}$. Similarly for $\mathcal{I}_A$. Consider now a vertex coloring of these graphs. All vertices (B, b) that are neighbors of (A, a) in $\mathcal{I}$ have a different color. Hence, A with input a may deduce the input of B from this color. This is what we call *input-colors*, and indeed these values are what it is *necessary* to transmit. The number of input-colors may be equal to the number of input values, e.g. when all pairs of input values are in $\mathcal{I}$ (a common case actually), but it may be much less. Let $c = \max\{chromNumb(\mathcal{I}_A), chromNumb(\mathcal{I}_B)\}$. We have the following result (that can be slightly improved, see Remark 2), assuming the protocol sends these colors instead of input values.

Theorem 5. *The average number of bits sent in an execution of the protocol in Fig. 4, over all executions, is at most $2k + 4 \log c$ bits. Namely, messages are of size at most $1 + \frac{2 \log c}{k}$ bits on average.*

Proof. To compute the number of bits that are sent by the protocol, the two first rounds have to be considered separately. In the first round A and B send at most $\log c + 1$ bits. This happens in the first round of 3^k executions. In total, $2(1 + \log c)3^k$.

We consider the label at the beginning of the round.

In the second round, A with label 0 sends 1 bit, and A with label 1 sends at most $\log c + 1$ bits. A with label 0 will be the second round of 3^{k-1} executions, A with label 1 will be the second round of $2 \cdot 3^{k-1}$ executions. In total $2(3^{k-1} + (\log c + 1) \cdot 2 \cdot 3^{k-1}) = 2 \cdot 3^k + 4(\log c)3^{k-1}$.

In the round i, A with label 0 will be in the i-th round of $3^{k-(i-1)}$ executions, the $(3^{i-1}-1)/2$ A with other labels will be in $2 \cdot 3^{k-(i-1)}$ executions. A with label $(3^{i-1} - 1/2)$ sends at most $(1 + \log c)$ bits. A with other labels sends 1 bit. In total $2(3^{k-(i-1)} + (1 + \log c) \cdot 2 \cdot 3^{k-(i-1)} + (3^{i-1} - 3/2) \cdot 3^{k-(i-1)}) = 2 \cdot 3^k + 4(\log c)3^{k-(i-1)}$.

If we have at least 2 rounds we have $(2 \cdot 3^k + 2(\log c)3^k) + (2 \cdot 3^k + 4(\log c)3^{k-1}) + \sum_{i=3}^{i=k}(2 \cdot 3^k + 4(\log c)3^{k-(i-1)}) = 2k \cdot 3^k + (\log c)(4 \cdot 3^k - 6)$. On average over the 3^k executions, at most $2k + 4 \log c$ bits per execution are exchanged.

4   Full-Information Protocol Using only Beeps

Here we consider the case where all messages sent by a protocol are identical: they consist of a unary signal equal to 1. A process can decide in each round to send a message or not. As before, in every round, the three events $EV = \{A \to B, B \to A, A \leftrightarrow B\}$ are possible. Thus, the message sent by a process will be delivered only if the corresponding delivery event is happening. Notice that if both A and B decide to send a beep in the same round, at least one of them receives a beep. But if only one of them sends, possibly no-one receives.

First, notice that it is possible to solve k-edge approximate agreement in $3k$ rounds, using the algorithm on the right. The figure above it depicts the 3 rounds, where 0 means no beep is sent.

A	B	A	B	A	B	A	B	A	B
0	4	1	3	2	2	3	1	4	0
0	1	1	0	1	1	0	1	1	0
1	0	1	1	0	1	1	0	1	1
1	1	0	1	1	0	1	1	0	1

CODE FOR PROCESSES A AND B
if $id = A$ **then** $d = 1$
    else $d = -1$
round $r = 0, 1, 2$ **do**
    if $\neg(label + d * r \equiv 0 \bmod 3)$ **then** $s_r = true$; send
    if receive in this round **then** $r_r = true$
if $(s_1 \wedge r_1) \vee (s_2 \wedge r_2) \vee (s_3 \wedge r_3)$ **then**
    let i such that $(s_i \wedge r_i)$
    if $i + label \equiv d * 1 \bmod 3$ **then** $label = 3 * label + 1$
    else $label = 3 * label - 1$
else $label = 3 * label$

Now we show that every wait-free solvable task is solvable in the strong beep model. Furthermore, it is still possible to solve any task with a number of rounds that is a constant away from the optimal. An emulation in the beep model, of the communication exchanges between A, B in the general model, is described in Fig. 5. More precisely, assume that V_A (resp. V_B) is the set of possible values sent by A (resp. by B). The emulation runs in rounds (x, y) for $x \in V_A$ and $y \in V_B$ enumerated in some predefined order. If A has to send v_A (resp. B has to send v_B) in the emulated round, A (resp. B) sends a beep at each round (v_A, y) for $y \in V_B$ (resp. round (x, v_B) for $x \in V_A$). Hence they both send beeps only in round (v_A, v_B), and the beep model ensures that at least one of A and B receives a beep. Processes that receive the bit use it in that round, and the emulated values of received bits, $(r_A$ and $r_B)$ are set to v_B for A, or to v_A for B. This implies the correctness of the emulation.

Lemma 4. *The protocol of Fig. 5 emulates a communication round of the wait-free dynamic model in which A sends $v_A \in V_A$ and B sends $V_B \in V_B$, using $|V_A| \times |V_B|$ rounds.*

Let c_a (resp. c_b) be the chromatic number of the graph $\mathcal{I}_A$ (resp. $\mathcal{I}_B$).

Theorem 6. *The k-approximate agreement task on $\mathcal{I}$ is solvable in the beep model in less than $4k \cdot c_a \cdot c_b$ rounds.*

It is possible to improve the constant 4 in the round complexity, using the 3-round emulation technique described above. The two first rounds are special, but after that it is possible to subdivide the edges in the middle using 3 rounds.

5   The Three-Process Case

Here we briefly discuss the case of 3 processes, A, B, C, while avoiding as much as possible technical topology notation. The goal is only to prove that for 3 processes, there is no protocol with fixed message size, which can solve all wait-free solvable tasks in an optimal number of rounds.

5.1   A Wait-Free Model for Three Processes

Interestingly, the impossibility happens already in the case of a single input, which we call *triangle-approximate agreement,* to contrast it with the edge-approximate agreement task.

There are several ways of generalizing our wait-free dynamic network model from 2 to 3 processes, see e.g. [13] or [1] (where characterizations are presented, using full-information protocols), for the results we want to present, it suffices to consider the set of possible communication graphs, for each round, that define a chromatic subdivision (if a full-information protocol is used) and hence, correspond to wait-free models.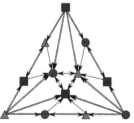
Namely, an *event* is a directed graph on three vertices, A, B, C, and the events EV for 3 processes we consider, are defined by each triangle in the figure. Notice that the boundary for each pair of processes of the chromatic subdivision consists of all events for our 2 processes model. Similarly, we have the following, illustrated in Fig. 6, where C is depicted as a blue corner vertex, that never gets messages from A, B.

---

emulation of send $v_A \in V_A$ for process A and send $v_B \in V_B$ for process B :

```
1    CODE FOR PROCESS A
2        r_A = ⊥
3        forall round (x, y) in some fixed order  do
4            if (x = v_A)
5                then send(beep)
6                if beep = receive() then r_A = y
7        /* r_A is the received message by A
8    CODE FOR PROCESS B
9        r_B = ⊥
10       forall round (x, y) in some fixed order  do
11           if (y = v_B)
12               then send(beep)
13               if beep = receive() then r_B = x
14       /* r_B is the received message by B
```

Fig. 5. Emulation of a round between A and B with beep-rounds.

Lemma 5. *In the k-round full-information protocol, there are 3^k executions, where C runs solo. These define a normal chromatic path for A, B of length 3^k, defined by all schedules where C runs solo.*

5.2   A Triangle Approximate Agreement Task for 3 Processes

The central task for three processes is *k-triangle approximate agreement,* where there is a single input configuration $\sigma = \{(A, 0), (B, 0), (C, 0)\}$ for the 3 processes in $\mathcal{I}$, and thus $\mathcal{I}$ consists of the simplex σ, together with all its subsets. The output complex consists of a chromatic subdivision, iterated k times. The case of $k = 2$ and $k = 3$ is in Fig. 6.

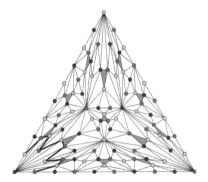
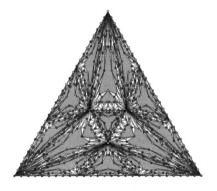

Fig. 6. Chromatically subdivided 2 times (left) and 3 times (right) for 3 processes

We don't give a formal definition of the k-triangle approximate agreement task, the intuition can be seen from the following chromatic subdivision figures. For the impossibility we prove, it is sufficient to define formally only one part of this task.

We focus on one part of the output complex, $\mathcal{O}$, corresponding to all the edges by A and B, incident to the output vertex $(C, 0)$ (blue corner), which is the one C has to output in executions where it runs solo. In these executions, A and B hear from C, but C never hears from them. There is a chromatic normal path in $\mathcal{O}$ of 3^k edges (yellow, green vertices)

$$P_{AB} = (A, 0), (B, (3^k - 1)/2), (A, 1), \ldots, (A, (3^k - 1)/2), (B, 0).$$

Each edge of P_{AB} is contained in the following two triangles. For each edge e of P_{AB}, there is a triangle by adding the vertex $(C, 0)$. Furthermore, for each edge e of P_{AB}, there is a vertex (C, e), and a triangle consisting of e together with (C, e). The meaning is that C should produce as output $(C, 0)$ in a solo execution where A and B decide the values in e, but if C does get a message in the last round of this execution, then C outputs (C, e).

5.3   The Impossibility for 3 Processes

We describe an impossibility result that shows that an optimal protocol cannot use messages of fixed size. At the end we discuss a more general version, similar to Theorem 4.

Consider all schedules of k rounds where C runs solo. Projecting out C, they are isomorphic to the schedules where only A and B participate. Recall that the full-information protocol is isomorphic to the k-edge approximate agreement task, which consists of a normal chromatic subdivision of length 3^k, with boundary that we denote $\{(A, 0), (B, 0)\}$ (Lemma 5).

Consider the $d = 3^{k-1}$ edges where A and B get messages from each other in the last round. For each such edge e, there are two triangles in the output complex of the k-triangle approximate agreement task. One where C runs solo, and another, where C gets a message from both A, B, in round k (the three get messages from each other). Of course, in that last round, A, B cannot tell if their messages were delivered to C or not, so they must decide the same value in both executions.

Let us enumerate the output values of C, as $e_1, e_2, \ldots, e_d$, for each one of the $d = 3^{k-1}$ edges where all get messages from each other in the last round. By definition of the task, in any execution where C runs solo, it has to decide $(C, 0)$. If C runs solo for $k - 1$ rounds, and gets messages from both A, B in round k, then C must decide vertex (C, e_i), if the values decided by A, B are in edge e_i. To see that C cannot decide $(C, 0)$ if it has received a message, an argument of distances similar to Theorem 4 can be used, or directly arguing in terms of emulating a full-information protocol. It follows that A, B must send messages that (together) allow C to distinguish $d = 3^{k-1}$ different edges.

Theorem 7. *Every k-round implementation of a full-information protocol for three processes has executions where A, B send in a round messages (together) of $\Omega(k)$ bits.*

We can get a stronger version, by using an argument similar to the argument of Theorem 4, to show that A, B must send longer and longer messages each round. But this version suffices to show that if instead of using full-information protocols, we use protocols that send messages of a fixed size, then there is a k-triangle approximate agreement task that is not solvable in k rounds (while it is solvable in k rounds using a full-information protocol).

6   Conclusions

We initiated the study of communication complexity of wait-free distributed computing, in a two-process dynamic graph model. In addition to being simple, this model is equivalent to other previously studied models in terms of task computability. We showed that it is possible to implement a full-information protocol using messages of constant size, which do not grow with the number of rounds. Then we showed that even sending only beeps it is possible to do so, at the cost of only a constant number of extra rounds.

Although we focused on the case of two processes, we showed that already for three processes, messages of constant size do not suffice to solve every task in an optimal number of rounds. To prove our main Theorem 5, we identified a parameter that determines communication complexity lower bounds, in the form of an *information chromatic number* of certain graphs related to the possible inputs, and showed that it can mostly be attained by our protocols; this notion seems to be implicit in many of our arguments, but further work is needed to explore this subject.

Many interesting avenues are open to explore in future work. It would be interesting to consider wait-free solvability in beeping models like in [5], which are weaker than ours, because in those models, when A beeps at the same round than B, neither knows if the other beeped. Our results for two processes apply directly to the iterated shared memory model [18], as in this case the models are essentially the same. Our lower bounds apply to non-iterated shared memory versions of our model, but not our algorithms, as the simulations of non-iterated models [11] do not try to optimize the communication complexity. We barely touched the case of $n \geq 3$ processes, much interesting work remains to be done. It is intriguing to consider [16], where an algorithm implementing a single-writer/multi-reader atomic register on an asynchronous message passing system using 2-bit messages is presented.

References

1. Afek, Y., Gafni, E.: A simple characterization of asynchronous computations. Theoret. Comput. Sci. **561**, 88–95 (2015). https://doi.org/10.1016/j.tcs.2014.07.022. http://www.sciencedirect.com/science/article/pii/S0304397514005659
2. Amdur, E.S., Weber, S.M., Hadzilacos, V.: On the message complexity of binary byzantine agreement under crash failures. Distrib. Comp. **5**(4), 175–186 (1992). https://doi.org/10.1007/BF02277665
3. Biran, O., Moran, S., Zaks, S.: A combinatorial characterization of the distributed 1-solvable tasks. J. Algorithms **11**(3), 420–440 (1990). https://doi.org/10.1016/0196-6774(90)90020-F
4. Casteigts, A., Flocchini, P., Quattrociocchi, W., Santoro, N.: Time-varying graphs and dynamic networks. Int. J. Parallel Emergent Distrib. Syst. **27**(5), 387–408 (2012). https://doi.org/10.1080/17445760.2012.668546
5. Casteigts, A., Métivier, Y., Robson, J.M., Zemmari, A.: Counting in one-hop beeping networks. Theor. Comput. Sci. **780**, 20–28 (2019). https://doi.org/10.1016/j.tcs.2019.02.009
6. Dinitz, Y., Moran, S., Rajsbaum, S.: Bit complexity of breaking and achieving symmetry in chains and rings. J. ACM **55**(1), 3:1–3:28 (2008). https://doi.org/10.1145/1326554.1326557
7. Dinitz, Y., Solomon, N.: Two absolute bounds for distributed bit complexity. In: Pelc, A., Raynal, M. (eds.) SIROCCO 2005. LNCS, vol. 3499, pp. 115–126. Springer, Heidelberg (2005). https://doi.org/10.1007/11429647_11
8. Dolev, D., Lynch, N.A., Pinter, S.S., Stark, E.W., Weihl, W.E.: Reaching approximate agreement in the presence of faults. J. ACM **33**(3), 499–516 (1986)
9. Fevat, T., Godard, E.: Minimal obstructions for the coordinated attack problem and beyond. In: 2011 International Parallel and Distributed Processing Symposium (IPDPS), pp. 1001–1011. IEEE, May 2011. https://doi.org/10.1109/IPDPS.2011.96
10. Gafni, E., Koutsoupias, E.: Three-processor tasks are undecidable. SIAM J. Comput. **28**(3), 970–983 (1999)
11. Gafni, E., Rajsbaum, S.: Distributed programming with tasks. In: Lu, C., Masuzawa, T., Mosbah, M. (eds.) OPODIS 2010. LNCS, vol. 6490, pp. 205–218. Springer, Heidelberg (2010). https://doi.org/10.1007/978-3-642-17653-1_17
12. Galil, Z., Mayer, A., Yung, M.: Resolving message complexity of byzantine agreement and beyond. In: 36th Annual Foundations of Computer Science, pp. 724–733. IEEE, Oct 1995. https://doi.org/10.1109/SFCS.1995.492674
13. Herlihy, M., Kozlov, D., Rajsbaum, S.: Distributed Computing Through Combinatorial Topology. Elsevier-Morgan Kaufmann, San Francisco (2013). https://doi.org/10.1016/C2011-0-07032-1

14. Herlihy, M., Rajsbaum, S.: The decidability of distributed decision tasks (extended abstract). In: Proceedings of the Twenty-ninth Annual ACM Symposium on Theory of Computing, STOC 1997, pp. 589–598. ACM, New York (1997). https://doi.org/10.1145/258533.258652

15. Herlihy, M., Shavit, N.: The topological structure of asynchronous computability. J. ACM **46**(6), 858–923 (1999)

16. Mostefaoui, A., Raynal, M.: Two-bit messages are sufficient to implement atomic read/write registers in crash-prone systems. In: Principles of Distributed Computing (PODC), pp. 381–389. ACM (2016). https://doi.org/10.1145/2933057.2933095

17. Pfleger, D., Schmid, U.: On knowledge and communication complexity in distributed systems. In: Lotker, Z., Patt-Shamir, B. (eds.) SIROCCO 2018. LNCS, vol. 11085, pp. 312–330. Springer, Cham (2018). https://doi.org/10.1007/978-3-030-01325-7_27

18. Rajsbaum, S.: Iterated shared memory models. In: López-Ortiz, A. (ed.) LATIN 2010. LNCS, vol. 6034, pp. 407–416. Springer, Heidelberg (2010). https://doi.org/10.1007/978-3-642-12200-2_36

19. Razborov, A.A.: Communication complexity. In: Dierk Schleicher, M.L. (ed.) An Invitation to Mathematics, pp. 97–117. Springer, Heidelberg (2011). https://doi.org/10.1007/978-3-642-19533-4

20. Winkler, K., Schmid, U.: An overview of recent results for consensus in directed dynamic networks. Bull. EATCS **128** (2019). http://bulletin.eatcs.org/index.php/beatcs/article/view/581/585

21. Yue, Y., Lei, F., Liu, X., Wu, J.: Asynchronous computability theorem in arbitrary solo models. Mathematics **8**(5), 757 (2020). https://doi.org/10.3390/math8050757

Distance Labeling Schemes for K_4-Free Bridged Graphs

Victor Chepoi, Arnaud Labourel$^{(\boxtimes)}$, and Sébastien Ratel

Aix Marseille Université, Université de Toulon, CNRS, LIS, Marseille, France
{victor.chepoi,arnaud.labourel,sebastien.ratel}@lis-lab.fr

Abstract. k-Approximate distance labeling schemes are schemes that label the vertices of a graph with short labels in such a way that the k-approximation of the distance between any two vertices u and v can be determined efficiently by merely inspecting the labels of u and v, without using any other information. One of the important problems is finding natural classes of graphs admitting exact or approximate distance labeling schemes with labels of polylogarithmic size. In this paper, we show that the class of K_4-free bridged graphs on n nodes enjoys 4-approximate distance labeling scheme with labels of $O(\log^3 n)$ bits.

Keywords: Bridged graphs · Distance labeling schemes

1 Introduction

A *(distributed) labeling scheme* is a scheme maintaining global information on a network using labels assigned to nodes of the network. Their goal is to locally store some useful information about the network in order to answer a specific query concerning a pair of nodes by only inspecting the labels of the two nodes. Motivation for such localized data structure in distributed computing is surveyed and widely discussed in [19]. The quality of a labeling scheme is measured by the size of the labels of nodes and the time required to answer queries. The predefined queries can be of various types such as distance, adjacency, or routing.

In this paper we investigate distance labeling schemes. A *distance labeling scheme* (*DLS* for short) on a graph family $\mathscr{G}$ consists of an *encoding* function $C_G : V \to \{0,1\}^*$ that gives binary labels to every vertex of a graph $G \in \mathscr{G}$, and of a *decoding* function $D_G : \{0,1\}^* \times \{0,1\}^* \to \mathbb{N}$ that, given the labels of two vertices u and v of G, computes the distance $d_G(u,v)$ between u and v in G. For $k \in \mathbb{N}^*$, we call a labeling scheme a *k-approximate distance labeling scheme* if given the labels of two vertices u and v, the decoding function computes an integer comprised between $d_G(u,v)$ and $k \cdot d_G(u,v)$.

By a result of Gavoille et al. [16], the family of all graphs on n vertices admits a distance labeling scheme using labels of $O(n)$ bits. This scheme is asymptotically optimal since simple counting arguments on the number of n-vertex graphs show that $\Omega(n)$ bits are necessary. Another important result is

© Springer Nature Switzerland AG 2020
A. W. Richa and C. Scheideler (Eds.): SIROCCO 2020, LNCS 12156, pp. 310–327, 2020.
https://doi.org/10.1007/978-3-030-54921-3_18

that trees admit a DLS with labels of $O(\log^2 n)$ bits. Quite recently, Freedman et al. [12] obtained such a scheme allowing constant time distance queries. Several graph classes containing trees also admit DLS with labels of length $O(\log^2 n)$: bounded tree-width graphs [16], distance-hereditary graph [14], bounded clique-width graphs [10], or planar graphs of non-positive curvature [8]. More recently, in [9] we designed DLS with labels of length $O(\log^3 n)$ for cube-free median graphs. Other families of graphs have been considered such as interval graphs, permutation graphs, and their generalizations [4,15] for which an optimal bound of $\Theta(\log n)$ bits was given, and planar graphs for which there is a lower bound of $\Omega(n^{\frac{1}{3}})$ bits [16] and an upper bound of $O(\sqrt{n})$ bits [17]. Other results concern approximate distance labeling schemes. For arbitrary graphs, Thorup and Zwick [22] proposed $(2k-1)$-approximate DLS, for each integer $k \geq 1$, with labels of size $O(n^{1/k} \log^2 n)$. In [13], it is proved that trees (and bounded tree-width graphs as well) admit $(1 + 1/\log n)$-approximate DLS with labels of size $O(\log n \log \log n)$, and this is tight in terms of label length and approximation. They also designed $O(1)$-additive DLS with $O(\log^2 n)$-labels for several families of graphs, including the graphs with bounded longest induced cycle, and, more generally, the graphs of bounded tree–length. Interestingly, it is easy to show that every exact DLS for these families of graphs needs labels of $\Omega(n)$ bits in the worst-case [13].

Finding natural classes of graphs admitting exact or approximate distance labeling schemes with labels of polylogarithmic size is an important and challenging problem. In this note we continue the line of research we started in [9] to investigate classes of graphs with rich metric properties, and we design approximate distance labeling schemes of polylogarithmic size for K_4-free bridged graphs. Together with hyperbolic, median, and Helly graphs, bridged graphs constitute the most important classes of graphs in metric graph theory [3,5]. They occurred in the investigation of graphs satisfying basic properties of classical Euclidean convexity: *bridged graphs* are the graphs in which the neighborhoods of convex sets are convex and it was shown in [11,21] that they are exactly the graphs in which all isometric cycles halve length 3. A local-to-global characterization of bridged graphs was found in [6]: they are exactly the graphs whose clique complexes are simply connected and the neighborhoods of vertices do not containing induced 4- and 5-cycles. This result was rediscovered in [18], where such graphs and complexes were called *systolic*. Bridged (alias systolic) graphs and complexes have been thoroughly investigated in graph theory and in geometric group theory. A K_4-*free bridged graph* is a bridged graph not containing 4-cliques (two examples are given on Fig. 1). Notice that topologically K_4-free bridged graphs are quite general: any graph of girth ≥ 6 may occur in the neighborhood of a vertex of a K_4-free bridged graph. The main result of this paper is the following:

Theorem 1. *The class $\mathscr{G}$ of K_4-free bridged graphs on n vertices admits a 4-approximate distance labeling scheme using labels of $O(\log^3 n)$ bits that can be decoded in constant time.*

The remaining part of this note is organized in the following way. The main ideas of our distance labeling scheme are informally described in Sect. 2.

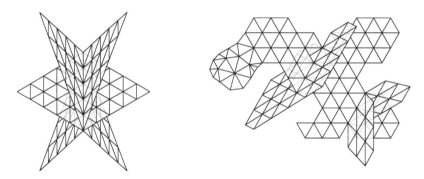

Fig. 1. Examples of K_4-free bridged graphs.

Section 3 introduces the notions used in this paper. The next three Sects. 4, 5, and 6 present the most important geometric and structural properties of K_4-free graphs, which are the essence of our distance labeling scheme. In particular, we describe a partition of vertices of G defined by the star of a median vertex. In Section 7 we characterize the pairs of vertices connected by a shortest path containing the center of this star. The distance labeling scheme and its performances are described in Sect. 8. Due to page limits, the missing proofs and the pseudocodes are provided in the full version of the paper and in Chapter 8 of the PhD thesis of S. Ratel [20].

2   Main Ideas of the Scheme

The global structure of our distance labeling scheme for K_4-free bridged graphs is similar to the one described in [9] for cube-free median graphs. Namely, the scheme is based on a recursive partitioning of the graph into a star and its fibers (which are classified as panels and cones). However, the stars and the fibers of K_4-free bridged graphs have completely different structural and metric properties than those of cube-free median graphs.

Let $G = (V, E)$ be a K_4-free bridged graph on n vertices. The encoding algorithm first searches for a *median vertex* m of G, i.e., m minimizing $x \mapsto \sum_{v \in V} d_G(x, v)$. It then computes a particular 2-neighborhood of m that we call the *star* $\mathrm{St}(m)$ of m. Every vertex of G can be associated to a unique vertex of $\mathrm{St}(m)$. This allows us to define the fibers of $\mathrm{St}(m)$: for a vertex $x \in \mathrm{St}(m)$, the *fiber* $F(x)$ of x corresponds to the set of all the vertices of G associated to x. The set of all the fibers of $\mathrm{St}(m)$ constitutes a partitioning of G. Moreover, choosing m as a median vertex ensures that every fiber contains at most the half of the vertices of G. The fibers are not convex. Nevertheless, they are connected and isometric, and thus induce bridged subgraphs of G. Consequently, we can apply recursively this partitioning to every fiber, without accumulating errors on distances at each step. Finally, we study the *boundaries* and the *total boundary* of those fibers, i.e., respectively, the set of all the vertices of a fiber having a neighbor in another particular fiber, and the union of every boundary of a fiber. We will see that those boundaries do not induce actual trees but something close that we call "starshaped trees". Unfortunately, these starshaped trees are not isometric. This explains why we obtain an approximate and not an exact distance labeling scheme. Indeed, distances computed in the total boundary can be twice as much as the distances in the graph.

We distinguish two types of fibers $F(x)$ depending on the distance between x and m: *panels* are fibers leaving from a neighbor x of m; *cones* are fibers associated to a vertex x at distance 2 from m. One of our main results to obtain a compact labeling scheme establishes that a vertex in a panel admits two "exits" on the total boundary of this panel, and that a vertex in a cone admits an "entrance" on each boundary of the cone (and it appears that every cone has exactly two boundaries). The median vertex m or those "entrances" and "exits" of a vertex u on a fiber $F(x)$ are guaranteed to lie on a path of length at most four times a shortest (u,v)-path for any vertex v outside $F(x)$. It follows that, at each recursive step, every vertex u has to store information relative only to three vertices (m, and the two "entrances" or "exits" of u). Since a panel can have a linear number of boundaries, without this main property, our scheme would use labels of linear length because a vertex in a panel could have to store information relative to each boundary to allow to compute distances with constant (multiplicative) error. Since we allow a multiplicative error of 4 at most, we will see that in almost every case, we can return the length of a shortest (u,v)-path passing through the center m of the star $\mathrm{St}(m)$ of the partitioning at some recursive step (to be determined). Lemmas 13 and 14 respectively indicate when this length corresponds to the exact distance between u and v, and when it is an approximation of the distance (up to factor 2). The case where u and v belong to distinct fibers that are "too close" is more technical and is the one leading to a multiplicative error of four in the worst case.

3   Preliminaries

All graphs $G = (V, E)$ in this note are finite, undirected, simple, and connected. We will write $u \sim v$ if two vertices u and v are adjacent. The *distance* $d_G(u, v)$ between two vertices u and v is the length of a shortest (u, v)-path in G, and the *interval* $I(u, v) := \{x \in V : d_G(u, x) + d_G(x, v) = d_G(u, v)\}$ consists of all the vertices on shortest (u, v)-paths. Let $H = (V', E')$ be a subgraph of G. Then H is called *convex* if $I(u, v) \subseteq H$ for any two vertices u, v of H. The *convex hull* of a subgraph H' of G is the smallest convex subgraph $\mathrm{conv}(H')$ containing H'. A connected subgraph H of G is called *isometric* if $d_H(u, v) = d_G(u, v)$ for any two vertices u, v of H. If an isometric subgraph H of G is a cycle, we call H an *isometric cycle*. The *metric projection* of a vertex $x \in V$ on H is the set $\mathrm{Pr}(x, H) := \{y \in V' : \forall y' \in V', d_G(y', x) \geq d_G(y, x)\}$. For a subset $S \subseteq V$, the *neighborhood of* S in G is the set $N[S] := S \cup \{v \in V \setminus S : \exists u \in S, v \sim u\}$. When S is a singleton s, then $N[s]$ is the closed neighborhood of s. The *ball* of radius k centered at s is the set $B_k(s) = \{v : d_G(s, v) \leq k\}$. Notice that $B_1(s) = N[s]$.

A graph G is *bridged* if any isometric cycle of G has length 3. As shown in [11, 21], bridged graphs are also characterized by one of the fundamental properties of CAT(0) spaces: *the neighborhoods of every convex subgraph of a bridged graph is also convex.* Consequently, balls in bridged graphs are convex. Bridged graphs constitute an important subclass of weakly modular graphs: a graph family that unifies numerous interesting classes of metric graph theory through "local-to-global" characterizations [5]. More precisely, *weakly modular graphs* are the graphs satisfying both the following *quadrangle (QC)* and *triangle (TC) conditions* [1, 7]:

(QC) $\forall u, v, w, z \in V$ with $k := d_G(u, v) = d_G(u, w)$, $d_G(u, z) = k + 1$, and $vz, wz \in E$, $\exists x \in V$ s.t. $d_G(u, x) = k - 1$ and $xv, xw \in E$.
(TC) $\forall u, v, w \in V$ with $k := d_G(u, v) = d_G(u, w)$, and $vw \in E$, $\exists x \in V$ s.t. $d_G(u, x) = k - 1$ and $xv, xw \in E$.

Bridged graphs are *exactly the weakly modular graphs containing no induced cycle of length 4 or 5* [7].

A *metric triangle* $u_1 u_2 u_3$ of a graph $G = (V, E)$ is a triplet u_1, u_2, u_3 of vertices such that for every $(i, j, k) \in \{1, 2, 3\}^3$, $I(u_i, u_j) \cap I(u_j, u_k) = \{u_j\}$ [7]. A metric triangle $u_1 u_2 u_3$ is *equilateral* if $d_G(u_1, u_2) = d_G(u_2, u_3) = d_G(u_1, u_3)$. Weakly modular graphs can be characterized via metric triangles in the following way: a graph G is weakly modular if and only if for any metric triangle $u_1 u_2 u_3$ of G and any $x, y \in I(u_1, u_2)$, the equality $d_G(u_1, x) = d_G(u_1, y)$ holds [7]. In particular, every metric triangle of a weakly modular graph is equilateral.

A metric triangle $u_1' u_2' u_3'$ is called a *quasi-median* of a triplet u_1, u_2, u_3 if for each pair $1 \le i < j \le 3$ there exists a shortest (u_i, u_j)-path passing via u_i' and u_j'. Notice that each triplet u_1, u_2, u_3 of vertices of any graph G admits at least one quasi-median: it suffices to take as u_1' a furthest from u_1 vertex from $I(u_1, u_2) \cap I(u_1, u_3)$, as u_2' a furthest from u_2 vertex from $I(u_2, u_1) \cap I(u_2, u_3)$, and as u_3' a furthest from u_3 vertex from $I(u_3, u_1) \cap I(u_3, u_2)$.

4   Metric Triangles and Intervals

From now on we suppose that G is a K_4-free bridged graph. A *flat triangle* is an equilateral triangle in the triangular grid; for an illustration see Fig. 2 (left). The interval $I(u, v)$ between two vertices u, v of the triangular grid induces a lozenge (see Fig. 2, right). A *burned lozenge* is obtained from $I(u, v)$ by iteratively removing vertices of degree 3; equivalently, a burned lozenge is the subgraph of $I(u, v)$ in the region of the plane bounded by two shortest (u, v)-paths. The vertices of a burned lozenge are naturally classified into *boundary* and *inner* vertices, and border vertices classified into *concave* and *convex corners*.

We denote the convex hull of a metric triangle uvw of K_4-free bridged graph G by $\Delta(u, v, w)$ and call it a *deltoid*. The following two lemmas were known before for K_4-free planar bridged graphs (see for example, [2, Proposition 3] for the first lemma) but their proofs remain the same.

Lemma 1. *Any deltoid $\Delta(u, v, w)$ of G is a flat triangle.*

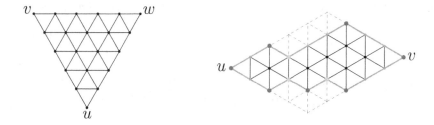

Fig. 2. A flat triangle (left) and a burned lozenge (right). The concave corners are in green and convex ones are drawn in blue. The border is in red and the inner vertices in black.

Lemma 2. *Any interval $I(u, v)$ of G induces a burned lozenge.*

We now introduce starshaped sets and trees, and we describe the structure of the intersection of an interval and a starshaped tree. Let T be a tree rooted at a vertex r. A path P of T is called *increasing* if it is entirely contained on a single branch of T, i.e., if $\forall u, v \in P$, either $I_T(u, r) \subseteq I_T(v, r)$, or $I_T(v, r) \subseteq I_T(u, r)$. A subset S of the vertices of an arbitrary graph G is said to be *starshaped* relatively to $s \in S$ if $I(s, s') \subseteq S$ for all $s' \in S$. If, additionally, any $I(s, s')$ induces a path, then S is called a *starshaped tree* (rooted at $s \in S$).

Lemma 3. *Let T be a starshaped tree of G (rooted at m), and let $u \in V \setminus T$. Then $I(u, m) \cap T$ is contained in the two increasing paths joining m to the two first non-degenerated convex corners of $I(u, m) \cap T$ (these corners are said to be extremal relatively to m) (Fig. 3).*

Fig. 3. Another burned lozenge illustrating Lemma 3. $I(u, m) \cap T$ appears in blue, and the red squares are the two extremal vertices.

5 Stars and Fibers

Let $z \in V$ be an arbitrary vertex of G. Since G is bridged, $N[z]$ is convex. Note that for any $u \in V \setminus N[z]$, z cannot belong to the metric projection $\mathrm{Pr}(u, N[z])$. Indeed, z necessarily has a neighbor z' on a shortest (u, z)-path. This z' is closer to u than z, and it belongs to $N[z]$. Since G is K_4-free and weakly modular, we obtain the following property of projections on $N[z]$.

Lemma 4. $Pr(u, N[z])$ *consists of a single vertex or of two adjacent vertices.*

Let $u \in V$ be a vertex with two vertices x, y in $Pr(u, N[z])$. By Lemma 4 and triangle condition, there exists a vertex $u' \sim x, y$ at distance $d_G(x, u) - 1$ from u. Moreover, $I(u', z) = \{u', z, x, y\}$. The *star* $St(z)$ of a vertex $z \in V$ consists of $N[z]$ plus all vertices $x \notin N[z]$ having two neighbors in $N[z]$, i.e., all x that can be derived by the triangle condition as above; see Fig. 4 (left) for an example.

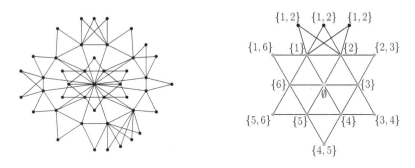

Fig. 4. Example of a star and of the encoding of the vertices of a star.

Let $x \in St(z)$. If $x \in N[z]$, we define the *fiber* $F(x)$ of x respectively to $St(z)$ as the set of all vertices of G having x as unique projection on $N[z]$. Otherwise (if $d_G(x, z) = 2$) $F(x)$ denotes the set of all vertices u such that $Pr(u, N[z])$ consists of two adjacent vertices v and w, and such that x is adjacent to v, w and is one step closer to u than v and w. A fiber $F(x)$ such that $x \sim z$ is called a *panel*. If $d_G(x, z) = 2$, then $F(x)$ is called a *cone*. Two fibers $F(x)$ and $F(y)$ are k-*neighboring* if $d_{St(z)\setminus\{z\}}(x, y) = k$. The two following lemmas are auxiliary.

Lemma 5. *A vertex u in a cone $F(x)$ can not be adjacent to a vertex v in a cone $F(y) \neq F(x)$. Similarly, a vertex $v \neq y$ in a panel $F(y)$ can not be adjacent to a vertex u in a panel $F(x) \neq F(y)$.*

Lemma 6. *Let $F(x)$ and $F(y)$ be two fibers, and let $u \in F(x)$ and $v \in F(y)$ be two adjacent vertices. We set $k := d_G(v, y)$. If $F(y)$ is a cone and if $F(x)$ is a panel, then $d_G(u, y) = d_G(u, x) \in \{k, k+1\}$ and $x \sim y$.*

Lemma 7. $\mathscr{F}_z := \{F(x) : x \in St(z)\}$ *defines a partition of G. Any fiber $F(x)$ is a bridged isometric subgraph of G and $F(x)$ is starshaped with respect to x.*

Proof. The fact that $\mathscr{F}_z$ is a partition follows from its definition. Since any isometric subgraph of a bridged graph is bridged, we have to prove that fiber $F(x)$ is starshaped and isometric. We begin by showing that $F(x)$ is starshaped with respect to x. For that, pick two vertices $u \in F(x)$ and $w \in I(u, x)$. If $F(x)$ is a cone, then $w \in I(u, x)$ and $x \in I(u, z)$ lead to $w \in I(u, z)$. This implies that $w \in F(x)$. Indeed, assume that $w \in F(y) \neq F(x)$ and consider two vertices $u' \in F(x) \cap I(u, w)$ and $w' \in F(y) \cap I(u, w)$ with $u' \sim w'$. By Lemma 6, $d_G(w', x) \in \{d_G(u', x), d_G(u', x) + 1\}$. This contradicts that $w' \in I(u, x)$. Now, if $F(x)$ is a panel and w belongs to a fiber $F(y) \neq F(x)$, then $F(y)$ has to be a cone 1-neighboring $F(x)$ (by Lemma 5). Consequently, y must belong to a shortest (u, x)-path passing via w, contrary to $d_G(u, y) = d_G(u, x)$.

It remains to show that each fiber $F(x)$ is isometric. Let u and v be two vertices of $F(x)$. We consider a quasi-median $u'v'x'$ of the triplet u, v, x. Since $F(x)$ is starshaped by Lemma 7, the intervals $I(u', x')$, $I(x, x')$, $I(u, u')$, $I(v, v')$, and $I(v', x')$ are all contained in $F(x)$. Consequently, to show that u and v are connected in $F(x)$ by a shortest path, it suffices to show that the unique shortest (u', v')-path in the deltoid $\Delta(u', v', x')$ belongs to $F(x)$. To simplify the notations, we can assume two things. First, since $I(u, u'), I(v, v') \subseteq F(x)$, we can let $u = u'$ and $v = v'$. Second, we can assume that $\Delta(u, v, x')$ is a minimal counterexample with $I(u, v) \not\subseteq F(x)$. We define this minimality according to two criteria: base length and height. Let w be the vertex closest to u on the (u, v)-shortest path of $\Delta(u, v, x')$ such that $w \notin F(x)$. The *base length* minimality means that we suppose u adjacent to w (otherwise, we replace u by the neighbor of w in $F(x) \cap \Delta(u, v, x')$). From the structure of deltoids described in Lemma 1, we know that there exists a vertex $u_1 \sim w, u$ at distance $k - 1$ from x, and that $u_1 \in F(x)$. Consider the neighbor w_1 of u_1 in $\Delta(u, v, x')$ at distance $k - 1$ from x. If $w_1 \notin F(x)$, then we replace $\Delta(u, v, x')$ by $\Delta(u_1, v_1, x')$, where $v_1 \sim v$ denotes the neighbor of v in $\Delta(u, v, x')$ at distance $k - 1$ from x. This defines the *height* minimality. Thus $u_1, w_1 \in F(x)$ and, if the path between u_1 and v_1 leaves $F(x)$, then we can also replace $\Delta(u, v, x')$ by $\Delta(u_1, v_1, x')$ and apply again the base length minimality. By iteratively applying the two criteria while it is possible, we can assume that the deltoid $\Delta(u_1, v_1, x')$ is entirely contained in $F(x)$ (see Fig. 5). In particular, $w_1 \in F(x)$. Two cases have to be considered.

Case 1. $F(x)$ is a panel. Then, by Lemma 5, w belongs to a cone 1-neighboring $F(y)$. Since $d_G(w, x) = k$, $d_G(w, y) = k - 1$. Moreover, by Lemma 6, we also have that $d_G(w_1, y) \in \{k - 1, k - 2\}$ and, since $d_G(w, y) = k - 1$, we can conclude that $d_G(w_1, y) = k - 1$. By the triangle condition applied to w, w_1, and y, there exists a vertex $t \sim w, w_1$ at distance $k - 2$ from y. Since $F(y)$ is starshaped, and since $t \in I(w, y)$, $t \in F(y)$. This also means that $d_G(t, x) = k - 1$. By the convexity of the ball $B_{k-1}(x)$, t must coincide with u_1 or with w_1 (otherwise, the quadruplet u_1, w_1, w, t would induce a K_4). By the minimality hypothesis, u_1 and w_1 must belong to $F(x)$, but then $t \in F(x)$, leading to a contradiction.

Case 2. $F(x)$ is a cone. Then w belongs to a panel 1-neighboring $F(y)$, and this case is quite similar to the previous one. By Lemma 6, $d_G(w, y) = d_G(w, x) = k$. We thus have $d_G(w, y) = d_G(w_1, y) = k$, and there exists a vertex $t \sim w, w_1$ at distance $k - 1$ from y. Still using Lemma 6, we deduce that $d_G(t, x) = k - 1$. Since $F(y)$ is starshaped, we obtain that $t \in F(y)$, and from the convexity of the ball $B_{k-1}(x)$ we conclude that $t = w_1$ or $t = u_1$. Finally, $t \in F(x)$. $\square$

If we choose the star centered at a median vertex of G, then the following lemma shows that the number of vertices in every fiber is bounded by $|V|/2$ (the proof is similar to the proof of [9, Lemma 8]).

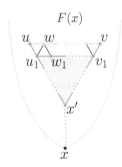

Fig. 5. To the proof of Lemma 7. The minimality hypothesis implies that the blue part belongs to the fiber $F(x)$. The proof then aims to show that the red shortest (u, v)-path also belongs to $F(x)$.

Lemma 8. *If z is a median vertex of G, then for all $x \in \mathrm{St}(z)$, $|F(x)| \leq |V|/2$.*

6   Boundaries and Total Boundaries of Fibers

Let x and y be two vertices of $\mathrm{St}(z)$. The *boundary* $\partial_{F(y)} F(x)$ of $F(x)$ *with respect to* $F(y)$ corresponds to the set of all vertices of $F(x)$ having a neighbor in $F(y)$. The *total boundary* $\partial^* F(x)$ of $F(x)$ is the union of all its boundaries.

Lemma 9. *The total boundary $\partial^* F(x)$ of any fiber $F(x)$ is a starshaped tree.*

Total boundaries of fibers are always starshaped trees, however they are not induced trees of G. The following result is a corollary of Lemma 9.

Corollary 1. *Let x be an arbitrary vertex of $\mathrm{St}(z)$. Then, for every pair u, v of vertices of $\partial^* F(x)$, $\mathrm{d}_G(u, v) \leq \mathrm{d}_{\partial^* F(x)}(u, v) \leq 2 \cdot \mathrm{d}_G(u, v)$.*

We now describe the structure of metric projections of the vertices on the total boundaries of fibers. We then justify that vertices in panels will have a constant number of "exits" on their total boundaries, even if their panel admits an arbitrary number of 1-neighboring cones. This is the purpose of Lemma 12.

Lemma 10. *Let $F(x)$ be a fiber and $u \in V \setminus F(x)$. Then the metric projection $\Pi := Pr(u, F(x)) = Pr(u, \partial^* F(x))$ is an induced tree of G.*

Lemma 11. *Let $F(x)$ be any fiber and $u \in V \setminus F(x)$. There exists a unique vertex u' in $\Pi := Pr(u, F(x))$ at minimum distance from x. Furthermore, $\mathrm{d}_{G(\Pi)}(u', v) \leq \mathrm{d}_G(u, u') = \mathrm{d}_G(u, v)$ for all $v \in \Pi$.*

Proof. The uniqueness of u' follows from the fact that Π is a rooted starshaped tree of $T := \partial^* F(x)$, which itself is a starshaped tree rooted at x. Indeed, every pair (a, b) of vertices of Π admits a nearest common ancestor in Π that coincides with the nearest common ancestor in T. This ancestor has to be closer to x than a and b (or at equal distance if $a = x$ or $b = x$).

Consider a vertex $v \in \Pi$. The equality $k := d_G(u, u') = d_G(u, v)$ holds since Π is the metric projection of u on $F(x)$. Assume by way of contradiction that $d_G(u', v) \geq k+1$. Then $I(v, u') =: P$ is an increasing path of length at least $k+1$ in a starshaped tree. By the triangle condition applied to u and to every pair of neighboring vertices of P, we derive at least k vertices. We then can show that each of these (at least k) vertices has to be distinct from every other (otherwise, P would contain a shortcut). We also can prove that those vertices create a path and, by induction on the length of this new shortest path, deduce that (u, v, t) forms a non-equilateral metric triangle, which is impossible. □

In particular, Lemma 11 establishes that every branch of the tree Π has depth smaller or equal to $d_G(u, u')$.

Lemma 12. *Let u be any vertex of G and let T be a starshaped tree rooted at $r \in V$. Let u_1 and u_2 be the two extremal vertices with respect to r in the two increasing paths of $I(u, r) \cap T$ (there are at most two of them by Lemma 3). Then, for all $v \in T$, the following inequality holds*

$$\min\{d_G(u, u_1) + d_T(u_1, v), d_G(u, u_2) + d_T(u_2, v)\} \leq 2 \cdot d_G(u, v).$$

Proof. Assume that the minimum $\min_{i \in \{1,2\}}\{d_G(u, u_i) + d_T(u_i, v)\}$ is obtained for $i = 1$. Let $x \in T$ be the nearest common ancestor of u_1 and v. Notice that, by Lemma 3, we know that $v \notin I(u, r)$, unless $v = x$. But we can assume that $v \neq x$, otherwise $d_G(u, v) = d_G(u, u_1) + d_T(u_1, v)$ would be shown already. Consider a quasi-median $u'v'r'$ of the triplet u, v, r (see on Fig. 6, left). We can make the following two remarks:

(1) Since T is a starshaped tree and $I(r, v)$ is one of its branches, v' necessarily belongs to this branch. If $v' \in I(r, x)$, then $v' \in I(u_1, r) \cap I(v, r)$ implies that $v' = x$ and that $d_G(u, u_1) + d_T(u_1, v') + d_T(v', v) = d_G(u, v)$. Indeed, if $v' = x$, then $v' \in I(u, r)$ (because $x \in I(u, r)$). So $d_G(u, u_1) + d_G(u_1, v') = d_G(u, v')$. Also, since T is starshaped, $d_G(u_1, v') = d_T(u_1, v')$ and $d_T(v', v) = d_G(v', v)$. Finally, since v' belongs to a quasi-median between u and v, it lies on a shortest (u, v)-path. So $d_G(u, u_1) + d_T(u_1, v') + d_T(v', v) = d_G(u, v') + d_G(v', v) = d_G(u, v)$. Therefore, we consider that $v' \in I(x, v)$.

(2) Since T is starshaped and $r' \in I(v, r)$, this implies that r' belongs to the branch $I(v, r)$ of T. Since $r' \in I(u, r)$, and since $I(x, v) \cap I(u, r) = \{x\}$, we conclude that r' is between r and x. Since $r' \in I(u', r) \cap I(u_1, r)$, and $u_1, u' \in I(u, r)$, we show that $d_G(u, u') + d_G(u', r') = d_G(u, u_1) + d_T(u_1, r')$. Indeed, $u' \in I(u, r')$ because it belongs to a quasi-median of u, v, r ; $u_1 \in I(u, r')$ by definition, so the distance between u and r' passing through u_1 and passing through u' are equal : $d_G(u, u') + d_G(u', r') = d_G(u, u_1) + d_G(u_1, r')$. Finally, since T is starshaped (and $r' \in T$), $d_G(u_1, r') = d_T(u_1, r')$.

Since the quasi-median $u'v'r'$ is an equilateral metric triangle, we have

$$2 \cdot d_G(u,v) \geq d_G(u,u') + d_G(u',r') + d_T(r',v)$$
$$= d_G(u,u_1) + d_G(u_1,r') + d_T(r',v)$$
$$= d_G(u,u_1) + d_T(u_1,v).$$

$\square$

Stated informally, Lemma 12 asserts that if $u \in F(x)$ and $T := \partial^* F(x)$, then the shortest paths from u to any vertex of T are "close" to a path passing via u_1 or via u_2. Thus, if u stores information relative to u_1 and u_2, approximate distances between u and all the vertices of T can be easily computed.

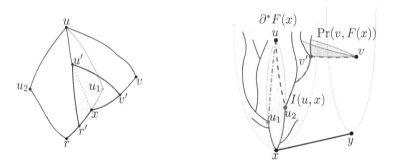

Fig. 6. Notations of Lemma 12 (left), and illustration of the entrance and exits used in Lemma 15 (right).

7   Shortest Paths and Classification of Pairs of Vertices

In this section, we characterize the pairs of vertices of G which are connected by a shortest path passing via the center z of the star $St(z)$ (Lemma 13). We then exhibit cases for which passing via z can lead to a multiplicative factor 2 (Lemma 14). Finally, we present the worst cases where our algorithm could make an error of at most 4 (Lemma 15). For this last case, let us point out that our analysis might not be tight.

Lemma 13. *Let $F(x)$ and $F(y)$ be two fibers, and let $u \in F(x)$ and $v \in F(y)$. Then $z \in I(u,v)$ iff $F(x)$ and $F(y)$ are distinct and either: (i) both are panels and are k-neighboring, for $k \geq 2$; (ii) one is a panel and the other is a k-neighboring cone, for $k \geq 3$; (iii) both are cones and are k-neighboring, for $k \geq 4$;*

Proof. Consider a quasi-median $u'v'z'$ of the triplet u, v, z. The vertex z belongs to a shortest (u,v)-path if and only if $u' = v' = z' = z$. In that case, let $s \in I(u,z)$ and $t \in I(v,z)$ be two neighbors of z. Since z belongs to a shortest (u,v)-path, s and t cannot be adjacent. It follows (see Fig. 7) that $F(x)$ and

$F(y)$ are k-neighboring with: (i) $k \geq 2$ if $F(x)$ and $F(y)$ are both panels; (ii) $k \geq 3$ if one of $F(x)$ and $F(y)$ is a cone, and the other a panel; (iii) $k \geq 4$ if $F(x)$ and $F(y)$ are both cones.

For the converse implication, we consider the cases where z does not belong to a shortest (u, v)-path. First notice that if $F(x) = F(y)$, then z cannot belong to such a shortest path because, then, $d_G(u, v) \leq d_G(u, x) + d_G(x, v) < d_G(u, z) + d_G(z, v)$. We now assume that $F(x) \neq F(y)$. Three cases have to be considered depending on the type of $F(x)$ and $F(y)$.

Case 1. $F(x)$ and $F(y)$ are both panels. If $z = z'$, then according to Lemma 1, x and y must be the two neighbors of z, respectively lying on the shortest (z, u')- and (z, v')-paths, and $x \sim y$, i.e., $F(x)$ and $F(y)$ are 1-neighboring. If $z \neq z'$, we consider a vertex $z'' \in I(z, z')$ adjacent to z. Then $z'', x \in I(u, z)$ and, since u belongs to a panel, $z'' = x$. With the same arguments, we obtain that $z'' = y$. Consequently, $F(x) = F(y)$, contrary to our assumption.

Case 2. $F(x)$ is a cone, and $F(y)$ is a panel (the symmetric case is similar). Let x' and x'' denote the two neighbors of x in the interval $I(x, z)$. If $z = z'$, then, still by Lemma 1, y and x' (or x'') must belong to the deltoid $\Delta(u', v', z')$, and then $x' \sim y$ (or $x'' \sim y$). It follows that $F(x)$ and $F(y)$ are 2-neighboring. If $z \neq z'$, we consider again a neighbor z'' of z in $I(z, z')$. Since $x', x'' \in I(u, z)$, z'' must coincide with x' or with x'', say $z'' = x'$. Also, $z'' = y$. Consequently, $F(x)$ and $F(y)$ are 1-neighboring.

Case 3. $F(x)$ and $F(y)$ are both cones. Let x' and x'' denote the two neighbors of x in $I(x, z)$, and let y' and y'' be those of y in $I(z, y)$. Again, if $z = z'$, then x' (or x'') and y' (or y'') belong to the deltoid $\Delta(u', v', z')$, leading to $x' \sim y'$ and to the fact that $F(x)$ and $F(y)$ are 3-neighboring. If $z \neq z'$, we consider $z'' \in I(z, z')$, $z'' \sim z$. By arguments similar to those used in previous cases, we obtain that $z'' = x' = y'$ (up to a renaming of the vertices x'' and y''). It follows that $F(x)$ and $F(y)$ are 2-neighboring.   □

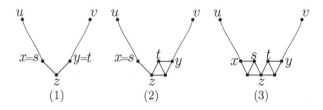

Fig. 7. To the proof of Lemma 13.

Let x and y be two vertices of $St(m)$ and let $(u, v) \in F(x) \times F(y)$. If $F(x) = F(y)$, then u and v are called *close*. If $F(x)$ and $F(y)$ are 1-neighboring, one of the fibers being a panel and the other a cone, then u and v are called *1pc-neighboring*. Finally, if $F(x)$ and $F(y)$ are as described in Lemma 13, i.e., if $z \in I(u, v)$, then u and v are called *separated*.

Lemma 14. *Let u and v be two vertices which are neither close, nor 1pc-neighboring, and nor separated. Then, $d_G(u, v) \leq d_G(u, z) + d_G(z, v) \leq 2 \cdot d(u, v)$.*

Proof. We denote by $F(x)$ and $F(y)$ the fibers respectively containing u and v, and we consider a quasi-median $u'v'z'$ of the triplet u, v, z. We have to show that $z = z'$. According to Lemma 13, four cases must be considered: $F(x)$ and $F(y)$ are two 1-neighboring or 2-neighboring fibers of distinct types and $F(x)$ and $F(y)$ are two 2-neighboring or 3-neighboring cones.

We first assume that $F(x)$ and $F(y)$ are 1-neighboring, one of them being a panel and the other a cone. If x and y belong to a shortest (u, v)-path, then $z = z'$. Let us assume that this is not the case. Then there exists a cone $F(w) \sim F(x), F(y)$ such that $I(u, v) \cap F(w) \neq \emptyset$. We claim that, if $z' \notin F(w)$, then $z = z'$. Indeed, this directly follows from the fact that $z' \in I(u, z) \cap I(v, z)$, $x \notin I(v, z)$ and $y \notin I(u, z)$. We now show that $z' \notin F(w)$. To do so, notice that $x \notin I(v, z)$, $y \notin I(u, z)$n and $x, y \in I(w, z)$ imply that $w \notin I(u, z) \cup I(v, z)$. By the definition of $u'v'z'$, we have in particular that $z' \in I(u, z)$. Moreover, $z' \in F(w)$ would lead to the contradiction $w \in I(z', z) \subseteq I(u, z)$.

We now assume that $F(x)$ and $F(y)$ are 2-neighboring, one of them being a panel and the other a cone. Suppose $F(y)$ to be the panel, and denote by x_1 and x_2 the two neighbors of x in $I(x, z)$. The same way as before, we show that $z = z'$. Indeed, $z' \in I(u, z)$ requires that $z' \in F(x) \cup F(x_1) \cup F(x_2) \cup \{z\}$, and $z' \in I(v, z)$ requires that $z' \in F(y) \cup \{z\}$. We conclude that $z = z'$.

The same arguments as in the two previous cases allow us to establish the result in the last two cases. □

Let $F(x)$ be a panel and let $F(y)$ be a cone 1-neighboring $F(x)$. We set $T := \partial^* F(x)$. Let $u \in F(x)$ and $v \in F(y)$. Recall that $\Pi := \Pr(v, T)$ induces a tree, according to Lemma 10. The vertex v' of Π the closest to x is called the *entrance of v* on the total boundary T. Similarly, two vertices u_1 and u_2 such as described in Lemma 12 are called the *exits of u* on the total boundary T. Confer to Fig. 6 (right) for an illustration of the notations of this paragraph.

Lemma 15. *Let $u \in F(x)$ and $v \in F(y)$ be two 1pc-neighboring vertices, where $F(x)$ is a panel and $F(y)$ a cone. Let T, u_1, u_2 and v' be such as described just above. Then,*

$$d_G(u, v) \leq \min\{d_G(u, u_1) + d_T(u_1, v'), d_G(u, u_2) + d_T(u_2, v')\} + d_G(v', v)$$
$$\leq 4 \cdot d_G(u, v).$$

The proof of Lemma 15 is more technical than previous proofs and makes use of tools unnecessary for the main purpose of this note.

8   Distance Labeling Scheme

We now describe our 4-approximate distance labeling scheme for K_4-free bridged graphs, based on the properties obtained in previous sections.

8.1 Encoding

We begin with a brief description of the encoding of stars of median vertices a K_4-free bridged graph $G = (V, E)$ (we explain in Section 8.2 how to use it to decode the distances). We then describe the labels L(u) given to vertices $u \in V$ by our encoding algorithm.

Encoding of the Star. Consider a median vertex m of G and the star St(m) of this vertex. The star-labeling of a vertex u of St(m) is denoted by $L_{St(m)}(u)$. We set $L_{St(m)}(m) := 0$ (where 0 will be considered as the empty set $\varnothing$). Each neighbor of m takes a distinct label in the range $\{1, \ldots, \deg(m)\}$ (interpreted as singletons). The label $L_{St(m)}(u)$ of a vertex u at distance 2 from m corresponds to the concatenation of the labels $L_{St(m)}(u')$ and $L_{St(m)}(u'')$ of the two neighbors u' and u'' of u in $I(u, m)$, i.e., $L_{St(m)}(u)$ is a set of size 2.

Remark 1. Note that vertices of St(m) not adjacent to m do not necessarily have unique identifiers. Moreover, this labeling of St(m) does not allow to determine adjacency of arbitrary pairs of vertices. Indeed, adjacency queries between vertices encoded by a singleton cannot be answered (such a singleton label only tells that the corresponding vertex is adjacent to m, see Fig. 4, right).

Algorithm 1: ENC_DIST $(G, L(V))$

Input: A K_4-free bridged graph $G = (V, E)$, and a list
$L(V) := \{L(u) := id(u) : u \in V\}$ of unique identifiers of its vertices.

1   **if** $V = \{v\}$ **then stop**;

2   Find a median vertex m of G ;

3   $L_{St(m)}(St(m)) \leftarrow$ Enc_Star $(St(m))$;

4   **foreach** panel $F(x) \in \mathcal{F}_m$ **do**

5      $L_{\partial^* F(x)}(\partial^* F(x)) \leftarrow$ Enc_Tree $(\partial^* F(x))$;

6   **foreach** $u \in F(x)$ **do**

7   Find the exits u_1 and u_2 of u on $\partial^* F(x)$;

8         $L_{St} \leftarrow (id(m), d_G(u, m), L_{St(m)}(x))$;

9         $L_{1st} \leftarrow (L_{\partial^* F(x)}(u_1), d_G(u, u_1))$;

10        $L_{2nd} \leftarrow (L_{\partial^* F(x)}(u_2), d_G(u, u_2))$;

11        $L(u) \leftarrow L(u) \circ (L_{St}, L_{1st}, L_{2nd})$;

12   ENC_DIST $(F(x), L(V))$;

13 **foreach** cone $F(x) \in \mathcal{F}_m$ **do**

14   **foreach** $u \in F(x)$ **do**

15   Find the panels $F(w_1)$ and $F(w_2)$ 1-neighboring $F(x)$;

16   Choose two entrances u_1^+ and u_2^+ of u on $F(w_1)$ and $F(w_2)$;

17         $L_{St} \leftarrow (id(m), d_G(u, m), L_{St(m)}(x))$;

18         $L_{1st} \leftarrow (L_{\partial^* F(w_1)}(u_1^+), d_G(u, u_1^+))$;

19         $L_{2nd} \leftarrow (L_{\partial^* F(w_2)}(u_2^+), d_G(u, u_2^+))$;

20        $L(u) \leftarrow L(u) \circ (L_{St}, L_{1st}, L_{2nd})$;

21   ENC_DIST $(F(x), L(V))$;

Encoding of the K_4-Free Bridged Graphs. Let u denote any vertex of G. We describe here the part $L_i(u)$ of the label of u built at step i of the recursion by the encoding procedure (see Enc_Dist). $L_i(u)$ consists of three parts: "St", "1st", and "2nd". The first part L_i^{St} contains information relative to the star $St(m)$ around the median m chosen in the corresponding step: the unique identifier $id(m) =: L_i^{St[Med]}(u)$ of m in G; the distance $d_G(u,m) =: L_i^{St[Dist]}(u)$ between u and m; and a star labeling $L_{St(x)}(m) =: L_i^{St[Root]}(u)$ of u in $St(m)$ (where $x \in St(m)$ is such that $u \in F(x)$). This last identifier is used to determine to which type of fibers the vertex u belongs, as well as the status (close, separated, 1pc-neighboring, or other) of the pair (u,v) for any other vertex $v \in V$.

The two subsequent parts, L_i^{1st} and L_i^{2nd}, contain information relative to the two entrances/exists u_1 and u_2 of u on (i) the total boundaries of the two 1-neighboring fibers $F(w_1)$ and $F(w_2)$ of $F(x)$, if $F(x)$ is a cone or on (ii) the total boundary of $F(x)$, if $F(x)$ is a panel. If $F(x)$ is a cone, then L_i^{1st} contains (1) an exact distance labeling $L_{\partial^* F(w_1)}(u_1) =: L_i^{1st[rep]}(u)$ of u_1 in the starshaped tree $\partial^* F(w_1)$ (the DLS described in [12], for example) and (2) the distance $d_G(u,u_1) =: L_i^{1st[Dist]}(u)$ between u and u_1 in G. If $F(x)$ is a panel, then L_i^{1st} contains (1) an exact distance labeling $L_{\partial^* F(x)}(u_1) =: L_i^{1st[rep]}(u)$ of u_1 in the total boundary of $F(x)$ and (2) the distance $d_G(u,u_1) =: L_i^{1st[Dist]}(u)$ between u and u_1 in G. Finally, the part L_i^{2nd} is the same as L_i^{1st} with u_2 replacing u_1 (and $\partial^* F(w_2)$ instead of $\partial^* F(w_1)$ if $F(x)$ is a cone).

8.2 Distance Queries

Given the labels $L(u)$ and $L(v)$ of two vertices u and v, the distance decoder (see DISTANCE below) starts by determining the state of the pair (u,v). To do so, it looks up for the first median m that separates u and v, i.e., such that u and v belong to distinct fibers with respect to the star $St(m)$. More precisely, it looks for the part i of the labels corresponding to the step in which m became a median. As noticed in [9], it is possible to find this median vertex m in constant time by adding particular $O(\log^2 n)$ bits information to the head of each label. Once the right part of label is found, the decoding function determines that two vertices are 1pc-neighboring if and only if the identifier (i.e., the star-label in $St(m)$ of the fiber of one of the two vertices u,v is included in the identifier of the other). In that case, the decoding function calls a procedure based on Lemma 15 (see Dist_1pc-neighboring below). More precisely, the procedure returns

$$\min\{d_G(u,u_1) + d_T(u_1,v'), d_G(u,u_2) + d_T(u_2,v')\} + d_G(v',v),$$

where we assume that u belongs to a panel (and v belongs to a cone), where u_1, u_2 and v' are contained in the label parts $L_i^{2nd}(u)$ (or $L_i^{1st}(u)$) and $L_i^{2nd}(v)$ (or $L_i^{1st}(v)$), and where the distances $d_T(u_1,v')$ and $d_T(u_2,v')$ are obtained by decoding the tree distance labels of u_1, u_2, and v' in T (also available in these label parts). We also point out that we assume that $L_i^{1st}(v)$ always contains the information to get to the panel whose identifier corresponds to the minimum of the two values identifying the cone of v. In each remaining case (i.e., when u and v are neither close, nor 1pc-neighboring), the decoding algorithm will return $d_G(u,m) + d_G(v,m)$. By Lemmas 13 and 14, this sum is sandwiched between

$d_G(u, v)$ and $2 \cdot d_G(u, v)$. The following procedure `Dist_1pc-neighboring` is based on Lemma 15 and assumes that its first argument u belongs to a panel, and the second v belongs to a cone.

> **function** `Dist_1pc-neighboring`($L_i(u)$, $L_i(v)$):
> $\quad$ dir $\leftarrow$ 2nd ; // If $L_i^{St[Root]}(u) = \max\{j : j \in L_i^{St[Root]}(v)\}$
> $\quad$ **if** $L_i^{St[Root]}(u) = \min\{j : j \in L_i^{St[Root]}(v)\}$ **then**
> $\qquad$ dir $\leftarrow$ 1st ;
> $\quad$ $v' \leftarrow L_i^{dir[rep]}(v)$;
> $\quad$ $u_1,\ u_2 \leftarrow L_i^{1st[rep]}(u),\ L_i^{2nd[rep]}(u)$;
> $\quad$ $d_1,\ d_2 \leftarrow$ `Dist_Tree`(v', u_1), `Dist_Tree`(v', u_2) ;
> $\quad$ **return** $L_i^{dir[Dist]}(v) + \min\left\{d_1 + L_i^{1st[Dist]}(u),\ d_2 + L_i^{2nd[Dist]}(u)\right\}$;

The following algorithm DISTANCE finds the first step where the given vertices u and v have belonged to distinct fibers for the first time. If they are 1pc-neighboring at this step, then DISTANCE calls procedure `Dist_1pc-neighboring`. Otherwise, it returns the sum of their distances to the median of the step.

Algorithm 2: DISTANCE $(L(u), L(v))$

Input: The labels $L(u)$ and $L(v)$ of two vertices u and v of G
Output: A value between $d_G(u, v)$ and $2 \cdot d_G(u, v)$

1 **if** $L_0(u) = L_0(v)$ /* $u = v$ */ **then return** 0 ;

2 Let i be the greatest integer such that $L_i^{St[Med]}(u) = L_i^{St[Med]}(v)$;
$\quad$ // If u is in a panel 1pc-neighboring the cone of v
3 **if** $L_i^{St[Root]}(u) \subsetneq L_i^{St[Root]}(v)$ **then**
4 $\quad$ **return** `Dist_1pc-neighboring`$(L_i(u), L_i(v))$;
$\quad$ // If v is in a panel 1pc-neighboring the cone of u
5 **if** $L_i^{St[Root]}(v) \subsetneq L_i^{St[Root]}(u)$ **then**
6 $\quad$ **return** `Dist_1pc-neighboring`$(L_i(v), L_i(u))$;
$\quad$ // In every other case
7 **return** $L_i^{St[Dist]}(u) + L_i^{St[Dist]}(v)$;

8.3 Correctness and Complexity

The number of recursive steps is $O(\log |V|)$ since by Lemma 8 the number of vertices in every part is at least divided by 2. At each recursive step, the vertices add to their label a constant number of information among which the longest consists in a distance labeling scheme for trees using $O(\log^2 |V|)$ bits. It follows that our scheme uses $O(\log^3 |V|)$ bits for each vertex.

The fact that the decoding algorithm returns distances with a multiplicative error at most 4 directly follows from Lemmas 13 and 14 for the non-1pc-neighbors cases and from Lemma 15 for 1pc-neighboring vertices. Those results are based on Lemmas 11 and 12 that respectively indicate the entrances and exits to store in total boundaries of panels. This concludes the proof of Theorem 1.

We would like to finish this paper with the following question: *Does there exist constants c and b such that any bridged graph G admits a c-approximate* *distance labeling scheme with labels of size* $O(\log^b n)$? The same question can be asked for bridged graphs of constant clique-size and for hyperbolic bridged graphs (i.e., bridged graphs in which all deltoids have constant size).

Acknowledgment. The work on this paper was supported by ANR project DISTAN-CIA (ANR-17-CE40-0015).

References

1. Bandelt, H.J., Chepoi, V.: A Helly theorem in weakly modular space. Discret. Math. **160**(1–3), 25–39 (1996)
2. Bandelt, H.J., Chepoi, V.: The algebra of metric betweenness ii: axiomatics of weakly median graphs. Euro. J. Combin. **29**, 676–700 (2008)
3. Bandelt, H.J., Chepoi, V.: Metric graph theory and geometry: a survey. Contemp. Math. **453**, 49–86 (2008)
4. Bazzaro, F., Gavoille, C.: Localized and compact data-structure for comparability graphs. Discret. Math. **309**, 3465–3484 (2009)
5. Chalopin, J., Chepoi, V., Hirai, H., Osajda, D.: Weakly modular graphs and non-positive curvature. Memoirs of AMS (2019)
6. Chepoi, V.: Graphs of some CAT(0) complexes. Adv. Appl. Math. **24**, 125–179 (2000)
7. Chepoi, V.: Classification of graphs by means of metric triangles. Metody Diskret. Analiz. **45**, 75–93 (1989)
8. Chepoi, V., Dragan, F.F., Vaxès, Y.: Distance and routing labeling schemes for non-positively curved plane graphs. J. Algorithms **61**, 60–88 (2006)
9. Chepoi, V., Labourel, A., Ratel, S.: Distance labeling schemes for cube-free median graphs. In: MFCS. vol. 138, pp. 15:1–15:14 (2019)
10. Courcelle, B., Vanicat, R.: Query efficient implementation of graphs of bounded clique-width. Discret. Appl. Math. **131**, 129–150 (2003)
11. Farber, M., Jamison, R.E.: On local convexity in graphs. Discret. Math. **66**(3), 231–247 (1987)
12. Freedman, O., Gawrychowski, P., Nicholson, P.K., Weimann, O.: Optimal distance labeling schemes for trees. In: PODC, pp. 185–194. ACM (2017)
13. Gavoille, C., Katz, M., Katz, N.A., Paul, C., Peleg, D.: Approximate distance labeling schemes. In: auf der Heide, F.M. (ed.) ESA 2001. LNCS, vol. 2161, pp. 476–487. Springer, Heidelberg (2001). https://doi.org/10.1007/3-540-44676-1_40
14. Gavoille, C., Paul, C.: Distance labeling scheme and split decomposition. Discret. Math. **273**, 115–130 (2003)
15. Gavoille, C., Paul, C.: Optimal distance labeling for interval graphs and related graph families. SIAM J. Discret. Math. **22**, 1239–1258 (2008)
16. Gavoille, C., Peleg, D., Pérennès, S., Raz, R.: Distance labeling in graphs. J. Algorithms **53**, 85–112 (2004)
17. Gawrychowski, P., Uznanski, P.: A note on distance labeling in planar graphs. CoRR abs/1611.06529 (2016). http://arxiv.org/abs/1611.06529
18. Januszkiewicz, T., Świątkowski, J.: Simplicial nonpositive curvature. Publications Mathématiques de l'Institut des Hautes Études Scientifiques, **104**(1), 1–85 (2006). https://doi.org/10.1007/s10240-006-0038-5
19. Peleg, D.: Distributed Computing: A Locality-Sensitive Approach. SIAM, Philadelphia (2000)

20. Ratel, S.: Densité. Aix-Marseille Université, VC-dimension et étiquetages de graphes (2019)
21. Soltan, V.P., Chepoi, V.: Conditions for invariance of set diameters under d-convexification in a graph. Cybernetics **19**(6), 750–756 (1983). https://doi.org/10.1007/BF01068561
22. Thorup, M.: Compact oracles for reachability and approximate distances in planar digraphs. J. ACM **51**, 993–1024 (2004)

Game Theory

Multi-winner Election Control via Social Influence

M. Abouei Mehrizi[(✉)] and Gianlorenzo D'Angelo

Gran Sasso Science Institute, L'Aquila, Italy
{mohammad.aboueimehrizi,gianlorenzo.dangelo}@gssi.it

Abstract. The extensive use of social media in political campaigns has motivated the recent study of *election control problem in social networks*. In an election, we are given a set of voters, each having a preference list over a set of candidates, that are distributed on a social network. The winners of the election are computed by aggregating the preference lists of voters according to a so-called voting rule. We consider a scenario where voters may change their preference lists as a consequence of the messages received by their neighbors in a social network. Specifically, we consider a political campaign that spreads messages in a social network in support or against a given candidate and the spreading follows a dynamic model for information diffusion. When a message reaches a voter, this latter changes its preference list according to an update rule. The election control problem asks to find a bounded set of nodes to be the starter of a political campaign in support (constructive problem) or against (destructive problem) a given target candidate c, in such a way that the margin of victory of c w.r.t. its most voted opponents is maximized. It has been shown that several variants of the problem can be solved within a constant factor approximation of the optimum, which shows that controlling elections by means of social networks is doable and constitutes a real problem for modern democracies. Most of the literature, however, focuses on the case of single-winner elections.

In this paper, we define the election control problem in social networks for *multi-winner elections* with the aim of modeling parliamentarian elections. Differently from the single-winner case, we show that the multi-winner election control problem is *NP*-hard to approximate within any factor in both constructive and destructive cases. We then study a relaxation of the problem where votes are aggregated on the basis of parties (instead of single candidates), which is a variation of the so-called *straight-party voting* used in some real parliamentarian elections. We show that the latter problem remains *NP*-hard but can be approximated within a constant factor.

1 Introduction

Nowadays, social media are extensively used and have become a crucial part of our life. Generating information and spreading in social media is one of the

This work has been partially supported by the Italian MIUR PRIN 2017 Project ALGADIMAR "Algorithms, Games, and Digital Markets".

A. W. Richa and C. Scheideler (Eds.): SIROCCO 2020, LNCS 12156, pp. 331–348, 2020.
https://doi.org/10.1007/978-3-030-54921-3_19

cheapest and most effective ways of advertising and sharing content and opinions. People feel free to share their opinion, information, news, or also gain something by learning or teaching in social media; on the other hand, they also use social media to get the latest news and information. Many people even prefer to check social media rather than news websites.

Social media are also exploited during election campaigns to support some party or a specific candidate. Many political parties diffuse targeted messages in social media with the aim of convincing users to vote for their candidates. Usually, these messages are posted by influential users and diffused on the network following a cascade effect, also called social influence. There are shreds of evidence of control election using the effect of social influence by spreading some pieces of information, including fake news or misinformation [21]. The presidential election of the United States of America is a real example. It has been shown that on average, ninety-two percent of Americans remembered pro-Trump false news, and twenty-three percent of them remembered the pro-Clinton fake news [2]. There are more real-life examples that have been presented in the literature [4,16,19,24].

This motivated the study of election control problems in social networks by using dynamic models for influence diffusion. We are given a social network of voters, a set of candidates, and a dynamic model for diffusion of information that models the spread of messages produced by political campaigns. The problem asks to find a bounded set of voters/nodes to be the starter of a political campaign in support of a given target candidate c, in such a way that the margin of victory of c w.r.t. its opponents is maximized. Each voter has its own preference list over the candidates and the winner of an election is determined by aggregating all preference lists according to some specific voting rule. Voters are autonomous, however their opinions about the candidates, and hence their preference lists, may change as a consequence of messages received by neighbors. When a message generated by a political campaign reaches a node, this latter changes its preference list according to some specific update rule. When the campaign aims to make the target candidate win, we refer to the *constructive* problem, while when the aim is to make c lose, we refer to the *destructive* problem. This problem recently received some attention. Most of the works in the area, however, focus on single-winner voting systems, while several scenarios require voting systems with multiple winners, e.g., parliamentarian elections.

In this paper, we consider the problem of *multi-winner election control via social influence*, where there are some parties, each with multiple candidates, and we want to find at most B nodes to spread a piece of news in the social network in such a way that a target party elects a large number of its candidates. In this model, more than one candidate will be elected as the winner, and parties try to maximize some function of the number of winners from their party. We considered this problem for some well-known objective functions in both constructive and destructive cases.

Related Work. There is an extensive literature about manipulation or control of elections, we refer to the survey in [14] for relevant work on election control without the use of social networks. In the following, we focus on election control problems where the voters are the nodes of a social network, which recently received some attention.

Finding strategies to maximize the spread of influence in a network is one of the main topics in network analysis. Given a network and a dynamic model for the diffusion of influence, find a bounded set of nodes to be the starters of a dynamic process of influence spread in such a way that the number of eventually influenced users is maximized. The problem, known as Influence Maximization (IM), has been introduced by Domingos and Richardson [12,22] and formalized by Kempe et al., who gave a $(1 - 1/e)$-approximation algorithm [18] for two of the most used dynamic models, namely Independent Cascade Model (ICM) and Linear Threshold Model (LTM). We point the reader to the book by Chen et al. [9] and to [18].

Wilder and Vorobeychik [25] started the study of election control by means of IM. They defined an optimization problem that combines IM and election control called *election control through influence maximization* that is defined as follows. We are given a set of candidates, a social network of voters, each having a preference list over the candidates, a budget B, and a specific target candidate $c_\star$. The network allows the diffusing influence of individuals according to ICM. When a node/voter v is influenced, it changes its preference list in such a way that the rank of $c_\star$ in the preference list of v is promoted (constructive) or demoted (destructive) by one position. At the end of a diffusion process, the voters elect a candidate according to the plurality rule [26]. The problem asks to find a set of at most B nodes to start a diffusion process in such a way that the chances for $c_\star$ to win (constructive) or lose (destructive) at the end of the diffusion are maximized. Wilder and Vorobeychik used the Margin of Victory (MoV) as an objective function and showed that there exists a greedy algorithm that approximates an optimal solution by a factor $1/3(1 - 1/e)$ for constructive and $1/2(1-1/e)$ for the destructive case. The same problem has been extended to LTM and general scoring rules [26] by Corò et al. [10,11]. They have shown that the problem can be approximated within the same bound. A similar problem has been studied in [13]. The authors consider a network where each node is a set of voters with the same preference list, and edges connect nodes whose preference lists differ by the ordering of a single pair of adjacent candidates. They use a variant of LTM for influence diffusion and show that the problem of making a specific candidate win is *NP*-hard and fixed-parameter tractable w.r.t. the number of candidates. Bredereck and Elkind [6] considered the following election control problem. Given a network where the influence spread according to a variant of LTM in which each node has a fixed threshold, and all edges have the same weight, find an optimal way of bribing nodes or add/delete edges in order to make the majority of nodes to vote for a target candidate. A different line of research investigates a model in which each voter is associated with a preference list over the candidates, and it updates its list according to the majority of opinions of its neighbors in the network [3,5,7]. All the previous works on election control through IM consider single-winner voting systems. Multi-winner voting systems raised recent and challenging research trends, we refer to a recent book chapter [15] and references therein.

Our Results. We introduce the multi-winner election control problem via social influence and show that it is *NP*-hard to approximate within any factor $\alpha > 0$, for two common objective functions known as *margin of victory* and *difference of winners* using a general scoring rule. This is in contrast with the previous work on single-winner election control through IM, in which it is possible to approximate the optimum within a constant factor. The hardness results hold for both constructive and destructive cases. Given the hardness result, we focus on a relaxed version of the problem, which is a variation of straight-party voting. We show that this latter remains *NP*-hard but admits a constant factor approximation algorithm for both constructive and destructive cases.

2   Multi-winner Election Control

In this section, we introduce the multi-winner election control problem. We consider elections with k winners and general scoring rule as a voting system, which includes many well-known scoring rules, such as plurality, approval, Borda, and veto [26]. We first introduce the models that we use for diffusion of influence and for updating the preference list of voters. Then we introduce the objective functions for the election control problem in both constructive and destructive cases.

Model for Influence Diffusion. We use the *Independent Cascade Model* (ICM) for influence diffusion [18]. In this model, we are given a directed graph $G = (V, E)$, where each edge $(u, v) \in E$ has a weight $b_{uv} \in [0, 1]$. The influence starts with a set of seed nodes S and keeps activating the nodes in at most $|V|$ discrete steps. In the first step, all the seed nodes S become active. In the next steps $i > 1$, all the nodes that were active in step $i - 1$ remain active, moreover, each node u that became active at step $i - 1$ tries to activate its outgoing neighbors at step i with probability b_{uv}, for each node $v \in N_u^o$. An active node will try to activate its outgoing neighbors independently and only once. The process stops when no new node becomes active. We denote by A_S the set of nodes that are eventually active by the diffusion started by the seeds S.

Model for Multi-winner Election Control. We consider a multi-winner election in which k candidates will be elected. Let $G = (V, E)$ be a directed social graph, where the nodes are the voters in the election, and the edges represent social relationships among users. The voters influence each other the same as ICM. We consider t parties $C_1, C_2, \ldots, C_t$, each having k candidates, $C_i = \{c_1^i, c_2^i, \ldots, c_k^i\}$, $1 \leq i \leq t$. Without loss of generality, we assume that C_1 is the target party. The set of all candidates is denoted by C, i.e. $C = \bigcup_{i=1}^t C_i$. Each voter $v \in V$ has a preference list π_v over the candidates. For each $c \in C$, we denote by $\pi_v(c) \in \{1, 2, \ldots, tk\}$ the rank (or position) of the candidate c in the preference list of node v.

Given a budget B, we want to select a set of B seed nodes that maximizes the number of candidates in C_1 who win the election after a political campaign that spread according to ICM starting from nodes S (see next section for a formal definition of the objective functions).[1]

After S, nodes in A_S will change the positions of candidates in C_1 in their preferences list. In contrast, nodes not in A_S will maintain their original preference list. The update rule for active nodes depends on the position of the target candidates and the goal of the campaign, i.e., if it is a constructive or a destructive one. We denote the preference list of node v after the process by $\tilde{\pi}_v$. If $v \notin A_S$, then $\tilde{\pi}_v = \pi_v$. In the following, we focus on nodes $v \in A_S$.

In the constructive case, like in the model in [25], the position of the target candidates in the list of active nodes will be decreased by one, if there is at least one opponent candidate in a smaller rank. The candidates who are overtaken will be demoted by the number of target candidates that were just after them. Formally, in the constructive case, the position of the candidates after the diffusion starting from seed S will change as follows. For each node $v \in A_S$ and for each target candidate $c \in C_1$, the new position of c in v is

$$\tilde{\pi}_v(c) = \begin{cases} \pi_v(c) - 1 & \text{if } \exists\, c' \in C \setminus C_1 \text{ s.t. } \pi_v(c') < \pi_v(c) \\ \pi_v(c) & \text{otherwise,} \end{cases}$$

while, for each opponent candidate $c \in C \setminus C_1$, if there exists a candidate $c' \in C_1$ s.t. $\pi_v(c') = \pi_v(c) + 1$ we have $\tilde{\pi}_v(c) = \pi_v(c) + |\{c'' \in C_1 \mid \pi_v(c'') > \pi_v(c) \wedge (\nexists\, \bar{c} \in C \setminus C_1 : \pi_v(c) < \pi_v(\bar{c}) < \pi_v(c''))\}|$, otherwise, we set $\tilde{\pi}_v(c) = \pi_v(c)$.

In the destructive case, we want to reduce the number of winners in C_1 and then each node $v \in A_S$ increases their position by one, if it is possible. Formally, after S the preferences list of the candidates will change as follows. For each node $v \in A_S$ and for each target candidate $c \in C_1$, the new position of c in v is

$$\tilde{\pi}_v(c) = \begin{cases} \pi_v(c) + 1 & \text{if } \exists\, c' \in C \setminus C_1 \text{ s.t. } \pi_v(c') > \pi_v(c) \\ \pi_v(c) & \text{otherwise,} \end{cases}$$

while for $c \in C \setminus C_1$, if there exists a candidate $c' \in C_1$ s.t. $\pi_v(c') = \pi_v(c) - 1$ we have $\tilde{\pi}_v(c) = \pi_v(c) - |\{c'' \in C_1 \mid \pi_v(c'') < \pi_v(c) \wedge (\nexists\, \bar{c} \in C \setminus C_1 : \pi_v(c'') < \pi_v(\bar{c}) < \pi_v(c))\}|$, otherwise, we set $\tilde{\pi}_v(c) = \pi_v(c)$.

As an example, if there are two parties with three candidates each, and the initial preferences list of a node is $(c_1^2, c_1^1, c_2^1, c_2^2, c_3^1, c_3^2)$, then if the node becomes active its preferences list in the constructive case will be $(c_1^1, c_2^1, c_1^2, c_3^1, c_2^2, c_3^2)$, i.e., all of the candidates c_i^1 will promote, and all the overtaken candidates will demote; while in the destructive case, it will be $(c_1^2, c_2^2, c_1^1, c_2^1, c_3^2, c_3^1)$, and all of the candidates in our target party demote, and all the overtaken candidates will promote.

The above rule for updating the preference lists is commonly used in the literature [10,25]. In this model, we consider just one message, which contains

---

[1] In the remainder of the paper, by *after S*, we mean *after the diffusion process started from the set of seed nodes S*.

some positive/negative information about the target party that will affect all the target candidates.

We consider a non-increasing scoring function $f(i)$, $1 \leq i \leq |C|$, such that for all $j > i > 0$ we have $f(j) \leq f(i)$. A candidate $c \in C$ gets $f(\pi_v(c))$ and $f(\tilde{\pi}_v(c))$ points from voter v before and after a diffusion, respectively. In other words, each voter will reveal his preferences list, and each candidate will get some score according to his position in the list and the scoring function. Also, we assume w.l.o.g. that there exist $1 \leq i < j \leq |C|$ such that $f(i) > f(j)$, i.e., the function does not return a fixed number for all ranks. The score of a candidate c is the sum of the scores received by all voters. The k candidates with the highest score will be elected.

We denote by $\mathcal{F}(c, S)$ the expected overall score received by candidate c after S, formally, $\mathcal{F}(c, S) = \mathbb{E}_{A_S}\left[\sum_{v \in V} f(\tilde{\pi}_v(c))\right]$ and $\mathcal{F}(c, \emptyset) = \sum_{v \in V} f(\pi_v(c))$.

Objective Functions. The objective function for the constructive election control problem in the single-winner case is maximizing the *margin of victory* (MoV) defined in [25]. Let us consider the difference between the votes for the target candidate and those for the most voted opponent candidate. MoV is the change of this value after S. Note that the most voted opponent before and after S might change. The notion of MoV captures the goal of a candidate to have the largest margin in terms of votes w.r.t. any other candidate. We extend the above definition of MoV in the case of *multi-winner* election control. Since the main goal is to elect more candidates from the target party, then we define the objective function in terms of the number of winning candidates in our target party before and after S.

Given a set A_S of nodes that are active at the end of a diffusion process started from S, we denote by $\mathcal{F}_{A_S}(c)$ the score that a candidate $c \in C$ receives if the activated nodes are A_S, and by $\mathcal{Y}_{A_S}(c)$ the number of candidates that have less score than the candidate c. As a tie-breaking rule, we assume that c_i^j has priority over $c_{i'}^{j'}$ if $j < j'$, or $j = j'$ and $i < i'$. In particular, the target candidates have priority over opponents when they have the same score. Then, for each $c_i^j \in C$, $i \in \{1, \ldots, k\}$, $j \in \{1, \ldots, t\}$, $\mathcal{Y}_{A_S}(c_i^j, S)$ is defined as $\mathcal{Y}_{A_S}(c_i^j) = \Big|\{c_{i'}^{j'} \in C \mid \mathcal{F}_{A_S}(c_i^j) > \mathcal{F}_{A_S}(c_{i'}^{j'}) \vee (\mathcal{F}_{A_S}(c_i^j) = \mathcal{F}_{A_S}(c_{i'}^{j'}) \wedge (j < j' \vee (j = j' \wedge i < i')))\}\Big|$.

For a party C_i, we define $\mathcal{F}(C_i, S)$ as the expected number of candidates in C_i that win the election after S; formally,

$$\mathcal{F}(C_i, S) = \mathbb{E}_{A_S}\left[\sum_{c \in C_i} \mathbb{1}_{\mathcal{Y}_{A_S}(c) \geq (t-1)k}\right]. \tag{1}$$

We denote by C_B and C_A^S the opponent party with the highest number of winners *before* and *after* S, respectively. For the constructive case, the *margin of victory* (MoV$_c$) for party C_1, w.r.t. seeds S, is defined as follows:

$$\mathrm{MoV}_c(C_1, S) = \mathcal{F}(C_1, S) - \mathcal{F}(C_A^S, S) - \big(\mathcal{F}(C_1, \emptyset) - \mathcal{F}(C_B, \emptyset)\big),$$

while for the destructive case, it is defined as:

$$\text{MoV}_d(C_1, S) = \mathcal{F}(C_1, \emptyset) - \mathcal{F}(C_\text{B}, \emptyset) - \big(\mathcal{F}(C_1, S) - \mathcal{F}(C_\text{A}^S, S)\big).$$

The *Constructive (Destructive, resp.) Multi-winner Election Control problem* (CMEC (DMEC, resp.)) asks to find a set S of B seed nodes that maximizes $\text{MoV}_c(C_1, S)$ ($\text{MoV}_d(C_1, S)$, resp.), where $B \in \mathbb{N}$ is a given budget.

In some scenarios, it is enough to maximize the difference between the number of our target candidates who win the election before and after S; we call this objective function the *difference of winners* (DoW), and for the constructive case we define it as follows:

$$\text{DoW}_c(C_1, S) = \mathcal{F}(C_1, S) - \mathcal{F}(C_1, \emptyset).$$

While for the destructive model it is defined as:

$$\text{DoW}_d(C_1, S) = \mathcal{F}(C_1, \emptyset) - \mathcal{F}(C_1, S).$$

The problems of finding a set of at most B seed nodes that maximize DoW_c and DoW_d, for a given integer B, are called *Constructive Difference of Winners* (CDW) and *Destructive Difference of Winners* (DDW), respectively.

3   Hardness Results

In this section, we show the hardness of approximation results for the problems defined in the previous section. We first focus on the constructive case and prove that CMEC and CDW are *NP*-hard to approximate within any approximation factor $\alpha > 0$. Then, we show that the same results hold for DMEC and DDW. All the results hold even when the instance is deterministic (i.e. $b_{uv} = 1$, for each $(u, v) \in E$) and when $t = 3$ and $k = 2$. Note that for $t = 1$ the problem is trivial and for $k = 1$ the problem reduces to the single-winner case.

Constructive Election Control. We first give an intuition of the hardness of approximation proof, which is formally given in Theorem 1. Consider an instance of the constructive case in which $t = k = 2$, $C_1 = \{c_1^1, c_2^1\}$, $C_2 = \{c_1^2, c_2^2\}$, and $C = C_1 \cup C_2$. The weight of all edges are equal to 1, that is, the diffusion is a deterministic process. Also, assume the scoring rule is plurality, i.e., $f(1) = 1$, $f(2) = f(3) = f(4) = 0$. Moreover, the nodes are partitioned into two sets of equal size, V_1 and V_2. In the preferences lists of all nodes in V_1, candidate c_1^2 is in the first position and c_1^1 is in the second position, while in the preferences lists of nodes in V_2, candidate c_2^2 is in first position and c_2^1 is in second position. In this instance, initially party C_1 does not have any elected candidate, that is, $\mathcal{F}(c_1^1, \emptyset) = \mathcal{F}(c_2^1, \emptyset) = 0$, $\mathcal{F}(c_1^2, \emptyset) = |V_1|$, $\mathcal{F}(c_2^2, \emptyset) = |V_2|$, $\mathcal{F}(C_1, \emptyset) = 0$, and $\mathcal{F}(C_2, \emptyset) = 2$.

Consider a diffusion process starting from seeds S that activate nodes A_S (note that, since the weights are all equal to 1, A_S is a deterministic set for any fixed S). The number of candidates that receives fewer votes than a target candidate c_i^1 after the diffusion process is $\mathcal{Y}_{A_S}(c_i^1) = \left| \{ c_{i'}^{j'} \in C \mid \mathcal{F}_{A_S}(c_i^1) > \mathcal{F}_{A_S}(c_{i'}^{j'}) \vee (\mathcal{F}_{A_S}(c_i^1) = \mathcal{F}_{A_S}(c_{i'}^{j'}) \wedge (j' = 2 \vee i < i')) \} \right|$.

Let us consider the case $i = 1$ and analyze the conditions that a seed set S must satisfy in order to include a candidate in the above set, i.e., make c_1^1 win. We analyze the three other candidates $c_{i'}^{j'}$ separately.

- If $j' = 2$ and $i' = 1$, then we must have $\mathcal{F}_{A_S}(c_1^1) \geq \mathcal{F}_{A_S}(c_1^2)$. Since the preferences list of each active nodes in V_1 is updated in a way that c_1^1 moves to the first position and c_1^2 moves to the second position, and the active nodes in V_2 do not affect the rankings of c_1^1 and c_1^2, we have that $\mathcal{F}_{A_S}(c_1^1) = |A_S \cap V_1|$ and $\mathcal{F}_{A_S}(c_1^2) = |V_1 \setminus A_S|$. Therefore, $\mathcal{F}_{A_S}(c_1^1) \geq \mathcal{F}_{A_S}(c_1^2)$ if and only if $|A_S \cap V_1| \geq |V_1 \setminus A_S| = |V_1| - |V_1 \cap A_S|$, which means that $|A_S \cap V_1| \geq |V_1|/2$.
- If $j' = 2$ and $i' = 2$, then we must have $\mathcal{F}_{A_S}(c_1^1) \geq \mathcal{F}_{A_S}(c_2^2)$. In this case, we still have $\mathcal{F}_{A_S}(c_1^1) = |A_S \cap V_1|$, and, since c_2^2 is moved down by one position for each active node in V_2, then $\mathcal{F}_{A_S}(c_2^2) = |V_2 \setminus A_S|$. This implies that $\mathcal{F}_{A_S}(c_1^1) \geq \mathcal{F}_{A_S}(c_2^2)$ if and only if $|A_S \cap V_1| \geq |V_2 \setminus A_S| = |V_2| - |V_2 \cap A_S|$, which means $|A_S \cap V_1| + |A_S \cap V_2| \geq |V_2|$.
- If $j' = 1$ and $i' = 2$, then we must have $\mathcal{F}_{A_S}(c_1^1) \geq \mathcal{F}_{A_S}(c_2^1)$. We again have $\mathcal{F}_{A_S}(c_1^1) = |A_S \cap V_1|$, and, since c_2^1 is moved by one position up for each active node in V_2, then $\mathcal{F}_{A_S}(c_2^1) = |A_S \cap V_2|$. Therefore, $\mathcal{F}_{A_S}(c_1^1) \geq \mathcal{F}_{A_S}(c_2^1)$ if and only if $|A_S \cap V_1| \geq |A_S \cap V_2|$.

Similar conditions hold for $i = 2$.

In order to elect candidate c_1^1 we should have $\mathcal{Y}_{A_S}(c_1^1) \geq (t-1)k = 2$, which means, we should find a seed set that satisfies at least two of the above conditions (or the corresponding conditions to elect c_2^1). Note that finding a seed set S of size at most B that satisfies any pair of the above conditions is a *NP*-hard problem since it requires to solve the IM problem, which is *NP*-hard even when the weight of all edges is 1 [18]. Let us assume that an optimal solution is able to elect both candidates in C_1 (e.g. by influencing $|V_1|/2$ nodes from V_1, and $|V_2|/2$ nodes from V_2), then the optimal MoV_c and DoW_c are equal to 4 and 2, respectively. Moreover, in this case $C_A^S = C_B = C_2$, then $\text{MoV}_c(C_1, S) = \mathcal{F}(C_1, S) - \mathcal{F}(C_1, \emptyset) + \mathcal{F}(C_2, \emptyset) - \mathcal{F}(C_2, S)$. Since $\mathcal{F}(C_2, \emptyset) - \mathcal{F}(C_2, S) = \mathcal{F}(C_1, S) - \mathcal{F}(C_1, \emptyset)$, i.e., for each candidate lost by C_2 there is a candidate gained by C_1, then $\text{MoV}_c(C_1, S) = 2(\mathcal{F}(C_1, S) - \mathcal{F}(C_1, \emptyset)) = 2\text{DoW}_c(C_1, S)$. Since $\mathcal{F}(C_1, \emptyset) = 0$, any approximation algorithm for CDW or CMEC must find a seed set S s.t. $\mathcal{F}(C_1, S) > 0$ and this requires to elect at least one candidate in C_1 (see Eq. (1)), which is *NP*-hard. It follows that it is *NP*-hard to approximate CMEC and CDW within any factor, as formally shown in the next theorem.

Theorem 1. *It is NP-hard to approximate* CMEC *and* CDW *within any factor* $\alpha > 0$.

Proof. We reduce the decision version of the deterministic IM problem, to CMEC and CDW, where *deterministic* refers to the weight of the edges in the graph, i.e., the weight of all edges is equal to 1. Let us define the decision version of the IM problem as follows: *Given a directed graph $G = (V, E)$ and budget $B \leq |V|$. Is there a set of seed nodes $S \subseteq V$ such that $|S| \leq B$ and $A_S = V$?*

Let $\mathcal{I}(G, B)$ be a deterministic instance of the decision IM problem (then, using a given seed set S, we can find the exact number of activated nodes in polynomial time). We create an instance $\mathcal{I}'(G', B)$ of CMEC and CDW, where $G' = (V \cup V', E \cup E')$. We use the same budget B for both problems. We first investigate the case where $t = 3, k = 2$, and consider two different cases as follows.

C1. If $f(1) = f(2) = f(3) = a, f(4) = f(5) = f(6) = b$ for $a, b \in \mathbb{R} \wedge a > b \geq 0$, we call this case *exceptional*, and do as follows.

For each $v \in V$ we add one more node in V' and it has just one incoming edge from v, i.e., $\forall v \in V : v_1 \in V', (v, v_1) \in E'$.

We set the preferences of all nodes $v \in V$ and its new outgoing neighbor as follows: $v = (c_1^2 \succ c_2^2 \succ c_1^3 \succ c_1^1 \succ c_2^1 \succ c_2^3), v_1 = (c_1^3 \succ c_2^3 \succ c_1^2 \succ c_2^1 \succ c_1^1 \succ c_2^2)$ where $a \succ b$ means a is preferred to b.

C2. For any non-increasing scoring function except the exceptional ones, we call it *general* and do as follows.

For each $v \in V$ we add three more nodes in V' and each of them has just one incoming edge from v, i.e., $\forall v \in V : v_1, v_2, v_3 \in V', (v, v_1), (v, v_2), (v, v_3) \in E'$.

We set the preferences of all nodes $v \in V$ and its new outgoing neighbors as follows: $v = (c_1^2 \succ c_1^1 \succ c_1^3 \succ c_2^2 \succ c_2^1 \succ c_2^3), v_1 = (c_2^2 \succ c_2^1 \succ c_2^3 \succ c_1^2 \succ c_1^1 \succ c_1^3), v_2 = (c_1^2 \succ c_1^3 \succ c_1^1 \succ c_2^2 \succ c_2^3 \succ c_2^1), v_3 = (c_2^2 \succ c_2^3 \succ c_2^1 \succ c_1^2 \succ c_1^3 \succ c_1^1)$.

In both cases, the weight of all edges is 1, i.e., $b_{uv} = 1$ for all $(u, v) \in E \cup E'$. The score of candidates before any diffusion is as follows.

C1. $\mathcal{F}(c_1^1, \emptyset) = \mathcal{F}(c_2^1, \emptyset) = |V|(f(4) + f(5)) = 2b|V|, \mathcal{F}(c_1^2, \emptyset) = \mathcal{F}(c_1^3, \emptyset) = |V|(f(1) + f(3)) = 2a|V|, \mathcal{F}(c_2^2, \emptyset) = \mathcal{F}(c_2^3, \emptyset) = |V|(f(2) + f(6)) = (a + b)|V|$. Since $a > b \geq 0$, it yields that $\mathcal{F}(C_2, \emptyset) = \mathcal{F}(C_3, \emptyset) = 1, \mathcal{F}(C_1, \emptyset) = 0$, and none of our target candidates win the election.

C2. $\mathcal{F}(c_1^1, \emptyset) = \mathcal{F}(c_2^1, \emptyset) = |V|(f(2) + f(3) + f(5) + f(6)), \mathcal{F}(c_1^2, \emptyset) = \mathcal{F}(c_2^2, \emptyset) = |V|(2f(1) + 2f(4)), \mathcal{F}(c_1^3, \emptyset) = \mathcal{F}(c_2^3, \emptyset) = |V|(f(2) + f(3) + f(5) + f(6))$. Since $f(\cdot)$ is a non-increasing function, it yields $\mathcal{F}(C_2, \emptyset) = 2$ and $\mathcal{F}(C_1, \emptyset) = \mathcal{F}(C_3, \emptyset) = 0$ and none of our target candidates win the election.

In $\mathcal{I}'(G', B)$, in both cases, all of the nodes $v \in V \cup V'$ become active if and only if all of the nodes $v \in V$ become active. Indeed, by definition, if $V \subseteq A_S$, then for each node $u \in V'$ there exists an incoming neighbor $v \in V$ s.t. $(v, u) \in E'$ and $b_{vu} = 1$, then if v is active, also u becomes active.

Suppose there exists an α−approximation algorithm called α-*appAlg* for CDW (resp. CMEC) and it returns $S \subseteq V \cup V'$ as a solution. We show that, by using the seed nodes S returned by the algorithm $\alpha\text{-}appAlg$, we can find the answer for the decision IM problem. We will show that $\mathrm{DoW}_c(C_1, S) > 0$ (resp. $\mathrm{MoV}_c(C_1, S) > 0$) if and only if S activates all of the nodes, i.e., $A_S = V \cup V'$. That is $\mathrm{DoW}_c(C_1, S) > 0$ (resp. $\mathrm{MoV}_c(C_1, S) > 0$) if and only if the answer to the decision IM problem is YES.

W.l.o.g., we assume $S \subseteq V$, because if there exists a node $u \in S \cap V'$, we can replace it with the node $v \in V$ s.t. $(v, u) \in E'$. Since $b_{uv} = 1$, this does not decrease the value of $\mathrm{DoW}_c(C_1, S)$ or $\mathrm{MoV}_c(C_1, S)$.

We now show that if $\mathrm{DoW}_c(C_1, S) > 0$ (resp. $\mathrm{MoV}_c(C_1, S) > 0$), then $A_S = V \cup V'$. By contradiction, assume that S will not activate all of the nodes, i.e., there exists a node v in $V \setminus A_S$. Then, the score of the candidates will be as follows.

C1. $\mathcal{F}(c_1^1, S) = \mathcal{F}(c_2^1, S) \leq (a + b)(|V| - 1) + 2b$, $\mathcal{F}(c_1^2, S) = \mathcal{F}(c_1^3, S) \geq (a + b)(|V| - 1) + 2a$, $\mathcal{F}(c_2^2, S) = \mathcal{F}(c_2^3, S) = |V|(f(2) + f(6)) = (a + b)|V|$. Since $a > b \geq 0$, then none of the target candidates will be among the winners, i.e., $\mathcal{F}(C_2, S) = \mathcal{F}(C_3, S) = 1$ and $\mathcal{F}(C_1, S) = 0$ and $\mathrm{DoW}_c(C_1, S) = \mathrm{MoV}_c(C_1, S) = 0$.

C2. $\mathcal{F}(c_1^1, S) = \mathcal{F}(c_2^1, S) \leq (|V| - 1)(f(1) + f(2) + f(4) + f(5)) + (f(2) + f(3) + f(5) + f(6))$, $\mathcal{F}(c_1^2, S) = \mathcal{F}(c_2^2, S) \geq (|V| - 1)(f(1) + f(2) + f(4) + f(5)) + (2f(1) + 2f(4))$, $\mathcal{F}(c_1^3, S) = \mathcal{F}(c_2^3, S) \geq (|V| - 1)(f(3) + f(4) + 2f(6)) + (f(2) + f(3) + f(5) + f(6))$. Since $f(\cdot)$ is a non-increasing function, then $\mathcal{F}(C_1, S) = \mathcal{F}(C_3, S) = 0$ and $\mathcal{F}(C_2, S) = 2$. Therefore $\mathrm{DoW}_c(C_1, S) = \mathrm{MoV}_c(C_1, S) = 0$.

In both cases we have a contradiction. To show the other direction, if all of the nodes become active, then the score of candidates will be as follows.

C1. For each $1 \leq i \leq 2$ and $1 \leq j \leq 3$, $\mathcal{F}(c_i^j, S) = (a + b)|V|$. Due to the tie-breaking rule it follows that both of our target candidates will be among the winners, i.e., $\mathcal{F}(C_1, S) = 2$ and $\mathcal{F}(C_2, S) = \mathcal{F}(C_3, S) = 0$.

C2. $\mathcal{F}(c_1^1, S) = \mathcal{F}(c_2^1, S) = \mathcal{F}(c_1^2, S) = \mathcal{F}(c_2^2, S) = |V|(f(1) + f(2) + f(4) + f(5))$, $\mathcal{F}(c_1^3, S) = \mathcal{F}(c_2^3, S) = |V|(f(3) + f(4) + 2f(6))$. Then, $\mathcal{F}(C_1, S) = 2$ and $\mathcal{F}(C_2, S) = \mathcal{F}(C_3, S) = 0$.

Therefore we have $\mathrm{DoW}_c(C_1, S) > 0$ (resp. $\mathrm{MoV}_c(C_1, S) > 0$), and it concludes the proof. The proof can be generalized for any $t, k > 2$, see [1]. □

Destructive Election Control. The following theorem shows the hardness of approximation of the destructive case. The proof is similar to that of Theorem 1, and hence it is omitted. Note that if we consider maximizing DoW_d, the destructive case can be reduced to the constructive model. We cannot apply the same reduction to the problem of maximizing MoV_d as the opponent party with the highest number of winners (i.e., C_B, C_A^S) may be different from that of the constructive case.

Theorem 2. *It is NP-hard to approximate* DMEC *and* DDW *within any factor* $\alpha > 0$.

4   Straight-Party Voting

Since all the variants of the multi-winner election control problems considered so far are NP-hard to be approximated within any factor, we now consider a relaxation of the problem in which, instead of focusing on the number of elected target candidates, we focus on the overall number of votes obtained by the target party. The rationale is that, even if a party is not able to (approximately) maximize the number of its winning candidates because it is computationally unattainable, it may want to maximize the overall number of votes, in the hope that these are not too spread among the candidates and still leads to a large number of seats in the parliament.

Moreover, the voting system that we obtain by the relaxation is used in some real parliamentary elections [23], and is called of *Straight-party voting* (SPV) or *straight-ticket voting* [8,20]. SPV was used very much until around the 1960s and 1970s in the United States. After that, the United States has declined SPV among the general voting; nevertheless, strong partisans are still voting according to SPV. Interestingly, the first time that every state voting for a Democrat for Senate also voted Democratic for president (and the same stability for Republicans) was the 2016 elections of the United states [17].

Note that in this model, if we consider that the controller targets a single candidate instead of a party and the preference lists are over candidates, then we can easily reduce the problem to the single-winner case. The same holds if the controller targets a party and the preference lists are over parties. Therefore, we assume that voters have preference lists over the candidates, but since the voting system is SPV and voters have to vote for a party, then they will cast a vote for each party based on the position of the candidates of the party in their preferences list, e.g., if the preferences list of a node $v \in V$ is $c_1^1 \succ c_1^2 \succ c_2^2 \succ c_2^1$, then the scores of v for party C_1 will be $f(1) + f(4)$, and $f(2) + f(3)$ for party C_2.

Let us define $\mathcal{F}_{spv}(C_i, \emptyset)$ and $\mathcal{F}_{spv}(C_i, S)$ as sum of the scores obtained by party C_i in SPV before and after S, respectively, as follows.

$$\mathcal{F}_{spv}(C_i, \emptyset) = \sum_{v \in V} \sum_{c \in C_i} f(\pi_v(c)), \quad \mathcal{F}_{spv}(C_i, S) = \mathbb{E}_{A_S}\Big[\sum_{v \in V} \sum_{c \in C_i} f(\tilde{\pi}_v(c))\Big].$$

As in the previous case, we denote by C_B and C_A^S the most voted opponents of C_1 before and after S, respectively. We define MoV_c and MoV_d for SPV as

$$\mathrm{MoV}_c^{spv}(C_1, S) = \mathcal{F}_{spv}(C_1, S) - \mathcal{F}_{spv}(C_A^S, S) - (\mathcal{F}_{spv}(C_1, \emptyset) - \mathcal{F}_{spv}(C_B, \emptyset)),$$
$$\mathrm{MoV}_d^{spv}(C_1, S) = \mathcal{F}_{spv}(C_1, \emptyset) - \mathcal{F}_{spv}(C_B, \emptyset) - (\mathcal{F}_{spv}(C_1, S) - \mathcal{F}_{spv}(C_A^S, S)).$$

Also, we define *difference of votes* for constructive (DoV_c) and destructive (DoV_d) as

$$\mathrm{DoV}_c^{spv}(C_1, S) = \mathcal{F}_{spv}(C_1, S) - \mathcal{F}_{spv}(C_1, \emptyset),$$
$$\mathrm{DoV}_d^{spv}(C_1, S) = \mathcal{F}_{spv}(C_1, \emptyset) - \mathcal{F}_{spv}(C_1, S).$$

Theorem 3. *Maximizing MoV and DoV in the constructive and destructive* SPV *problems is NP-hard.*

Proof. As in Theorem 1, we use the decision version of the deterministic IM problem. Let $\mathcal{I}(G, B)$ be a deterministic instance of the decision IM problem. We create an instance $\mathcal{I}'(G', B)$ of SPV, where $G' = G$ and B is the same budget for both problems. Assume $t = k = 2$, and we are using a non-increasing scoring function. Consider the minimum j such that $1 \le i < j \le |C|$ and $f(i) > f(j)$, i.e., j is the minimum rank that has less score than rank $i = j - 1$. Note that $2 \le j \le 4$. We set the preferences list of each voter $v \in V$ in graph G' as follows.

C1. If $j = 2$. $v : c_1^2 \succ c_1^1 \succ c_2^1 \succ c_2^2$. In this case, $\mathcal{F}_{spv}(C_1, \emptyset) = (f(2) + f(3))|V|$.
C2. If $j = 3$. $v : c_1^2 \succ c_2^2 \succ c_1^1 \succ c_2^1$. In this case, $\mathcal{F}_{spv}(C_1, \emptyset) = (f(3) + f(4))|V|$.
C3. If $j = 4$. $v : c_1^1 \succ c_1^2 \succ c_2^2 \succ c_2^1$. In this case, $\mathcal{F}_{spv}(C_1, \emptyset) = (f(1) + f(4))|V|$.

By this preferences assignment, if all of the nodes become active after S, then the score and DoV_c^{spv} for party C_1 will be as following.

C1. $\mathcal{F}_{spv}(C_1, S) = (f(1) + f(2))|V|$,

$\text{DoV}_c^{spv}(C_1, S) = (f(1) + f(2))|V| - (f(2) + f(3))|V| = (f(1) - f(3))|V|$;

and, if at least one node is not active $\text{DoV}_c^{spv}(C_1, S) < (f(1) - f(3))|V|$.
C2. $\mathcal{F}_{spv}(C_1, S) = (f(2) + f(3))|V|$,

$\text{DoV}_c^{spv}(C_1, S) = (f(2) + f(3))|V| - (f(3) + f(4))|V| = (f(2) - f(4))|V|$;

also, if at least one node is not active $\text{DoV}_c^{spv}(C_1, S) < (f(2) - f(4))|V|$.
C3. $\mathcal{F}_{spv}(C_1, S) = (f(1) + f(3))|V|$,

$\text{DoV}_c^{spv}(C_1, S) = (f(1) + f(3))|V| - (f(1) + f(4))|V| = (f(3) - f(4))|V|$;

moreover, if at least one node is not active $\text{DoV}_c^{spv}(C_1, S) < (f(3) - f(4))|V|$.

Then by this reduction, we can distinguish between the case that all of the nodes become active or not, which is the answer of IM problem. In this case, since there are just two parties, whatever C_2 looses will go for C_1. Then, $\text{MoV}_c^{spv}(C_1, S) = 2\text{DoV}_c^{spv}(C_1, S)$, which means we also can answer to IM problem by maximizing MoV_c^{spv}. The generalized version of this proof, $t, k > 0$, is available in the extended version [1].

Regarding the destructive case, the reduction is similar to the constructive one, except that we set the preferences of the voters s.t. at least one of the candidates $c \in C_1$ can decrease the score of C_1. The same approach gets the NP-hardness result. □

We now give an approximation algorithm for the problems of maximizing DoV_c^{spv} and DoV_d^{spv} that is based on a reduction to the node-weighted version of the IM problem. We construct an instance of this problem where the weight to each node $v \in V$, which is equal to the increase in the score of C_1 when v becomes active. The node-weighted IM problem can be approximated by a factor of $1 - \frac{1}{e} - \epsilon$, for any $\epsilon > 0$, by using the standard greedy algorithm [18].

Theorem 4. *There exists an algorithm that approximates DoV_c^{spv} and DoV_d^{spv} within a factor $(1 - \frac{1}{e}) - \epsilon$ from the optimum, for any $\epsilon > 0$.*

Proof. We first consider the constructive case, i.e., DoV_c^{spv}. Let us define $\bar{C}_1^v \subseteq C_1$ as a set of candidates in our target party whose rank is decreased if v become active; in other words, $\bar{C}_1^v = \{c \in C_1 : \exists c' \in C \setminus C_1, \pi_v(c') < \pi_v(c)\}$. In this case, a node $v \in V$ can increase the score of C_1 by $\sum_{c \in \bar{C}_1^v} f(\pi_v(c) - 1) - f(\pi_v(c))$.[2] Given an instance of $\mathcal{I}(G, B)$ of the DoV_c^{spv} maximization problem, we define an instance $\mathcal{I}'(G, B, w)$ of the node-weighted IM problem, where w is a node-weight function defined as $w(v) = \sum_{c \in \bar{C}_1^v} (f(\pi_v(c) - 1) - f(\pi_v(c)))$, for all $v \in V$. Given a set S of nodes, we denote by $\sigma(S)$ the expected weight of active nodes in G, when the diffusion starts from S. We will show that $DoV_c^{spv}(C_1, S) = \sigma(S)$ for any set $S \subseteq V$, since the standard greedy algorithm guarantees an approximation factor of $1 - \frac{1}{e} - \epsilon$, for the node-weight IM problem, for any $\epsilon > 0$, this shows the statement.

Given a set S, $\sigma(S)$ can be computed as follows:

$$\sigma(S) = \mathbb{E}_{A_S} \left[\sum_{v \in A_S} w(v) \right] = \sum_{A_S \subseteq V} \sum_{v \in A_S} w(v) \mathbb{P}(A_S),$$

where $\mathbb{P}(A_S)$ is the probability that $A_S \subseteq V$ is the set of nodes activated by S.

By definition $DoV_c^{spv}(C_1, S) = \mathcal{F}_{spv}(C_1, S) - \mathcal{F}_{spv}(C_1, \emptyset)$, where $\mathcal{F}_{spv}(C_1, \emptyset) = \sum_{v \in V} \sum_{c \in C_1} f(\pi_v(c))$ and

$$\mathcal{F}_{spv}(C_1, S) = \mathbb{E}_{A_S} \left[\sum_{v \in V} \sum_{c \in C_1} f(\tilde{\pi}_v(c)) \right] = \sum_{v \in V} \sum_{c \in C_1} \mathbb{E}_{A_S} [f(\tilde{\pi}_v(c))]$$

$$= \sum_{v \in V} \left(\sum_{c \in \bar{C}_1^v} \mathbb{E}_{A_S} [f(\tilde{\pi}_v(c))] + \sum_{c \in C_1 \setminus \bar{C}_1^v} f(\pi_v(c)) \right),$$

where the last equality is due to the fact that, a node v doesn't change the positions of candidates in $C_1 \setminus \bar{C}_1^v$. Let us focus on the first term of the above formula,

---

[2] We assume function $f(\cdot)$ is defined in such away that $f(i-1) - f(i)$, for $i = 2, \ldots, m$, does not depend exponentially on the graph size (e.g. it is a constant). The influence maximization problem with arbitrary node-weights is still an open problem [18].

$$\sum_{v \in V} \sum_{c \in \bar{C}_1^v} \mathbb{E}_{A_S}\left[f(\tilde{\pi}_v(c))\right]$$

$$= \sum_{v \in V} \sum_{c \in \bar{C}_1^v} \sum_{A_S \subseteq V} \left(f(\pi_v(c) - 1)\mathbb{1}_{v \in A_s} + f(\pi_v(c))\mathbb{1}_{v \notin A_s}\right)\mathbb{P}(A_S)$$

$$= \sum_{A_S \subseteq V} \left(\sum_{v \in A_S} \sum_{c \in \bar{C}_1^v} f(\pi_v(c) - 1) + \sum_{v \notin A_S} \sum_{c \in \bar{C}_1^v} f(\pi_v(c))\right)\mathbb{P}(A_S).$$

It follows that

$$\mathrm{DoV}_c^{spv}(C_1, S) = \sum_{v \in V} \sum_{c \in \bar{C}_1^v} \mathbb{E}_{A_S}\left[f(\tilde{\pi}_v(c)) - f(\pi_v(c))\right]$$

$$\sum_{A_S \subseteq V} \sum_{v \in A_S} \sum_{c \in \bar{C}_1^v} \left(f(\pi_v(c) - 1) - f(\pi_v(c))\right)\mathbb{P}(A_S) = \sigma(S),$$

since the term related to candidates in $C_1 \setminus \bar{C}_1^v$ and to nodes not in A_S are canceled out.

The destructive case is similar to the constructive one except that a node $v \in V$ can decrease the score of C_1 by $\sum_{c \in C_1 : \exists c' \in C \setminus C_1, \pi_v(c') > \pi_v(c)} f(\pi_v(c)) - f(\pi_v(c) + 1)$. Therefore the same approach, where the weights are set to the above value, yields the same approximation factor for DoV_d^{spv}.   □

In the following theorems, we show that using Theorem 4, we get a constant approximation factor for the problem of maximizing MoV. Specifically, we show that by maximizing DoV_c^{spv} we get an extra $1/3$ approximation factor for the problem of maximizing MoV_c^{spv}. For the destructive case, the extra approximation factor is $1/2$. It follows that, by using the greedy algorithm for maximizing DoV_c^{spv} and DoV_d^{spv}, we obtain approximation factors of $\frac{1}{3}(1 - \frac{1}{e}) - \epsilon$ and $\frac{1}{2}(1 - \frac{1}{e}) - \epsilon$, for any $\epsilon > 0$, of the maximum MoV_c^{spv} and MoV_d^{spv}, respectively.

Theorem 5. *There exists an algorithm that approximates MoV_c^{spv} within a factor $\frac{1}{3}(1 - \frac{1}{e}) - \epsilon$ from the optimum, for any $\epsilon > 0$.*

Proof. Let S and S^* be the solution returned by the greedy algorithm for DoV_c^{spv} maximization and a solution that maximizes MoV_c^{spv}, respectively. For each party $C_i \neq C_1$, we denote by $\mathrm{DoV}_c^-(C_i, S)$ the score lost by C_i after S, that is $\mathrm{DoV}_c^-(C_i, S) = \mathcal{F}(C_i, \emptyset) - \mathcal{F}(C_i, S) \geq 0$. Let $\alpha_\epsilon := (1 - \frac{1}{e}) - \epsilon$. Since S is a factor α_ϵ from the optimum DoV_c^{spv}, the following holds.

$$\mathrm{MoV}_c^{spv}(C_1, S) = \mathcal{F}_{spv}(C_1, S) - \mathcal{F}_{spv}(C_A^S, S) - \left(\mathcal{F}_{spv}(C_1, \emptyset) - \mathcal{F}_{spv}(C_B, \emptyset)\right)$$

$$= \mathrm{DoV}_c^{spv}(C_1, S) + \mathrm{DoV}_c^-(C_A^S, S) - \mathcal{F}_{spv}(C_A^S, \emptyset) + \mathcal{F}_{spv}(C_B, \emptyset)$$

$$\geq \alpha_\epsilon \mathrm{DoV}_c^{spv}(C_1, S^*) - \mathcal{F}_{spv}(C_A^S, \emptyset) + \mathcal{F}_{spv}(C_B, \emptyset)$$

$$\overset{(a)}{\geq} \frac{1}{3}\alpha_\epsilon \left[\mathrm{DoV}_c^{spv}(C_1, S^*) + \mathrm{DoV}_c^-(C_A^{S^*}, S^*) + \mathrm{DoV}_c^-(C_A^S, S^*)\right]$$

$$- \mathcal{F}_{spv}(C_A^S, \emptyset) + \mathcal{F}_{spv}(C_B, \emptyset)$$

$$\overset{(b)}{\geq} \frac{1}{3}\alpha_\epsilon \left[\mathrm{DoV}_c^{spv}(C_1, S^*) + \mathrm{DoV}_c^-(C_A^{S^*}, S^*) + \mathrm{DoV}_c^-(C_A^S, S^*)\right.$$

$$\left. - \mathcal{F}_{spv}(C_A^S, \emptyset) + \mathcal{F}_{spv}(C_B, \emptyset) + \mathcal{F}_{spv}(C_A^{S^*}, \emptyset) - \mathcal{F}_{spv}(C_A^{S^*}, \emptyset)\right]$$

$$= \frac{1}{3}\alpha_\epsilon \left[\mathrm{MoV}_c^{spv}(C_1, S^*) + \mathrm{DoV}_c^-(C_A^S, S^*) + \mathcal{F}_{spv}(C_A^{S^*}, \emptyset) - \mathcal{F}_{spv}(C_A^S, \emptyset)\right]$$

$$\overset{(c)}{\geq} \frac{1}{3}\alpha_\epsilon \mathrm{MoV}_c^{spv}(C_1, S^*) \geq \left(\frac{1}{3}\left(1 - \frac{1}{e}\right) - \epsilon\right) \mathrm{MoV}_c^{spv}(C_1, S^*),$$

for any $\epsilon > 0$. Inequality (a) holds because, by definition, the score lost by C_A^S and $C_A^{S^*}$ will be added to the score of C_1. Inequality (b) holds since, by definition of C_B, $\mathcal{F}_{spv}(C_B, \emptyset) \geq \mathcal{F}_{spv}(C_A^S, \emptyset)$ and then $-\mathcal{F}_{spv}(C_A^S, \emptyset) + \mathcal{F}_{spv}(C_B, \emptyset) \geq 0$. Inequality (c) holds because

$$\mathrm{DoV}_c^-(C_A^{S^*}, S^*) - \mathrm{DoV}_c^-(C_A^S, S^*)$$

$$= \mathcal{F}_{spv}(C_A^{S^*}, \emptyset) - \mathcal{F}_{spv}(C_A^{S^*}, S^*) - \mathcal{F}_{spv}(C_A^S, \emptyset) + \mathcal{F}_{spv}(C_A^S, S^*)$$

$$\overset{(d)}{\leq} \mathcal{F}_{spv}(C_A^{S^*}, \emptyset) - \mathcal{F}_{spv}(C_A^S, \emptyset),$$

which implies that

$$\mathrm{DoV}_c^-(C_A^S, S^*) + \mathcal{F}_{spv}(C_A^{S^*}, \emptyset) - \mathcal{F}_{spv}(C_A^S, \emptyset) \geq \mathrm{DoV}_c^-(C_A^{S^*}, S^*) \geq 0.$$

Inequality (d) holds since, by definition of $C_A^{S^*}$, $\mathcal{F}(C_A^S, S^*) \leq \mathcal{F}(C_A^{S^*}, S^*)$. $\qquad \square$

Theorem 6. *There exists an algorithm that approximates MoV_d^{spv} within a factor $\frac{1}{2}(1 - \frac{1}{e}) - \epsilon$ from the optimum, for any $\epsilon > 0$.*

Proof. Let S and S^* be the solution returned by the greedy algorithm for DoV_d^{spv} maximization and a solution that maximizes MoV_d^{spv}, respectively. For each party $C_i \neq C_1$, we denote by $\mathrm{DoV}_c^+(C_i, S)$ the score gained by C_i after S, that is $\mathrm{DoV}_c^+(C_i, S) = \mathcal{F}(C_i, S) - \mathcal{F}(C_i, \emptyset) \geq 0$. Let $\alpha_\epsilon := (1 - \frac{1}{e}) - \epsilon$. Since S is a factor α_ϵ from the optimum DoV_c^{spv}, the following holds.

$$\text{MoV}_d^{spv}(C_1, S) = \mathcal{F}_{spv}(C_1, \emptyset) - \mathcal{F}_{spv}(C_{\text{B}}, \emptyset) - \left(\mathcal{F}_{spv}(C_1, S) - \mathcal{F}_{spv}(C_{\text{A}}^S, S)\right)$$

$$= \text{DoV}_d^{spv}(C_1, S) + \text{DoV}_c^+(C_{\text{A}}^S, S) + \mathcal{F}_{spv}(C_{\text{A}}^S, \emptyset) - \mathcal{F}_{spv}(C_{\text{B}}, \emptyset)$$

$$\geq \alpha_\epsilon \text{DoV}_d^{spv}(C_1, S^*) + \text{DoV}_c^+(C_{\text{A}}^S, S) + \mathcal{F}_{spv}(C_{\text{A}}^S, \emptyset) - \mathcal{F}_{spv}(C_{\text{B}}, \emptyset)$$

$$\overset{(a)}{\geq} \frac{1}{2}\alpha_\epsilon \left[\text{DoV}_d^{spv}(C_1, S^*) + \text{DoV}_c^+(C_{\text{A}}^{S^*}, S^*)\right] + \text{DoV}_c^+(C_{\text{A}}^S, S) +$$

$$\mathcal{F}_{spv}(C_{\text{A}}^S, \emptyset) - \mathcal{F}_{spv}(C_{\text{B}}, \emptyset)$$

$$\overset{(b)}{\geq} \frac{1}{2}\alpha_\epsilon \left[\text{DoV}_d^{spv}(C_1, S^*) + \text{DoV}_c^+(C_{\text{A}}^{S^*}, S^*) + \text{DoV}_c^+(C_{\text{A}}^S, S) +\right.$$

$$\left. \mathcal{F}_{spv}(C_{\text{A}}^S, \emptyset) - \mathcal{F}_{spv}(C_{\text{B}}, \emptyset) + \mathcal{F}_{spv}(C_{\text{A}}^{S^*}, \emptyset) - \mathcal{F}_{spv}(C_{\text{A}}^{S^*}, \emptyset)\right]$$

$$= \frac{1}{2}\alpha_\epsilon \left[\text{MoV}_d(C_1, S^*) + \text{DoV}_c^+(C_{\text{A}}^S, S) - \mathcal{F}_{spv}(C_{\text{A}}^{S^*}, \emptyset) + \mathcal{F}_{spv}(C_{\text{A}}^S, \emptyset)\right]$$

$$\overset{(c)}{\geq} \frac{1}{2}\alpha_\epsilon \text{MoV}_d(C_1, S^*) \geq \left(\frac{1}{2}\left(1 - \frac{1}{e}\right) - \epsilon\right) \text{MoV}_d(C_1, S^*),$$

for any $\epsilon > 0$. Inequality (a) holds because $C_{\text{A}}^{S^*}$ can gain at most all of the scores lost by C_1. Inequality (b) holds since we have

$$\text{DoV}_c^+(C_{\text{A}}^S, S) + \mathcal{F}_{spv}(C_{\text{A}}^S, \emptyset) - \mathcal{F}_{spv}(C_{\text{B}}, \emptyset)$$

$$= \mathcal{F}_{spv}(C_{\text{A}}^S, S) - \mathcal{F}_{spv}(C_{\text{A}}^S, \emptyset) + \mathcal{F}_{spv}(C_{\text{A}}^S, \emptyset) - \mathcal{F}_{spv}(C_{\text{B}}, \emptyset)$$

$$= \mathcal{F}_{spv}(C_{\text{A}}^S, S) - \mathcal{F}_{spv}(C_{\text{B}}, \emptyset),$$

and, by definition of C_{A}^S, $\mathcal{F}_{spv}(C_{\text{A}}^S, S) \geq \mathcal{F}_{spv}(C_{\text{B}}, S) \geq \mathcal{F}_{spv}(C_{\text{B}}, \emptyset)$. Inequality (c) holds because

$$\text{DoV}_c^+(C_{\text{A}}^S, S) - \mathcal{F}_{spv}(C_{\text{A}}^{S^*}, \emptyset) + \mathcal{F}_{spv}(C_{\text{A}}^S, \emptyset)$$

$$= \mathcal{F}_{spv}(C_{\text{A}}^S, S) - \mathcal{F}_{spv}(C_{\text{A}}^S, \emptyset) - \mathcal{F}_{spv}(C_{\text{A}}^{S^*}, \emptyset) + \mathcal{F}_{spv}(C_{\text{A}}^S, \emptyset)$$

$$= \mathcal{F}_{spv}(C_{\text{A}}^S, S) - \mathcal{F}_{spv}(C_{\text{A}}^{S^*}, \emptyset)$$

and, by definition of C_{A}^S, $\mathcal{F}_{spv}(C_{\text{A}}^S, S) \geq \mathcal{F}_{spv}(C_{\text{A}}^{S^*}, S) \geq \mathcal{F}_{spv}(C_{\text{A}}^{S^*}, \emptyset)$.   □

5   Conclusions and Future Work

Controlling elections through social networks is a significant issue in modern society. Political campaigns are using social networks as effective tools to influence voters in real-life elections. In this paper, we formalized the multi-winner election control problem through social influence. We proved that finding an approximation to the maximum margin of victory or difference of winners, for both constructive and destructive cases, is *NP*-hard for any approximation factor. We relaxed the problem to a variation of straight-party voting and showed that this case is approximable within a constant factor in both constructive and destructive cases. To our knowledge, these are the first results on multi-winner election control via social influence.

The results in this paper open several research directions. We plan to study the problem in which the adversary can spread a different (constructive/destructive) message for each candidate, using different seed nodes. In these cases, a good strategy could be that of sending a message regarding a third party (different from the target one and the most voted opponent), and our results cannot be easily extended.

References

1. Abouei Mehrizi, M., D'Angelo, G.: Multi-winner election control via social influence. arXiv e-prints, arXiv:2005.04037, May 2020
2. Allcott, H., Gentzkow, M.: Social media and fake news in the 2016 election. J. Econ. Perspect. **31**(2), 211–36 (2017)
3. Auletta, V., Caragiannis, I., Ferraioli, D., Galdi, C., Persiano, G.: Minority becomes majority in social networks. In: Markakis, E., Schäfer, G. (eds.) WINE 2015. LNCS, vol. 9470, pp. 74–88. Springer, Heidelberg (2015). https://doi.org/10.1007/978-3-662-48995-6_6
4. Bond, R.M., et al.: A 61-million-person experiment in social influence and political mobilization. Nature **489**, 295 (2012)
5. Botan, S., Grandi, U., Perrussel, L.: Proposition wise opinion diffusion with constraints. In: 4th AAMAS Workshop on Exploring Beyond the Worst Case in Computational Social Choice (EXPLORE) (2017)
6. Bredereck, R., Elkind, E.: Manipulating opinion diffusion in social networks. In: IJCAI 2017 pp. 894–900 (2017)
7. Brill, M., Elkind, E., Endriss, U., Grandi, U.: Pairwise diffusion of preference rankings in social networks. In: IJCAI 2016, pp. 130–136 (2016)
8. Campbell, B.A., Byrne, M.D.: Straight-party voting: what do voters think? IEEE Trans. Inf. Forensics Secur. **4**(4), 718–728 (2009)
9. Chen, W., Castillo, C., Lakshmanan, L.V.S.: Information and Influence Propagation in Social Networks. Morgan & Claypool, San Rafael (2013)
10. Corò, F., Cruciani, E., D'Angelo, G., Ponziani, S.: Exploiting social influence to control elections based on scoring rules. In: IJCAI2019, pp. 201–207 (2019)
11. Corò, F., Cruciani, E., D'Angelo, G., Ponziani, S.: Vote for me! Election control via social influence in arbitrary scoring rule voting systems. In: AAMAS 2019, pp. 1895–1897 (2019)
12. Domingos, P., Richardson, M.: Mining the network value of customers. In: KDD 2001, pp. 57–66. ACM (2001)
13. Faliszewski, P., Gonen, R., Koutecký, M., Talmon, N.: Opinion diffusion and campaigning on society graphs. In: IJCAI 2018, pp. 219–225 (2018)
14. Faliszewski, P., Rothe, J.: Control and bribery in voting. In: Handbook of Computational Social Choice, pp. 146–168. Cambridge University Press (2016)
15. Faliszewski, P., Skowron, P., Slinko, A., Talmon, N.: Multiwinner voting: a new challenge for social choice theory. In: Endriss, U. (ed.) Trends in Computational Social Choice, chap. 2, pp. 27–48. AI Access (2017)
16. Ferrara, E.: Disinformation and social bot operations in the run up to the 2017 french presidential election. First Monday **22**(8) (2017)
17. Hershey, M.R.: Party Politics in America. Taylor & Francis, London (2017)
18. Kempe, D., Kleinberg, J.M., Tardos, É.: Maximizing the spread of influence through a social network. Theory Comput. **11**, 105–147 (2015)

19. Kreiss, D.: Seizing the moment: the presidential campaigns' use of twitter during the 2012 electoral cycle. New Media Soc. **18**(8), 1473–1490 (2016)

20. Kritzer, H.M.: Roll-off in state court elections: change over time and the impact of the straight-ticket voting option. J. Law Courts **4**(2), 409–435 (2016)

21. Pennycook, G., Cannon, T., Rand, D.G.: Prior exposure increases perceived accuracy of fake news. J. Exp. Psychol.: Gener. **147**, 1865 (2018)

22. Richardson, M., Domingos, P.: Mining knowledge-sharing sites for viral marketing. In: KDD 2002, pp. 61–70. ACM (2002)

23. Ruhl, J.M., Mcdonald, R.H.: Party Politics and Elections in Latin America. Taylor & Francis, London (2019)

24. Stier, S., Bleier, A., Lietz, H., Strohmaier, M.: Election campaigning on social media: politicians, audiences, and the mediation of political communication on Facebook and Twitter. Polit. Commun. **35**(1), 50–74 (2018)

25. Wilder, B., Vorobeychik, Y.: Controlling elections through social influence. In: AAMAS 2018, pp. 265–273 (2018)

26. Zwicker, W.S.: Introduction to the theory of voting. In: Handbook of Computational Social Choice, pp. 23–56. Cambridge University Press (2016)

Network Creation Games with Local Information and Edge Swaps

Shotaro Yoshimura[1] and Yukiko Yamauchi[2(✉)]

[1] Graduate School of Information Science and Electrical Engineering,
Kyushu University, Fukuoka, Japan
yoshimura@tcs.inf.kyushu-u.ac.jp
[2] Faculty of Information Science and Electrical Engineering, Kyushu University,
Fukuoka, Japan
yamauchi@inf.kyushu-u.ac.jp

Abstract. In the *swap game (SG)*, selfish players, each of which is associated with a vertex, form a graph by edge swaps, i.e., a player changes its strategy by simultaneously removing an adjacent edge and forming a new edge (Alon et al. 2013). The cost of a player considers the average distance to all other players or the maximum distance to other players. Any SG by n players starting from a tree converges to an equilibrium with a constant Price of Anarchy (PoA) within $O(n^3)$ edge swaps (Lenzner 2011). We focus on SGs where each player knows the subgraph induced by players within distance k. Therefore, each player cannot compute its cost nor a best response. We first consider *pessimistic* players who consider the worst-case global graph. We show that any SG starting from a tree (i) always converges to an equilibrium within $O(n^3)$ edge swaps irrespective of the value of k, (ii) the PoA is $\Theta(n)$ for $k = 1, 2, 3$, and (iii) the PoA is constant for $k \geq 4$. We then introduce *weakly pessimistic* players and *optimistic* players and show that these less pessimistic players achieve constant PoA for $k \leq 3$ at the cost of best response cycles.

Keywords: Network creation game · Local information · Price of Anarchy · Dynamics

1 Introduction

Static and dynamic properties of networks not controlled by any centralized authority attracts much attention in last two decades as self-organizing large-scale networks play a critical role in a variety of information systems, for example, the Internet, Peer-to-Peer networks, ad-hoc networks, wireless sensor networks, social networks, viral networks, and so on. In these networks, participants selfishly and rationally change a part of the network structure to minimize their cost and maximize their gain. Controlling such networks is essentially impossible and many theoretical and empirical studies have been conducted; stochastic

This work was supported by JSPS KAKENHI Grant Number JP18H03202.

A. W. Richa and C. Scheideler (Eds.): SIROCCO 2020, LNCS 12156, pp. 349–365, 2020.
https://doi.org/10.1007/978-3-030-54921-3_20

network construction models such as the Barabási–Albert model were proposed, and key structural properties such as the small world networks [21] and the scale-free networks [3] have been discovered. Stochastic communication models such as the voting models [20], the random phone call model [13], and the rewiring model [11] were proposed and many phase transition phenomena have been reported. Many problems related to broadcasting, gossiping, and viral marketing were also proposed [4,10,15].

In this paper, we take a game-theoretic approach to analyze dynamics and efficiency of the network structure resulting from local reconstruction by selfish agents. The *network creation game (NCG)* considers n players forming a network [12]. Each player is associated with a vertex of the network, can construct a communication edge connecting itself to another player at the cost of α, and can remove an adjacent edge for free. The cost of a player is the sum of the *construction cost* for edges and the *communication cost*, which is the sum of distances to all other players in the current network, i.e., the average distance to other players. Each player selfishly changes its strategy to minimize its cost and the social cost of a network is the sum of all players' costs. The *Price of Anarchy* (PoA) of NCG is constant for almost all values of α [1,9,17,18], yet the PoA is not known for some values of α. However, computing the best response in NCG is NP-hard [12], and this fact makes the NCG unrealistic in large-scale networks. The NCG with another type of communication cost is proposed in [9], where the cost of a player is the maximum distance to other players. We call this game the *Max Network Creation Game (MAX-NCG)* and the original NCG the *Sum Network Creation Game (SUM-NCG)*. However, the SUM-NCG and the MAX-NCG ignores one of the most critical limitations in large-scale networks; each player cannot obtain "global" information. This type of locality is a fundamental limitation in distributed computing, although players can neither compute its cost nor the best response without global information.

In this paper, we focus on games in such a distributed environment where each player cannot obtain the current strategy of all players nor have enough local memory to store the global information. Rather, players can access only local information. The NCG by players with local information is first proposed in [6]. Each player can observe a subgraph of the current graph induced by the players within distance k. We call this information the k-*local information*. The players are *pessimistic* in the sense that they consider the worst-case global graph when they examine a new strategy. Computing the best response for MAX-NCG is still NP-hard because k-local information may contain the entire network. For small k, more specifically, for $1 \leq k \leq \alpha + 1$, PoA $= \Omega(\frac{n}{1+\alpha})$ for MAX-NCG and for $k \leq c\sqrt[3]{\alpha}$ PoA $= \Omega(n/k)$ for SUM-NCG. These results contrast global information with local information. The SUM-NCG and MAX-NCG by players with global trace-route based information is proposed, yet PoA $= \Theta(n)$ for some values of α [5]. The NCG for more powerful players with k-local information is considered in [8], where the players can probe the cost of a new strategy. Computing the best response is NP-hard for any $k \geq 1$ while there exists tree equilibrium that achieves PoA $= O(\log n)$ and PoA $= \Omega(\frac{\log n}{k})$ for $2 \leq k \leq \log n$

and PoA $= \Theta(n)$ for $k = 1$. For non-tree networks, depending on the values of α and k, we have PoA $= O(n)$.

The *swap game* (SG) restricts strategy changes to *edge swaps*, i.e., simultaneously removing an edge and creating a new edge [2]. Thus, any strategy change does not change the number of edges in the network and the best response can be computed in polynomial time. Additionally, when we restrict initial networks to trees, a star achieves the minimum social cost. Above mentioned cost functions were adopted and these SGs are called the SUM-SG and the MAX-SG, respectively. The aim of SG is to omit parameter α from NCG with keeping the essence of NCG. The authors showed that the diameter of a tree equilibrium is two for the SUM-SG and at most three for the MAX-SG, while there exists an equilibrium with a large diameter in general networks. Thus, PoA of a tree equilibrium is always constant. Moreover, any SUM-SG and MAX-SG starting from a tree converges to an equilibrium within $O(n^3)$ edge swaps while they admit best response cycles starting from a general graph [14,16]. Consequently, local search at players with global information achieves efficient network construction for initial tree networks. The SUM-SG and MAX-SG with "powerful" players with k-local information is investigated in [8]. For $k \geq 2$, the SUM-SG and MAX-SG starting from general networks admits best response cycles while convergence within $O(n^3)$ moves is guaranteed for tree networks. However, to the best of our knowledge, SG with k-local information has not been considered.

1.1 Our Results

In this paper, we investigate the convergence property and PoA of SGs by players with local information. First, we consider pessimistic players and demonstrate that starting from an initial tree, any SUM-SG and MAX-SG converge to an equilibrium within $O(n^3)$ edge swaps in the same manner as [16], i.e., we present a generalized ordinal potential function for the two games. We also show that convergence from a general network is not always guaranteed. Then, we present a clear phase transition phenomenon caused by the locality.

- When $k = 1, 2$, pessimistic players never perform any edge swap in the SUM-SG and in MAX-SG. Any network is an equilibrium of the two games, thus PoA $= \Theta(n)$.
- When $k = 3$, in the SUM-SG and MAX-SG, there exists an equilibrium of diameter $\Theta(n)$, thus PoA $= \Theta(n)$.
- When $k \geq 4$, in the SUM-SG and MAX-SG, the diameter of every equilibrium is constant, thus PoA is constant.

We then introduce *weakly-pessimistic* players and *optimistic* players to obtain a better PoA for $k \leq 3$. A weakly pessimistic player performs an edge swap even when its cost does not decrease. This relaxation results in a constant PoA of the MAX-SG when $k = 3$ at the cost of best response cycles. An optimistic player assumes the best-case global graph for an edge swap and this optimism results in a constant PoA of the SUM-SG and MAX-SG for any value of k. Consequently, the combination of k-locality for $k \geq 4$ and pessimism enables distributed construction of efficient trees by selfish players.

1.2   Related Works

We briefly survey existing results of the NCG and SG for players with global information. Regarding the SUM-NCG, when $\alpha \leq n^{1-\varepsilon}$ for $\varepsilon \geq 1/\log n$ the PoA is $O(3^{1/\varepsilon})$ [9]. Thus, when n is sufficiently large, the PoA is bound by a constant. When $\alpha > 4n + 13$, the PoA is at most $3 + 2n/(2n + \alpha)$ [7]. In addition, any constant upper bound of PoA for $n \leq \alpha \leq 4n + 13$ is not known and the best upper bound is $O(2^{\sqrt{\lg n}})$ [9]. If every equilibrium is a tree, then PoA < 5 and an interesting conjecture is that every equilibrium is a tree for sufficiently large α [12]. Regarding the MAX-NCG, the PoA is $2^{O(\sqrt{\lg n})}$ and it is constant when $\alpha = O(n^{-1/2})$ or $\alpha > 129$ [18].

Regarding the SUM-SG, there exists an equilibrium with diameter $2^{O(\sqrt{\lg n})}$ while the diameter of any equilibrium is at most two (thus, a star) if an initial graph is a tree [2]. Regarding the MAX-SG, there exists an equilibrium with diameter $\Theta(\sqrt{n})$ while the diameter of any equilibrium is at most three if an initial graph is a tree [2].

1.3   Organization

Section 2 introduces the SGs and pessimistic players with local information. We analyze the dynamics of SGs by pessimistic players in Sect. 3 and PoA in Sect. 4. In Sect. 5, we introduce less pessimistic players and present best response cycles and equilibria with small diameter. Finally, we conclude this paper with open problems.

2   Preliminaries

A *swap game (SG)* by players with k-local information is denoted by (G_0, k), where $G_0 = (V, E_0)$ is an initial network and integer k is the size of each player's "visibility". G_0 is a simple undirected connected graph, where $|V| = n$ and $|E_0| = m$. We say $u \in V$ is adjacent to $v \in V$ if edge $\{u, v\}$ is an element of E. Each player is associated with a vertex in V and the *strategy* of a player $u \in V$ is the set of its incident edges.

Each player can change its strategy by an *edge swap*, i.e., removing one incident edge and creating a new edge. Starting from G_0, a sequence of edge swaps generates a network evolution $G_0, G_1, G_2, \ldots$.

Let $N_G(u)$ be the set of adjacent vertices of $u \in V$ in G and $d_G(u, v)$ be the distance between $u, v \in V$ in G. When G is not connected and v is not reachable from u, $d_G(u, v) = \infty$. The cost of a player depends on the current graph G. We consider two different types of cost functions, $c_{\text{SUM},u}(G)$ and $c_{\text{MAX},u}(G)$ defined as follows:

$$c_{\text{SUM},u}(G) = \sum_{v \in V} d_G(u, v)$$

$$c_{\text{MAX},u}(G) = \max_{v \in V} d_G(u, v).$$

When G is not connected, $c_{\mathrm{SUM},u}(G) = \infty$ and $c_{\mathrm{MAX},u}(G) = \infty$. We call a swap game where each player u uses $c_{\mathrm{SUM},u}$ the *sum swap game (SUM-SG)* and a swap game where each player u uses $c_{\mathrm{MAX},u}$ the *max swap game (MAX-SG)*. When it is clear from the context, we omit the name of the game and use c_u.

Each player u can access local information determined by G. Let $V_{G,k}(u)$ denote the set of vertices within distance k from u in G (thus, the k-neighborhood of u). Player u can observe the subgraph of G induced by $V_{G,k}(u)$ and we call this subgraph the *view* of u. We say the information at u is *k-local* and we call its view the *k-local information* of u. We assume that each player does not know any global information such as the values of n and m.

In a transition from G_t to G_{t+1}, a single player performs an edge swap. Consider the case where a player u performs an edge swap $(v, w) \in N_{G_t}(u) \times (V_{G_t,k}(u)\backslash(N_{G_t}(u)\cup\{u\}))$ in G_t. We call u the *moving player* in G_t. The resulting graph is $G_{t+1} = (V, E_t\backslash\{\{u,v\}\}\cup\{\{u,w\}\})$. Note that the number of edges does not change in a SG.

Due to local information, each player cannot compute its current cost nor the improvement by a strategy change. We first consider *pessimistic* players that consider the worst-case improvement for each possible edge swap and select one that achieves positive improvement. A player u is *unhappy* if it has an edge swap that decreases its cost in the worst-case global graph. In other words, there exists at least one edge swap (v, w) at u that satisfies

$$\Delta_u(v, w) = \min_{H \in \mathcal{G}_u} (c_u(H) - c_u(H')) > 0,$$

where $\mathcal{G}_u$ is the set of simple undirected connected graphs consisting of finite number of vertices and compatible with u's local view, and H' is a graph obtained by the edge swap (v, w) at u in $H \in \mathcal{G}_u$. We assume that a moving player always performs an edge swap (v, w) with $\Delta_u(v, w) > 0$. When every player u is not unhappy with respect to $c_{\mathrm{SUM},u}(G)$ in graph G, we call G a *sum-swap equilibrium*. When every player u is not unhappy with respect to $c_{\mathrm{MAX},u}(G)$ in graph G, we call G a *max-swap equilibrium*. When a graph is a sum-swap equilibrium and a max-swap equilibrium we simply call the graph *swap equilibrium*.

We define the *social cost* $SC(G)$ of a graph G as the sum of all players' costs, i.e., $SC(G) = \sum_{u \in V} c_u(G)$. Let $\mathcal{G}(n, m)$ be the set of simple undirected connected graphs of n players and m edges and $\overline{\mathcal{G}}_{\mathrm{SUM}}(n, m, k)$ be the set of sum-swap equilibrium graphs of n players with k-local information and m edges. The *Price of Anarchy (PoA)* of the SUM-SG is defined as follows:

$$\mathrm{PoA}_{\mathrm{SUM}}(n, m, k) = \frac{\max_{G \in \overline{\mathcal{G}}_{\mathrm{SUM}}(n,m,k)} SC(G)}{\min_{G' \in \mathcal{G}(n,m)} SC(G')}.$$

In the same way, the PoA of the MAX-SG is defined for the set $\overline{\mathcal{G}}_{\mathrm{MAX}}(n, m, k)$ of max-swap equilibrium graphs of n players with k-local information and m edges. The PoA of the SUM-SG (and MAX-SG) starting from a tree is denoted by $\mathrm{PoA}_{\mathrm{SUM}}(n, n-1, k)$ ($\mathrm{PoA}_{\mathrm{MAX}}(n, n-1, k)$, respectively).

A strategic game has the *finite improvement property* (FIP) if every sequence of improving strategy changes is finite [19]. Thus, from any initial state, any sequence of finite improving strategy changes reaches an equilibrium. Monderer and Shapley showed that a strategic game has the FIP if and only if it has a *generalized ordinal potential function*. Regarding a swap game, a function $\Phi : \mathcal{G}_{n,m} \rightarrow \mathbb{R}$ is a generalized ordinal potential function if we have the following property for every graph $G \in \mathcal{G}_{n,m}$, every unhappy player u, and every edge swap (v, w) that makes u unhappy,

$$c_u(G) - c_u(G') > 0 \Rightarrow \Phi(G) - \Phi(G') > 0,$$

where G' is a graph obtained by the edge swap (v, w) at u. That is, any transition in the SUM-SG and MAX-SG satisfies the above property for the moving player.

The *best response* of a player u in G_t is an edge swap (v, w) that maximizes $\Delta_u(v, w)$. We call an evolution $G_0, G_1, G_2, \ldots, G_i(= G_0)$ a *best response cycle* when each moving player in G_t performs a best response for $t = 0, 1, 2, \ldots, i-1$.

We further introduce some notations for graph $G = (V, E)$. For a set of vertices $V' \subseteq V$ the graph obtained by removing vertices in V' and their incident edges is denoted by $G \setminus V'$. Additionally, for a set of edges $E' \subseteq E$ the graph obtained by removing edges in E' is denoted by $G \setminus E'$. The vertex set and the edge set of a graph G' is denoted by $V(G')$ and $E(G')$, respectively.

3   Convergence Properties for Pessimistic Players

In this section, we investigate the dynamics of the SUM-SG and MAX-SG by pessimistic players with local information. We first consider general settings where the initial graph is not a tree and multiple players perform edge swaps simultaneously. We show that the two games admit best response cycles. We then demonstrate that when the initial graph is a tree, the SUM-SG and MAX-SG have the FIP and converge to an equilibrium within polynomial number of edge swaps.

3.1   Impossibility in General Settings

We first present several necessary conditions for an evolution of the SUM-SG and MAX-SG by players with local information to reach an equilibrium. We first present the necessary visibility for each player to change their strategies.

Theorem 1. *In the SUM-SG and MAX-SG, when $k \leq 2$, no player is unhappy in an arbitrary graph. Thus, any graph is a swap equilibrium.*

Proof. When $k = 1$, no player can perform an edge swap because $V_{G_t,1}(u) \setminus (N_{G_t}(u) \cup \{u\}) = \emptyset$ at any $u \in V$.

When $k = 2$, we first consider the SUM-SG. Assume player u is unhappy because of edge swap (v, w) in graph G. Let G' be the graph obtained by this edge swap. Thus, $d_G(u, w) = 2$ and $d_G(u, w) - d_{G'}(u, w) = 1$. In a worst-case global graph, w has no adjacent vertex other than those in $V_{G,2}(u)$ and the cost of u decreases by at most one by this edge swap. In G', v must be reachable from u. There exists at least one player that is in $V_{G,2}(u)$ and adjacent to v, otherwise v is not reachable from u in a worst-case global graph. Hence, $d_G(u,v) - d_{G'}(u,v) = -1$. Additionally, $d_G(u,x) - d_{G'}(u,x) \leq 0$ for any $x \in V_{G,2}(u) \setminus \{v,w\}$. Consequently, $\Delta_u(v,w) \leq 0$ and u is not unhappy in G.

Next, we consider the MAX-SG. Assume player u is unhappy because of edge swap (v,w) in graph G. Thus, $d_G(u,w) = 2$ and w is the only player at distance 2 from u in $V_{G,2}(u)$ otherwise u is not unhappy because of the edge swap (v,w) in G. Let G' be the graph obtained by this edge swap. In a worst-case global graph, w has no adjacent vertex other than those in $V_{G,2}(u)$ and the cost of u is expected to be reduced to 1. By the same discussion above, v is reachable from u in G', however $d_{G'}(u,v) = 2$. Hence, the maximum distance from u to players in $V_{G',2}(u)$ is still two, thus $\Delta_u(v,w) \leq 0$. Hence, u is not unhappy in G.   □

The following theorem justifies our assumption of a single edge swap in each transition.

Theorem 2. *When $k \geq 3$, if multiple players change their strategies simultaneously, the SUM-SG and MAX-SG admit best response cycles.*

Proof. We first consider the SUM-SG. Consider a path of four players u, v, w, and x aligned in this order. When $k \geq 3$, the two endpoint players u and x are unhappy because of the edge swap (v,w) and (w,v), respectively. If the two players perform the edge swaps simultaneously, the resulting graph is again a path graph, where u and x are unhappy.

The above example is also a best response cycle in the MAX-SG.   □

Finally, we consider dynamics of SGs starting from an arbitrary initial graph. Lenzner presented a best response cycle for the SUM-SG by players with global information [16]. During the evolution, the distance to any player from a moving player is always less than four and we can apply the result to the SUM-SG by pessimistic players with k-local information for $k \geq 3$. In addition, the edge swaps are also best responses in the MAX-SG. Hence, we can also apply the result to the MAX-SG by pessimistic players with k-local information for $k \geq 3$. We have the following theorem.

Theorem 3. *When $k \geq 3$, there exists an initial graph from which the SUM-SG and MAX-SG admit a best response cycle.*

In the following, we concentrate on the SUM-SG and MAX-SG by pessimistic players with k-local information for $k \geq 3$ starting from a tree. As defined in the preliminary, a single player changes its strategy in each transition.

3.2   Convergence from an Initial Tree

In this section, we show that the SUM-SG and MAX-SG have the FIP. For players with global information, generalized ordinal potential functions for the SUM-SG [16] and MAX-SG [14] have been proposed. We can use these generalized ordinal potential functions for pessimistic players with local information.

Theorem 4. *If G_0 is a tree, any SUM-SG (G_0, k) has the FIP and reaches a sum-swap equilibrium within $O(n^3)$ edge swaps.*

Proof. We show that $\Phi_{\text{SUM}} = SC(G)$ is a generalized ordinal potential function for the SUM-SG irrespective of the value of k. Consider a tree G_t where an arbitrary unhappy player u performs an edge swap (v, w) that yields a new graph G_{t+1}. We have $\Delta_u(v, w) > 0$.

Lenzner showed that for players with global information $SC(G_t) - SC(G_{t+1}) \geq 2$ holds if $c_u(G_t) - c_u(G_{t+1}) > 0$ [16]. Since $\Delta_u(v, w)$ considers the worst case graph, $c_u(G_t) - c_u(G_{t+1}) \geq \Delta_u(v, w) > 0$ holds. Consequently, Φ_{SUM} is a generalized ordinal potential function for the SUM-SG.

Lenzner also showed that when the graph is a path of n vertices, Φ_{SUM} achieves the maximum value of $\Theta(n^3)$, and if the graph is a star of n vertices, Φ_{SUM} achieves the minimum value of $\Theta(n^2)$. Hence, the number of edge swaps is $O(n^3)$. □

We next show the FIP of the MAX-SG. Kawald and Lenzner presented a generalized ordinal function for the MAX-SG by players with global information [14]. Their generalized ordinal function is an n-tuple of players' costs, where the players are sorted in the descending order of their costs. We apply their function to the MAX-SG by pessimistic players with local information.

Consider the case where an unhappy player u performs an edge swap (v, w) in G_t and a new graph G_{t+1} is formed. Graph $G_t \setminus \{\{u, v\}\}$ consists of two trees and let G_t^u be the tree containing vertex u and G_t^v be the tree containing vertex v. We have the following two lemmas. We omit the proof of Lemma 2 due to page restriction.

Lemma 1 [14]. *Any player $x \in V(G_t^u)$ satisfies $c_x(G_t) > c_x(G_{t+1})$.*

Lemma 2. *Any player $y \in V(G_t^v)$ satisfies at least one of the following two conditions; (i) there exists a player $x \in V(G_t^u)$ that satisfies $c_x(G_t) > c_y(G_{t+1})$ and (ii) $c_y(G_t) \geq c_y(G_{t+1})$.*

We define $\Phi_{\text{MAX}}(G)$ for a graph G as an n-tuple $(c_{u_1}(G), c_{u_2}(G), \ldots, c_{u_n}(G))$ where $c_{u_i}(G) \geq c_{u_{i+1}}(G)$ for $i = 1, 2, \cdots, n - 1$. We assume that ties are broken arbitrarily. We then consider lexicographic ordering of n-tuples. For two n-tuples $C = (c_1, c_2, \ldots, c_n)$ and $C' = (c'_1, c'_2, \ldots, c'_n)$ where $c_i, c'_i \in \mathbb{Z}$ for $i = 1, 2, \cdots, n$, when $\Delta = (c_1 - c'_1, c_2 - c'_2, \ldots, c_n - c'_n) \neq \mathbf{0}$ and the leftmost non-zero entry of Δ is positive, we say C is lexicographically larger than C', denoted by $C >_{lex} C'$. We can demonstrate that any transition from G_t to G_{t+1} satisfies $\Phi_{\text{MAX}}(G_t) >_{lex} \Phi_{\text{MAX}}(G_t + 1)$. We have the following theorem.

Theorem 5. *If G_0 is a tree, a MAX-SG (G_0, k) has the FIP and reaches a max-swap equilibrium within $O(n^3)$ edge swaps.*

Proof. We demonstrate that any transition from G_t to G_{t+1} satisfies

$$\Phi_{\text{MAX}}(G_t) >_{lex} \Phi_{\text{MAX}}(G_t + 1).$$

Let u be the moving player in G_t and x be the player with the maximum cost in G_t^u. Then, in G_t^v, there may exist a player with larger cost than $c_x(G_t)$. We sort these players in the descending order of their costs and let $y_1, y_2, \ldots, y_p$ be the obtained sequence of players and $y_{p+1}, \ldots, y_q$ be the remaining players in G_t^v.

We first show that any y_j ($1 \le j \le p$) satisfies $c_{y_j}(G_t) \ge c_{y_j}(G_{t+1})$. If the second condition of Lemma 2 holds for all $y_1, y_2, \ldots, y_p$, we have the property. Otherwise, there exists y_j that does not satisfy the second condition but the first condition. However, we have $c_{y_j}(G_t) \ge c_x(G_t)$ and $c_x(G_t) > c_{y_j}(G_{t+1})$. This is a contradiction and all y_j satisfies $c_{y_j}(G_t) \ge c_{y_j}(G_{t+1})$.

Then we consider a player $v \in V(G_t^u) \cup \{y_{p+1}, y_{p+2}, \ldots, y_q\}$. By Lemma 1 and Lemma 2 such player v satisfies $c_v(G_{t+1}) < c_x(G_t)$. Consequently, we have $|\{v \mid c_v(G_t) \ge c_x(G_t)\}| > |\{v \mid c_v(G_{t+1}) \ge c_x(G_t)\}|$, and $\Phi_{\text{MAX}}(G_i) >_{lex} \Phi_{\text{MAX}}(G_{t+1})$.

We can bound the number of edge swaps in the same manner as [14]. $\square$

By Theorem 4 and 5, when an initial graph is a tree, the SUM-SG and MAX-SG by pessimistic players with local information converge to a sum-swap equilibrium and max-swap equilibrium, respectively within $O(n^3)$ edge swaps.

4   PoA for Pessimistic Players

In this section, we analyze PoA of the SUM-SG and MAX-SG by pessimistic players with local information. Alon et al. showed that for players with global information, the diameter of a tree swap equilibrium is constant for the two cost functions, thus PoA is also constant [2]. On the other hand, our results show the clear contrast by the value of k. When $k = 3$, there exists a sum-swap equilibrium of diameter $\Theta(n)$ and a max-swap equilibrium of diameter $\Theta(n)$. Thus, PoA is $\Theta(n)$ for both games. When $k \ge 4$, the diameter of any sum-swap equilibrium is at most two and that of any max-swap equilibrium is at most three. Thus, the PoA is bounded by a constant for both games.

In the following, we consider a path in a graph. A path P of length ℓ is denoted by a sequence of vertices on it, i.e., $P = v_0 v_1 \ldots v_\ell$. The set of vertices that appear on P is denoted by $V(P)$ and the set of edges of P is denoted by $E(P)$. Given a tree $G = (V, E)$ and a path $P = v_0 v_1 v_2 \ldots v_\ell$ in G, consider the forest $G' = (V, E \setminus E(P))$ and let $T_{G,P}(v_i)$ denote the connected component (thus, a tree) containing v_i. We consider v_i as the root of $T_{G,P}(v_i)$ when we address the depth of $T_{G,P}(v_i)$. The following lemma provides a basic technique to check the existence of an unhappy player.

Lemma 3. *In the SUM-SG, when $k = 3$, a player u in a tree G is unhappy if and only if there exists a path $P = uvw$ that satisfies the following two conditions; (i) the depth of $T_{G,P}(v)$ is at most one, and (ii) $|V(T_{G,P}(v))| < |N_{T_{G,P}(w)}(w)|$.*

Proof. We first show that u is unhappy if the two conditions hold. Assume that there is a path $P = uvw$ satisfying the two conditions. See Fig. 1. Let G' be the graph obtained by the edge swap (v, w) at u in G. For every $x \in V(T_{G,P}(v))$,

$d_{G'}(u, x) = d_G(u, x) + 1$ and for every $y \in N_{T_{G,P}(w)}(w)$ $d_{G'}(u, y) = d_G(u, y) - 1$. By condition (i), u knows that the edge swap (v, w) increases the distance to $x \in V(T_{G,P}(v))$. By condition (ii), u knows that the edge swap (v, w) decreases its cost by at least $|N_{T_{G,P}(w)}(w)|$. In the worst-case global graph, w has no adjacent players other than $N_{T_{G,P}(w)}(w)$. Hence,

$$\Delta_u(v, w) = |N_{T_{G,P}(w)}(w)| - |V(T_{G,P}(v))| > 0,$$

and u is unhappy because of this edge swap (v, w).

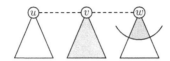

Fig. 1. $T_{G,P}(u)$, $T_{G,P}(v)$, and $T_{G,P}(w)$.

Next, we show that u is unhappy in G only if the two conditions hold. Consider the case where for any path $P = uvw$, (i') the depth of $T_{G,P}(v)$ is larger than one, or (ii') $|V(T_{G,P}(v))| \geq |N_{T_{G,P}(w)}(w)|$ holds. We show that any player $u \in V$ is not unhappy. We check an arbitrary edge swap (v', x') at u. Thus, $v' \in N_G(u)$ and $x' \in V_{G,3}(u) \setminus (N_G(u) \cup \{u\})$. G must have a path between v' and x', otherwise the edge swap (v', x') disconnects the players. If u cannot see this path, in the worst case global graph, v' is not reachable from u. Hence, G contains a path $uv'w'x'$ or $uv'x'$.

If G contains a path $uv'w'x'$, the edge swap (v', x') satisfies $d_G(u, x') - d_{G'}(u, x') = 2$, $d_G(u, v') - d_{G'}(u, v') = -2$, and $d_G(u, w') - d_{G'}(u, w') = 0$. The worst-case global graph for the edge swap (v', x') is a graph where x' is not adjacent to any other vertex in $V(G) \setminus V_{G,3}(u)$. Thus, $\Delta_{c_u}(v', x') \leq 0$. Hence, u is not unhappy with respect to the edge swap (v', x').

If G contains a path $P' = uv'x'$ and condition (i') holds, there exist vertices $v'_1, v'_2 \in T_{G,P'}(v')$ that form a path $uv'v'_1v'_2$. In the worst case global graph for the edge swap (v', x'), v'_2 has many children whose distance from u increases by one in G'. Hence, $\Delta_{c_u}(v', x') \leq 0$ and u is not unhappy with respect to the edge swap (v', x').

If G contains a path $P' = uv'x'$ and condition (ii') holds, in the worst-case global graph for the edge swap (v', x'), the number of players whose distance from u decreases by one with the edge swap (v', x') is $|N_{T_{G,P}(w)}(w)|$ and the number of players whose distance from u increases by one is lower bounded by $|V(T_{G,P}(v))|$. Thus, $\Delta_{c_u}(v', x') \geq |N_{T_{G,P}(w)}(w)| - |V(T_{G,P}(v))| \leq 0$ holds and player u is not unhappy with respect to the edge swap (v', x').

Consequently, u is not unhappy with respect to any edge swap.   □

We present a sum-swap equilibrium of diameter $\Theta(n)$. We define a tree $TS(p)$ with a spine path of length p as follows: For $i = 1, 2, \cdots, p$, H_i is a tree, where

a_i has four children b_i, c_i, d_i, and e_i. For $i = 0, p+1$, H_i is a tree rooted at a_i with three children b_i, c_i, and d_i. $TS(p)$ is a tree defined by

$$V(TS(p)) = \bigcup_{i=0}^{p+1} V(H_i)$$

$$E(TS(p)) = \bigcup_{i=0}^{p+1} E(H_i) \cup \{\{a_0, e_1\}, \{e_p, a_{p+1}\}\} \cup \bigcup_{i=1}^{p-1} \{\{e_i, e_{i+1}\}\}.$$

Figure 2 shows $TS(7)$ as an example.

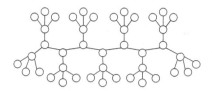

Fig. 2. $TS(7)$

For each $u \in V(TS(p))$ ($p \geq 3$), any path $P = uvw$ does not satisfy the two conditions of Lemma 3. Thus, every player is not unhappy and $TS(p)$ is a sum swap equilibrium. The social cost in $TS(p)$ is $\Theta(n^3)$. Since a star is a sum-swap equilibrium with the minimum social cost, the PoA of $TS(p)$ is $\Theta(n)$. By adding extra vertices to some a_i of $TS(p)$, we obtain a sum-swap equilibrium of diameter $\Theta(n)$ for any $n \geq 13$. We have the following theorem.

Theorem 6. When $n \geq 13$ and $k \leq 3$, $PoA_{SUM}(n, n-1, k) = \Theta(n)$.

We now demonstrate that when $k \geq 4$, sum-swap equilibrium for pessimistic players with k-local information achieves the same PoA as that with global information. We first show the following lemma.

Lemma 4. In an arbitrary tree G whose diameter is larger than two, there exists a path $P = vabw$ that satisfies the following two conditions; (i) $|V_{G_{T,P}(a),2}(a)| \leq |V_{G_{T,P}(b),2}(b)|$, and (ii) the depth of $T_{G,P}(a)$ is at most two.

Proof. We prove the lemma by induction. There exists at least one path of length at least three in G. We choose a path $P = vabw$ arbitrarily. Let d_a and d_b be the depth of $T_{G,P}(a)$ and $T_{G,P}(b)$, respectively.

(Base Case). Consider the case where $\max\{d_a, d_b\} \leq 2$. Let $P_{rev} = wbav$. The two paths P and P_{rev} satisfies the second condition and either P or P_{rev} satisfies the first condition. Thus, the statement holds when $\max\{d_a, d_b\} \leq 2$.

(Induction Step). Assume the statement holds when $\max\{d_a, d_b\} \leq d-1$ for $d \geq 3$. Consider the case where $\max\{d_a, d_b\} = d$. Without loss of generality,

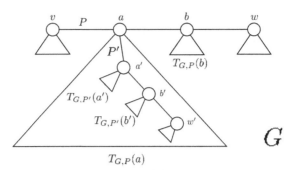

Fig. 3. Induction step of Lemma 4.

we assume $d_a = d \geq 3$. Hence, there exists at least one path $P' = aa'b'w'$ in $T_{G,P}(a)$. See Fig. 3. Let $d_{a'}$ and $d_{b'}$ be the depth of $T_{G,P'}(a')$ and $T_{G,P'}(b')$, respectively. Clearly, $d_a > d_{a'}$ and $d_{a'} > d_{b'}$ hold and the statement holds by the induction hypothesis. □

By Lemma 4, in any graph G whose diameter is larger than two there exists an unhappy player.

Theorem 7. *When G_0 is a tree and $k \geq 4$, any sum-swap equilibrium is a star and* $\mathrm{PoA}_{SUM}(n, n-1, k) = 1$.

Proof. Assume that there exists a sum-swap equilibrium G whose diameter is larger than two. By Lemma 4, there exists a path $P = vabw$ that satisfies the two conditions. Hence, v is unhappy because

$$\Delta c_v(a, b) \geq |V_{T_{G,P}(b),2}(b)| + 1 - |V_{T_{G,P}(a),2}(a)| > 0.$$

This is a contradiction and G is not a sum-swap equilibrium. Hence, the diameter of a sum-swap equilibrium is smaller than or equal to two and we have the statement. □

Consequently, the "visibility" of pessimistic players has a significant effect on the PoA of the SUM-SG. We then demonstrate that this is also the case for the MAX-SG.

The tree shown in Fig. 4 is a max-swap equilibrium for $k = 3$. By adding inner vertices (with its child), we have the similar equilibrium for any even $n \geq 6$. For odd $n \geq 6$, we attach an extra player to an inner vertex and obtain a max swap equilibrium.

Since the star graph is an max-swap equilibrium with the minimum social cost, we have PoA $= \Theta(n)$. We have the following theorem.

Theorem 8. *When $n \geq 6$ and $k = 3$,* $\mathrm{PoA}_{MAX}(n, n-1, k) = \Theta(n)$.

We now demonstrate that when $k \geq 4$, any MAX-SG by pessimistic players with k-local information achieves the same PoA as that of players with global information. The following lemma shows that in any tree of diameter larger than three, there is at least one unhappy player.

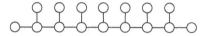

Fig. 4. A max-swap equilibrium of 16 players.

Lemma 5. *In any tree G whose diameter is larger than three, there exists a path $P = vabcw$ that satisfies the following two conditions; (i) P starts from a leaf v, and (ii) the depth of $T_{G,P}(a)$ is at most one.*

Proof. There exists at least one path of length at least four in G. We arbitrarily choose a path $P = vabcw$ that starts from some leaf v. If the depth of $T_{G,P}(a)$ is smaller than two, the statement holds. If the depth of $T_{G,P}(a)$ is larger than one, choose a leaf v' in $T_{G,P}(a)$ and its parent vertex, say a'. There exists at least one path $P' = v'a'b'c'w'$ and P' satisfies the second condition. □

Theorem 9. *When G_0 is a tree and $k \geq 4$, the diameter of any max-swap equilibrium is at most three and $\mathrm{PoA}_{MAX}(n, n-1, k) \leq 3/2$.*

Proof. Assume that there exists a max-swap equilibrium G whose diameter is larger than three. By Lemma 5, there exists a path $P = vabcw$ such that v is a leaf and the depth of $T_{G,P}(a)$ is at most one. Player v is unhappy because $\Delta c_v(a, b) \geq 1$. This is a contradiction and G is not a max-swap equilibrium. Thus, the diameter of any max-swap diameter is at most three.

Because a equilibrium with the minimum cost is a star, the PoA is bounded by $3/2$. □

5   Swap Games with Non-pessimistic Players

We demonstrated that when $k = 2, 3$, the PoA for pessimistic players is $\Theta(n)$ in the SUM-SG and MAX-SG. In this section, we introduce less pessimistic players to obtain smaller PoA for these cases. We consider two types of non-pessimistic players: A player u is *weakly pessimistic* if u is unhappy when there exists an edge swap (v, w) at u such that $\Delta c_u(v, w) \geq 0$.

A player u is *optimistic* if its $\Delta c_u(v, w)$ is defined as

$$\Delta_{c_u}(v, w) = \max_{H \in \mathcal{G}_u} (c_u(H) - c_u(H')),$$

where H' is a graph obtained by an edge swap (v, w) at u in $H \in \mathcal{G}$.

Weakly pessimistic players and optimistic players do not perform any edge swap in the SUM-SG and MAX-SG when $k = 1$. However, weakly pessimistic players change their strategies when $k = 2$. However, when $k > 2$, weakly pessimistic players cause a cycle of edge swaps from an initial path graph: Let $P = u_0 u_1 u_2 \ldots$ be a path of n ($\geq 2k$) weakly pessimistic players with k-local information. In the SUM-SG and MAX-SG, player u_k is unhappy because of the edge swap (u_{k-1}, u_0). However after u_k performs this edge swap, the graph is

$u_{k-1}u_{k-2}\ldots u_0 u_k u_{k+1}\ldots u_n$ and u_k is again unhappy because of the edge swap (u_0, u_{k-1}). By selecting u_k forever, the graph never reach an equilibrium.

We now consider a more restricted *round robin scheduling*. In a round-robin scheduling, players have a fixed ordering and at each time step a moving player is selected according to this ordering. Consider n players $u_1, u_2, \ldots, u_n$ and let the subscript i indicate the order of player u_i. In G_0 if u_1 is unhappy, u_1 is selected as the moving player. Otherwise, we check $u_2, u_3, \ldots$ until we find an unhappy player. Thus, the unhappy player with the smallest order, say j, is selected as a moving player. In G_1 if u_{j+1} is unhappy, u_{j+1} is selected as the moving player. Otherwise, we check $u_{j+2}, u_{j+3}, \ldots$ until we find an unhappy player. If u_n is selected in G_t, the check start with u_1 in G_{t+1}. We show that there exists a best response cycle from an initial tree.

Theorem 10. *When $k \geq 3$, for weakly pessimistic players with k-local information, there exist infinitely many trees from which the SUM-SG and MAX-SG admit best response cycles under the round-robin scheduling.*

Proof. We present an initial tree G_0 for the SUM-SG by $n = 2m + 3$ weakly pessimistic players $u_1, u_2, \ldots, u_{2m+3}$ ($m = 2, 3, \ldots$). The players are divided into two subtrees rooted at u_{2m+3} and u_{2m+2}, respectively; u_{2m+3} has $m + 1$ leaves $u_1, u_3, \ldots, u_{2m+1}$ and u_{2m+2} has m leaves $u_2, u_4, \ldots, u_{2m}$. Additionally, G_0 contains an edge connecting u_{2m+3} and u_{2m+2}. See Fig. 5 for $n = 9$.

Fig. 5. An example of G_0 for nine players. The two inner vertices are u_9 with four leaves (u_1, u_3, u_5, and u_7) and u_8 with three leaves (u_2, u_4, and u_6).

In G_0, u_1 is unhappy because of the edge swap (u_{2m+3}, u_{2m+2}); u_{2m+3} and u_{2m+2} have m leaves except u_1. When u_1 performs this edge swap, a new graph G_1 is formed, where u_{2m+3} has m leaves and u_{2m+2} has $m + 1$ leaves. In G_1, u_2 is unhappy because of the edge swap (u_{2m+2}, u_{2m+3}). When u_2 performs this edge swap, a new graph G_2 is formed, where u_{2m+3} has $m + 1$ leaves and u_{2m+2} has m leaves. In this way, the leaves of a graph keep on changing their parent under the round robin scheduling. The seesaw game continues until G_{2m} where u_{2m+3} has m leaves with even subscripts and u_{2m+2} has $m + 1$ leaves with odd subscripts. Player u_{2m+2} is not unhappy in G_{2m}; the only possibility is an edge swap $(u_{2m+3}, u_{2\ell})$ for some $\ell \in \{2, 4, \ldots, m\}$, but it increases its cost. Player u_{2m+3} is not unhappy in G_{2m} with the same reason. Player u_1 is unhappy in G_{2m}, and the leaf players start the seesaw game again.

This cycle is also a best response cycle in the MAX-SG.   □

Next lemma shows the PoA for weakly pessimistic players with 2-local information. We omit the proof due to page restriction.

Lemma 6. *In the SUM-SG and MAX-SG by weakly pessimistic players with 2-local information, a player u is unhappy if and only if there is a path uvw where the degree of v is two.*

By Lemma 6, a graph is a swap equilibrium if it does not have any vertex of degree two. The graph shown in Fig. 4 is a swap equilibrium with diameter $\Theta(n)$. We have the following theorem.

Theorem 11. *For weakly pessimistic players with 2-local information, when $n \geq 4$, $\mathrm{PoA}_{SUM}(n, n-1, 2) = \Theta(n)$ and $\mathrm{PoA}_{MAX}(n, n-1, 2) = \Theta(n)$.*

We then present the PoA of the SUM-SG by weakly pessimistic players with 3-local information in the same manner as Theorem 6. For $i = 1, 2, \cdots, p$, H'_i is a tree in which a_i has five children b_i, c_i, d_i, e_i, and f_i. For $i = 0, p+1$, H'_i is a tree rooted at a_i with four children b_i, c_i, d_i, and e_i. Then $TS'(p)$ is is a tree defined by

$$V(TS(p)) = \bigcup_{i=0}^{p+1} V(H_i)$$

$$E(TS(p)) = \bigcup_{i=0}^{p+1} E(H_i) \cup \{\{a_0, f_1\}, \{f_p, a_{p+1}\}\} \cup \bigcup_{i=1}^{p-1} \{\{f_i, f_{i+1}\}\}.$$

Figure 6 shows $TS'(7)$ as an example.

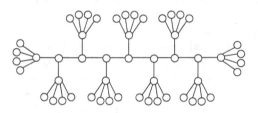

Fig. 6. $TS'(7)$

Theorem 12. *For weakly pessimistic players with 3-local information, when $n \geq 16$, $\mathrm{PoA}_{SUM}(n, n-1, 3) = \Theta(n)$.*

When a pessimistic player u is unhappy in graph G, u is unhappy in G when u is a weakly pessimistic player. By Theorem 7, we have the following theorem.

Theorem 13. *When $k \geq 4$, for weakly pessimistic players with k-local information, the diameter of a sum-swap equilibrium is at most two and $\mathrm{PoA}_{SUM}(n, n-1, k) = 1$.*

On the other hand, the diameter of any max-swap equilibrium is smaller than three for weakly pessimistic players with 3-local information.

Theorem 14. *When $k \geq 3$, for weakly pessimistic players with k-local information, the diameter of a max-swap equilibrium is at most two and* $\text{PoA}_{MAX}(n, n - 1, k) = 1$.

We omit the proof due to page restriction.

Finally, we consider optimistic players. An optimistic player u with k-local information expects that a player at distance k has a long path that u cannot observe. Thus, u always perform an edge swap to create an edge connecting itself to another player at distance k if any.

Lemma 7. *For optimistic players with k-local information, the diameter of any swap equilibrium is smaller than k.*

Consequently, in a sum-swap equilibrium and a max-swap equilibrium, all optimistic players can observe the entire graph. We have the following theorem.

Theorem 15. *For optimistic players with k-local information,* $\text{PoA}_{SUM}(n, n - 1, k) = 1$ *and* $\text{PoA}_{MAX}(n, n - 1, k) < 3/2$.

6   Conclusion

In this paper, we introduced swap games with k-local information and investigated their dynamics and PoA. First, we showed that when $k \geq 4$, starting from a tree, the SUM-SG and MAX-SG by pessimistic players with k-local information promise convergence to an equilibrium with constant PoA. In other words, in a distributed environment, rational participants can construct a tree of small diameter without global information.

We then introduced weakly pessimistic players to obtain a tree equilibrium with small PoA for $k \leq 3$. When $k = 3$, the MAX-SG achieves a constant PoA, at the cost of best response cycles. Thus, relaxing pessimism does not promise distributed graph construction. Finally, we introduced optimistic players and presented the constant PoA of the SUM-SG and MAX-SG for any value of k.

There are many interesting future directions. One is a better upper bound and a lower bound of the number of edge swaps during convergence. The dynamics of non-pessimistic players is also an open problem.

Although games with imperfect information have been investigated in game theory, to the best of our knowledge, there are few games where each player knows the existence of only a part of players. We hope games in this form open up new vistas for game theory and distributed computing.

References

1. Albers, S., Eilts, S., Even-Dar, E., Mansour, Y., Roditty, L.: On Nash equilibria for a network creation game. ACM Trans. Econ. Comput. **2**(1), 2:1–2:27 (2014)
2. Alon, N., Demaine, E.D., Hajiaghayi, M.T., Leighton, T.: Basic network creation games. SIAM J. Discrete Math. **27**, 656–668 (2013)

3. Barabási, A.L., Bonabeau, E.: Scale-free networks. Sci. Am. **288**, 50–59 (2003)
4. Berenbrink, P., Czyzowicz, J., Elsässer, R., Gąsieniec, L.: Efficient information exchange in the random phone-call model. In: Abramsky, S., Gavoille, C., Kirchner, C., Meyer auf der Heide, F., Spirakis, P.G. (eds.) ICALP 2010. LNCS, vol. 6199, pp. 127–138. Springer, Heidelberg (2010). https://doi.org/10.1007/978-3-642-14162-1_11
5. Bilò, D., Gualà, L., Leucci, S., Proietti, G.: Network creation games with traceroute-based strategies. In: Halldórsson, M.M. (ed.) SIROCCO 2014. LNCS, vol. 8576, pp. 210–223. Springer, Cham (2014). https://doi.org/10.1007/978-3-319-09620-9_17
6. Bilò, D., Gualà, L., Leucci, S., Proietti, G.: Locality-based network creation games. ACM Trans. Parallel Comput. **3**(1), 6:1–6:26 (2016)
7. Bilò, D., Lenzner, P.: On the tree conjecture for the network creation game. In: Proceedings of STACS 2018, pp. 14:1–14:15 (2018)
8. Cord-Landwehr, A., Lenzner, P.: Network creation games: think global – act local. In: Italiano, G.F., Pighizzini, G., Sannella, D.T. (eds.) MFCS 2015. LNCS, vol. 9235, pp. 248–260. Springer, Heidelberg (2015). https://doi.org/10.1007/978-3-662-48054-0_21
9. Demaine, E.D., Hajiaghayi, M., Mahini, H., Zadimoghaddam, M.: The price of anarchy in network creation games. ACM Trans. Algorithms **8**(2), 13:1–13:13 (2012)
10. Domingos, P., Richardson, M.: Mining the network value of customers. In: Proceedings of KDD 2001, pp. 57–66 (2001)
11. Durrett, R., et al.: Graph fission in an evolving voter model. Proc. Natl. Acad. Sci. **109**(10), 3682–3687 (2012)
12. Fabrikant, A., Luthra, A., Maneva, E., Papadimitriou, C.H., Shenker, S.: On a network creation game. In: Proceedings of PODC 2003, pp. 347–351 (2003)
13. Karp, R.M., Schindelhauer, C., Shenker, S., Vocking, B.: Randomized rumor spreading. In: Proceedings of FOCS 2000, pp. 565–574 (2000)
14. Kawald, B., Lenzner, P.: On dynamics in selfish network creation. In: Proceedings of SPAA 2013, pp. 83–92 (2013)
15. Kempe, D., Kleinberg, J., Tardos, E.: Maximizing the spread of influence through a social network. In: KDD 2003, pp. 137–146 (2003)
16. Lenzner, P.: On dynamics in basic network creation games. In: Persiano, G. (ed.) SAGT 2011. LNCS, vol. 6982, pp. 254–265. Springer, Heidelberg (2011). https://doi.org/10.1007/978-3-642-24829-0_23
17. Mamageishvili, A., Mihalák, M., Müller, D.: Tree nash equilibria in the network creation game. In: Bonato, A., Mitzenmacher, M., Prałat, P. (eds.) WAW 2013. LNCS, vol. 8305, pp. 118–129. Springer, Cham (2013). https://doi.org/10.1007/978-3-319-03536-9_10
18. Mihalák, M., Schlegel, J.C.: The price of anarchy in network creation games is (mostly) constant. Theor. Comput. Syst. **53**(1), 53–72 (2013)
19. Monderer, D., Shapley, L.S.: Potential games. Games Econ. Behav. **14**, 124–143 (1996)
20. Nakata, T., Imahayashi, H., Yamashita, M.: Probabilistic local majority voting for the agreement problem on finite graphs. In: Asano, T., Imai, H., Lee, D.T., Nakano, S., Tokuyama, T. (eds.) COCOON 1999. LNCS, vol. 1627, pp. 330–338. Springer, Heidelberg (1999). https://doi.org/10.1007/3-540-48686-0_33
21. Watts, D.J., Strogatz, S.H.: Collective dynamics of 'small-world' networks. Nature **393**, 440–442 (1998)

The Value of Information in Selfish Routing

Simon Scherrer[1]([⊠]), Adrian Perrig[1], and Stefan Schmid[2]

[1] Department of Computer Science, ETH Zürich, Zürich, Switzerland
simon.scherrer@inf.ethz.ch
[2] Faculty of Computer Science, University of Vienna, Vienna, Austria

Abstract. Path selection by selfish agents has traditionally been studied by comparing social optima and equilibria in the Wardrop model, i.e., by investigating the Price of Anarchy in selfish routing. In this work, we refine and extend the traditional selfish-routing model in order to answer questions that arise in emerging path-aware Internet architectures. The model enables us to characterize the impact of different degrees of congestion information that users possess. Furthermore, it allows us to analytically quantify the impact of selfish routing, not only on users, but also on network operators. Based on our model, we show that the cost of selfish routing depends on the network topology, the perspective (users versus network operators), and the information that users have. Surprisingly, we show analytically and empirically that less information tends to lower the Price of Anarchy, almost to the optimum. Our results hence suggest that selfish routing has modest social cost even without the dissemination of path-load information.

Keywords: Price of Anarchy · Selfish routing · Game theory · Information

1 Introduction

If selfish agents are free to select communication paths in a network, their interactions can produce sub-optimal traffic allocations. A long line of research relating to *selfish routing* [27,28,30] has quantified many effects of distributed, uncoordinated path selection by selfish individuals in networks. While seminal work on such game-theoretic analyses dates back to Wardrop [38], especially the notion of *Price of Anarchy*, coined by Koutsoupias and Papadmitriou [17], has received much attention: the Price of Anarchy compares the worst possible outcome of individual decision making, i.e., the worst Nash equilibrium, to the global optimum, by taking the corresponding cost ratio. The Price of Anarchy in network path selection is typically measured in terms of *latency*.

In this paper, we revisit these concepts to investigate two key aspects which have been less explored in the literature so far and are highly relevant for newly emerging path-aware network architectures (cf. Sect. 1.1):

© Springer Nature Switzerland AG 2020
A. W. Richa and C. Scheideler (Eds.): SIROCCO 2020, LNCS 12156, pp. 366–384, 2020.
https://doi.org/10.1007/978-3-030-54921-3_21

- **Impact of information:** A fundamental design question of network architectures concerns which information about the network state should be shared with end-hosts, beyond the latency information that can be observed by the end-hosts directly.
- **Impact on network operators:** While game-theoretic analyses usually revolve around the cost experienced by users, it is also important to understand the impact of selfish routing on the network operators' cost.

1.1   Practical Motivation

The traditional question studied in the selfish-routing literature, namely the efficiency of uncoordinated path selection by selfish agents, has recently received new relevance in the context of emerging Internet architectures relying on source-based path selection [2,9,25,39]. In particular, the already deployed SCION architecture [3,23] offers extensive path-selection control to users.

Today's Internet infrastructure is based on a forwarding mechanism that grants almost exclusive control to the network and almost no control to users (or *end-hosts*). In fact, all communication from a given end-host to another end-host takes place over the *single* AS-level path that BGP (Border Gateway Protocol) converged on. In the upcoming paradigm of *source-based path selection* [37], network operators supply end-hosts with a pre-selected set of paths to a destination, enabling end-hosts to select a forwarding path themselves.

Source-based path selection allows end-hosts to select paths with superior performance to the BGP-generated path [12,16,32] or to quickly switch to an alternative path upon link failures. However, a widely shared concern about source-based path selection regards the loss of control by network operators, which fear that the traffic distribution resulting from individual user decisions might impose considerable cost on both themselves and their customers. Another concern is that end-hosts require path-load information in order to perform path selection effectively, necessitating complex systems for the dissemination of network-state information. We refine and extend concepts from the selfish-routing literature to investigate the validity of these concerns.

1.2   Our Contributions

We present a game-theoretic model (Sect. 2) which allows us to quantify not only the Price of Anarchy experienced by end-hosts, but also to account for the network operators. Furthermore, we use our model to explore how end-host information about the network state affects the Price of Anarchy.

We find that different levels of information indeed lead to different Nash equilibria and thus also to different Prices of Anarchy. Intriguingly, we find that while more information can improve the efficiency of selfish routing in networks with few end-hosts (Sect. 3), more information tends to induce a *higher* Price of Anarchy in more general settings (Sect. 4). Indeed, near-optimal outcomes are typically achieved if end-hosts select paths based on simple latency measurements of different paths. These theoretical results suggest that source-based path selection cannot only achieve a good network performance in selfish contexts, but can be realized in a fairly light-weight manner, avoiding the need to distribute much information about the network state. This insight is validated with a case study on the Abilene topology (Sect. 5).

2   Model and First Insights

2.1   Model

As in previous work on selfish routing [10,30], our model is inspired by the classic Wardrop model [38]. In this model, the network is abstracted as a graph $G = (A, L)$, where the edges $\ell \in L$ between the nodes $A_i \in A$ represent links. Every link $\ell \in L$ is described by a link-cost function $c_\ell(f_\ell)$, where f_ℓ is the amount of load on link ℓ, i.e., a *link flow*. Typically, link-cost functions are seen as describing the latency behavior of a link. To reflect queuing dynamics, link cost functions are convex and non-decreasing. For every node pair (A_i, A_j), there is a set of paths $P(A_i, A_j)$ that contains all non-circular paths between A_i and A_j. Between any node pair (A_i, A_j), there is a demand d shared by infinitely many agents, where each agent is controlling an infinitesimal share of traffic.

However, the traditional Wardrop model is not suitable to analyze traffic dynamics in an Internet context. We thus adapt the Wardrop model into a more realistic model as follows. First, we introduce the concept of *ASes* and *end-hosts*, which allows us to perform a more fine-grained analysis of traffic in an inter-domain network. An AS $A_i \in A$ is represented by a node in the network graph G. The AS contains a set of end-hosts, which are the players in the path-selection game. Differently than in the Wardrop model, we allow for non-negligible, heterogeneous demand between end-host pairs in order to accommodate the variance of demand in the Internet. For example in origin-destination pair $od = (e_s, e_t) \in OD$ (short: (s, t)), an end-host $e_s \in A_i$ can have a demand $d_{s,t} \geq 0$ towards another end-host $e_t \in A_j$. We also deviate from the Wardrop model by considering a multi-path setting, where the demand $d_{s,t}$ of one agent can be arbitrarily distributed over all paths $p \in P(A_i, A_j)$. The amount of flow from end-host e_s to end-host e_t on path $p \in P(A_i, A_j)$ is denoted as a *path flow* $F_{(s,t),p}$, which must be non-negative, with $\sum_{p \in P(A_i, A_j)} F_{(s,t),p} = d_{s,t}$. The set $\Pi(e_s, e_t) \subseteq \Pi$ contains all end-host paths of the form $\pi = [(s, t), p]$, where e_s, e_t are end-hosts connected by the AS-level path p. All path flows $F_{(s,t),p}$ for an origin-destination pair (e_s, e_t) are collected in a path-flow vector $\mathbf{F}_{s,t} \in \mathbb{R}^{|\Pi(e_s, e_t)|}$. All such path-flow vectors $\mathbf{F}_{s,t}$ are collected in the global *path-flow pattern* $\mathbf{F} \in \mathbb{R}^{|\Pi|}$. A link flow f_ℓ for link $\ell \in L$ is the sum of the path flows in $\mathbf{F}$ that refer to end-host paths π containing link ℓ, i.e., $f_\ell = \sum_{\pi \in \Pi : \ell \in \pi} F_\pi$.

The cost C_π of an end-host path π given a certain path-flow pattern $\mathbf{F}$ is the sum of the cost of all links in the path: $C_\pi(\mathbf{F}) = \sum_{\ell \in \pi} c_\ell(f_\ell)$. The cost to end-hosts $C^*(\mathbf{F})$ from a path-flow pattern $\mathbf{F}$ is the latency experienced by all end-hosts on all the paths to all of their destinations, weighted by the amount of traffic that goes over a given path. This term can be simplified as follows:

$$C^*(\mathbf{F}) = \sum_{(s,t)\in OD} \sum_{\pi\in\Pi(s,t)} F_\pi \cdot C_\pi(\mathbf{F}) = \sum_{\pi\in\Pi} F_\pi \cdot \sum_{\ell\in\pi} c_\ell(f_\ell) = \sum_{\ell\in L} f_\ell \cdot c_\ell(f_\ell)$$

Existing work on selfish routing [28, 30] usually defines total cost in the above sense. However, when analyzing source-based path selection architectures, the network-operator perspective on social cost is essential. Therefore, we also introduce a social cost function relating to the perspective of network operators.

The basic idea of the network-operator cost function $C^\#$ is to treat links as investment assets. Thus, the business performance of a link ℓ is given by a function $p_\ell^\#(f_\ell) = b_\ell^\#(f_\ell) - c_\ell^\#(f_\ell)$, where $b_\ell^\#$ and $c_\ell^\#$ are the benefits and costs of a link, respectively. As we investigate effects on the aggregate of network operators, we model the network-operator cost function as follows:

$$C^\#(\mathbf{F}) = \sum_{\ell\in L} -p_\ell^\#(f_\ell) = \sum_{\ell\in L} c_\ell^\#(f_\ell) - b_\ell^\#(f_\ell) = \sum_{\ell\in L} c_\ell(f_\ell)$$

We justify this formulation as follows. Concerning link costs $c_\ell^\#$, a central insight is that network-operator costs mostly stem from heavily used links. In volume-based interconnection agreements, excessive usage of a link induces high charges, whereas in peering agreements, excessive usage violates the agreement and triggers expensive renegotiation. Moreover, heavy usage necessitates expensive capacity upgrades. As the latency function $c_\ell(f_\ell)$ indicates the congestion level on link ℓ, we approximate $c_\ell^\# \approx c_\ell$. The link benefit $b_\ell^\#$ captures the link revenue, both revenue from customer ASes and customer end-hosts. In the aggregate, the monetary transfers between ASes (charges paid and received) sum up to zero. Given a fixed market size, the revenue from end-hosts sums up to a constant in the aggregate. Hence, the global benefit $\sum_{\ell\in L} b_\ell^\#$ is constant and can be dropped, as the absolute level of the network-operator cost is irrelevant for our purposes. This convex formulation of $C^\#$ allows theoretical analysis.

2.2   Social Optima

According to Wardrop [6, 38], a socially optimal traffic distribution is reached iff the total cost cannot be reduced by moving traffic from one path to another. In the optimum, the cost increase on an additionally loaded path at least outweighs the cost reduction from a relieved path. Because the cost functions are convex and non-decreasing, it suffices that this condition holds for an infinitesimal traffic share. Adding an infinitesimal amount to the argument of a cost function imposes a *marginal cost*, given by the derivative of the cost function. A socially optimal traffic distribution is thus reached iff the marginal cost of every alternative path is not smaller than the marginal cost of the currently used paths [6]:

Social optimum. A path-flow pattern $\mathbf{F}$ represents a social optimum w.r.t. cost function C if and only if for every origin-destination pair $od \in OD$, the paths $\pi_1, ..., \pi_i, \pi_{i+1}, ..., \pi_{|\Pi(od)|} \in \Pi(od)$ stand in the following relationship:

$$\frac{\partial}{\partial F_{\pi_1}}C(\mathbf{F}) = ... = \frac{\partial}{\partial F_{\pi_i}}C(\mathbf{F}) \leq \frac{\partial}{\partial F_{\pi_{i+1}}}C(\mathbf{F}) \leq ... \leq \frac{\partial}{\partial F_{\pi_{|\Pi(od)|}}}C(\mathbf{F})$$

$$F_\pi > 0 \quad \text{for} \quad \pi = \pi_1, ..., \pi_i, \qquad\qquad F_\pi = 0 \quad \text{for} \quad \pi = \pi_{i+1}, ..., \pi_{|\Pi(od)|}.$$

In this work, we refine the conventional notion of the social optimum by distinguishing two different perspectives on social cost: The *end-host optimum* $\mathbf{F}^*$ satisfies the above conditions with respect to the function C^*, whereas the *network-operator optimum* $\mathbf{F}^\#$ satisfies the above conditions with respect to function $C^\#$.

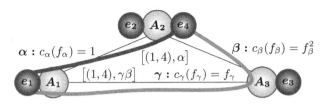

Fig. 1. Example network illustrating the source-based path selection model.

Interestingly, the end-host optimum $\mathbf{F}^*$ and the network-operator optimum $\mathbf{F}^\#$ can differ substantially. Assume that end-host e_1 in Fig. 1 has a demand of $d_{1,4} = 1$ towards end-host e_4 and that there is no other traffic in the network. The network-operator cost function $C^\#(\mathbf{F})$ is $1 + F_{\gamma\beta}^2 + F_{\gamma\beta}$ and is minimized by $\mathbf{F}^\# = (1,0)^\top$, i.e., by sending all traffic over link α. In contrast, the end-host cost function is $F_\alpha + F_{\gamma\beta}^3 + F_{\gamma\beta}^2$ and is minimized by $\mathbf{F}^* = (2/3, 1/3)^\top$, i.e., by sending two thirds of traffic over link α and the remaining third over path $\gamma\beta$.

2.3   Degrees of Information

In this paper, we consider the following two assumptions on the network information possessed by end-hosts:

- **Latency-only information (LI):** End-hosts know the latency of every path to a destination.
- **Perfect information (PI):** End-hosts know not only the latency of different paths, but also how the latency of the network links depends on the current load, i.e., the *latency functions*. Moreover, the end-hosts know the current link utilization, i.e., the background traffic.

The LI assumption hence reflects a scenario where end-hosts have to rely solely on latency measurements of paths, i.e., through RTT measurements from their own device. The LI assumption is the standard model traditionally considered in the selfish routing literature [11,17,30].

In this work, we extend the standard model by introducing the concept of *perfect information* (PI). The PI assumption reflects a scenario where end-hosts can always take the *best* traffic-allocation decision in selfish terms. More specifically, the PI assumption allows end-hosts to compute the *marginal cost* of a path. In path-aware networking, supplying end-hosts with perfect information is possible, as such information is known by network operators and can be disseminated along with path information.

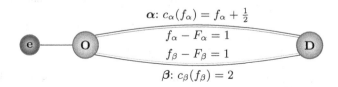

Fig. 2. Example illustrating the different degrees of end-host information.

Figure 2 illustrates the difference between the LI assumption and the PI assumption. Assume that end-host e, residing in AS A, has a demand of $d = 1$ to a destination in AS D. End-host e can split its traffic between two paths α and β, both consisting of a single link with the cost functions c_α (linear) and c_β (constant). The background traffic (traffic not from end-host e) is 1 on both paths. Assuming the traffic allocation of end-host e is $(F_\alpha, F_\beta) = (0.5, 0.5)$, the path-latency values are given by $c_\alpha(0.5 + 1) = 2$ and $c_\beta(0.5 + 1) = 2$. Given the LI assumption, end-host e performs no traffic reallocation, as there is no lower-cost alternative path which traffic could be shifted to. Moreover, there is no method for predicting the path costs for a different traffic allocation. However, such a prediction is possible with perfect information (PI): under the PI assumption, end-host e knows the cost functions and the background traffic such that it can optimize the objective $F_\alpha \cdot (F_\alpha + 1 + \frac{1}{2}) + (1 - F_\alpha) \cdot 2$. As a result, end-host e discovers the optimal traffic assignment $(0.25, 0.75)$. Intriguingly, the more detailed perfect information (PI) enables end-host e to detect an optimization that it cannot directly observe when confronted with latency values only (LI).

2.4   Nash Equilibria

In general, uncoordinated actions of selfish end-hosts do not result in socially optimal traffic allocations. Instead, the only stable states that arise in selfish path selection are *Nash equilibria*, i.e., situations in which no end-host perceives an opportunity to reduce its selfish cost by unilaterally reallocating traffic. However, as shown in Sect. 2.3, the degree of available information (LI or PI) strongly influences the optimization opportunities that an end-host perceives. Therefore, different information assumptions induce different types of Nash equilibria:

LI Equilibrium. An end-host restricted to latency measurements will shift traffic from high-cost paths to low-cost paths whenever there is a cost discrepancy between paths, and will stop reallocating traffic whenever there is no lower-cost path anymore which the traffic could be shifted to. In the latter situation, an end-host under the LI assumption cannot perceive any way of reducing its selfish cost. A Nash equilibrium under the LI assumption (short: LI equilibrium) can thus be defined as follows:

LI equilibrium. A path-flow pattern $\mathbf{F}$ represents an LI equilibrium $\mathbf{F}^0$ if and only if for every origin-destination pair $od \in OD$, the paths $\pi_1, ..., \pi_i, \pi_{i+1}, ..., \pi_{|\Pi(od)|} \in \Pi(od)$ have the following relationship:

$$C_{\pi_1}(\mathbf{F}) = ... = C_{\pi_i}(\mathbf{F}) \leq C_{\pi_{i+1}}(\mathbf{F}) \leq ... \leq C_{\pi_{|\Pi(od)|}}(\mathbf{F})$$

$$F_\pi > 0 \quad \text{for} \quad \pi = \pi_1, ..., \pi_i \qquad\qquad F_\pi = 0 \quad \text{for} \quad \pi = \pi_{i+1}, ..., \pi_{|\Pi(od)|}$$

Traditionally, selfish-routing literature [11,27,30] considers a Nash equilibrium in the sense of the LI equilibrium, namely an equilibrium defined by the cost equality of all used paths to a destination. Under this classical definition, selfish routing is an instance of a *potential game* [31].

PI Equilibrium. We contrast the classical equilibrium (LI equilibrium) with a different equilibrium definition that builds on our new concept of perfect information (PI). As explained in Sect. 2.3, the PI assumption states that end-hosts do not only possess cost information of available paths to a destination, but are informed about the cost *functions* of all links in the available paths, as well as the background traffic on these links, i.e., the arguments to the cost functions. An end-host can thus calculate the selfish cost of a specific traffic reallocation and find the path-flow pattern that minimizes the end-host's selfish cost.

The selfish cost $C_{(e)}^*(\mathbf{F})$ of end-host e is given by the cost of all paths to all desired destinations, weighted by the amount of flow relevant to end-host e:

$$C_{(e)}^*(\mathbf{F}) = \sum_{\ell \in L} f_{\ell,(e)} \cdot c_\ell(f_\ell)$$

where $f_{\ell,(e)}$ is the flow volume on link ℓ for which e is origin or destination.

Similar to the end-host social cost function C^* of which it is a partial term, $C_{(e)}^*$ has a minimum that is characterized by a marginal-cost equality. An equilibrium under the PI assumption is thus given if and only if all end-hosts are at the minimum of their respective selfish cost functions, given the traffic by all other end-hosts:

PI equilibrium. A path-flow pattern $\mathbf{F}$ represents a PI equilibrium $\mathbf{F}^+$ if and only if for every origin-destination pair $od = (e, \_) \in OD$, the paths $\pi_1, ..., \pi_i, \pi_{i+1}, ..., \pi_P \in \Pi(od)$ stand in the following relationship:

$$\frac{\partial}{\partial F_{\pi_1}} C_{(e)}^*(\mathbf{F}) = ... = \frac{\partial}{\partial F_{\pi_i}} C_{(e)}^*(\mathbf{F}) \leq \frac{\partial}{\partial F_{\pi_{i+1}}} C_{(e)}^*(\mathbf{F}) \leq ... \leq \frac{\partial}{\partial F_{\pi_{|\Pi(od)|}}} C_{(e)}^*(\mathbf{F})$$

$$F_\pi > 0 \quad \text{for} \quad \pi = \pi_1, ..., \pi_i \qquad\qquad F_\pi = 0 \quad \text{for} \quad \pi = \pi_{i+1}, ..., \pi_{|\Pi(od)|}$$

2.5 Price of Anarchy

A natural way of analyzing the efficiency of selfish routing is to compare the social optima and the equilibria in a network. Typically, such a comparison involves computing the *Price of Anarchy (PoA)*, i.e., the ratio of the equilibrium cost and the optimal cost. By definition of the optimal cost, this ratio is always larger or equal to 1.

In our model, the classical Price of Anarchy from the existing literature reflects a comparison of the end-host cost C^* of the LI equilibrium $\mathbf{F}^0$ and the end-host cost C^* of the end-host optimum $\mathbf{F}^*$. With the additional versions of social optima and equilibria established in the preceding sections, a total of four different variants of the Price of Anarchy are possible, one for each combination of equilibrium (LI or PI) and perspective (end-hosts or network operators) (Table 1):

Table 1. Different versions of the Price of Anarchy.

	LI equilibrium	PI equilibrium
End-host perspective	$PoA^{*0} = \frac{C^*(\mathbf{F}^0)}{C^*(\mathbf{F}^*)}$	$PoA^{*+} = \frac{C^*(\mathbf{F}^+)}{C^*(\mathbf{F}^*)}$
Network-operator perspective	$PoA^{\#0} = \frac{C^\#(\mathbf{F}^0)}{C^\#(\mathbf{F}^\#)}$	$PoA^{\#+} = \frac{C^\#(\mathbf{F}^+)}{C^\#(\mathbf{F}^\#)}$

2.6 Value of Information

To compare different equilibria for different information assumptions, we introduce the *Value of Information (VoI)*. For a given perspective, the Value of Information is the difference between the Prices of Anarchy under the LI and PI assumptions, denominated by the Price of Anarchy under the LI assumption:

$$VoI^* = \frac{PoA^{*0} - PoA^{*+}}{PoA^{*0}} \qquad VoI^\# = \frac{PoA^{\#0} - PoA^{\#+}}{PoA^{\#0}}$$

A positive Value of Information reflects a situation where the equilibrium under the PI assumption is closer to the social optimum than the equilibrium under the LI assumption. We identify and analyze scenarios with a positive impact of information in Sect. 3. A negative Value of Information reflects the counter-intuitive scenario where additional information makes the equilibrium more costly (cf. Sect. 4).

3 The Benefits of Information

In this section, we will show that information is beneficial in the artificial network settings traditionally considered in the literature [27]. More precisely, we show that in this setting, the PI equilibrium induces a lower Price of Anarchy than the

LI equilibrium such that the Value of Information is positive. This is intuitive: if end-hosts possess more information, source-based path selection is more efficient.

In the network of Fig. 3, K end-hosts e_1, ..., e_K reside in AS O. Each end-host has a demand of d/K towards a destination in AS D. ASes O and D are connected by m links α_1, ..., α_m with a constant cost function $c_{\alpha_i}(f_{\alpha_i}) = d^p$ and one link β with a load-dependent cost function $c_\beta(f_\beta) = f_\beta^p$, where $p \geq 1$.

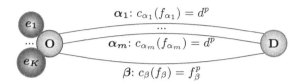

Fig. 3. Example network with beneficial impact of end-host information.

Such networks of parallel links are of special importance in the theoretical selfish-routing literature. In particular, Roughgarden [27] proved that the network in Fig. 3 reveals the worst-case Price of Anarchy for any network with link cost functions limited to polynomials of degree p. The intuition behind this result is that the Price of Anarchy relates to a difference of steepness between cost functions of competing links: the link β allows to reduce the cost of traffic from AS O to AS D if used modestly, but loses its advantage over the links α_i if fully used. However, in selfish routing, end-hosts will use link β until the link is fully used, as it is always a lower-cost alternative path if not fully used. Therefore, the end-hosts overuse link β compared to the optimum. Intuitively, the parallel-links network represents a network where end-hosts have a choice between paths with different latency behavior.

Roughgarden's result refers to the classical Price of Anarchy, i.e., the Price of Anarchy PoA^{*0} to end-hosts under the LI assumption. In this section, we will show how this result is affected by additionally introducing the network-operator perspective and the PI assumption. In particular, we will prove the following theorem:

Theorem 1. *In a network of parallel links, a higher degree of information (PI assumption) is always more socially beneficial compared to a lower degree of information (LI assumption), both from the perspective of end-hosts and network operators:*

$$PoA^{*+} \leq PoA^{*0} \qquad PoA^{\#+} \leq PoA^{\#0}$$

3.1 Social Optima

The end-host optimum in the network of parallel links can be shown to have social cost $C^*(\mathbf{F}^*) = d^{p+1}[1 - p/(p+1)^{(p+1)/p}]$. As the derivation is relatively similar to Roughgarden [27], it has been moved to Appendix A.1 in the full version of the paper [33].

The network-operator optimum $\mathbf{F}^\#$ is simple to derive: Since the cost of the links α_i is independent of the flow on these links in contrast to the cost of link β, any flow on link β increases the cost $C^\#$ to network operators. The minimal cost to network operators is thus simply $C^\#(\mathbf{F}^\#) = m \cdot d^p$.

3.2   LI Equilibrium

Under the LI assumption, a network is in equilibrium if for every end-host pair, all used paths have the same cost and all unused paths do not have a lower cost. Applied to the simple network in Fig. 3, this condition is satisfied if and only if $f_\beta^0 = d$ and $f_{\alpha_i}^0 = 0 \; \forall f_{\alpha_i}$, implying $c_\beta(f_\beta^0) = d^p = c_{\alpha_i}(f_{\alpha_i}^0)$. The path-flow pattern $\mathbf{F}^0$ with $F_{(k,D),\beta} = d/K$ and $F_{(k,D),\alpha_i} = 0$ represents the LI equilibrium. The cost C^* of the LI equilibrium $\mathbf{F}^0$ to end-hosts is simply $C^*(\mathbf{F}^0) = d^{p+1}$. The Price of Anarchy to end-hosts under the LI assumption is thus $PoA^{*0} = C^*(\mathbf{F}^0)/C^*(\mathbf{F}^*) = [1 - p/(p+1)^{(p+1)/p}]^{-1}$.

The cost $C^\#$ of the LI equilibrium $\mathbf{F}^0$ to network operators is given by $C^\#(\mathbf{F}^0) = d^p + \sum_{\alpha_i} d^p = (m+1) \cdot d^p$. The Price of Anarchy to network operators under the LI assumption is thus $PoA^{\#0} = C^\#(\mathbf{F}^0)/C^\#(\mathbf{F}^\#) = (m+1)/m$, which is maximal for the number $m = 1$ of links α_i. The Price of Anarchy to network operators in networks of parallel links is thus upper-bounded by $PoA^{\#0}_{m=1} = 2$ whereas the Price of Anarchy to end-hosts is unbounded for arbitrary p.

3.3   PI Equilibrium

If the end-hosts $e_1, ..., e_K$ are equipped with perfect information, they are in equilibrium if and only if the *selfish* marginal cost of every path to AS D is the same for every end-host. Under this condition, the cost term C^* of the PI equilibrium $\mathbf{F}^+$ to end-hosts can be derived to be $C^*(\mathbf{F}^+) = d^{p+1}\big(1 - (p/K)/(p/K + 1)^{(p+1)/p}\big)$ (cf. Appendix A.2 [33]). The Price of Anarchy to end-hosts under the PI assumption is

$$PoA^{*+} = \left(1 - \frac{p/K}{(p/K + 1)^{(p+1)/p}}\right) \cdot PoA^{*0} \leq PoA^{*0}.$$

The cost $C^\#$ of the PI equilibrium $\mathbf{F}^+$ to network operators is $C^\#(\mathbf{F}^+) = (m + 1/(p/K + 1)) \cdot d^p$ and the corresponding Price of Anarchy to network operators is

$$PoA^{\#+} = \frac{m + 1/(p/K + 1)}{m} \leq \frac{m+1}{m} = PoA^{\#0}.$$

Based on the Prices of Anarchy in Table 2, Theorem 1 holds. However, the Prices of Anarchy PoA^{*+} and $PoA^{\#+}$ under the PI assumption are dependent on K, which is the number of end-hosts in the network. If K is very high, as it is in an Internet context, the Prices of Anarchy under the PI assumption approximate the Prices of Anarchy under the LI assumption. Thus, for scenarios of heterogeneous parallel paths to a destination, the benefit provided by perfect information is undone in an Internet context. In fact, the effect of additional information can even turn negative when considering more general networks, as we will show in the next section.

Table 2. Price of Anarchy for different perspectives and different equilibrium definitions in the network of parallel links (Fig. 3).

	LI equilibrium	PI equilibrium
End-host perspective	$\frac{1}{1-p/(p+1)^{(p+1)/p}}$	$\frac{1-(p/K)/(p/K+1)^{(p+1)/p}}{1-p/(p+1)^{(p+1)/p}}$
Network-operator perspective	$\frac{m+1}{m}$	$\frac{m+1/(p/K+1)}{m}$

4   The Drawbacks of Information

We will now show that in more general settings, more information for end-hosts can *deteriorate* outcomes of selfish routing. Such a case is given by the general *ladder network* in Fig. 4, a natural generalization of the simple topology considered above and a traditional ISP topology [19].

A ladder network of height H contains H horizontal links $h_1,..., h_H$, which represent the rungs of a ladder and have the cost function $c_{h_i}(f_{h_i}) = f_{h_i}^p$. Each horizontal link h_i connects an AS A_{i1} to AS A_{i2}, which accommodate the end-hosts e_{i1} and e_{i2}, respectively. Every end-host e_{i1} has the same demand d towards the corresponding end-host e_{i2}. Neighboring rungs of a ladder are connected by vertical links v_{ij}, $i \in \{1,...,V = H-1\}$, $j \in \{1,2\}$, where the vertical link v_{ij} connects the ASes A_{ij} and $A_{i+1,j}$ and has the linear cost function $c_{v_{ij}}(f_{v_{ij}}) = t \cdot f_{v_{ij}}$ with $t \geq 0$. We denote a ladder network with this structure and a choice of parameters H, p, d, and t by $\mathcal{L}(H,p,d,t)$.

By comparing optima and equilibria, we will prove the following theorem in the following subsections:

Theorem 2. *For any ladder network $\mathcal{L}(H,p,d,t)$, the Value of Information for both end-hosts and network operators is **negative**, i.e., $VoI^* < 0$ and $VoI^\# < 0$.*

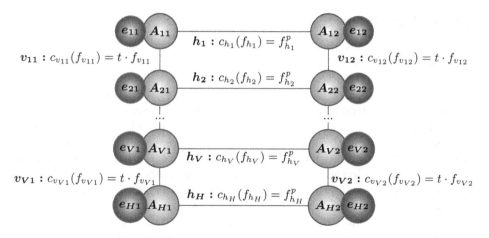

Fig. 4. Example network illustrating the harmful impact of end-host information (Read: $V = H - 1$).

4.1 Social Optima

Both the end-host optimum $\mathbf{F}^*$ and $\mathbf{F}^\#$ are equal to the *direct-only* path-flow pattern $\mathbf{F}^\sim$ that is defined as follows: For every end-host e_{i1}, $F^\sim_{(i1,i2),h_i} = d$ and $F^\sim_{(i1,i2),q} = 0$ where q is any other path between A_{i1} and A_{i2} than the direct path over link h_i.

Simple intuition already confirms the optimality of this path-flow pattern. The social cost from the horizontal links is minimized for an equitable distribution of the whole-network demand Hd onto the H horizontal links. In contrast, the cost from vertical links v_{ij} can be minimized to 0 by simply abstaining from using vertical links. In fact, every use of the vertical links is socially wasteful.

More formally, if $f_{h_i} = d$ for $i \in \{1, ..., H\}$ and $f_{v_{i1}} = f_{v_{i2}} = 0$ for $i \in \{1, ..., V\}$, the marginal costs of the direct path and every indirect path can be easily shown to equal $(p + 1)d^p$, given end-host cost function C^*. Concerning network-operator cost $C^\#$, the direct and indirect paths have marginal costs $p \cdot d^{p-1}$ and $p \cdot d^{p-1} + 2yt \ \forall y \in \mathbb{N}_{\geq 1}$, respectively. The used direct paths thus do not have a higher marginal cost than the unused indirect paths.

4.2 LI Equilibrium

Also the LI equilibrium path-flow pattern $\mathbf{F}^0$ is equal to the direct-only path-flow pattern $\mathbf{F}^\sim$. For $\mathbf{F}^\sim$, the cost of a direct path π is $C_\pi(\mathbf{F}^\sim) = F^p_{(i1,i2),\pi} = d^p$ and the cost of an indirect path π' is $f^p_{h'} + \sum_{v \in W_{\pi'}} f_v = d^p + 0 = d^p$, where π' contains the remote horizontal link h' and the vertical links $v \in W_{\pi'}$. Thus, the LI equilibrium conditions of cost equality are satisfied by $\mathbf{F}^\sim$.

As the LI equilibrium is equal to the social optimum both from the end-host perspective and the network-operator perspective, both variants of the Price of Anarchy under the LI assumption are optimal, i.e., $PoA^{*0} = PoA^{\#0} = 1$.

4.3 PI Equilibrium

Differently than under the LI assumption, the direct-only flow distribution $\mathbf{F}^{\sim}$ is not stable under the PI assumption. An end-host e_i can improve its individual cost by allocating some traffic to an indirect path π_k (involving the horizontal link h_k) and interfering with another end-host e_k. This reallocation decision will increase the social cost for end-hosts and network operators. In particular, the end-host e_k that previously used the link h_k exclusively will see its selfish cost increase. In turn, the harmed end-host e_k will reallocate some of its traffic to an indirect path in order to reduce its selfish cost $C_{(e_k)}$, leading to a process where all end-hosts in the network interfere with each other until they reach a PI equilibrium with a suboptimal social cost for end-hosts and network operators.

Similar to Sect. 3.3, we use the condition of marginal selfish cost equality in order to derive the Price of Anarchy under the PI assumption for a ladder network with $H = 2$. This derivation, as performed in Appendix A.3 [33], yields the following results for the Price of Anarchy to end-hosts and network operators:

$$PoA^{*+}_{H=2}(p) = 1 + p/12 \qquad PoA^{\#+}_{H=2}(p) = 1 + p/3$$

Since the LI equilibrium is optimal and the PI equilibrium is generally suboptimal on the considered ladder networks, Theorem 2 holds. This finding is confirmed by a case study of the Abilene network (cf. Sect. 5), which structurally resembles a ladder topology. The case study also reveals that the negative impact of information is especially pronounced if path diversity is high.

Interestingly, there is an upper bound of the Price of Anarchy to network operators for a *general* ladder network. This bound is given by the following theorem and proven in Appendix A.4 in the full-paper version [33]:

Theorem 3. *For every ladder network $\mathcal{L}(H, p, d, t)$, the Price of Anarchy $PoA^{\#+}$ to network operators is lower than the following upper bound $PoA^{\#+}_{\max}$:*

$$PoA^{\#+} \leq PoA^{\#+}_{H,\max} = 1 + \frac{2(H-1)}{3H}p \leq PoA^{\#+}_{\max} = 1 + \frac{2}{3}p$$

5 Case Study: Abilene Network

To verify and complement our theoretical insights, we conducted a case study with a real network: we consider the well-known Abilene network, for which topology and workload data is publicly available [14,15]. We accommodate the Abilene topology into our model as follows. For the demand d between the 11 points-of-presence, which we consider ASes, we rely on the empirical traffic matrix from the dataset. Concerning the link-cost functions c_ℓ, we model the latency behavior of a link by a function $c_\ell(f_\ell) = f_\ell^2 + \delta_\ell$, where f_ℓ^2 captures the queueing delay and δ_ℓ is a constant quantity depending on the geographical distance between the two end-points of link ℓ, approximating the link's propagation delay.

In order to study the effect of both end-host information and multi-path routing on the Price of Anarchy, we perform the following simulation experiment. First, we compute the social optima $\mathbf{F}^*$ and $\mathbf{F}^\#$ for the Abilene network. Second, we simulate the convergence to the Nash equilibria $\mathbf{F}^0$ and $\mathbf{F}^+$ for different degrees of multi-path routing, represented by the maximum number of shortest paths that end-hosts consider in their path selection. Once converged, we compute the social cost of the equilibrium traffic distributions and the corresponding Prices of Anarchy.

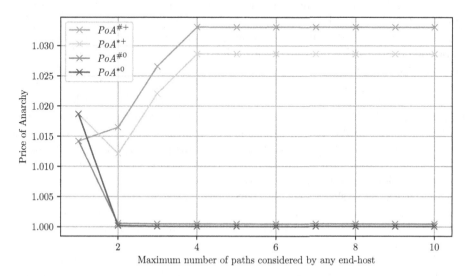

Fig. 5. Abilene network results.

The experiment results in Fig. 5 offer multiple interesting insights. Most prominently, if simple shortest-path routing represents the baseline of network-controlled path selection, source-based path selection with latency-only information improves the performance of the network (up to a near-optimum), which confirms findings of prior work [24]. In contrast, path selection with perfect information deteriorates performance, especially for a higher degree of multi-path routing. Therefore, the potential performance benefits of source-based path selection with multi-path routing are conditional on the amount of information possessed by end-hosts, where a higher degree of information is associated with lower performance. However, while an increasing degree of multi-path routing is associated with worse performance under perfect information, the resulting inefficiency is bounded at a modest level of less than 4% for both end-hosts and network operators. The near-optimality of latency-only information in terms of performance and the bounded character of the Price of Anarchy under perfect information reflect the findings from Sect. 4 about ladder topologies, which resemble the Abilene topology. Thus, the experiment results not only show that source-based path selection can be a means to improve the performance of a network but also confirm the practical relevance of our theoretical findings.

6   Related Work

Inefficiency arising from selfish behavior in networks is well-known to exist in transportation networks and has been thoroughly analyzed with the framework of the Wardrop model [6,38]. The most salient expressions of this inefficiency is given by the Braess Paradox [4].

Literature on *selfish routing* is often concerned with the discrepancy between optimum and Nash equilibrium: the *Price of Anarchy* [8,17]. The Price of Anarchy was initially studied for network models (see Nisan et al. [22] for an overview), but literature now covers a wide spectrum, from health care to basketball [29]. Our work has a closer connection to more traditional research questions, such as bounds on the Price of Anarchy for selfish routing. An early result has been obtained by Koutsoupias and Papadimitriou [17], who formulated routing in a network of parallel links as a multi-agent multi-machine scheduling problem.

A different model has been developed by Roughgarden and Tardos [30] who build on the Wardrop model [38] for routing in the context of computer networks. The Price of Anarchy in the proposed routing game is the ratio between the latency experienced by all users in the Wardrop equilibrium and the minimum latency experienced by all users. For different classes of latency functions, the authors derive explicit high bounds on the resulting Price of Anarchy. In a different work, they show that the worst-case Price of Anarchy for a function class can always be revealed by a simple network of parallel links and that the upper bound on the Price of Anarchy depends on the growth rate of the latency functions [27].

The relatively loose upper bounds on the Price of Anarchy of previous works [17,30] have been qualified by subsequent research. It was found that problem instances with high Prices of Anarchy are usually artificial. By introducing plausible assumptions to make the routing model more realistic, upper bounds on the Price of Anarchy can be reduced substantially. For instance, Friedman [11] shows that the Price of Anarchy is lower than the mentioned worst-case derived by Roughgarden and Tardos [30] if the Nash equilibrium cost is not sensitive to changes in the demand of agents. By computing the Price of Anarchy for a variety of different latency functions, topologies, and demand vectors, Qiu et al. even show that selfish routing is nearly optimal in many cases [24].

Convergence to Nash equilibria has been studied in the context of congestion games [26] and, in a more abstract form, in the context of potential games [21,31]. Sandholm [31] showed that selfish player behavior in potential games leads to convergence to the Nash equilibrium and, under some conditions, even to convergence to the social optimum. As the question of equilibrium convergence is traditionally studied separately from the question of equilibrium cost, we do not address convergence issues in this paper.

The study of the effect of incomplete information also has a long tradition [13], but still poses significant challenges [29]. Existing literature in this area primarily focuses on scenarios where players are uncertain about each others' payoffs, studying alternative notions of equilibria such as Bayes-Nash equilibria [36], which also leads to alternative definitions of the price of anarchy such as the Bayes-Nash Price of Anarchy [18, 29] or the price of stochastic anarchy [5]. A common observation of many papers in this area is that less information can lead to significantly worse equilibria [29]. There is also literature on the impact on the Price of Anarchy in scenarios where interacting players only have *local* information, e.g., the evolutionary price of anarchy [34].

However, much less is known today about the role of information in games related to *routing*. In this context, one line of existing literature is concerned with the recentness of latency information. Most prominently, research on the damage done by stale information in load-balancing problems [7, 20] has been applied to routing games by Fischer and Vöcking [10]. This work investigates whether and how rerouting decisions converge onto a Wardrop equilibrium if these rerouting decisions are based on obsolete latency information. Other recent work about the role of information in routing games investigates how the amount of topology information possessed by agents affects the equilibrium cost [1].

Existing work on the subject of source-routing efficiency differs from our work in two important aspects. First, to the best of our knowledge, all existing work on the subject defines the social optimum as the traffic assignment that minimizes the total cost experienced by users, which is indeed a reasonable metric. However, our work additionally investigates the total cost experienced by *links*, i.e., the network operators. Since cost considerations by network operators are a decisive factor in the deployment of source-based path selection architectures, the Price of Anarchy to network operators is an essential metric. Second, although existing work on the topic has investigated the role played by the *recentness* of congestion information or the degree of *topology* information, it does not investigate the role played by the *degree of congestion information* that agents possess. Indeed, a major contribution of our work is to highlight the effects of perfect information, i.e., information that allows agents to perfectly minimize their selfish cost. Latency-only information, which agents are assumed to have in existing work, does not enable agents to perform perfect optimization.

7 Conclusion

Motivated by emerging path-aware network architectures, we refined and extended the Wardrop model in order to study the implications of source-based path selection. Our analysis provides several interesting insights with practical relevance. First, the cost of selfish routing to network operators differs from the cost experienced by users. Since network operators are central players in the adoption of path-aware networking, research on the effects of selfish routing thus needs to address the network-operator perspective separately. However, we proved upper bounds on the Price of Anarchy which suggest that selfish routing imposes a low cost on network operators. Second, we found that basic latency information, which can be measured by the end-hosts themselves, leads to near-optimal traffic allocations in many cases. Selfish routing thus causes modest ineffiency even if end-hosts have only imperfect path information and network operators do not disseminate detailed path-load information.

Our model and first results introduce several exciting avenues for future research. First, we note that while we have focused on path-aware network architectures, we hope to apply our model to other practical applications where source routing has currently received much attention, e.g., in the context of segment routing [9] and multi-cast [35]. Furthermore, we aim to obtain a more general understanding of the interactions between the network topology structure and the Price of Anarchy in selfish routing. Moreover, our focus in this paper was on rational players, and it is important to extend our model to account for other behaviors, e.g., players combining altruistic, selfish and Byzantine behaviors.

Acknowledgement. We gratefully acknowledge support from ETH Zurich, from SNSF for project ESCALATE (200021L_182005), and from WWTF for project WHATIF (ICT19-045, 2020–2024). Moreover, we thank Markus Legner, Jonghoon Kwon, and Juan A. García-Pardo for helpful discussions that supported and improved this research. Lastly, we thank the anonymous reviewers for their constructive feedback.

References

1. Acemoglu, D., Makhdoumi, A., Malekian, A., Ozdaglar, A.: Informational braess' paradox: the effect of information on traffic congestion. Oper. Res. **66**(4), 893–917 (2018)
2. Andersen, D., Balakrishnan, H., Kaashoek, F., Morris, R.: Resilient overlay networks, vol. 5 (2001)
3. Barrera, D., Chuat, L., Perrig, A., Reischuk, R.M., Szalachowski, P.: The scion internet architecture. Commun. ACM **60**(6), 56–65 (2017)
4. Braess, D.: Über ein Paradoxon aus der Verkehrsplanung. Unternehmensforschung **12**(1), 258–268 (1968)
5. Chung, C., Ligett, K., Pruhs, K., Roth, A.: The price of stochastic anarchy. In: Monien, B., Schroeder, U.-P. (eds.) SAGT 2008. LNCS, vol. 4997, pp. 303–314. Springer, Heidelberg (2008). https://doi.org/10.1007/978-3-540-79309-0_27
6. Dafermos, S.C., Sparrow, F.T.: The traffic assignment problem for a general network. J. Res. Natl. Bureau Stand. B **73**(2), 91–118 (1969)
7. Dahlin, M.: Interpreting stale load information. IEEE Trans. Parallel Distrib. Syst. **11**(10), 1033–1047 (2000)
8. Dubey, P.: Inefficiency of nash equilibria. Mathematics of Operations Research **11**(1), (1986)
9. Filsfils, C., Nainar, N.K., Pignataro, C., Cardona, J.C., Francois, P.: The segment routing architecture. In: IEEE Global Communications Conference (GLOBECOM) (2015)
10. Fischer, S., Vöcking, B.: Adaptive routing with stale information. Theor. Comput. Sci. **410**(36), 3357–3371 (2009)
11. Friedman, E.J.: Genericity and congestion control in selfish routing. In: 43rd IEEE Conference on Decision and Control (CDC), vol. 5 (2004)
12. Gupta, A., et al.: SDX: a software defined internet exchange. ACM SIGCOMM Comput. Commun. Rev. **44**(4), 551–562 (2015)
13. Harsanyi, J.C.: Games with incomplete information played by "Bayesian" players, I–III part I. The basic model. Manag. Sci. **14**(3), 159–182 (1967)
14. Knight, S., Nguyen, H.X., Falkner, N., Bowden, R., Roughan, M.: The internet topology zoo. IEEE J. Sel. Areas Commun. **29**, 1765–1775 (2011)

15. Kolaczyk, E.D.: Statistical Analysis of Network Data (Datasets). Springer, Heidelberg (2009). http://math.bu.edu/people/kolaczyk/datasets.html
16. Kotronis, V., Kloti, R., Rost, M., Georgopoulos, P., Ager, B., Schmid, S., Dimitropoulos, X.: Stitching inter-domain paths over IXPs. In: Proceedings of ACM Symposium on SDN Research (SOSR) (2016)
17. Koutsoupias, E., Papadimitriou, C.: Worst-case equilibria. In: Annual Symposium on Theoretical Aspects of Computer Science (STACS) (1999)
18. Leme, R.P., Tardos, E.: Pure and Bayes-Nash price of anarchy for generalized second price auction. In: IEEE 51st Annual Symposium on Foundations of Computer Science (FOCS) (2010)
19. Luizelli, M.C., Bays, L.R., Buriol, L.S., Barcellos, M.P., Gaspary, L.P.: Characterizing the impact of network substrate topologies on virtual network embedding. In: Proceedings of the 9th International Conference on Network and Service Management (CNSM). IEEE (2013)
20. Mitzenmacher, M.: How useful is old information? IEEE Trans. Parallel Distrib. Syst. **11**(1), 6–20 (2000)
21. Monderer, D., Shapley, L.S.: Potential games. Games Econ. Behav. **14**, 124–143 (1996)
22. Nisan, N., Roughgarden, T., Tardos, E., Vazirani, V.V.: Algorithmic Game Theory. Cambridge University Press, Cambridge (2007)
23. Perrig, A., Szalachowski, P., Reischuk, R.M., Chuat, L.: SCION: A Secure Internet Architecture. Springer, Heidelberg (2017). https://doi.org/10.1007/978-3-319-67080-5. https://netsec.ethz.ch/publications/papers/SCION-book.pdf
24. Qiu, L., Yang, Y.R., Zhang, Y., Shenker, S.: On selfish routing in internet-like environments. In: Proceedings of the ACM SIGCOMM (2003)
25. Raghavan, B., Snoeren, A.C.: A system for authenticated policy-compliant routing. In: ACM SIGCOMM Computer Communication Review, vol. 34 (2004)
26. Rosenthal, R.W.: A class of games possessing pure-strategy nash equilibria. Int. J. Game Theory **2**(1), 65–67 (1973)
27. Roughgarden, T.: The price of anarchy is independent of the network topology. J. Comput. Syst. Sci. (JCSS) **67**(2), 341–364 (2003)
28. Roughgarden, T.: Routing games. Algorithmic Game Theory **18**, 459–484 (2007)
29. Roughgarden, T.: The price of anarchy in games of incomplete information. ACM Trans. Econ. Comput. (TEAC) **3**(1), 1–20 (2015)
30. Roughgarden, T., Tardos, É.: How bad is selfish routing? J. ACM (JACM) **49**(2), 236–259 (2002)
31. Sandholm, W.H.: Potential games with continuous player sets. J. Econ. Theory **97**(1), 81–108 (2001)
32. Savage, S., Collins, A., Hoffman, E., Snell, J., Anderson, T.: The end-to-end effects of internet path selection. In: ACM SIGCOMM Computer Communication Review, vol. 29 (1999)
33. Scherrer, S., Perrig, A., Schmid, S.: The value of information in selfish routing (2020). https://arxiv.org/abs/2005.05191
34. Schmid, L., Chatterjee, K., Schmid, S.: The evolutionary price of anarchy: locally bounded agents in a dynamic virus game. In: Proceedings of 33rd International Symposium on Distributed Computing (OPODIS) (2019)
35. Shahbaz, M., Suresh, L., Rexford, J., Feamster, N., Rottenstreich, O., Hira, M.: Elmo: source-routed multicast for cloud services. arXiv preprint arXiv:1802.09815 (2018)

36. Singh, S., Soni, V., Wellman, M.: Computing approximate Bayes-Nash equilibria in tree-games of incomplete information. In: Proceedings of the 5th ACM conference on Electronic Commerce (EC) (2004)
37. Trammell, B., Smith, J.P., Perrig, A.: Adding path awareness to the internet architecture. IEEE Internet Comput. **22**(2), 96–102 (2018). https://doi.org/10.1109/MIC.2018.022021673
38. Wardrop, J.G.: Some theoretical aspects of road traffic research. Proc. Inst. Civ. Eng. **1**(3), 325–362 (1952)
39. Xu, W., Rexford, J.: MIRO: multi-path Interdomain Routing. In: Proceedings of the SIGCOMM (2006)

Author Index

Printed in the United States
By Bookmasters